Modern Methods of Plant Analysis

New Series Volume 13

Editors
H. F. Linskens, Erlangen/Nijmegen/Amherst
J. F. Jackson, Adelaide

Volumes Already Published in this Series:

Volume 1: Cell Components
1985, ISBN 3-540-15822-7

Volume 2: Nuclear Magnetic Resonance
1986, ISBN-3-540-15910-X

Volume 3: Gas Chromatography/
Mass Spectrometry
1986, ISBN 3-540-15911-8

Volume 4: Immunology in Plant Sciences
1986, ISBN 3-540-16842-7

Volume 5.: High Performance Liquid Chromatography
in Plant Sciences
1987, ISBN 3-540-17243-2

Volume 6: Wine Analysis
1988, ISBN 3-540-18819-3

Volume 7: Beer Analysis
1988, ISBN 3-540-18308-6

Volume 8: Analysis of Nonalcoholic Beverages
1988, ISBN 3-540-18820-7

Volume 9: Gases in Plant and Microbial Cells
1989, ISBN 3-540-18821-5

Volume 10: Plant Fibers
1989, ISBN 3-540-18822-3

Volume 11: Physical Methods in Plant Sciences
1990, ISBN 3-540-50332-3

Volume 12: Essential Oils and Waxes
1991, ISBN 3-540-51915-7

Volume 13: Plant Toxin Analysis
1992, ISBN 3-540-52328-6

Forthcoming:

Volume 14: Seed Analysis
1992, ISBN 3-540-52737-0

Volume 15: Alkaloids
1992, ISBN 3-540-52738-9

Plant Toxin Analysis

Edited by
H. F. Linskens and J. F. Jackson

Contributors

J. T. Arnason R. R. Aucion B. Beißman W. Bors A. Breiman
A. M. Brinker E. F. Elstner R. Fall E. Galun A. J. Hills
H. M. Kalisz M. E. Kalisz K.-H. Kogel K. Kohmoto
R. K. Monson M. R. A. Morgan S. D. Murphy P. L. Pfahler
M. Saran E. Schneider D. S. Seigler L. M. V. Soares
A. Stierle D. Stierle G. Strobel F. Sugawara K. Rudolph
C. M. Ward A. P. Wilkinson K. Wydra

With 33 Figures

Springer-Verlag Berlin Heidelberg GmbH

Prof. Dr. HANS FERDINAND LINSKENS
Goldberglein 7
W-8520 Erlangen, FRG

Prof. Dr. JOHN F. JACKSON
Department of Viticulture, Enology and Horticulture
Waite Agricultural Research Institute
University of Adelaide
Glen Osmond, S. A. 5064
Australia

ISBN 978-3-642-08090-6 ISBN 978-3-662-02783-7 (eBook)
DOI 10.1007/978-3-662-02783-7

The Library of Congress Card Number 87-659239 (ISSN 0077-0183)

Originally published by Springer-Verlag Berlin Heidelberg New York in 1992
Softcover reprint of the hardcover 1st edition 1992

Production Editor: Herta Böning, Heidelberg
Reproduction of the figures: Gustav Dreher GmbH, Stuttgart
Typesetting: Best-set, Hong Kong
31/3145-5 4 3 2 1 0 – Printed on acid-free paper

Introduction

Modern Methods of Plant Analysis

When the handbook *Modern Methods of Plant Analysis* was first introduced in 1954 the considerations were:

1. the dependence of scientific progress in biology on the improvement of existing and the introduction of new methods;
2. the difficulty in finding many new analytical methods in specialized journals which are normally not accessible to experimental plant biologists;
3. the fact that in the methods sections of papers the description of methods is frequently so compact, or even sometimes so incomplete that it is difficult to reproduce experiments.

These considerations still stand today.

The series was highly successful, seven volumes appearing between 1956 and 1964. Since there is still today a demand for the old series, the publisher has decided to resume publication of *Modern Methods of Plant Analysis.* It is hoped that the New Series will be just as acceptable to those working in plant sciences and related fields as the early volumes undoubtedly were. It is difficult to single out the major reasons for success of any publication, but we believe that the methods published in the first series were up-to-date at the time and presented in a way that made description, as applied to plant material, complete in itself with little need to consult other publications.

Contribution authors have attempted to follow these guidelines in this New Series of volumes.

Editorial

The earlier series *Modern Methods of Plant Analysis* was initiated by Michel V. Tracey, at that time in Rothamsted, later in Sydney, and by the late Karl Paech (1910–1955), at that time at Tübingen. The New Series will be edited by Paech's successor H.F. Linskens (Nijmegen, The Netherlands) and John F. Jackson (Adelaide, South Australia). As were the earlier editors, we are convinced "that there is a real need for a collection of reliable up-to-date methods for plant analysis in large areas of applied biology ranging from agriculture and horticultural experiment stations to pharmaceutical and technical institutes concerned with raw material of plant origin".

The recent developments in the fields of plant biotechnology and genetic engineering make it even more important for workers in the plant sciences to become acquainted with the more sophisticated methods, which sometimes come from biochemistry and biophysics, but which also have been developed in commercial firms, space science laboratories, non-university research institutes, and medical establishments.

Concept of the New Series

Many methods described in the biochemical, biophysical, and medical literature cannot be applied directly to plant material because of the special cell structure, surrounded by a tough cell wall, and the general lack of knowledge of the specific behavior of plant raw material during extraction procedures. Therefore all authors of this New Series have been chosen because of their special experience with handling plant material, resulting in the adaptation of methods to problems of plant metabolism.

Nevertheless, each particular material from a plant species may require some modification of described methods and usual techniques. The methods are described critically, with hints as to their limitations. In general it will be possible to adapt the methods described to the specific needs of the users of this series, but nevertheless references have been made to the original papers and authors. While the editors have worked to plan in this New Series and made efforts to ensure that the aims and general layout of the contributions are within the general guidelines indicated above, we have tried not to interfere too much with the personal style of each author.

There are several ways of classifying the methods used in modern plant analysis. The first is according to the technological and instrumental progress made over recent years. These aspects were used for the first five volumes in this series describing methods in a systematic way according to the basic principles of the methods.

A second classification is according to the plant material that has to undergo analysis. The specific application of the analytical method is determined by the special anatomical, physiological, and biochemical properties of the raw material and the technology used in processing. This classification will be used in Volumes 6 to 8, and for some later volumes in the series.

A third way of arranging a description of methods is according to the classes of substances present in the plant material and the subject of analytic methods. The latter will be used for later volumes of the series, which will describe modern analytical methods for alkaloids, drugs, hormones, etc.

Naturally, these three approaches to developments in analytical techniques for plant materials cannot exclude some small overlap and repetition; but careful selection of the authors of individual chapters, according to their expertise and experience with the specific methodological technique, the group of substances to be analyzed, or the plant material which is the subject of chemical and physical analysis, guarantees that recent developments in analytical methodology are described in an optimal way.

Volume Thirteen – Plant Toxin Analysis

In producing a volume of *Modern Methods of Plant Analysis* on plant toxins, it is at once apparent that we are dealing with one of the most important aspects of plant defence mechanisms. Chemical defence by means of toxins poisonous to other organisms, be they animals or plants, is widespread amongst the plant kingdom, and that includes microorganisms as well.

According to the Encyclopedia Brittanica, a toxin is properly a poisonous protein; however, non-proteinaceous poisons such as fungal aflotoxins and plant alkaloids are also referred to as toxins, and it is this wider usage of the term which we will adopt here.

The toxicity of these chemical compounds is related to the dose, sensitivity of target organism, mechanism of uptake, mechanism of action, mode of excretion, whether or not the toxin is degraded or detoxified and the possibility that through adoption certain organisms are not harmed by the toxin.

Regarding dose, the amount of toxin which is lethal can be quite small; for example, the dose of the plant protein toxin ricin, which is lethal to humans, is as little as 1 mg/kg body weight, and many other toxins are even more toxic than this. In the case of ricin, we now know that the mechanism of action is highly specific, in that the A-chain of the protein toxin is an N-glycosidase which removes a single base from r-RNA, thereby inactivating the 60-s ribosomal unit. Other cytotoxins made by bacteria involved in enteric and renal diseases are functionally and structurally related to ricin.

An example of a non-proteinaceous toxin is the neurotoxin β-N-oxalyl aminovaline, which is found in the food grain *Lathyrus sativus*. In India it is known that some people eat kesari dhal made with *L. sativus*, leading to a crippling condition of the nervous system.

Not all of the so-called toxins are injurious to man and other animals; some can even be beneficial. For example, certain killer strains of yeast secrete a complex protein toxin which is active against other yeasts from several genera. While the mechanism of action of this toxin is not well understood, it is known that its production is associated with the presence, in for example a killer strain of *Kluyveromyces lactis*, of two double-stranded DNA plasmids, one of which encodes the toxin. Yeasts susceptible to this particular toxin are arrested in the G1 phase of the cell cycle. Wine makers are endeavouring to turn the phenomenon of killer yeasts to good advantage by using wine yeasts containing these "killer" encoding plasmids, which could help discourage the growth of other undesirable yeasts in fermenting wines.

In this thirteenth volume of *Modern Methods of Plant Analysis*, a wide variety of plant toxins are dealt with.

Thus, a description is given of the way in which the easily carried out TLC (thin layer chromatography) is used to detect such specific toxins as aflotoxin, ochrotoxin A, zearalenone and sterigmatocystin in food commodities like wheat, barley, beans, corn, oats, rye, peanuts, rice, sorghum, cheese, olives, almonds and so on.

At the other extreme, consideration is given to the effect of those potentially toxic agents released as the result of agricultural or industrial activity. This represents an extremely diverse spectrum of toxins chemically, so that a specific chemical test is out of question.

Instead, a chapter by Professor Pfahler puts forward the proposition that pollen systems can be used to detect such toxins that could have effect on the seed and fruit commodity production of the world through pollination and fertilization malfunction. Another very chemically diverse class of biologically active compounds is that represented by the so-called elicitors. These compounds lead to the induction of defence responses in plants subject to microbial attack, usually in the form of induction of phyto-alexin production. A vast number of fungal and bacterial metabolites, including carbohydrates, glycoproteins, peptides, fatty acids and hydrolytic enzymes are known to act as elicitors; a chapter is devoted to this aspect of plant analysis.

Another chapter looks at fungal plant disease from the standpoint of phytotoxins isolated from culture filtrates of pathogens and host-selective toxins. Various bioassays for this class of plant toxins are described, as well as chemical analysis in the form of HPLC (high performance liquid chromatography, see Vol. 5, *Modern Methods of Plant Analysis*).

The very diversity of host-selective toxins illustrates the flexibility of HPLC as a handy tool in plant analysis. The subject of bioassays is taken up also in another chapter dealing with quantitative assays of phytotoxins using plant protoplasts and isolated cells. These bioassays are of course restricted to certain sensitive cells; thus, when setting up a bioassay for fusicoccin, one must use whole plants or leaves, since this phytotoxin causes opening of stomata and mediates uncontrolled water loss from treated plants.

Gas chromatography lends itself easily to the analysis of monoterpenes such as isoprene, and it is this that is taken up in a chapter in this Volume dealing with the analysis of isoprene emission from leaves. Such monoterpenes have been the subject of numerous studies in chemical analysis, the motivation being the role of monoterpenes in protecting plants from herbivory.

It is well known that many plant secondary metabolites are rendered toxic by light (visible or ultra-violet); several classes of compounds are involved, e.g. coumarins, quinones, furans, flavonones and some others. Upon activation, they interact with and inactivate various important macromolecules such as DNA, proteins and lipids. A chapter is devoted to these light-activated plant phototoxins, an outline of bioassays is given which have been used to evaluate phototoxicity and photogenotoxicity.

The toxins dealt within this Volume and the methods used to evaluate them are not meant to be exhaustive. However, they are illustrative of the wide range of analytical procedures needed for plant toxins.

Acknowledgements. The editors express their thanks to all contributors for their efforts in keeping to production schedules, and to Dr. Dieter Czeschlik, Ms S. Mees and Ms H. Böning of Springer-Verlag for their cooperation in preparing this and other volumes in the Series *Modern Methods of Plant*

Analysis. A very special thanks is due to Jose Broekmans in Nijmegen, The Netherlands, for her continuing assistance in the preparation of these volumes.

Nijmegen and Adelaide, Christmas 1990

H. F. LINSKENS
J. F. JACKSON

Contents

Evaluating the Phototoxicity and Photogenotoxicity of Plant Secondary Compounds
R. R. AUCOIN, E. SCHNEIDER, and J. T. ARNASON (With 1 Figure)

Toxic Extracellular Enzymes
H. M. KALISZ and M. E. KALISZ

Multi-Toxin TLC Methods for Aflatoxins, Ochratoxin A, Zearalenone and Sterigmatocystin in Foods
L. M. V. SOARES (With 1 Figure)

Isolation and Characterization of Elicitors
K.-H. KOGEL and B. BEISSMANN (With 5 Figures)

Identification and Characterization of Suppressors
B. BEISSMANN and K.-H. KOGEL (With 4 Figures)

Screening for Plant Antioxidants
W. BORS, M. SARAN, and E.F. ELSTNER

Methods for the Analysis of Isoprene Emission from Leaves
A.J. HILLS, R. FALL, and R.K. MONSON (With 6 Figures)

Analysis of Ecotoxic Agents Using Pollen Tests
P. L. PFAHLER (With 1 Figure)

The Determination of the Allelopathic Potential of Pollen and Nectar
S. D. MURPHY (With 1 Figure)

Determination of Cyanide and Cyanogenic Glycosides from Plants
A. M. BRINKER and D. S. SEIGLER

List of Contributors

ARNASON, JOHN T., Biology Department, University of Ottawa, Ottawa, Canada K1N 6N5

AUCOIN, RICHARD R., Ottawa-Carleton Institute of Biology, University of Ottawa, Ottawa, Canada K1N 6N5

BEISSMANN, BERND, Institut für Biologie III, Worringer Weg, RWTH Aachen, W-5100 Aachen, FRG

BORS, WOLF, GSF Forschungszentrum für Umwelt und Forschung, W-8042 Neuherberg, FRG

BREIMAN, ADINA, Department of Botany, The George S. Wise Faculty of Life Sciences, Tel Aviv University, Tel Aviv 69978, Israel

BRINKER, ANITA M., Department of Plant Biology, University of Illinois, Urbana, IL 61801, USA

ELSTNER, ERICH F., Institut für Botanik und Mikrobiologie, Technische Universität, Arcisstraße 21, W-8000 München, FRG

FALL, RAY, Department of Chemistry and Biochemistry, Cooperative Institute for Research in Environmental Sciences, University of Colorado, Boulder, CO 80309, USA

GALUN, ESRA, Department of Plant Genetics, The Weizmann Institute of Science, Rehovot 76100, Israel

HILLS, ALAN J., Atmospheric Chemistry Division, National Center for Atmospheric Research, Boulder, CO 80307, USA

KALISZ, HENRYK M., GBF – Gesellschaft für Biotechnologische Forschung, Mascheroder Weg 1, W-3300 Braunschweig, FRG

KALISZ, MARIOLA E., Department of Botany and Microbiology, University College London, Gower St., London, UK

KOGEL, KALLE, Institut für Biologie III, Worringer Weg, RWTH Aachen, W-5100 Aachen, FRG

KOHMOTO, KEISUKA, Faculty of Agriculture, Tottori University, Koyama, Tottori 680, Japan

MONSON, RUSSELL K., Department of Environmental, Population and Organismic Biology, University of Colorado, Boulder, CO 80309, USA

MORGAN, MICHAEL R. A., Department of Food Research, AFRC Institute of Food Research, Norwich Research Park, Colney Lane, Norwich, Norfolk, NR4 7UA, UK

MURPHY, STEPHEN D., Department of Biology, Queen's University, Kingston, Ontario, Canada K7L 3N6

PFAHLER, PAUL L., Department of Agronomy, University of Florida, Gainesville, FL 32611-0500, USA

SARAN, MANFRED, GSF Forschungszentrum für Umwelt und Forschung, W-8042 Neuherberg, FRG

SCHNEIDER, EDWARD, Plant Research Center, Agriculture Canada, Ottawa, Ontario, Canada

SEIGLER, DAVID S., Department of Plant Biology, University of Illinois, Urbana, IL 61801, USA

SOARES, LUCIA M. V., Faculdade de Engenharia del Alimentos, Universidade Estadual de Campinas, Caixa Postal 6121, 13081 Campinas, São Paulo, Brazil

STIERLE, ANDREA, Department of Plant Pathology, Montana State University, Bozeman, Montana, USA

STIERLE, DONALD, Department of Chemistry and Geochemistry, Montana State University, Bozeman, Montana, USA

STROBEL, GARY, Department of Plant Pathology, Montana State University, Bozeman, Montana, USA

SUGAWARA, FUMIO, Riken, The Institute of Physical and Chemical Research, Wako Saitama 351-01, Japan

RUDOLPH, KLAUS, Institute for Plant Pathology and Plant Protection, University of Göttingen, Grisebachstr. 6, W-3400 Göttingen, FRG

WARD, COLIN M., Department of Food Research, AFRC Institute of Food Research, Norwich Research Park, Colney Lane, Norwich, Norfolk, NR4 7UA, UK

WILKINSON, ANDREW P., Department of Food Research, AFRC Institute of Food Research, Norwich Research Park, Colney Lane, Norwich, Norfolk, NR4 7UA, UK

WYDRA, KERSTIN, Institute for Plant Pathology and Plant Protection, University of Göttingen, Grisebachstr. 6, W-3400 Göttingen, FRG

Analytical Methods for Phytotoxins

A. STIERLE, G. STROBEL, D. STIERLE, and F. SUGAWARA

1 Introduction

Diseases of plants are caused by many of the same classes of agents responsible for the diseases of man and animals. However, fungi and bacteria are the most important in terms of distribution, diversity, and total damage to plants in the field as well as in storage. Most of the plant pathogenic fungi examined to date produce one or more phytotoxins. Phytotoxin, for the purpose of this chapter, refers to a compound produced by a plant parasitic organism that is toxic to plants and may play some role in symptom expression. Phytotoxins are commonly found in those organisms that are necrotrophic, versus the biotrophic or obligate parasitic microbes. The term phytotoxin does not refer to the phytohormones and is to be contrasted with plant toxins, which are compounds produced by plants that adversely affect man or animals such as the castor bean toxin (ricin) or the death cap mushroom toxin (amanitin).

Prior to 1970, reports on phytotoxins appeared only sporadically. Many of these early reports focused on observations of biological activity of phytotoxins rather than their chemistry or mode of action. Recently, new and improved techniques of separation science and spectroscopy have facilitated the isolation and characterization of phytotoxins. The importance of phytotoxins to the welfare of mankind was sharply brought into focus in 1970 with the advent of the Southern corn leaf blight disease in the USA and Canada. The disease, caused by *Helminthosporium maydis* race T, was the most devastating plant disease occurring in the shortest time in world history (Tatum 1971). Phytotoxins were subsequently shown to be involved in this disease (Karr et al. 1974); determining the nature of these compounds was the subject of several studies (Karr et al. 1974; Kono and Daly 1979; Strobel 1982; Sugawara et al. 1987). Understanding the chemistry and biology of phytotoxins may provide important insights into the mechanisms of plant disease and possible control measures. Thus, for the benefit of pathologists, plant breeders, and others interested in the health and well being of plants, a study of phytotoxinology is warranted.

One of the major difficulties in studying and utilizing a given phytotoxin is its availability. This problem was overcome when European groups in the 1970s succeeded in isolating and characterizing fusicoccin, a phytotoxin produced by *Fusicoccum amygdali* (Ballio et al. 1964; Ballio 1977). This unleashed an unprecedented number of physiologists, biologists, chemists, pathologists, and agronomists in an attack on the mode of action, usefulness, and general biology of this phytotoxin (Marrè 1979). Each of the many new phytotoxins that has since been described provides a new target for a thorough, concerted chemical and biological investigation.

Besides their obvious role in the development of symptoms of certain plant diseases, phytotoxins also possess some unusual and intriguing chemical and biological properties. Approximately 15 of the phytotoxins isolated to date exhibit extremely high host specificity; at low concentrations they affect only those plant species or cultivars that are similarly affected by the producing pathogen. On the other hand, some exhibit host specificity that differs from that of the producing pathogen. At the other extreme are those compounds that have no specificity whatsoever. Included in this group are certain macromolecular compounds that cause wilting in plants not by interfering with metabolic function or cell integrity, but by virtue of their size: they physically impair water movement in infected plants.

In a toxin-sensitive host, phytotoxins interact with a specific target site, which ultimately leads to some outward manifestation by the plant. Thus, toxins help us understand the mechanisms of plant disease development, including pathogen specificity, disease susceptibility and resistance. Furthermore, phytotoxins have proven useful tools for probing the normal physiological and biochemical activities of plants. Some have even provided a practical means for selecting seeds, tissues, or seedling seeds for disease resistance (Strobel 1982).

In recent years, a number of phytotoxins from pathogens of weedy plants have been isolated and characterized (Kenfield et al. 1988a). In several cases, these phytotoxins represent novel chemistry and have been considered as new herbicide models. There is a potentially endless list of pathogens from weed species that should be studied for phytotoxin production. Knowledge of the phytotoxinology of a given weed pathogen may eventually lead to the development of genetically engineered pathogens as weed biocontrol agents.

Phytotoxins may have other uses than as weed biocontrol agents. By expanding bioassay techniques to include other organisms, we have recently found excellent antifungal activity in the phytotoxin cercosporamide (Sugawara et al. 1991). Earlier, others made similar observations with other phytotoxins, including the syringomycins (Ballio et al. 1988).

The tropical regions of the world represent a virtually untapped reservoir of pathogenic microbes for phytotoxinology. These phytopathogens affect untold numbers of forest herb species as well as grasses and flowering plants. These rapidly disappearing higher plant species are most certainly accompanied by their associated pathogens into the oblivion of extinction; and with these pathogens, pass all the secrets and potentially unique biology and chemistry their phytotoxins might represent. Such compounds could have important pharmacological, biochemical or agrochemical uses. Many of the world's common food and fiber crops have associated pathogens that have never been investigated for their phytotoxin-producing ability. These include banana, cassava, plantain, rice, tropical fruit species, and countless other plant species.

Understanding the biology and chemistry of phytotoxins provides important insights into potential uses of these compounds. But it is important for other reasons as well. With the advent of modern methods of genetic engineering technology, it is possible that plants may be genetically manipulated to produce enzymes that degrade or derivatize phytotoxins to inactive forms. This type of work requires knowledge of phytotoxin structure, of the enzyme(s) that use it as a substrate, and the source of such an enzyme. The work of Anzai (1989) on tabtoxin illustrates this point. Tabtoxin is produced by *Pseudomonas syringae* pv. tabaci and causes chlorotic symptoms associated with wildfire disease of tobacco. The tabtoxin resistance gene (ttr) coding for an acyltransferase (detoxifying enzyme) of *P. syringae* pv. tabaci was fused to the promoters of the cauliflower mosaic virus to construct a chimeric gene, which was introduced into tobacco cells by *Agrobacterium*-mediated transformation. The transgenic tobacco plants expressed the ttr gene, which detoxifies tabtoxin.

Although many plant pathogens exhibit phytotoxic activity, relatively few phytotoxins have been isolated and completely characterized, and even fewer have been synthesized. Some have been subjected to biosynthetic studies, but a complete pathway with all intermediates and enzymes identified is known for very few phytotoxins. The interaction of the host plant with a phytotoxin also requires additional study. Certain plants modify phytotoxins to either activate or deactivate them. The mechanisms involved in this interaction demand further attention.

As a group, phytotoxins have no common structural features. Unlike the proteinaceous toxins isolated from snake and insect venoms, and most other poisonous animals, the majority of phytotoxins characterized to date are relatively low molecular weight compounds. Although some proteinaceous phytotoxins are known, this chapter is primarily concerned with the low molecular weight organic molecules that constitute the bulk of characterized phytotoxins. Phytotoxins represent most of the various classes of organic molecules, including polyketides, terpenes, phenolics, amino acid derivatives, alkaloids, and glycosides.

Phytopathogens usually produce a series of related compounds, of which only one or two may be phytotoxic (Stierle et al. 1988). Organisms may also produce highly host-specific phytotoxins in conjunction with compounds that adversely affect a wide variety of plants. And there is some evidence that phytotoxins may act synergistically (Stierle et al. 1989).

2 Culturing Considerations

Determining the phytotoxin-producing potential of a pathogenic microorganism can be facilitated by observing its affect on an infected plant. Generally, if it induces flecks, spots, necrotic areas, discolored "runners', localized chlorosis, or other abnormal coloration, then it is a good candidate for phytotoxin study.

Investigation of phytotoxins from a phytopathogen is usually initiated with fermentation of the microbe in 1–2 l of an appropriate medium. Generally, a defined medium is used, such as M-1-D (Sugawara et al. 1987). More complex media may support growth of the pathogen, but complicate the isolation procedure. An effusion made from the host plant tissues is usually added to the medium; many pathogens do not produce phytotoxins without this inclusion. Culture conditions may be critical to toxin production. Temperature, pH, duration, light intensity, and whether or not a culture is shaken may have a profound affect on the production of phytotoxins. Certain microorganisms must be maintained on infected plant parts, on a selective medium, or on agar plugs (at 4 °C in sterile, distilled water) to retain their ability to produce phytotoxins.

3 Bioassay Techniques

A phytotoxin cannot be isolated unless a specific bioassay procedure has been devised to guide each step in the purification process. A number of phytotoxin assays have been developed that toxicity information in a relatively short period of time. One of the most common is a leaf puncture-wound overlay technique that was instrumental in the isolation of bipolaroxin (F. Sugawara et al. 1985). A leaf, preferably of the host species, is first punctured with a 10–25 gauge Hamilton syringe. The puncture is overlayed with a droplet of toxin solution (5–10 µl). The inoculated leaf is then incubated in a moist, sterile chamber. The toxin solution is prepared by dissolving the compound (1–20 µg) in 2% methanol or ethanol. Water-insoluble toxins are first dissolved in methanol, then diluted to the appropriate concentration with water. Control tests are run concurrently with the solvent. The leaves are usually tested 12 h day/12 h night, although etiolation may be necessary for green island formation. Leaves are checked periodically over 72 h for symptom development.

Several other assay techniques have been devised which measure a number of physiological responses, including water conductivity, CO_2 fixation, mitochondrial effects, seed germination inhibition, root or pollen growth inhibition, effects on stomatal opening, and various membrane effects. Many of these tests are summarized by Strobel (1982). One of the most quantitative bioassay methods involves the use of flow cytometry to measure phytotoxin effects on protoplasts (Berglund et al. 1988). Protoplasts exposed to a phytotoxin preparation are labeled with a fluorescent dye, such as fluorescein diacetate, and then passed through a flow cytometer. Toxin concentration effects, time course dependence, LD_{50} values, and other information can be quickly and accurately obtained using this method. It may also provide some insight into the mode of action of the phytotoxin.

Before any chemical or biological studies can be initiated on a phytotoxin, however, it must be isolated and characterized. The bioassays mentioned above can be used to guide any isolation scheme. But before an effective separation protocol can be devised, it is important to know and understand the various separation techniques available.

4 Separation Techniques

Microbial fermentation usually results in a complex aqueous mixture of salts and dissolved or suspended organic metabolites. The isolation of phytotoxic metabolites from the filtrate usually involves separating pure, active compounds from this mixture of both organic and inorganic material. Two concerns are critical: the purification process must be thorough and it must be nondestructive. It is essential to choose a gentle separation strategy that does not alter metabolites in the purification process. To facilitate isolation of the phytotoxic components, we have adopted a bioassay-guided fractionation strategy that includes a variety of separation techniques.

Once an organism is grown, it is removed from the filtrate either by centrifugation or gravity filtration. The phytotoxin isolation procedure is initiated with a liquid-liquid partition process: a simple solvent extraction. The phytotoxic components in the filtrate may vary in polarity, and the polarity of desired metabolites dictates the separation protocol. The microbial filtrate is partitioned between water and an immiscible, moderately polar solvent (i.e., ethyl acetate or methylene chloride). Both extracts are then reduced to dryness; the water-soluble extract is lyophilized, and organic-soluble extract is rotoevaporated. The two extracts are then tested for phytotoxicity (100 and 1000 ppm in 2% aqueous ethanol) using an appropriate bioassay. Determining whether the phytotoxic components are water-soluble or organic-soluble determines the type of separation strategy followed.

Most of the phytotoxins isolated have been moderately polar, organic-soluble compounds with molecular weights less than 2000 amu. However, phytotoxins are not limited to this physical description. Regardless of the nature of the active metabolite, it must be isolated from either the organic-soluble or water-soluble extract and purified. Resolution of the organic-soluble extract into pure, active metabolites usually involves a combination of chromatographic techniques. These include column chromatography (gravity, low and high pressure) and centrifugal countercurrent chromatography. Aqueous components can be isolated with similar methodologies using more polar solvent systems. If the toxin is proteinaceous, gel filtration and gel electrophoresis can also be employed.

4.1 Column Chromatography

Column chromatography encompasses a variety of separation techniques and employs numerous stationary supports. The separation mechanism can exploit different physical properties of the components of a mixture, depending on the type of stationary support used. Chromatographic separations are all based on a similar phenomenon: the balance of the driving force of the mobile phase and the retarding force of the staionary phase, to optimize separation of a mixture into individual components. The mechanisms involved in achieving this balance, however, may vary greatly in different techniques. Compounds can be separated based on size, polarity, functional groups, charge, or specific affinities. Certain techniques utilize one or more of these properties to effect separation of mixtures. Purification is greatly enhanced when different types of chromatographic procedures are used in tandem.

It is not our intention to catalog and describe all of the chromatographic techniques currently available. We will only discuss those techniques that we use routinely, techniques that afford excellent resolution power and gentleness. Our stategy includes size exclusion (gel filtration) gravity columns and bonded phase silica low and high pressure columns. All of our gravity and HPLC columns are interfaced with UV monitors and chart recorders. Fractions are designated based on UV peak correlations. We also use thin-layer chromatography (TLC), both silica gel and bonded phase silica gel, generally not as a purification technique, but as indicator of purity and as an aid in the selection of appropriate solvent systems.

4.1.1 Gel Filtration Chromatography

The terms "size exclusion", "gel filtration", and "gel permeation" are often used interchangeably to describe a chromatographic technique that separates molecules based on size. The advent of gel filtration chromatography is generally attributed to P. Flodin and J. Porath, of the Biochemical Institute in Uppsala, Sweden. In 1959, it was demonstrated that columns packed with cross-linked polydextran gels, swollen in aqueous solvents, could size-separate various water-soluble macromolecules (Porath and Flodin 1959). The advent of gel permeation chromatography followed in 1964, when J.C. Moore of the Dow Chemical Company disclosed the use of cross-linked polystyrene "gels" to separate organic-soluble synthetic polymers (Moore 1964; Moore and Hendrickson 1965). Gel filtration and gel permeation are synonymous, although the latter usually refers to molecular weight determination and separation of synthetic polymers on polystyrene gels (Yau et al. 1979). Gel filtration chromatography is an important component of our separation protocols. The products that we utilize routinely with consistently satisfactory results are described below.

The original Sephadex G-type gels are prepared by cross-linking dextran with epichlorohydrin. These bead-formed gels are extremely hydrophilic due to the large number of hydroxyl groups; consequently, they swell readily in aqueous solutions. Sephadex G-type gels are available with varying degrees of cross-linking. Each type has a different molecular weight range over which molecules can be fractioned. The highly cross-linked gels, G-10 and G-15, have exclusion limits of 700 and 1500 amu, respectively, and are useful for polar organic compounds below these limits (typical of most phytotoxins). Proteins and polydextrans can be separated on gels with a small degree of cross-linking.

Pharmacia Fine Chemicals modified the Sephadex G-type gels in 1966 (Joustra et al. 1967). The resultant LH-type gels facilitate gel filtration chromatography in organic solvents. Sephadex LH-20, ideal for compounds <2500 amu, is prepared by hydroxypropylation of Sephadex G-25. The introduction of hydroxypropyl groups does not alter the number of hydroxyl groups, but does increase the ratio of carbon to hydroxyl. Sephadex LH-20 has both hydrophilic and lipophilic properties. It swells in water, aqueous solvent mixtures, polar-organic solvents, and mixtures of organic solvents. In solvent mixtures containing both polar and nonpolar organic solvents, the gel preferentially includes the polar component.

In a classical sense, separation in gel filtration is dependent on the interaction of solute molecules with the mobile and stationary phases, strictly as a function of size. Molecules small enough to diffuse into the stationary phase (solvent within the gel pores) are retained. Large molecules which exceed the exclusion limit of a particular gel are not retained, and elute rapidly from the column. Molecules are eluted in order of decreasing molecular size, as long as molecular sieving effects predominate. In practice, however, gel filtration is far more complex.

Reversible adsorption and partition phenomena also figure strongly in elution order (Janson 1967; Williams 1972). Both the Sephadex G-type and LH-20 gels have an affinity for aromatic and polar compounds. In Sephadex LH-20 these affinities can be used to enhance separation by increasing retention of specific types of molecules. Reversible aromatic adsorption is most pronounced when lower alcohols are used in the solvent mixture. Reversible adsorption of polar compounds is most pronounced when chloroform is used as a solvent. Since Sephadex LH-20 swells only in polar and moderately polar organic solvents, its use in partition chromatography is limited to "straight-phase" partition. It specifically retains the more polar-organic solvent, creating a marked polarity difference between the mobile and stationary phases. The resultant stationary phase has a large volume and is quite stable, with highly reproducible results. Partition effects are also observed with two polar solvents, like water-MeOH. Water is retained within the gel pores, enhancing the retention of very polar compounds.

Bio-Rad also offers a variety of gel filtration media. Their Bio-Gel P series, hydrophilic polyacrylamide gels, can be used in place of the Sephadex G-type gels. They are compatible with dilute organic acids, detergents, buffers, distilled water, and aqueous methanol or ethanol solutions <20% (v/v). The extremely hydrophilic Bio-Gel P series is an excellent support for the separation of peptides, proteins and oligosaccharides, and other water-soluble compounds.

For nonaqueous or nonpolar systems, Bio-Rad offers S-X beads, neutral, nonporous styrene-divinylbenzene copolymer beads intended for gel permeation chromatography of lipophilic organic solutes. The beads are usually used in toluene, chloroform, xylene, and in mixtures of relatively nonpolar organic solvents. The resolving power of these gels is excellent, but the lipophilicity of the gels limits their application in many separation schemes.

We have used Sephadex LH-20 and Bio-Rad S-X beads in tandem in many isolation protocols. In keeping with the basic philosophy that each step in a separation scheme should exploit a different mechanism, these two gels offer disparate retention characteristics. Each of these gels can be used in different solvent systems which can dramatically alter separation. The isolation of maculosin, a host-specific toxin of spotted knapweed, *Centaurea maculosa* (Stierle et al. 1988), is a good example of their resolving power. The ethyl acetate-soluble extract of *Alternaria alternata* exhibited phytotoxicity toward spotted knapweed. This extract was chromatographed on Bio-Beads S-X4 (hexane/methylene chloride/ethyl acetate, 4:3:1). Twelve fractions eluted: fraction five exhibited selective phytotoxicity, fractions four and six general phytotoxicity, toward spotted knapweed. Size exclusion chromatography of fraction five on Sephadex LH-20 (methylene chloride/*tert*-butyl methyl ether/isopropanol, 1:1:1) yielded crystalline maculosin **1**. Further gel permeation of fraction four on Bio-Beads S-X8 (methylene chloride/cyclohexane, 3:2) yielded the crystalline diketopiperazine, *cyclo*Pro–*homo*Leu **2**.

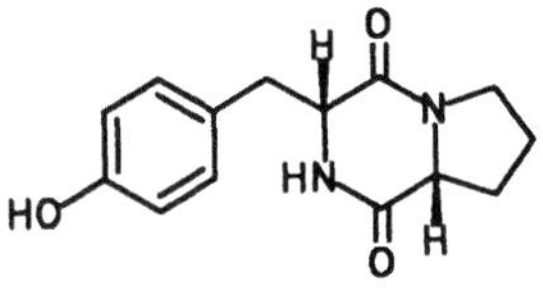

1

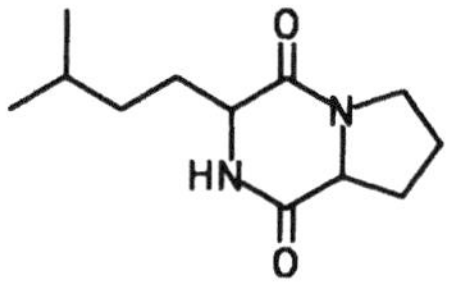

2

Most of our separations begin with an initial chromatography using Sephadex LH-20, in either chloroform/methanol (1:1); methanol; or methanol/water (4:1), depending on the solubility of the extract. The polarity of phytotoxic fractions determines what chromatographic protocol is employed. Sephadex LH-20 has proven successful as the sole chromatography support in the isolation of certain phytotoxins. The pure phytotoxin cyperine **3** was isolated from the ethyl acetate extract of the fungus *Ascochyta cypericola* in our study of weed pathogens of purple nutsedge, *Cyperus rotundus* L (Stierle et al. 1991).

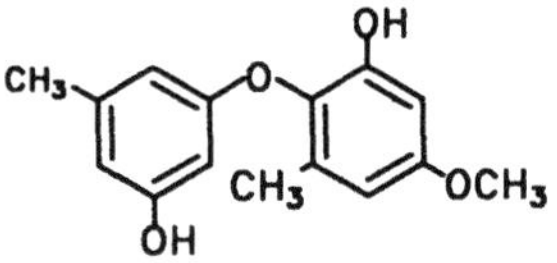

3

Not all of our separation problems have been resolved with Sephadex LH-20 or Bio-Beads S-X gels. LH-20 does not always effectively separate polar components, and S-X gels cannot be used in polar-solvent systems. More and more of our phytotoxic mixtures have defied the use of Bio-Beads S-X and its excellent resolution power, because they were too polar to dissolve in the appropriate solvent systems. Professor Yuzuru Shimizu, University of Rhode Island, recommended a size exclusion gel to deal with this dilemma. TSK-gel Toyopearl is a porous, semirigid spherical gel manufactured by Toya Soda of Japan. The base material of this gel is a hydrophilic vinyl polymer. TSK-gel Toyopearl HW-type gels combine the affinities associated with unsaturated resins with the hydrophilicity of a Sephadex LH-20 or G-type gel. It can be used with a wide range of solvents, and shows similar swelling properties with water, methanol, acetone, and chloroform. Because of its extreme stability, it is an excellent gel for gradient elution gel filtration chromatography. Because of its swelling characteristics, it can be used in a gradient system that ranges from water to methanol/water to methanol/chloroform to methanol to chloroform. In different column chromatographic separations of crude phytotoxic extracts using this gradient system, the resolution was phenomenal. The bands were discreet and well separated. Many of these mixtures were poorly resolved in prior attempts at separation on Sephadex LH-20, G-15, and C_{18}-bonded phase silica gel HPLC. TSK-gel is a new addition to our laboratory, but is rapidly gaining favor as it repeatedly solves difficult separation problems.

4.1.2 Reverse Phase Chromatography

Compound purification can be facilitated by alternating between different separation mechanisms. Gel filtration chromatography separates molecules based primarily on their size, their partition coeffients, and on their reversible adsorption to the gel matrix. Techniques that emphasize adsorption phenom-ena provide a good alternative to gel filtration. When used in tandem, compound isolation can be rapid and effective.

Classical adsorption chromatography is the oldest of the chromatographic techniques. It is based on the retention of a solute by surface interactions with a solid support and is one of the simplest and most effective methods for separating mixtures of nonpolar compounds. It can be used in both column and thin layer applications. Columns can be run with gravity flow, low pressure, or high pressure.

Silica gel is widely used for adsorption chromatography, particularly for the separation and purification of lipids and other nonpolar compounds. It is a polar, weakly acidic matrix that interacts with adsorbed molecules via specific forces such as electrostatic attraction and hydrogen bonding. These interactions are much stronger than the reversible adsorbent forces in gel filtration chromatography. Unfortunately, these stronger adsorbent effects can lead to irreversible binding of solute molecules to the silica gel support. This is particularly true of polar or basic molecules. Adsorption chromatography with silica gel can also lead to alteration of labile compounds. Chromatography on silica gel can promote oxidation of sensitive materials if oxygen is not excluded (March 1985). This problem is most pronounced with adsorption chromatography on thin layer plates exposed to air. Alteration phenomena are mitigated in high pressure systems (HPLC): rapid flow rates reduce compound exposure to the adsorbent and aerobic effects are minimized.

Despite its excellent resolving power, silica gel (or any other strong adsorption support) is rarely used in our initial compound isolations. The risk of molecular alteration or irreversible binding is too great with small, sensitive, bioactive samples. Only if a compound is very nonpolar and readily available, will we use low pressure (flash) silica gel chromatography. It can be a powerful separation tool, and is actually preferable to HPLC silica chromatography. Flash columns are disposable, so use of methanol as a deactivating solvent is acceptable. Deactivating silica gel mitigates adsorptivity, and the use of methanol as a solvent improves elution of moderately polar compounds. Careful examination of the eluted material for spectral modifications or loss of activity will obviate the use of silica gel in a separation scheme.

With the advent of bonded phase column packings, however, adsorption chromatography can be used effectively with even the most sensitive samples. Bonded phase packings typically consist of a silica gel matrix with covalently bound side chains. The bonded phase can be either polar or nonpolar, and can be used in either normal phase or reversed phase mode. The most versatile of the bonded phases are the alkyl-substituted packings, with either C_8 or C_{18} substituent groups. These relatively nonpolar bonded phases are used with polar mobile phases to separate a variety of mderately polar to polar solutes. Alkyl-substituted C_8 or C_{18} bonded phases can be used most effectively in a gradient elution mode.

Reversed phase HPLC (RP-HPLC) has proven an effective separation tool in a number of our studies. It can be used either isocratically or in a gradient elution scheme. In our investigation of the phytotoxic components of *Exserohilum holmii*, the phytotoxin exserohilone **4** and its analog 9,10-dihydro exserohilone **5** were cleanly resolved using C_{18} (Merck, RP-18) with acetonitrile/water (63:35) (D. Sugawara et al. 1985).

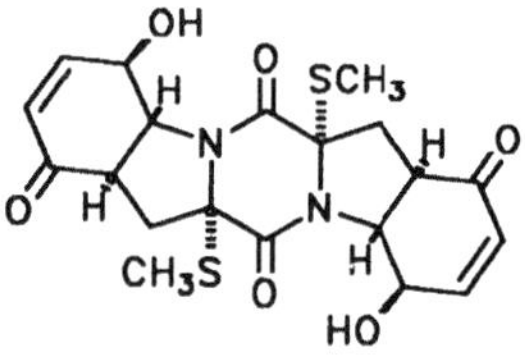

4

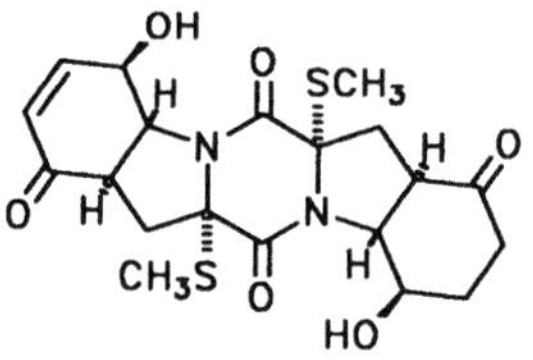

5

Gigantenone **6**, a highly oxidized eremophilane, was isolated from the ethyl acetate extract of *Drechslera gigantea* using flash chromatography on silica gel with chloroform/methanol (9:1), followed by reversed phase HPLC on RP-18 with acetonitrile/water (1:1) (Kenfield et al. 1989).

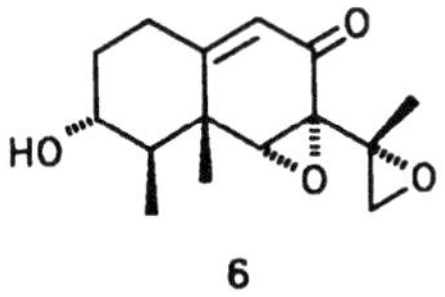

6

A series of phytotoxic sesterterpenes were isolated from the pathogenic fungi *Drechslera maydis* and *Drechslera sorghicola*. *D. maydis* was the causal agent of the Southern corn leaf blight epidemic of 1970. The phytotoxins associated with this disease have been investigated with contradictory results (Karr et al. 1974; Strobel 1982; Kono et al. 1985). We isolated a series of phytotoxic ophiobolins from the ethyl acetate extracts of both *D. maydis* and *D. sorghicola*. The extracts were flash chromatographed on silica gel with chloroform/methanol (20:1) and resolved with reversed phase HPLC on RP-18 with acetonitrile/water (65:35). The retention times of the six ophiobolins **7–12** varied from 4.3 min for 25-hydroxyophiobolin I **10** to 30.5 min for ophiobolin C **9** (Sugawara et al. 1987).

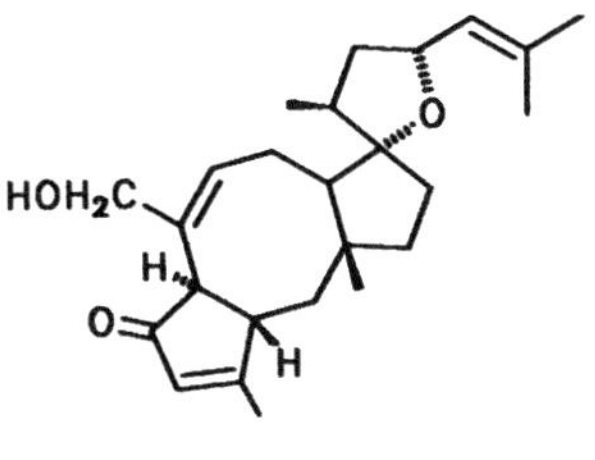

7

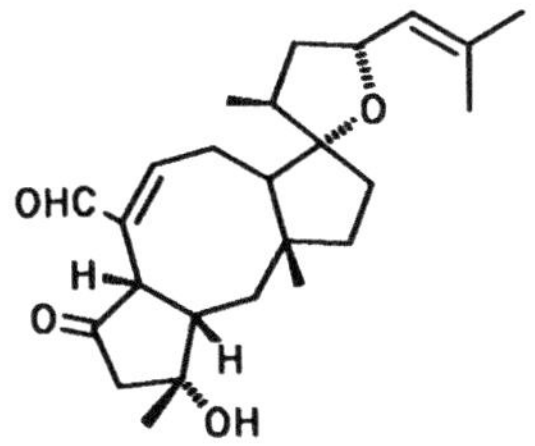

8

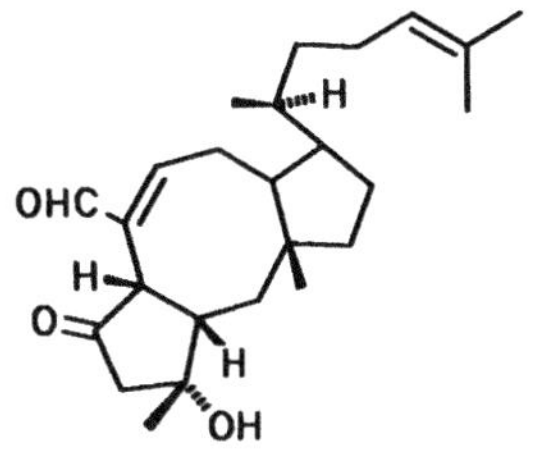

9

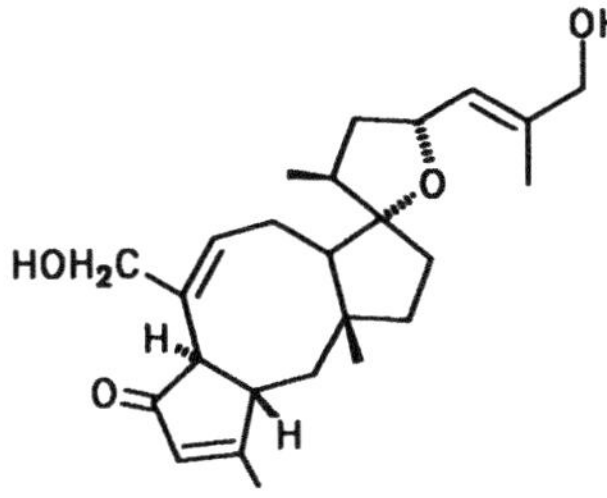

10

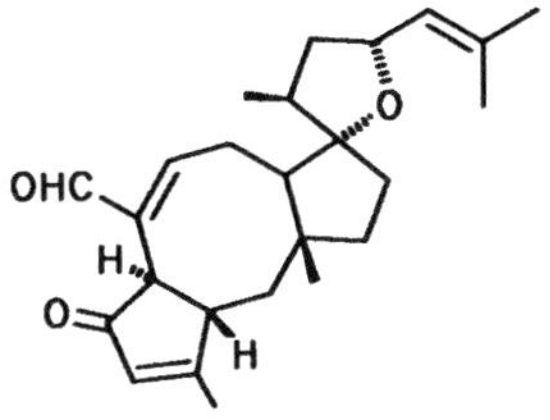

11

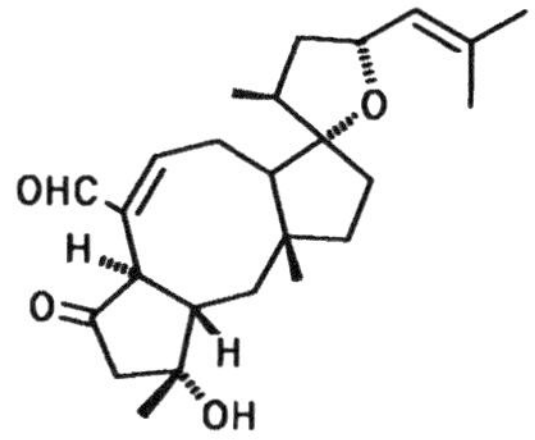

12

Although isocratic systems can be most effective, finding the optimum solvent system may be a time-consuming, trial and error proposition. Unlike silica gel TLC plates, reversed phase TLC plates are not always reliable as solvent system models. The use of gradient elution HPLC provides a real solution to this dilemma. The appropriate gradient can separate a complex mixture of components in a relatively short period of time. Because a single gradient covers all possible binary solvent compositions within its polarity range, it can be an effective scouting technique for isocratic systems as well.

A gradient system is chosen with a polarity range that flushes all material from the column. In the investigation of phytotoxins of bananas and plantains from *Mycosphaerella fijiensis*, gradient elution HPLC provided rapid, effective separation. The ethyl acetate extract was chromatographed on Sephadex LH-20; eight fractions eluted, with activity concentrated in fraction four. Fraction four could not be resolved with gel filtration, and there was TLC evidence that silica gel altered some of the components in the mixture.

Preparative HPLC was attempted using a gradient from acetonitrile/water (25:75) to acetonitrile, on an Altex Ultrasphere RP-18 column. Three main clusters of peaks eluted over 25 min. Phytotoxicity was concentrated in the peak cluster eluting between 27 and 31% acetonitrile. This fraction was reinjected using a gradient between 25 and 50% acetonitrile. The pure phytotoxin 2,4,8-trihydroxytetralone **13** was isolated (Stierle et al. 1991).

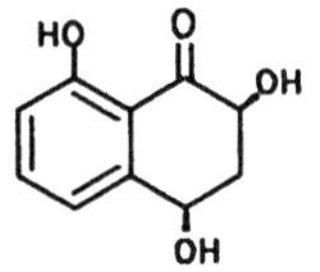

13

Thin layer chromatography (TLC) with a silica gel adsorbent is an excellent isolation tool. Although it can be used to effect actual separations of mixtures, its primary utility is diagnostic. TLC can indicate both the relative purity of samples and appropriate solvent choices for further separations. After every step in an isolation scheme, each eluant is spotted on silica TLC plates. Polar samples are also spotted on C^{18}-bonded phase TLC plates. The plates are developed in a variety of solvent systems, usually consisting of chloroform/methanol or ethyl acetate/hexane. Multiple development serves two purposes. If a compound is resolved to one discreet spot on a TLC plate, the presence of a single compound is suggested. If this apparent purity is repeated in three or four different solvent systems, then the presence of one compound is strongly indicated. The differential rates of migration (R_f) of individual compounds in a mixture in various TLC solvent systems can be used to select an appropriate HPLC solvent system. This is particularly true in normal phase silica gel HPLC, which, unlike reverse phase HPLC, is generally run in an isocratic mode. Selecting a solvent system that results in well-resolved solutes with R_f values between 0.3 and 0.6, generally provides good separation without tailing. Of course, this is an ideal situation. If a single solvent system cannot be found for a given mixture, flash silica gel chromatography can provide preliminary separation. The eluants can be tested for phytotoxicity, and the active fraction can be further resolved with HPLC.

4.2 Centrifugal Countercurrent Chromatography

Centrifugal countercurrent chromatography (CCC) has proven to be a powerful tool for the separation of complex mixtures of natural products. CCC is actually a reincarnation of countercurrent chromatography, a classic separation technique pioneered by Lyman Craig in the 1950s (Craig and Craig 1956). The earliest instruments were capable of resolving complex mixtures with great facility and without the problem of irreversible adsorption. Unfortunately, a typical experiment could consume the greater part of a week, although the resolution achieved surpassed other separation techniques available at that time. The advent of HPLC, which promised rapid separations, supplanted countercurrent techniques with their exhorbitant run times.

In the last 10 years, however, new instrument designs capable of reducing experiments from days to hours have renewed interest in CCC (Ito 1981). The advent of more efficient, faster instruments coincided with increased demand for new agrochemicals and pharmaceutical agents from natural sources. Many bioactive compounds are quite polar or labile, and pose a real isolation challenge. Countercurrent chromatography has several inherent advantages designed to meet even the toughest separation challenges. It is extremely versatile: any immiscible two-phase solvent system can be used in either a normal phase or reversed phase mode. There is no solid support, so there is no danger of adsorption. Solvents can be chosen which mitigate or eliminate the chance of chemical alteration. If the perfect solvent system is found, CCC is an ideal separation technique.

Countercurrent chromatography is a liquid-liquid partition technique using two immiscible liquid phases. One phase is designated as the stationary phase and is retained either by gravitational (DCC) or centrifugal force (CCC). The mobile phase is passed through the stationary phase, usually as fine droplets. Sample mixture is dissolved and introduced with the mobile phase at one end of the stationary phase. The various components of the sample mixture have different partition coefficients in each of the two solvents, and therefore migrate through the stationary phase at different rates. Compounds that favor the mobile phase will migrate more rapidly than compounds that favor the stationary phase (Ito 1986).

In modern centrifugal countercurrent chromatography, a multilayer coiled column is subjected to synchronous planetary and revolutionary motions, generating a centrifugal force field that effects retention of the stationary phase. The dual rotatory motion of the CCC creates a hydrodynamic equilibrium between the two liquid phases, dominated by the stationary phase. The introduced mobile phase moves rapidly through the column in a dynamic continuous flow fashion, analogous to other chromatography systems. Mandava and Ito achieved baseline resolution of abscisic acid and indole acetic acid from a crude extract of zoysia grass. Resolution was achieved in slightly more than 1 h (Mandava and Ito 1982). The authors also resolved four indole auxins: indole-3-acetamide, indole-3-acetic acid, indole-3-butyric acid, and indole-3-acetonitrile, with baseline precision in 2.5 h, using the multilayer coil CCC (Ito et al. 1982; Mandava and Ito 1984). The same separation required 15 h using an earlier instrument design (Ito 1981) (Mandava and Ito 1982).

The only potential difficulty in using CCC is selecting the appropriate solvent system for maximal separation. Numerous tables of solvent schemes have been published that deal with specific classes of compounds. Systems employed in droplet countercurrent chromatography work well in CCC, and can be used without modification (Hostettmann 1980). Ito lists solvent systems for tannins, plant hormones, steroids, glycosides, lignans, and a wide array of other compounds in his excellent review of countercurrent chromatography (Ito 1986). If the class of compounds to be isolated is known, then a recommended solvent system can be used directly.

If the sample is a complex mixture, or if the nature of the components is unknown, selection of an ideal solvent system can be more involved. A simple approach is to select a system and modify it. Chloroform/methanol/water systems can accommodate a broad polarity range and constitute a good starting point. Choose a moderately polar system such as 7:13:8 (v/v) to begin. Allow the solvents to equilibrate, then put 2 ml of each phase into a test tube. Dissolve a small amount of sample in the solvent mixture and mix thoroughly. Equilibrate the solvent mixture and allow the two phases to reestablish themselves. Remove an aliquot of each phase, evaporate to dryness, and weigh the residue. The ratio of the weights of the two phases is a measure of the partition coefficient. If the desired component absorbs at a particular wavelength in the ultraviolet region, then the partition coefficient can be determined using a spectrometer. An aliquot of each phase is measured on a spectrometer, and the ratio of these absorbances determines the partition coefficient.

An ideal partition coefficient is between 0.5 and 2, although values between 0.2 and 5 are acceptable. Partition coefficients can be adjusted by altering the hydrophilicity of the solvent system. If most of the sample is dissolved in the lower, nonaqueous phase, then the system can be made more hydrophobic by increasing the ratio of chloroform to water. If most of the sample is dissolved in the upper, aqueous phase, then the hydrophilicity of the system can be increased. Chloroform/methanol/water systems can range from moderately nonpolar (13:7:8) to polar (5:16:6) and still maintain two distinct phases. If the sample partition coefficient is unacceptable in any variant of this mixture, greater hydrophilicity can be achieved in butanol/water systems, and greater hydrophobicity in nonaqueous systems, such as heptane/methylene chloride/acetonitrile.

Solvent selection on the CCC actually becomes more straightforward with time. We have found that a few systems work quite well for most separations, and we work almost exclusively with these systems. They might not be the perfect system, but they usually afford baseline resolution of many of the components within a mixture. Each system can be used in either a normal (polar stationary phase) or reversed mode (polar mobile phase). Moderately polar compounds separate well with chloroform/methanol/water (25:34:20). If the components favor the polar phase, the chromatography should be run in a normal mode with the lower, organic phase mobile. If the components favor the organic phase, the chromatography should be run in a reversed mode with the upper, aqueous phase mobile.

This solvent system effected baseline resolution of several complex mixtures of phytotoxins. In our study of the phytotoxic components of *Alternaria alternata* against spotted knapweed, CCC solved two difficult separation challenges. The ethyl acetate extract was chromatographed on Bio-Beads S-X4. One of the moderately phytotoxic eluants could not be further resolved by gel filtration or RP-HPLC. CCC with chloroform/methanol/water (25:34:20, lower phase mobile) cleanly separated this mixture in less than 1 h, as is shown in Fig. 1. Two of the isolates were the diastereomeric diketopiperazines *cyclo*L-Pro-L-Phe **14** and *cyclo*L-Pro-D-Phe **15** (Stierle et al. 1988).

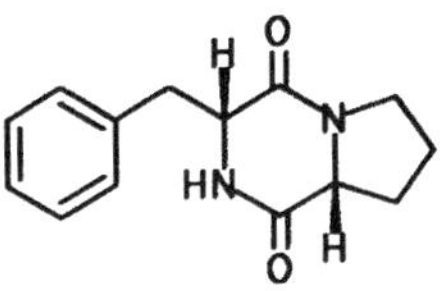

14

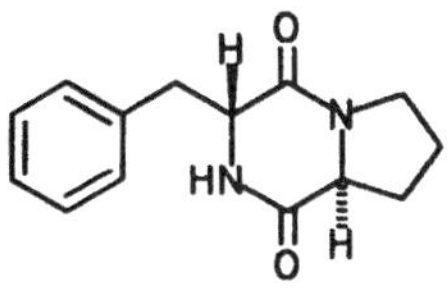

15

The methylene chloride soluble extract of *A. alternata* was also phytotoxic against spotted knapweed. It was chromatographed on Sephadex LH-20 with methylene chloride/methanol (1:1). All of the phytotoxicity was concentrated in fraction three, which defied all further attempts at separation with gel filtration chromatography. Centrifugal countercurrent chromatography with chloroform/methanol/water (25:34:20, upper phase mobile) provided baseline separation of four perylenequinones **16–19** (Fig. 1), including the phytotoxin alterlosin II **16** (Stierle et al. 1989).

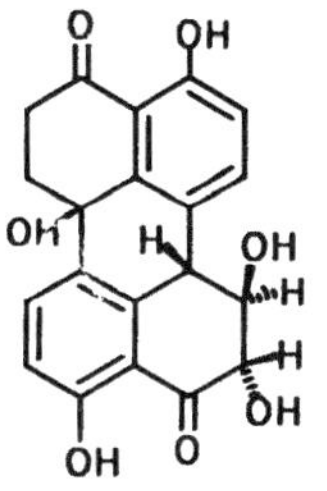

16

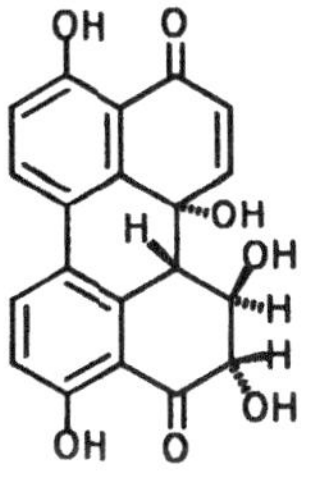

17

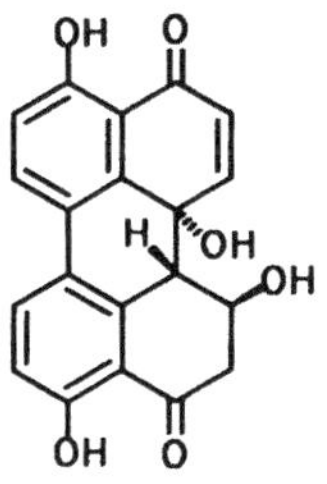

18

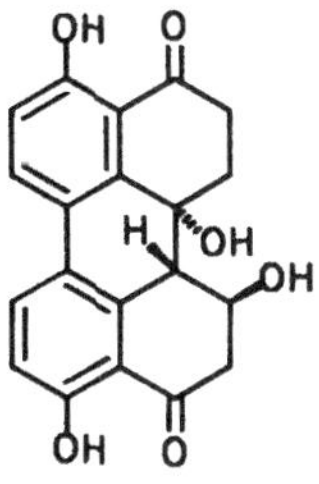

19

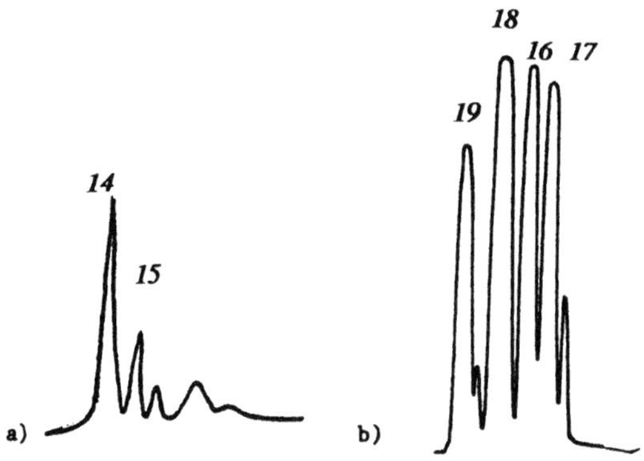

Fig. 1. CCC trace of separation of **a** diastereomers of *cyclo* Pro-Phe; **b** perylenequinones

5 Structure Elucidation Techniques

Once a phytotoxic metabolite has been isolated and purified, it must be characterized. Structure elucidation is a science in itself. It relies on several spectroscopic methods including nuclear magnetic resonance, mass spectrometry, ultraviolet spectroscopy, infrared spectroscopy, and X-ray crystallography. Individual natural products chemists may favor one technique over the other, but each of these techniques contributes to the process of characterization. No one technique, except X-ray analysis, supplies sufficient information for unequivocal structure elucidation. Even X-ray analysis should be supplemented with additional spectral data.

Numerous texts have been written to describe the various spectral techniques, in far more detail and with far greater authority than we can claim here. Rather than rewrite what is already written elsewhere, we hope to introduce the plant pathological community with some of the recent advances in structure determination techniques. Most of our attention will center on nuclear magnetic resonance and mass spectrometry. This emphasis is not necessarily a reflection of the relative importance of these techniques, but does reflect the rapid developments in their technologies in recent years.

5.1 Nuclear Magnetic Resonance

All of the available spectroscopic techniques provide unique information about the structure of sample molecules. Infrared spectroscopy indicates which functional groups are present. Ultraviolet spectroscopy provides information about extended chromophores. Mass spectrometry can provide the molecular weight and formula of a compound. Nuclear magnetic resonance (NMR) indicates the numbers and types of individual atoms (proton, carbon, nitrogen, etc.) present in a molecule. Each of these is important. If this were the extent of their capabilities, all spectroscopic techniques would be comparable. Nuclear magnetic resonance, however, supersedes other techniques because it can indicate *connections* between the various atoms in a molecule. Relationships between nuclei, often several bonds apart, can be determined using different NMR experiments.

Over the last 10 years, the advantages of pulsed-Fourier transform NMR have been investigated, leading to the development of a number of experiments designed to exploit internuclear connectivities. A list of these new experiments can be overwhelming, particularly if a natural products chemist attempts to understand the mechanics involved. Fortunately, a few key experiments can provide most of the information needed to characterize the structure of a phytotoxin. It is beyond the scope of this chapter to describe the principles of NMR spectroscopy, and the mechanisms inherent in various experiments. Several textbooks are available that provide insight into these phenomena.

All of our NMR experiments are run on Bruker AC300 and AM500 spectrometers. Routine analyses are run on the lower field instrument; the AM500 is reserved for very small samples (<1 mg) and very complex, congested spectra. ^{1}H-NMR analysis is run on every fraction generated by our separation protocol. This serves two purposes. First, phytotoxic compounds are often produced in conjunction with inactive but closely related metabolites. Plant pathologists are usually interested in isolating and characterizing both phytotoxins and any related, inactive analogs. NMR spectroscopy is uniquely suited to this goal. Structural similarities can be readily deduced from the chemical shifts patterns and coupling interactions presented in ^{1}H-NMR spectra. Second, NMR spectroscopy can be used to determine whether artifacts are generated during separation. Each fraction is chronicled from the beginning. All of the peaks present in the NMR spectrum of a pure compound should be present in the spectrum of the crude extract. Any extraneous peaks may be indicative of artifact formation.

Pure phytotoxic compounds and their analogs are subjected to more rigorous NMR analysis, which includes both one- and two-dimensional ^{1}H-NMR and ^{13}C-NMR experiments. Comprehensive one-dimensional ^{1}H-NMR analysis includes the chemical shift assignment of each peak in a spectrum, with full expansion of congested regions. Obvious coupling patterns may be recognized at this time, and supported by direct measurement of coupling constants and selective decoupling experiments. Peaks may also be integrated at this point, with attention given to delay time, to insure adequate relaxation of all nuclei. These preliminary analyses provide an idea of functional groups, and the number and types of protons present in a molecule.

Preliminary ^{1}H-NMR analysis is followed by ^{13}C-NMR analysis, which includes the indispensable broad-band proton decoupled (or composite pulse decoupled) experiment and the DEPT experiment. The broad-band decoupled experiment provides ^{13}C-NMR chemical shifts devoid of ^{13}C-H couplings. The loss of coupling information is offset by signal-to-noise enhancement in the decoupled experiment and reduced experimental time. This experiment indicates both the number and types of carbons present in a molecule. Adequate relaxation time must be provided to insure that all carbons are represented by the spectrum.

The DEPT experiment (distortionless enhancement by polarization transfer) is an excellent counterpart to the broad-band decoupled experiment. Through varying parameters, it is possible to ascertain the number of protons attached to a particular carbon. These one-dimensional experiments provide insight into the electronic environment of the various nuclei within a molecule, although little connectivity information can be inferred. Connectivity information is the domain of two-dimensional NMR.

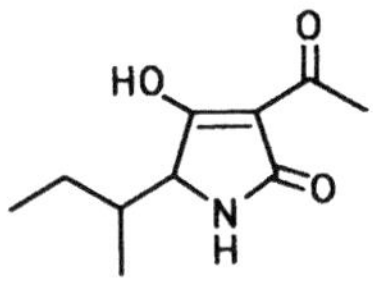

20

The use of 1D-NMR to solve the structure of the phytotoxin tenuazonic acid **20** should illustrate the importance of these techniques. Tenuazonic acid, a phytotoxin with a wide host range, was originally isolated from *Alternaria tenuis* (Stickings 1958) and characterized using 1D-NMR techniques, mass spectrometry, and infrared spectroscopy. Mass spectrometry provided a molecular weight of m/z 197 corresponding to $C_{10}H_{15}NO_3$, which indicated four sites of unsaturation. The infrared spectrum suggested the presence of a lactam, and a β-diketone moiety. The CPD spectrum of tenuazonic acid showed ten carbons. DEPT experiments indicated that the three highest field peaks were methyl groups, and that there was a single methylene carbon and two methine carbons (Fig. 2). This accounted for 13 protons; the remaining protons were attached to heteroatoms. The β-diketone suggested by IR was supported by the three carbon absorbances at δ195, δ184, and δ102 ppm. The methyl absorbance at δ24 ppm indicated proximity to a ketone (Silverstein et al. 1981). The carbon absorbance at δ175 ppm supported the presence of a γ lactam.

The proton NMR of tenuazonic acid verified the preliminary structural features suggested by carbon NMR. Three discrete absorbances integrated to three protons: a singlet at δ2.42 ppm, a doublet at δ0.98 ppm, and a triplet at δ0.86 ppm. The methyl at δ2.42 ppm suggested proximity to a carbonyl carbon, and corresponded to the carbon absorbance at δ24 ppm. Selective decoupling experiments indicated that the triplet methyl group was adjacent to the methylene, which was coupled, in turn, to the methine adjacent to the doublet methyl group. This methine was also coupled to the methine at δ3.76 ppm. These data generated an isopentyl moiety reminiscent of

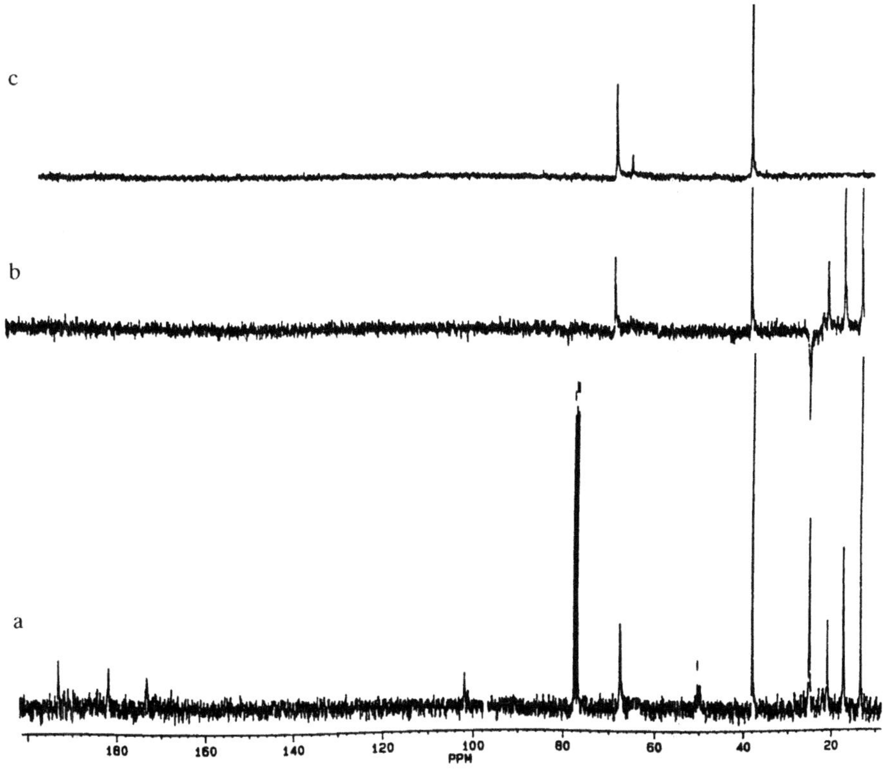

Fig. 2. **a** Composite pulse decouple experiment; comparison of DEPT experiments exhibiting; **b** methine and methyl carbons above baseline; **c** methine carbons

isoleucine. Piecing the fragments together, utilizing the γ lactam framework, generated the structure indicated for tenuazonic acid.

Although structure characterization is possible with 1D-NMR, the natural products chemist is no longer restricted to 1D-NMR experiments. There have been numerous changes in NMR technology since the first observations of "nuclear induction" 45 years ago (Bloch 1946; Bloch et al. 1946). For the natural products chemist, the last decade has produced the most exciting development: two-dimensional NMR spectroscopy (2D-NMR). Two-dimensional NMR was first conceived in 1971 by Jeener, although his early work was not published. Conventional 1D-NMR experiments portray response intensity strictly as a function of the observation frequency. In 2D-NMR, however, response intensity is a function of two frequencies. This second frequency domain increases the information content of the spectrum by several orders of magnitude. Experimentally, data is collected as a function of two independent time domains, denoted as $S(t_1, t_2)$. By applying double Fourier transformation to the data matrix, the independent time domains are converted to the frequency domain, denoted as $S(F_1, F_2)$ (Martin and Zektzer 1988).

Displaying the information obtained in a 2D-NMR experiment poses a problem. The data obtained from the experiment must be displayed in a two-dimensional array. One-dimensional NMR spectra are actually two-dimensional. The first dimension is chemical shift frequency and the second dimension is signal intensity. Two-dimensional NMR spectra are actually three dimensional, but must be portrayed in two dimensions. The simplest representation of the data is a contour plot, with intensity as the third dimension. Contour lines are generated by intersecting the third dimension of the spectrum with lines parallel to the F_1, F_2 plane. The planar F_1, F_2 dimensions of the spectrum may represent chemical shift frequencies, either of the same or different nuclides (scalar-correlated spectra). Alternatively, scalar couplings may be displayed in the first dimension and chemical shifts in the second (J-resolved spectra) (Duddeck et al. and Dietrich 1989). It is important to realize that 2D-NMR is not simply an alternative display strategy. The information obtained in 2D-NMR frequently cannot be gleaned from a conventional 1D-NMR experiment. The second frequency domain provides valuable connectivity data and correlations that greatly facilitate structure elucidation (Martin and Zektzer 1988).

Numerous two-dimensional experiments have been developed in recent years. These have been described at great length and considerable depth by a number of authors including Derome (1988) and Martin and Zektzer (1988). Most of the commonly used experiments are available as micro-programs. Implementation of a program requires some knowledge of the mechanisms involved in the experiment itself to allow parameter optimization. It is assumed that either the natural products chemist or a colleague has some NMR expertize to successfully use 2D-NMR. Experiments that have consistently proven useful in our investigations include the H-H COSY (homonuclear COrrelation SpectroscopY) and the H-C COSY or HETCOR (HETeronuclear CORrelation spectroscopy). Both of these experiments provide connectivity data that cannot be obtained through most other spectral methods.

The H-H COSY experiment is the exact experiment originally proposed by Jeener (Martin et al. 1988). The COSY experiment displays ^1H chemical shifts in both the F_1 and F_2 dimensions. Two types of peaks appear in the H-H COSY spectrum. Those along the diagonal represent the original spectrum. The off-diagonal peaks, or cross-peaks, indicate scalar (through bond) couplings between nuclei. Coupled nuclei can be found by drawing horizontal and vertical lines that intersect a cross-peak and associated off-diagonal peaks. The information obtained in a H-H COSY experiment is equivalent to that obtained by a series of selective decouplings. Selective decouplings can be laborious and inconclusive, because of signal overlap.

There are two caveats to consider when using the simple COSY experiment. The first concerns the information derived from the experiment:

a simple COSY indicates the existence of scalar coupling, but not the magnitude. An important aspect of H-H coupling is the magnitude of the coupling constant. Scalar couplings can be used to establish relative stereochemistry in a molecule, as well as to establish connectivity. Coupling constant measurements cannot be obtained from a simple COSY experiment, although phase-sensitive COSY techniques can provide these values. In simple COSY spectra, the size of the cross-peak is not an indication of the size of the coupling constant. Direct measurement of chemical shift differences in the proton spectrum is the simplest method to determine coupling constants.

The second caveat applies to most 2D-NMR techniques. Extraneous signals may sometimes appear in spectra. These are usually noise artifacts and can be difficult to distinguish from real chemical shift correlations. If these erroneous data are considered, they may either lead to an incorrect structure or render elucidation impossible. Fortunately, because both axes represent proton chemical shifts, H-H COSY spectra are symmetrical. If frequency domains are intentionally selected to generate a square data matrix, extraneous peaks can be mitigated through symmetrization. The operation of symmetrizing a spectrum applies a simple algorithm that examines the symmetry of the data file. Signals are symmetrically located; noise is strictly random, and is eliminated from the spectrum. In any symmetrical 2D-NMR experiment that generates a square data matrix, symmetrization can greatly facilitate data analysis (Martin and Zektzer 1988).

The H-C COSY experiment is extremely important because it connects proton signals in the F_1 dimension with carbon signals in the F_2 dimension. Cross-peaks illustrate which carbons are directly attached to which protons. This information can be extremely valuable in the determination of molecular structure. Unfortunately, problems may arise with extraneous peaks which cannot be corrected with symmetrization. The H-C COSY is asymmetrical because the two axes represent different nuclei. Dubious peak assignments can be clarified with 1D-NMR single frequency irradiation experiments. One of the greatest drawbacks of the H-C COSY is the time factor involved. It can require large blocks of instrument time, and may not be effective with small sample sizes. A typical H-C COSY experiment on the Bruker AM 500 requires approximately 12 h for a 20-mg sample. And we rarely have 20 mg of a single phytotoxin. Fortunately, H-C connectivity can also be obtained through the proton channel rather than the carbon channel, if the NMR spectrometer is equipped with an inverse probe. What this means to the natural products chemist is that the same information available from the H-C COSY experiment can be obtained in much less time, even on small samples. The Bruker BIRDPH.AU experiment requires 4–6 h for 1 mg of sample. Both experiments provide one bond H-C connectivity information, but inverse detection makes this information available even with small sample sizes.

Several of the phytotoxins elucidated in recent years illustrate the importance of various NMR spectral techniques. These include cyperine **3** (Stierle et al. 1991), a fairly simple molecule, and the more complex ophiobolins,

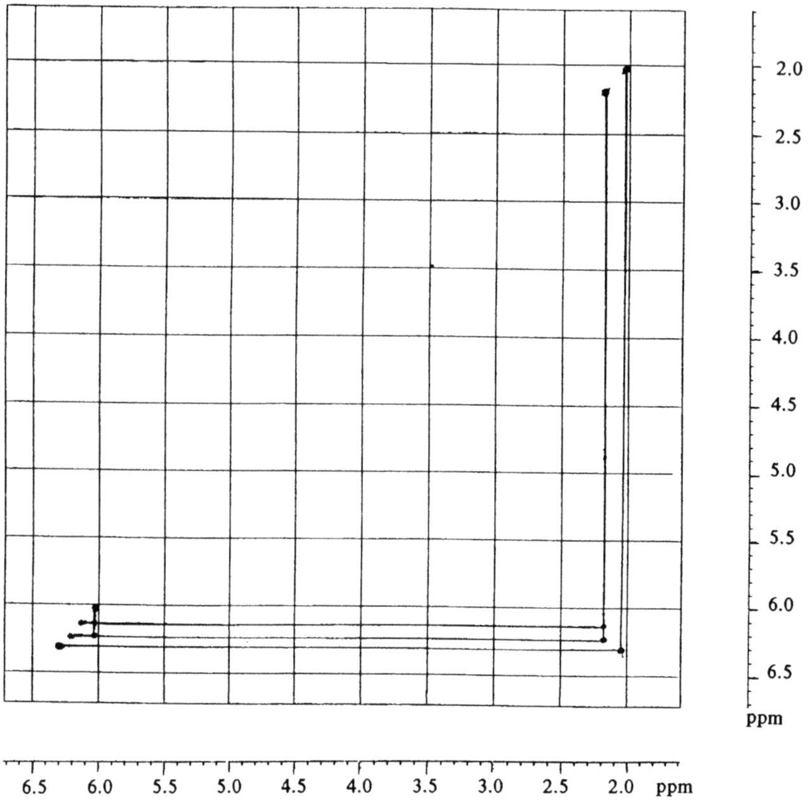

Fig. 3. H-H COSYPHDQ of cyperine

isolated from several different species of *Drechslera* (Sugawara et al. 1987). We will describe in detail the structure elucidation of cyperine, including the 1D- and 2D-NMR techniques that were instrumental in this process. This should provide insight not only into the methods available for elucidation, but also how these methods interact.

Mass spectral analysis of cyperine provided a molecular weight of m/z 260, corresponding to a molecular formula of $C_{15}H_{16}O_4$, and eight sites of unsaturation. Examination of the proton spectrum of cyperine provided a number of structural clues. The proton absorbances between δ6.0 and δ6.4 ppm, which integrated to five protons, were indicative of shielded aromatic protons. Selective decoupling experiments indicated that the peaks at δ6.35 and δ6.29 ppm were metacoupled to each other (J = 2.7 Hz), as were the three higher field signals (δ6.22, δ6.12 and *d*6.02 ppm, J = 2.1 Hz). No two protons were adjacent to each other. Five nonadjacent aromatic protons cannot possibly be accommodated by one benzene ring. These assignments suggested the presence of both a trisubstituted and a tetrasubstituted benzene ring, which satisfied the unsaturation data. The chemical shifts suggested that all of the aromatic protons were adjacent to at least one hydroxyl or methoxyl substituent.

The chemical shift at δ3.72 ppm, which integrated to three protons, is typical of a methoxyl group. The two chemical shifts at δ2.17 and δ2.02 ppm, each of which integrated to three protons, typify aromatic methyl groups. This one experiment indicated the nature of 14 protons. It showed the presence of three methyl groups, one attached to oxygen, and suggested the presence of two benzene rings.

The information obtained from the carbon spectrum of cyperine (composite pulse decoupled experiment with a delay time of 3 s) complemented the information acquired in the proton spectrum. There were clearly 15 carbons in cyperine, with 12 in the aromatic region between δ110 and δ165 ppm. DEPT parameters were chosen to show methines and methyl groups above the baseline, and methylene groups below the baseline. Quaternary carbons were not shown. The DEPT experiment was then rerun using parameters that selected solely for methines. These two experiments verified the presence of three methyl groups and five aromatic methines. These groups accounted for 14 of the 16 protons. This left two protons and three oxygens, which could be readily accommodated by two hydroxyl moieties and an ether linkage. No other assumptions could be made from the 1D-NMR data. No other conclusions could be drawn concerning the structure of cyperine without additional data.

Piecing the molecule together required connectivity information. Two-dimensional NMR analysis was instrumental in solving this structure. H-H COSYPHDQ (double quantum phase-sensitive COSY), which mitigates diagonal peaks resulting from uncoupled methyl groups, clearly delineated the aromatic couplings described above (Fig. 3).

The most important information derived from this experiment, however, was the benzylic coupling between the methyl singlet at δ2.18 and the aromatic protons at δ6.12 and δ6.22 ppm, and between the methyl singlet at δ2.03 and the aromatic protons at δ6.29 and δ6.34 ppm. Benzylic couplings are usually measurable only for ortho (0.6–0.9 Hz) or para (0.5–0.6 Hz) interactions (Jackman and Sternhell 1969). From this information we were able to propose a structure for cyperine.

One-dimensional nOe experiments supported this structure. Irradiating the molecule at δ3.72 ppm to determine absolutely the position of the methoxyl group. This resulted in the enhancement of the two aromatic protons at δ6.34 and δ6.29 (3.5% and 4.2% respectively), which is consistent with the proposed structure.

5.2 Mass Spectrometry

An important aspect of structure elucidation is the ability to determine the molecular weight and molecular formula of an unknown compound. Mass spectrometry provides that ability, and a great deal more. The appropriate experiment can provide not only the correct molecular weight but can also indicate fragments present in a molecule. This fragmentation information can provide valuable clues in structure elucidation.

Mass spectrometry is a microanalytical technique to determine molecular weight, and in the case of high resolution mass spectrometry, actual molecular formulae of samples. Although it is a destructive technique, only a few micrograms of sample are required. In recent years, a number of new mass spectral techniques have been developed. We have adopted a few of these techniques and will describe the advantages and disadvantages of each. It is not our intention to provide an exhaustive overview of the methods available, merely a "user-friendly" review of techniques that have worked well in the analyses of phytotoxic compounds. A working knowledge of mass spectrometry is helpful, however, in selecting the technique appropriate for a particular natural product.

For conventional mass spectrometry, molecules must be introduced in the vapor state, with a vapor pressure greater than 10^{-2} mm Hg. Molecules migrate by diffusion from the inlet system (either direct probe or gas-chromatographic) into the ionization chamber. As these neutral molecules randomly diffuse throughout the ion source, only a few hundredth of a percent of them reach the center of the ionization chamber and are ionized.

The ionization process involves the transfer of energy to neutral molecules in the vapor state. The most common ionization process, electron-impact ionization (EI), requires electron bombardment of a molecule with 70 eV of energy. Consequently, the neutral molecule has enough energy to eject one of its own electrons, acquiring a residual positive charge in the process. This positively charged species, the molecular ion (M^+), has considerable excess energy, which can be dissipated through fragmentation processes. If a molecule produces a stable molecular ion there may be a substantial molecular ion peak, which provides the molecular weight of a compound (Watson 1985).

The electrically charged ions are forced by a system of ion repellers, ion-acceleration plates, and ion-focusing plates into an ion-acceleration region. The ions are then directed through a slit system as a well-collimated beam, and accelerated out of the ionization chamber and into the mass analyzer. There are several different types of mass analyzers available: magnetic, time of flight, quadrupole, ion cyclotron resonance, etc.). Each type of mass analyzer serves the same function: to disperse a fast-moving beam of charged particles into a spectrum, according to the mass-to-charge ratio of the ions (Watson 1985).

The choice of ionization technique is a critical one for the natural products chemist. Electron-impact ionization (EI) is not always the method of choice to determine organic structures. Certain problems may arise with the use of this "hard" ionization process:

1. The possibility of pyrolysis is increased for molecules forced into a vapor phase, particularly if a compound has low volatility.
2. The use of 70 eV ionization voltage often destroys the molecular ion.

The ionization potential of most molecules is about 10 eV. The much higher voltage used in EI insures that there is enough energy to break any bond in the sample molecule. Unfortunately, if the molecular ion is unstable, it may fragment so extensively that the molecular ion is mitigated to the point of undetectability. Such destruction can pose a serious problem because an extremely weak molecular ion peak confuses molecular weight determination. "Soft" ionization techniques (chemical ionization, field desorption, field ionization, microprobe ionization, fast atom bombardment, etc.) have been developed in recent years. These various ionization techniques, many of which are readily available, provide the natural products chemist with a number of analytical options. "Hard" and "soft" ionization techniques are complementary processes: if the latter provides an intense molecular ion peak, the former provides fragmentation patterns that can be invaluable in structure determination (Watson 1985). We will describe in detail the techniques we have used, and compare our results with selected phytotoxins.

5.2.1 Chemical Ionization

In chemical ionization, an analyte is ionized without the transfer of excessive energy. The result of this "soft" technique is the formation of abundant adduct ions that contain the intact molecular species of the analyte. Molecular adduct ions undergo very little fragmentation, as compared with electron impact ionization. In CI, analyte molecules are ionized by collisions with highly reactive reagent gas ions, rather than electrons. If alkyl reagent gases (i.e., methane) are used to ionize an analyte, protonated adduct ions usually result. Protonation generally occurs on a heteroatom with the greatest proton affinity. Although some loss of the protonated heteroatom may result, fragmentation involving breakage of C-C bonds rarely occurs (Watson 1985).

Chemical ionization is initiated by electron-impact ionization of the reagent gas. Although both analyte and reagent gas are present in the ionization chamber, only reagent ions are produced because of their large excess. Typically, the partial pressure of reagent gas is 1 torr, that of the analyte is 10^{-5} torr. Electrom impact of methane follows a reaction pathway that results in the formation of several ionic species. The most important reactions, however, are those shown below: they result in the formation of the reagent ions (CH_5^+, $C_2H_2^+$):

$$CH_4 + e^- \rightarrow CH_4^{+\cdot} + 2e^-$$
$$\hookrightarrow CH_3^+ + H_2.$$
$$CH_3^+ + CH_4 \rightarrow C_2H_5 + H_2;$$
$$CH_4^{+\cdot} + CH_4 \rightarrow CH_5^+ + {}^{\cdot}CH_3.$$

Reagent ions do not continue reacting with methane, because methane lacks sufficient proton affinity to abstract a proton from these species. They accumulate in the ionization chamber and interact with analyte molecules, producing protonated analyte ions:

$$M + CH_5^+ \rightarrow MH^+ + CH_4.$$

The major high mass peak in this type of CI experiment corresponds to $(M + 1)^+$ rather than M^+ (Watson 1985).

Different reaction pathways occur with different reagent gases. Commonly used gases include He/H_2, H_2, and NH_3. NH_3 is particularly effective because it has a high proton affinity and a low fragmentation potential. The predominant reactant ion generated by EI of NH_3 is NH_4^+, which has a proton affinity of 207 kcal/mol. Most organic molecules have proton affinities closer to 200 kcal/mol. When proton affinities are approximately the same, analyte molecules and reagent ions associate to form adduct ions. The predominant high mass peak in this case is $(M + 17)^+$ rather than M^+ (Watson 1985).

A high mass peak that corresponds to the protonated molecular ion is not guaranteed in CI, although it is more likely than in EI. This is especially true if reagent ion species have low proton affinities. Such reagent ions interact with the analyte to produce predominantly protonated species that tend to fragment, losing their protonated heteroatom. It is beneficial to run more than one CI expertiment with different gases to compare the high mass peaks generated. We use CH_4 and NH_3, gases that generate ions with low and high proton affinities, respectively. It is then possible to determine whether the high mass peak in a CI experiment with CH_4 is actually the protonated molecular ion or a fragment ion.

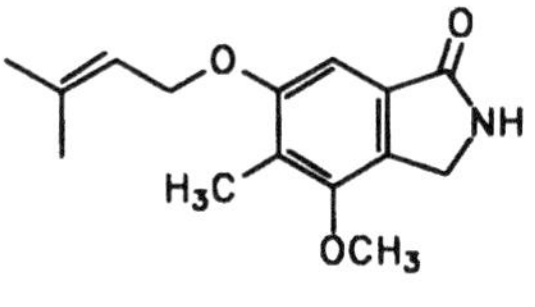

21

Both EI and CI mass spectrometry were used in the structure elucidation of a series of compounds isolated from *Alternaria cichorii* (Fig. 4). Mass spectral comparisons of one of these compounds will illustrate differences in the two techniques. Cichorine II **21** consists of a substituted phthalimidine moiety attached through an ether linkage to an isoprene unit (Stierle et al. unpubl. results). Original mass spectral data, obtained through electron-impact ionization at 70 eV, indicated a molecular weight of m/z 193. It could not accommodate the number of carbons (15) and protons (19) present in the NMR spectra. It was obvious that the molecular ion was not the parent peak. To determine the molecular ion, we tried chemical ionization, which gen-

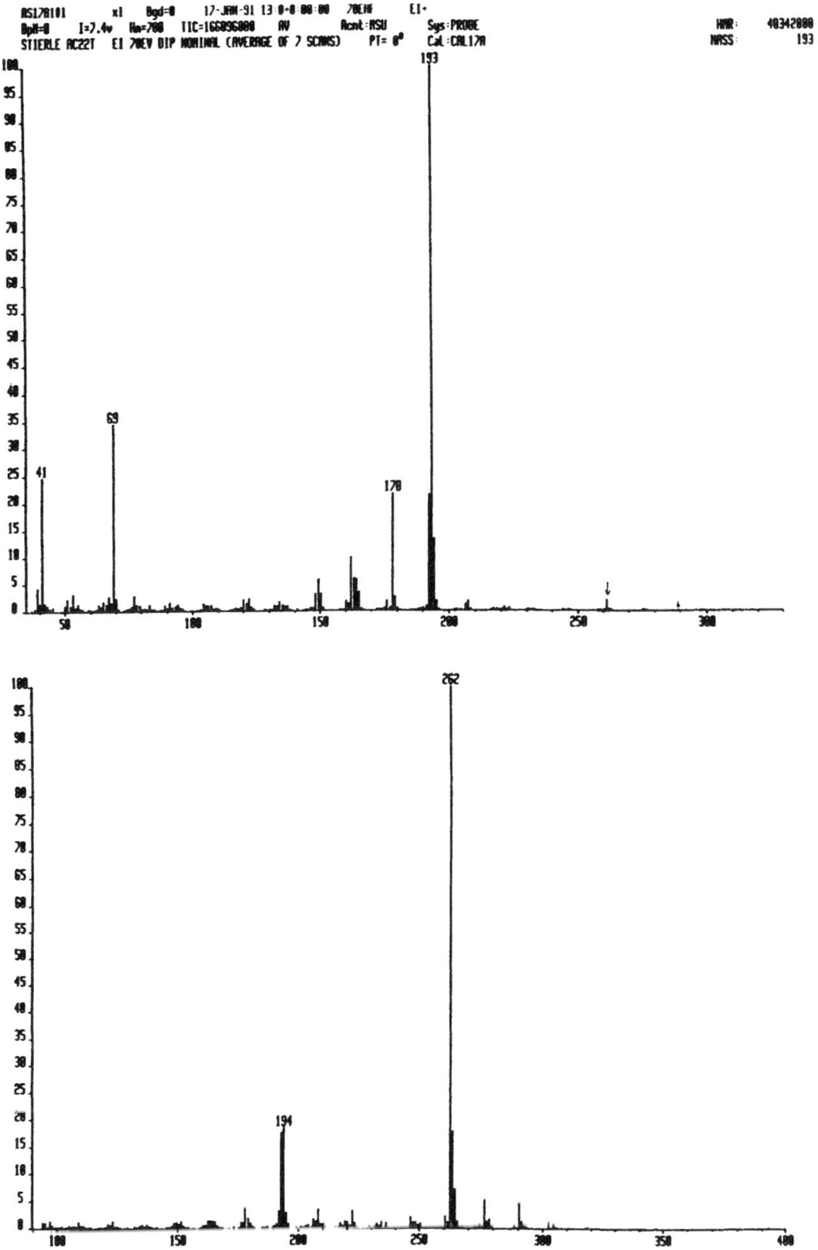

Fig. 4. Comparison of **a** EI MS and **b** CI MS of cichorine IV

erated $(M + 1)^+$ at m/z 262. The molecular weight of 261 corresponded to a molecular formula of $C_{15}H_{19}NO_3$.

It is good practice to view every piece of data with a small "grain of salt", and to consider how each spectral analysis relates to data already accrued. The discrepancy between the information contained in the carbon NMR spectrum and the EI mass spectral analysis indicated that one was erroneous. Because the high energy associated with EI often fragments the molecular ion, it is good practice to supplement this technique with additional mass spectral analysis.

5.2.2 Fast Atom Bombardment

In the past, mass spectral analysis of many biologically active compounds has been complicated by the low volatility of more polar metabolites. Until recently, adequate sample volatility was a prerequisite for mass spectrometry, because high energy electron impact was exclusively a gas phase phenomenon. Polysaccharides and proteins are virtually impossible to analyze by traditional techniques. Even simple sugars and fairly small peptides may defy analysis, because these molecules are difficult to volatilize. Desorption ionization (DI) circumvents this dilemma by eliminating the need for volatile samples. Samples are analyzed in the condensed phase either as pure solids or as solutions in a viscous solvent.

Fast atom bombardment (FAB) is a DI technique that has been used to characterize high molecular weight molecules with low volatility, including peptides, polysaccharides, and proteins. Desorption energy is provided by a beam of atoms of xenon, argon, or other suitable gas with energy in the range of 6 to 10 keV. The sample is dissolved in a viscous, low boiling solvent such as glycerol prior to bombardment. The solvent is important in promoting ionization of the analyte. The sample is then mounted on a probe in such a way that the sample surface rests at the focal point of the instrument. The beam of fast atoms is generated by ionizing the corresponding gas to produce cations. The ions are accelerated into an appropriate medium where the fast ions capture an electron, and become fast atoms. The beam of fast atoms bombards the sample with a large incident angle, which promotes desorption of molecular species. Particles desorbed from the surface are spewed over a wide angle. Positive and negative ions are captured by the ion optics. They include molecular species of the analyte as well as ions emanating from glycerol or common salts. These ions form a variety of adduct ions with the analyte and glycerol (Watson 1985).

FAB was an important tool in the structure elucidation of a host-specific toxin isolated from *Helminthosporium sacchari*, a pathogen of sugarcane. The phytotoxin, helminthosporoside, defied conventional mass spectral techniques. FAB analysis, however, provided a molecular weight of 884 amu. ^{13}C NMR gave evidence of two sugar moieties, each consisting of an internal and

Table 1. Examples of phytotoxins isolated from plant pathogens

Phytotoxins	Isolation	Characterization	Reference
Bipolaroxin	Flash silica; RP-HPLC	X-ray	F. Sugawara et al. (1985)
Gigantenone	Flash silica; LH-20; RP-18 HPLC	X-ray, NMR; MS	Kenfield et al. (1989)
Triticones A and B	Crystallization	X-ray	Sugawara et al. (1988a)
Ophiobolins	Flash silica; LH-20; RP-18 HPLC	X-ray	Sugawara et al. (1987)
Ophiobolins	Flash silica; RP-18 HPLC	NMR; MS	Sugawara et al. (1988b)
Maculosin	SX-4 and LH-20	NMR; MS	Stierle et al. (1988)
Tryptophol	Flash silica; LH-20; RP-18 HPLC	NMR; MS	Sugawara et al. (1987)
Exserohilone	Flash silica; RP-18 HPLC	X-ray	D. Sugawara et al. (1985)
Alterlosin I and II	S-X4; LH-20; CCC	NMR; MS	Stierle et al. (1989)
Cercosporamide	Crystallization	X-ray; NMR; MS	Sugawara et al. (1991)
Pseudomycins	XAD-2; Amicon C-8 HPLC	Amino acid analysis, protein Sequencer, plasma desorption MS	Harrison et al. (1991)
Curvulin	prep-TLC; RP-HPLC	X-ray	Kenfield et al. (1988)
Methylcurvulinic acid	prep-TLC; RP-HPLC	X-ray	Kenfield et al. (1988)
De-o-methyldiaporthin	LH-20; prep-TLC; RP-HPLC	NMR; MS	Hallock et al. (1988)
Fijiensin	prep-TLC; RP-HPLC	X-ray; NMR; MS	Upadhyay et al. (1990)

a terminal β-galactofuranoside (Beier et al. 1982). The remainder of the molecule consisted of a $C_{15}H_{22}$. The definitive structure was proposed by Macko et al. (1983).

This is a brief overview of the techniques we depend upon for phytotoxin isolation and structure characterization. Several of the references cited in Table 1 will provide more detailed understanding of the techniques described. We have listed many of the phytotoxins isolated in our lab over the years, indicating the techniques used in the separation and the method of characterization. Many of these compounds were identified by X-ray crystallography, in J. Clardy's lab at Cornell University. We have not described the technique simply because we do not run the analyses ourselves. Should Prof. Clardy ever choose to write a "user-friendly" approach to X-ray crystallography, well . . . that remains to be seen. For the natural products chemist it is usually a question of obtaining crystals and mailing them off to another laboratory. The results are conclusive and sometimes absolute. But we always supplement X-ray data with additional spectral analyses because this adds to the body of knowledge relevent to structure elucidation.

References

Anzai H (1989) Transgenic tobacco resistant to a bacterial disease by the detoxification of a pathogenic toxin. Mol Gen Genet 219:492–494

Ballio A (1977) Regulation of cell membrane activities in plants. In: Marrè E, Ciferri O (eds) North Holland, Amsterdam, pp 217–223

Ballio A, Chain EB, Deheo P, Erlanger BF, Mauri M, Tonolo A (1964) Fusicoccin: a new wilting toxin produced by *Fusicoccum amygdali*. Nature 203:297

Ballio A, Barra D, Bassa F, DeVay JF, Grguring I, Iacobellis NS, Marino G, Pucci P, Simmaco M, Surio G (1988) Multiple forms of syringomycin. Phys Mol Plant Pathol 32:1–4

Beier RC, Mundy BP, Strobel GA (1982) Helminthosporoside, a host-specific toxin from *Helminthosporium sacchari*: a structure revision and a new partial structure. Experientia 38:1312–1314

Berglund D, Strobel S, Sugawara F, Strobel G (1988) Flow cytometry as as method for assaying the biological activity of phytotoxins. Plant Science 56:183–186

Bloch F (1946) Nuclear induction. Phys Rev 70:460

Bloch F, Hansen WW, Packard M (1946) Nuclear induction experiments. Phys Rev 69:127

Craig LC, Craig D (1956) Countercurrent distribution. In: Weissberger A (ed) Techniques of organic chemistry, vol III. Separation and purification. Interscience, New York, pp 247–254

Derome AE (1988) Modern NMR techniques for chemistry research. In: Baldwin JE (ed) Organic chemistry series, vol 6. Pergamon, Oxford

Duddeck H, Dietrich W (1989) Structure elucidation by modern NMR. Springer Berlin Heidelberg New York

Hallock Y, Clardy J, Kenfield D, Strobel G (1988) De-O-methyldiaporthin, a new phytotoxin from *Drechslera siccans*. Phytochemistry 27:3123–3125

Harrison L, Teplow DB, Rinaldi M, Strobel G (1991) Pseudomycins, a family of novel peptides from *Pseudomonas syringae*, possessing broad spectrum antifungal activity. Proc Natl Acad Sci USA (submitted)

Hostettmann K (1980) Droplet countercurrent chromatography and its application to the preparative scale separation of natural products. Planta Med 39:1–18
Ito Y (1981) Efficient preparative counter-current chromatography with a coil planet centrifuge. J Chromatogr 214:122–125
Ito Y (1986) High speed countercurrent chromatography. CRC Crit Rev Anal Chem 17:65–143
Ito Y, Sandlin J, Bowers WG (1982) High-speed preparative counter-current chromatography with a coil planet centrifuge. J Chromatogr 244:247–258
Jackman LM, Sternhell S (1969) Applications of nuclear magnetic resonance spectroscopy in organic chemistry. In: Barton DHR, Doering W (eds) International series of monographs on organic chemistry, vol 5. Pergamon, Oxford
Janson JC (1967) Adsorption phenomena on Sephadex. J Chromatogr 28:12–20
Joustra M, Söderqvist B, Fischer L (1967) Gel filtration in organic solvents. J Chromatogr 28:21–25
Karr AL, Karr DB, Strobel G (1974) Isolation and partial characterization of four host-specific toxins of *Helminthosporium maydis* (race T). Plant Physiol 53:250–257
Kenfield D, Bunkers G, Strobel G, Sugawara F (1988a) Potential new herbicides-phytotoxins from plant pathogens. Weed Tech 2:519–524
Kenfield DS, Hallock YF, Clardy J, Strobel GA (1988b) Curvulin and o-methylcurvulinic acid: phytotoxic metabolites of *Drechslera indica* which causes necroses on purslane and spiny amaranth. Plant Sci 60:123–127
Kenfield D, Bunkers G, Wu Ya-Hua, Strobel G (1989) Gigantenone-a novel phytohormone mimic. Experientia 45:900–902
Kono Y, Daly JM (1979) Characterization of the host-specific pathotoxin produced by *Helminthosporium maydis*, race T, affecting corn with Texas male sterile cytoplasm. Bioorg Chem 8:391–397
Kono Y, Suzuki Y, Takeuchi S, Knoche HW, Daly JM (1985) Studies on the host-specific pathotoxins produced by *H. maydis*, race T and *P. maydis*: absolute configuration of PM-toxins and HMT-toxins. Agric Biol Chem 49:559–562
Macko V, Acklin W, Hildebrand C, Weibel F, Arigoni D (1983) Structure of three isomeric host-specific toxins from *Helminthosporium sacchari*. Experientia 39:343–347
Mandava NB, Ito Y (1982) Separation of plant hormones by countercurrent chromatography. J Chromatogr 247:315
Mandava NB, Ito Y (1984) Plant hormone analysis by countercurrent chromatography. J Liq Chromatogr 7:303
March J (1985) Advanced organic chemistry, 3rd edn. John Wiley & Sons, New York
Marrè E (1979) Fusicoccin-a tool in plant physiology. Annu Rev Plant Physiol 30:273–288
Martin GE, Zektzer AS (1988) Two-dimensional NMR methods for establishing molecular connectivity. VCH Publ, New York
Moore JC (1964) Gel permeation chromatography. I. New method for molecular-weight distribution of high polymers. J Polymer Sci Part A 2:835–843
Moore JC, Hendrickson JG (1965) Gel permeation chromatography. II. Nature of the separation. J Polymer Sci Part C 8:233–241
Porath J, Flodin P (1959) Gel filtration: a method for desalting and group separation. Nature 183:1657–1659
Silverstein RM, Bassler GC, Morrill TC (1981) Spectroscopic identification of organic compounds, 4th edn. John Wiley & Sons, New York
Stickings CE (1958) Biochemistry of Microorganisms. CVIII. Metabolites of *Alternaria tenuis*. Biochem J 72:332–340
Stierle AC, Cardellina JH, Strobel GA (1988) Maculosin, a host-specific phytotoxin for spotted knapweed from *Alternaria alternata*. Proc Natl Acad Sci USA 85:8008–8011
Stierle AC, Cardellina JH, Strobel GA (1989) Phytotoxins from *Alternaria alternata*, a pathogen of spotted knapweed. J Nat Prod 52:42–47
Stierle A, Upadhyay R, Strobel G (1991) Cyperine, a novel phytotoxin produced by *Ascochyta cypericola*. Phytochemistry, (in press)
Strobel G (1982) Phytotoxins. Annu Rev Biochem 51:309–333

Sugawara D, Sugawara F, Strobel GA, Yali Fu, He Cun-Heng, Clardy J (1985) Exserohiline: a novel phytotoxin produced by *Exserohilum holmii*. J Org Chem 50:5631–5633

Sugawara F, Strobel GA (1987) Tryptophol, a phytotoxin produced by *Drechslera nodulosum*. Phytochemistry 26:1349–1351

Sugawara F, Strobel G, Fisher LE, Van Duyne GD, Clardy J (1985) Bipolaroxin, a selective phytotoxin produced by *Bipolaris cynodontis*. Proc Natl Acad Sci USA 82:8291–8294

Sugawara F, Strobel G, Strange R, Siedow J, Van Duyne GD, Clardy J (1987) Phytotoxins from the pathogenic fungi *Drechslera maydis* and *Drechslera sorghicola*. Proc Natl Acad Sci USA 84:3081–3085

Sugawara F, Samsoedin R, Takahashi N, Liu H, Fu Y, Clardy J, Strobel S, Berglund DL, Strobel G (1988a) Triticones, novel spirocyclic lactam compounds isolated from the plant pathogenic fungus, *Drechslera tritici-repentis*. J Am Chem Soc 110:4086

Sugawara F, Takahashi N, Strobel G, Yun CH, Gray G, Fu Y, Clardy J (1988b) Some new phytotoxic ophiobolins produced by *Drechslera oryzae*. J Org Chem 53:2170–2172

Sugawara F, Strobel S, Strobel G, Larsen RD, Berglund DL, Gray G, Takahashi N, Coval SJ, Stout TJ, Clardy J (1991) The Structure and biological activity of cercosporamide from *Cercosporidium henningsii*. J Org Chem 56:909–910

Tatum LA (1971) The southern corn leaf blight epidemic. Science 171:1113–1115

Upadhyay R, Strobel GA, Coval SJ, Clardy J (1990) Fijiensin, the first phytotoxin from *Mycosphaerella fijiensis*, the causative agent of Black Sigatoka disease. Experientia 46:982–984

Watson JT (1985) Introduction to mass spectrometry, 2nd edn. Raven Press, New York

Williams KW (1972) Solute-gel interactions in gel filtration. Lab Pract 21:667–670

Yau WW, Kirkland JJ, Bly DD (1979) Modern size-exclusion liquid chromatography. John Wiley & Sons, New York

Quantitative Assays of Phytotoxins Using Plant Protoplasts and Isolated Cells

E. GALUN and A. BREIMAN

1 Introduction

In the absence of a chemical definition, an extract (or a fraction of it) from a phytopathogen is characterized as a phytotoxin by the effect of the former in bioassays (Yoder 1980, 1981). Conceptually, the most appropriate bioassay should involve a system that directly expresses the interaction of the assayed phytotoxin with its immediate target. Unfortunately, such systems are rare in plants. Even in well-studied phytotoxin systems such as the interaction of tentoxin with the β-subunit of the CF_1 component of the chloroplast ATPase (Steele et al. 1976; Avni 1990), changes in electrical properties of oat-cell membranes by victorin C (Ullrich and Novacky 1991), and the effect of *Helminthosporium maydis* race T toxin (*Hm* T toxin) on maize mitochondria (Earle 1982; Holden and Sze 1989), the immediate target and the exact mode of action of the respective phytotoxins were not unravelled. We therefore face a dilemma: a chemically defined toxin is the preferable tool for the study of the mode of action of a phytotoxin, but once a phytotoxin is purified, its chemical analysis, rather than its biological impact, may provide the preferable means for quantitative determination. On the other hand, bioassays are imperative for the chemical purification of putative phytotoxins. Moreover, some phytotoxins impart rather typical and well-defined symptoms on the appropriate susceptible hosts (e.g., phaseotoxin from *Pseudomonas syringae* pv. *phaseolicola* on beans; Patil et al. 1972); using such symptoms for bioassay could thus be a useful step in the study of a given pathogen-phytotoxin-host system. Another point in favor of bioassays is that some (chemically defined) phytotoxins seem to be synergetic with less-defined extracts from a given phytopathogen. For the evaluation of such synergistic effects, bioassays could be rather useful.

Before focusing on phytotoxin bioassays with isolated cells and protoplasts, one should be aware of the usefulness and the limitations of bioassays of various kinds, for biological, biochemical, chemical and genetic studies as well as for the selection of pathogen-resistant cultivars. These were discussed in several excellent reviews (e.g., Strobel 1977; Yoder 1980, 1981; Durbin 1981; Earle 1983; Daub 1986; Gengenbach and Rines 1986; Durbin and Graniti 1989). It is therefore recommended to consult these reviews before utilizing bioassays in phytotoxin studies.

In this Chapter, our attention shall be focussed on a specific subject, namely the utilization of isolated cells and protoplasts in the study of phytotoxins and for breeding pathogen resistant plants. In respect to utilization, we would like to emphasize some general points.

Cells and protoplasts can serve to assay only those phytotoxins that have a direct impact on such isolated cells. While this appears as a superfluous statement, some investigators neglected to consider this limitation. Moreover, when the mode of action of a phytotoxin is not yet known, insensitivity of isolated cells and protoplasts to this phytotoxin can be expected. A few examples will clarify this situation. Fusicoccin is a phytotoxin with a defined chemical identity; it causes opening of stomata and consequently mediates uncontrolled water loss from treated plants (Turner and Graniti 1969). Thus whole plants or leaves can serve to bioassay fusicoccin but not isolated cells and protoplasts. Similarly a toxin specifically involved in the photosynthetic apparatus, like tentoxin (Steele et al. 1976) that binds to a CF_1 subunit of sensitive chloroplasts' ATPase (as noted above) will affect only cells (e.g., mesophyll cells) that are engaged in photosynthesis or are differentiating towards this activity. Therefore cell-suspension cultures in a sugar-containing medium will not be affected by tentoxin. Moreover, even protoplasts isolated from the mesophyll will not be affected, since in such protoplasts the photosynthetic activity is shut down. Present assays for tentoxin are therefore based on seedlings (e.g., Galun 1982), however, in the future could be based on photoautotrophically cultured cell suspensions. Some antibiotics, although commonly not categorized as phytotoxins (e.g., streptomycin, oligomycin) have a toxic effect on plants since they interfere with the normal functioning of cellular organelles (chloroplasts and mitochondria, respectively). Cultured plant cells and isolated protoplasts may survive and divide despite the malfunctioning of chloroplasts and/or mitochondria. Therefore streptomycin will arrest the differentiation of green shoots from susceptible cell cultures (Maliga et al. 1973) and oligomycin will arrest the growth of cultured roots from sensitive plants (Perl et al. 1990), but neither will have an impact on cultured cells of the respective plants.

On the other hand, several toxicity parameters can conveniently be assessed in isolated plant cells and protoplasts. Details shall be provided below with the specific protocols. Here we should note that the evaluation of cell vitality, cell-division capability, enzymatic-activity tests, changes in cell-membrane permeability and changes in cell organelles can all be assessed with rather reliable and accurate procedures. Moreover, such assessments can be handled in parallel with numerous samples; consequently, isolated cells and protoplasts are useful for several types of phytotoxin bioassays. Briefly, they can serve for the following purposes:

1. Evaluation of the toxicity-rate of a defined phytotoxin or an extract-fraction from a cultured phytopathogen;
2. To assist in the purification of phytotoxins from their respective crude extracts by the use of extract fractions in bioassays;
3. To evaluate the relative toxicity of phytotoxins from various pathovars of a given pathogenic species;

4. To determine the susceptibility of a plant genotype (or of a cell-line derived from this genotype) to a phytotoxin;
5. To assess the impact of specific additives (e.g., synergists, suppressors) to an assay system, on phytotoxin sensitivity;
6. To investigate the effect of physical and chemical factors (e.g., light fluence, temperature, gas composition, pH) on the toxicity of phytotoxins.

It is beyond the capacity of this chapter to provide procedural details for all available phytotoxin bioassays that are based on isolated and cultured plant cells and protoplasts. Moreover, it is our intention to provide information that is useful for the establishment of respective bioassays that were not yet provided by the existing literature. Consequently, we shall first present basic principles for the isolation and culture of plant cells and protoplasts. In a subsequent section, we shall detail several examples of phytotoxin bioassays.

2 Isolation and Culture of Cells and Protoplasts for Phytotoxin Bioassays

Notably, plant pathologists were among the first to establish reproducible procedures for the isolation and culture of plant cells (e.g., Hildebrandt et al. 1946) and the isolation of protoplasts having sustained division capability was originally a part of an effort to study viral infection in *Nicotiana* (Takebe and Otsuki 1969). Apart from some species of the Gramineae family, there are well-established procedures to obtain cell cultures from most annual crop plants and from an increasing number of perennials. Protoplasts having division capability can likewise be obtained from crop plants of important dicotyledoneous families such as Compositae, Cruciferaceae, Legumosaceae, Rutaceae, Solanaceae and Umbellifera as well as from several speceis (e.g., rice, maize, barley and wheat) of the Gramineae. Principles and methods for the isolation and culture of plant cells and protoplasts were presented in several text books and treatises (e.g., Reinert and Bajaj 1977; Street 1977; Wetter and Constabel 1982; Evans et al. 1983 and subsequent volumes; Vasil 1984 and subsequent volumes; Pierik 1987; Bajaj 1989).

2.1 Establishment and Maintenance of Cultured Plant Cells for Phytotoxin Bioassays

Phytotoxin bioassays can be performed with cells that were directly isolated from intact plant organs or from in vitro cultured cells (calli or cell suspensions). Although methods to obtain cells directly from intact organs such as leaves were reported (Takebe et al. 1968), these methods were developed for only a few plant species and the "primary" cells do not constitute a convenient plant material for routine experimentation. The much more commonly employed means to obtain large quantities of viable isolated cells is to release them from in vitro cultures, namely from either calli or cell suspensions. Calli constitute the most prevalent source of sustained cell-suspension cultures. Consequently, in the majority of the cases, the first step in obtaining isolated cells will be to establish callus cultures.

Callus initiation is usually started by the removal of an organ, or part of it (explants), from the intact plant. Hence, in practice sections of shoots, leaf-blades, inflorescences, petioles, storage organs, hypocotyls, epicotyls and roots are common sources for callus initiation. Under appropriate culture conditions these explants will initiate cell proliferation from the cut surfaces. Callus may be initiated also from non-wounded surfaces of the nucellus (e.g., in *Citrus*, see Sect. 3), leaf thin layers, and the scutellum of immature embryos of Gramineae. In addition, division of isolated cells and protoplasts will result in cell colonies, and further cell division in these colonies may lead to calli.

Unless the plants from which the explants are removed are grown under strictly axenic conditions (e.g., in vitro cultured seedlings from surface sterilized seeds), a surface sterilization step must preclude the culture. Such a sterilization should be effective enough to kill all epiphytic microorganisms (e.g., bacteria and fungi) but should not be lethal to the plant tissue. One standard procedure employs a short detergent wash followed by a dip in 70% ethanol and a subsequent immersion (5–15 min) in sodium hypochlorite (about 1%, namely about 30% of a commercial solution). The explants are then washed with several changes of sterile water. Other common sterilizing agents are silver nitrate, mercury chloride and hydrogen peroxide. The use of antibiotics is practiced in some cases, but it should be noted that for extended cultures and/or when the plant material has a high microbial contamination, the addition of antibiotics is useless.

Various basic media were devised for the culture of explants as listed in the above-mentioned text books (e.g., Wetter and Constabel 1982). One of the most commonly used media is the Murashige and Skoog medium that contains a relatively high nitrogen concentration. Recently ready-made plant-cell culture media (e.g., Murashige and Skoog, Gamborg, Nitsch) are offered by commercial companies (e.g., Sigma). The media may be composed of only "synthetic" (chemically defined) compounds; ill-defined components, such as malt extract, yeast extract, casein hydrolyzate or coconut milk, are also added to the media. The chemicals used should be of "analytical" grade and the water should be deionized and glass distilled.

For callus initiation the explant is usually cultured in solidified medium containing 0.6–2% agar. Most plant tissues and cells will successfully grow in the pH range of 5–6. Hence, the pH of the medium is adjusted accordingly.

Most explants will not react to the transfer to culture medium unless one or more plant hormones are added. Of these auxins, 2,4-dichlorophenoxyacetic acid (2,4-D), β-naphthaleneacetic acid (NAA) and β-indoleacetic acid (IAA) are the most important ones in respect to callus initiation. The choice of the auxin and its concentration is specific for the type of the explant. Very generally speaking, monocots will require a higher dose of 2,4-D than dicots, but there are notable exceptions even to this generalization. Moreover, certain explants do not require any auxin for callus initiation (e.g., *Citrus* ovules, tumorous tissues of *Agrobacterium tumefaciens* infected plants). Kinins (e.g., kinetin, zeatin) are required in some cases for callus initiation and very commonly are needed for continuous culture. The size of the explant may be crucial for callus initiation; too small a piece may not initiate callus, and it is not practical to use too large an explant because, as noted above, callus usually starts only from the cut surfaces.

Once callus is established, it may be transferred (with or without the explant) to another culture vessel (plate, flask, vial or tube) for further growth. The medium for the subculture may be the same as during callus initiation, or it may be changed completely or partially (e.g., change of hormone composition). The frequency of subcultures depends on the rate of callus growth; in most cases, calli are transferred at least once in 6 weeks, and very rarely less than once a week. Between transfers the callus mass increases between about two- to tenfold. Many calli, when cultured under appropriate conditions, can be subcultured indefinitely, while in some cases there is a gradual decline in growth or even a degeneration and final death. The size of the callus piece to be used for subculture should not be too small, otherwise the callus will either degenerate in the new culture medium or it will have a long lag phase before the renewal of growth.

Soft and friable calli are usually a good indication of healthy and vigorous growth. It is rather common that from the same or parallel explants, cultured under identical conditions, lines of subcultures with different friability and vigor can be selected. One should therefore be observant to the callus growth, especially during the early subcultures. Light is essential for the growth of some types of calli, but bright light may be harmful to certain calli. Therefore calli are commonly maintained in dim light or in the dark.

Once a callus is established, it may serve to initiate a suspension culture. One of the prerequisites for establishing and maintaining suspension cultures is that the "cell density" should be above a certain limit (which is typical for any specific case). Therefore, unless a minimal quantity of cells is added to a given volume of culture solution, no cell division will occur. Likewise, during subculture of suspension cultures, the total cell volume (relative to the volume of culture solution used for inoculation of the subculture) should be above a certain threshold ("packed volume"). Various factors affect the level of this threshold. By certain changes in the medium (especially when "conditioned" medium is used) the packed volume can be lowered. Thus a "rich" medium including several organic acids, specific sugars and vitamins may allow division and growth of cells at a relatively low density. Following inoculation, several growth phases can be recognized in the suspension culture. There is first a *lag* phase (before the cells start to divide in the new medium). This phase is followed by an exponential phase, a linear phase, a progressive deceleration phase, and finally a stationary phase. The length of the first two phases is strongly affected by the initial cell density.

Plant cell suspensions require aeration. This is achieved normally in the laboratory by rotating the culture flasks on an appropriate shaker. Erlenmeyer flasks filled up to about one-fifth of their volume with liquid medium are most commonly used for routine suspension cultures. Other vessels can also be used (such as glass carboys with internal rotators and fermentors with rotators or with air lift systems). The use of fermentors enables culture by the chemostat (continuous) system rather than by batch culture.

Suspension cultures may be composed of mostly single cells (e.g., *Haplopappus gracilis*), big lumps of cells, or of cell groups of various sizes. The type of suspension very much depends on two factors: the plant source of the suspension and the hormonal composition. A relatively high auxin content tends to cause a shift towards fewer lumps and more single cells or small cell groups.

2.2 Isolation and Culture of Protoplasts for Phytotoxin Bioassays

Isolated protoplasts may be exposed to phytotoxins for a short duration to assess mortality, metabolic arrest and structural changes. Otherwise they may be cultured, for extended periods, in the presence of a phytotoxin to evaluate long-term effects such as the reduction of cell-division capability.

Due to relatively high solute concentrations in the vacuoles of the plant cells, intact cells exert a constant pressure (turgor) on their walls. Therefore, removal of the cell walls and exposure of the protoplasts to solutions with low osmotic pressures will cause a sudden inflow of water, resulting, in most cases, in the rupture of the plasmalemma and killing of the cell. Thus, the removal of cell walls should take place in solutions with osmotic pressures that are similar to those in the vacuoles. The vacuolar osmotic pressure is characteristic to the specific tissue and plant species and varies with certain environmental conditions (e.g., it is higher in mesophyll cells of leaves in optimal photosynthetic condition than in shaded leaves). Practice indicated that incipial plasmolysis (i.e., an outside osmotic pressure higher than inside the vacuoles) improves the removal of the cell wall by enzymatic degradation. Generally, the isolation is practiced at 0.4–0.7 M of sugar or sugar-alcohol (e.g., mannitol) solutions. When inorganic salts (or seawater) are utilized the molarity is obviously maintained at a lower range (e.g., 0.2–0.3 M).

Protoplasts may be isolated from either of two main sources: leaves or cultured cells. In either case the cells are surrounded by a cellulosic primary wall and a secondary wall composed mainly of pectic compounds. The latter wall is the outermost one. Accordingly, cell-wall degradation is commonly accomplished by pectinases, cellulases and hemicellulases. The tissues are thus exposed to these enzymes (dissolved in solutions with the appropriate osmotic pressure and filter-sterilized) either sequentially or to a mixture of these enzymes. The enzymes that are in actual use are not biochemically pure, but rather are crude protein extracts obtained from microorganisms having high pectinic and/or cellulotic degrading ability. These are supplied under a variety of commercial names which can be misleading, e.g., the so-called cellulases also contain several other enzymatic activities. For a variety of reasons, the exposure of tissues and cells to these commercial enzyme preparation can be rather harmful. Therefore, one should use the minimal concentrations which will still result in good yields of protoplasts.

When leaves serve as source for protoplasts they are derived preferably from axenic plants ("shoot cultures") grown in containers with solidified (1% agar) culture medium (e.g., Nitsch medium with reduced sugar concentration). The leaves are harvested alternatively from plants grown either in growth chambers or greenhouses. The latter leaves require surface sterilization. This can be accomplished by dipping in a low concentration (0.1% or less) of detergent, followed by thorough washing with tap water. The washed leaves are dipped (1–2 s) in 70% ethanol, transferred for about 15 min to a 1% sodium hypochlorite solution, and finally washed several times with sterile deionized water.

The following procedure to isolate and culture leaf protopalsts is applicable to many plant species; it was found efficient especially in *Nicotiana* species (e.g., Aviv and Galun 1985, 1988) and some other plants of the Solanaceae family. The leaves are pricked by a multi-needle pricker and incubated in the maceration solution. The incubation may be short: up to 3 h with continuous agitation or relatively long (overnight) with little or no agitation. It is normally performed at room temperature (22–27 °C), but low temperature with several days incubation was also successfully employed. The release of protoplasts is usually quite abrupt but may proceed for several hours. It is a matter of experience to decide when the incubation should be stopped for maximal yield of viable protoplasts: extension of the incubation may release more protoplasts, but at the same time existing protoplasts may be harmed, killed, or made otherwise unfit for phytotoxin bioassay or for future culture. Some tissues (e.g., potato leaves) are induced during maceration to produce damaging levels of ethylene and/or free radicals. To avoid such damage, either or both silver thiosulphate and salicylic acid can be added.

The incubation is stopped by precipitating the protopalsts at low gravity forces (50–200 g), followed by several washes with either the final culture medium or with a special "wash medium". The protoplasts may be cleaned from debris with a higher specific density by flotation in a solution of 0.6-M sucrose. There are also protoplast purification procedures that employ Percoll gradients to separate the protoplasts from the debris. Alternatively, the protoplasts are passed through appropriate sieves.

The purified protoplasts may either be plated directly in the culture vessels or maintained for several hours, at a relatively high density (i.e., over $10^5\,ml^{-1}$) in a salt medium and then plated out.

Plant protoplasts, similarly to plant cells in general, require an optimal cell density for successful culture. The optimum may vary for specific types of protoplasts and probably depends on the volume of individual protoplasts. The density range for tobacco protoplasts is about 5×10^3 to $10^5\,ml^{-1}$.

The protoplasts are plated out in either liquid or solid medium. When cultured in liquid medium, either microculture or mass-culture is employed. For microculture, the protoplasts are diluted to the required density into the liquid culture medium and drops (1–50 μl according to the specific technique used) are inserted into the appropriate culture vessels (normally petri dishes). Special precautions are taken to keep the humidity within the culture vessel high in order to avoid evaporation of the drops. For mass culture in a liquid medium, the purified protoplasts are diluted to the required density and the volume of protoplast suspension per vessel (petri dish) is calibrated to finally give a 2–4 mm depth of suspension.

When cultured in solidified medium, the protoplasts are suspended in liquid medium at a density that is twice the final one and mixed at a 1:1 ratio with culture-medium containing twice the final agar concentration. The agar medium, before being mixed with the concentrated protoplast suspension, is kept liquid by a 42 °C water bath. Agarose or other solidifiers may be utilized in cases where the protoplasts are sensitive to agar.

Freshly isolated protopalsts are very sensitive to a variety of harsh treatments. The following warnings should therefore be considered:

1. Strong shaking and agitation should be avoided;
2. Most freshly isolated protopalsts are light-sensitive: therefore their isolation, purification and early culture should be performed in dim light;
3. The minimal *g* force sufficient for precipitation and/or flotation should be employed;
4. Pipettes with wide open ends should be used and the protoplast suspension should not be forced to enter or exit the pipette quickly;
5. One should be careful in using only thoroughly rinsed glassware in order to eliminate trace amounts of soap or detergent;
6. Protoplasts are heat-sensitive; therefore the agar containing part of the culture medium should not be kept at a too-high temperature (usually not above 43 °C) and one should always add the warm agar medium to the concentrated protoplast suspension and not vice versa (as in English tea preparation: never add the milk to the hot tea, but add tea to milk already in the cup, otherwise the milk will have a "burned" taste).

When calli or cell suspensions are the source of protoplasts, these are incubated directly in the macerating solution. Calli are thus harvested from the petri plates and added to the latter solution. Suspended cells are precipitated and the maceration solution is exchanged by a culture medium. Generally, young cultures are a better source for protoplasts than mature cultures. Hence, calli are usually used 2–3 weeks after transfer and cell suspensions are collected at the early log phase. In order to separate the protoplasts from calli and cell groups that are not affected by the enzyme treatment, a stepwise screening is usually employed (e.g., sequential screening through sieves having a pore diameter of ca. 350, 150 and 40 μm). The further steps of protoplast isolation from callus and cell suspension are the same as those mentioned above for leaf mesophyll-derived protoplasts.

For bioassays requiring a short phytotoxin exposure, the isolated protoplasts are dispersed in liquid media. When a long exposure is required, such as in assays where cell-division capability is monitored, the protoplasts can either be cultured in an appropriate liquid medium, or they are embedded in a solidified culture medium as indicated above.

3 Specific Bioassays

The actual procedures of quantitative phytotoxin bioassays shall be demonstrated for five specific cases. The respective protocols are based on studies performed by several investigators, including ourselves. The cases to be described include bioassays of phytotoxin culture filtrates and of defined phytotoxins from pathogenic bacteria and fungal organisms; the bioassays are based on protoplasts or on isolated cells. Our choice of examples includes phytotoxins affecting only sensitive species or cultivars as well as phytotoxins lacking such a specificity.

3.1 Bioassay of Toxicity Imposed by a Culture Filtrate from *Xanthomonas campestris* on Peach Mesophyll Cells

3.1.1 General Considerations and Background

Xanthomonas campestris pv. *pruni* is pathogenic to peach trees causing bacterial leaf spot in susceptible cultivars. The virulence of *X. campestris* pv. *pruni* strains differs considerably. Low molecular mass components from culture filtrates of this pathogen were found to be toxic to cells of susceptible peach cultivars but the phytotoxin(s) was not yet chemically identified. Hammerschlag (1984) found that culture filtrate toxicity is correlated to *X.c.* pv. *pruni* virulence that the culture filtrate of this bacterial pathogen is active at the cellular level, and that leaf cells of susceptible peach cultivars respond to the culture filtrate differentially. The bioassay of the *X.c.* pv. *pruni* was based on changes in the transmembrane electrical potential (PD) of mesophyll cells as revealed by the fluorescence probe MC-540. A reduction of fluorescence relates to increase of hyperpolarization of the cell membranes.

3.1.2 Protocol

Culture Filtrate. Erlenmeyer flasks with 100 ml medium (0.8% Difco nutrient, 5 g NaCl, 2 g glucose) are inoculated with a standard inoculum (e.g., 2×10^8 colony forming units) of *X.c.* pv. *pruni* and shaken for 48 h (30 °C, dark). After centrifugation (5000 g, 10 min) the supernatant is filtered (0.22 μm) and the filtrate is passed through an ultrafiltrator (e.g., Amicon, with a UMO5 membrane) having a 500-dalton cutoff. The volume of the final filtrate is restored to the volume prior to filtration by the addition of water. As control, the noninoculated medium is filtered and restored in the same manner.

Isolation of Cells. Leaf tissue (e.g., of peach trees) is surface sterilized (5 g/l sodium hypochlorite, 0.01% Tween 20), washed with sterile distilled water, blotted and cut into stripes. The stripes are incubated overnight in several dishes, each with 5 ml filter-sterilized maceration fluid (2% cellulysin and 1% macerase, both from Calbiochem-Behring, 0.45 M sucrose, pH 5.7). The suspension is then filtered (150 and 90 μm), the filtrate if centrifuged (100 g, 2 min) and the pellet of cells is resuspended in 5 ml of medium (Murashige Skoog salts, 20 g/l sucrose, 0.4 μg/l thiamine HCl, 100 mg/l myo-inositol, 0.05 mg/l nicotinic acid, 0.05 mg/l pyridoxine HCl, 0.01 mg/l p-aminobenzoic acid, 10 mg/l indolebutyric acid, 0.1 mg/l 6-benzylamino purin, pH 5.7) and centrifuged again. While the original protocol calls for the various vitamins, these are probably not essential for the bioassay that lasts for only a few hours.

Fluorometry. The pellet of cells is suspended in 1 ml of 5.7 μM MC-540 (Eastman Kodak, dissolved in 0.01 M MES, 0.6 M sorbitol, pH 6) and incubated for 10 min. Then the cells are washed (by low-speed centrifugations as above) and the volume is adjusted by the medium to 10^5 cells/ml. Fluorescence measurements are performed with an epifluorescence microscope with appropriate exitation and barrier filters (e.g., Zeiss BP 546/10 and LP590, respectively) and quantitized by measuring the fluorescence emission with a photomultiplier. The loss of fluorescence during a few hours provides a measure of the toxicity rate.

3.2 The Evaluation of Toxicity of a *Phytophthora citrophthora* Culture Filtrate by *Citrus* Protoplasts

3.2.1 General Considerations and Background

Phytophthora citrophthora is a fungal pathogen causing trunk gummosis, collar and root rot, damping-off and brown rot of *Citrus* trees (see Breiman and Barash 1981). *Citrus* cultivars are considered to have a differential sensitivity to this pathogen, but the culture filtrates derived from *P. citrophthora* is toxic to cells of *Citrus* as well as to cells of unrelated genera. Furthermore, recent results (Vardi et al. 1986) questioned the usefulness of such culture filtrates as a tool for the selection of resistant cell lines. The procedures presented below based primarily on Breiman and Galun (1981) are therefore intended to provide information on three bioassay systems for the evaluation of a filtrate's toxicity to plant cells, rather than to advocate the use of these procedures to select resistant cells. Thus, neither details of culture filtrate preparation nor details of the protocols for the isolation of protoplasts shall be given. The protocol is based primarily on our own studies (Breiman and Galun 1981; Vardi et al. 1986).

3.2.2 Protocol

Culture Filtrate. Details of culture filtrate preparation are as described (Breiman and Barash 1981; Breiman and Galun 1981). Briefly, a pathogenic *P. citrophthora* isolate is inoculated into 1-l flasks, each containing 100 ml fungal growth medium and the flasks are maintained for several weeks at 25 °C in dim light. The suspension from several flasks is then combined and filtered and the filtrate is flash-evaporated to about 1/25 of its original volume. After acetone precipitation, dispersion of the pellet and further Sephadex G-25 and DEAE Sepharose factionation, a concentrated, "partially purified toxin" (PPT), is obtained that has a volume of less than 1/100 of the original culture medium. It is quantitized as µg protein per ml.

Protoplasts. The following assays can be performed with protoplasts from various sources. Procedures for obtaining *Citrus*, carrot, tobacco and *Nicotiana sylvestris* protoplasts were described by Vardi and Galun (1989), Wetter and Constabel (1982), Aviv and Galun (1985, 1988). General references for the preparation of other protoplasts were provided in the previous section.

Evaluation of Vitality by Fluorescein Diacetate. This procedure is based on the transformation of fluorescein diacetate (FDA) staining (Widholm 1972). The monopolar molecules enter into the cells, where active esterases probably cleave off the acetate residue, leaving the fluorescenting fluorescein in the cell. Protoplasts are suspended in the appropriate densities (e.g., 2×10^4 ml of *Nicotiana* protoplasts, 10^5 ml of *Citrus* protoplasts) in culture medium. The PPT is added in the amount between 50 to 1000 $\mu g\,ml^{-1}$ and the protoplasts are incubated for a few hours or overnight. Samples of protoplast suspensions are then stained by FDA and the results are expressed as percentage of fluorescent (i.e., vital) protoplasts.

Uptake and Incorporation of Radioactive Precursors. Protoplast are incubated in their appropriate liquid medium for 1 day before the assay. Then aliquots (e.g., 5 ml) of protoplast suspensions are transferred to 5-cm petri dishes and the radiolabeled precursors are added. For amino acid incorporation, 1.25 μCi of [^{14}C] protein hydrolysate (5.7 Ci g-atom^{-1} specific activity) may be used, but specific radiolabelled amino-acids are also appropriate. The suspensions are then incubated for up to 1 or 2 h and samples are removed at desired time intervals (e.g., every 15 min). For uptake, samples (0.5 ml) of protoplasts are filtered through a glass filter washed with cold medium and the filters are dried (70 °C overnight), placed in scintillation fluid and the radioactivity is evaluated in a scintillation counter. For incorporation, the aliquots of protoplasts are transferred into chilled tubes containing 10% trichloroacetic acid. The latter suspension is then filtered as described above, then washed four times with cold 5% trichloroacetic acid and twice with 5 ml 96% ethanol. Radioactivity is measured as for uptake. Similar procedures (Zelcer and Galun 1976) can be followed to evaluate precursors incorporation into the RNA and DNA of cultured protoplasts.

Colony Formation as Affected by PPT. Colony formation can be assayed by protoplasts incubated continuously in PPT or by protoplasts exposed to PPT for a limited duration (e.g., 2 h, 24 h) and then cultured in media without PPT. Obviously for short exposure (2 h) the PPT concentration should be higher than for continuous culture in the presence of PPT. Colony counts are usually performed when in control cultures the number of cells per colony exceeds 10–20. Thus, *Nicotiana* and *Citrus* colonies may be counted 2 and 4 weeks after plating the protoplasts, respectively.

3.3 Bioassay of Cercosporin by Tobacco Cells

3.3.1 General Consideration and Background

Cercosporin is a chemically defined compound with a molecular mass of 534 daltons, produced by members of the fungal genus *Cercospora*. It is a nonspecific toxin and will also kill cells of plants that are not hosts of pathogenic *Cercospora* species. Moreover, cercosporin is also toxic to mammalian cells and to bacteria. Its killing capability is light dependent; illuminated cercosporin appears to generate both singlet oxygen and superoxide. The latter are probably the more direct killing agents, exerting detrimental effects on cellular membranes. The phytotoxicity of cercosporin was amply studied and reviewed by M.E. Daub and her collaborators (e.g., Daub 1982, 1984, 1987; Daub and Briggs 1983; Daub and Hangarter 1983). The following protocol for cercosporin toxicity assays is based on these publications as well as on personal communication (with M.E. Daub) and on our own laboratory procedures. The evaluation of toxicity is based on staining of viable cells (see Sect. 3.2.2). The enzymatic cleavage of the acetate takes a few minutes; then, vitality can be determined microscopically. This protocol will describe the procedure for protoplasts isolated from tobacco (*Nicotiana tabacum*) cell cultures, but it should be applicable to various other cell-culture derived protoplasts.

3.3.2 Protocol

Plant Material and Protoplasts. The sources of protoplasts for this bioassay are cell suspension cultures. Tobacco cell suspensions as those from cv. Xanthi and cv. Wisconsin 38 are availabe from many laboratories. Otherwise such suspensions can be established from tobacco cultivars as well as from other dicot species by the procedure outlined in Section 2.2 above. Cell cultures in their early log phase (2 to 3 days after transfer of a healthy cell suspension culture) are sedimented, washed once with a wash solution (CPW: 0.55 M mannitol, 0.2 mM KH_2PO_4, 10 mM $CaCl_2$, 1 mM $MgSO_4$, 1 mM KNO_3) and then incubated overnight in a filter-sterilized maceration fluid containing 1% Driselase in CPW. Incubation is best performed in 9-cm petri dishes containing 10-ml cells in maceration suspension, at 25 °C, in the dark. For isolation of protoplasts from the maceration fluid, the suspension is filtered through a nylon sieve (~70-μm pores) and precipitated by low speed centrifugation (100–150 g) for 3–10 min. The pelleted protoplasts are then suspended and washed three times with a protoplast buffer (5 mM MES, 5 mM $CaCl_2$, 450 mM mannitol, pH 6.0) by centrifugations as described above.

Bioassay with Fluorescein Diacetate (FDA). FDA is stored as a 5 mg per 1 ml acetone stock solution. The diluted FDA in aqueous solution should be prepared shortly before use. Cercosporin can be dissolved in acetone and up to 0.5% acetone can be added to protoplasts with little or no damage. This compound is also not stable in aqueous solutions. Its concentration in acetone solution can be determined spectrophotometrically (extinction coefficient is 23 600 at 473 nm). The range of cercosporin concentration that is usually used is 0.1–1.0 μM.

For the assay, aliquotes of protoplasts (2×10^4/ml) are placed in petri dishes and the various quantities of cercosporin are added. Cercosporin is active only after light excitation. Therefore a standard fluence of light should be provided such as 2×10^4 erg $cm^{-2} s^{-1}$, by exposure to fluorescent light for various time periods. After incubation in cercosporin the protoplasts may be assayed by FDA or washed, pelleted and redispersed in protoplast buffer or in culture medium. After extended culture in culture medium, cell or colony counts can be performed. For earlier determinations of cell vitality, FDA up to 0.01% is added to the protoplast suspensions and the percentage of fluorescenting protoplasts is evaluated with an epifluorescence microscope equipped with appropriate excitation and barrier filters (e.g., Zeiss filter set G436); fluorescenting cells are considered vital. In order to count the dead cells also, a low level of regular light should be maintained during observation.

3.4 Bioassay of Sirodesmin by *Brassica* Protoplasts and Cells

3.4.1 General Considerations and Background

The fungal organism *Phoma lingam* is pathogenic to crops such as cabbage and rape of Cruciferaceae. A number of investigations (see Ferezou et al. 1977; Sjödin E.G. et al. 1988; Sojöin E.G. and Glimelius 1989) showed that culture filtrates of *P. lingam* contain several toxic components, one of which is sirodesmin PL. While the exact correlation between the pathogenecity of *P. lingam* and sirodesmin phytotoxicity to host and non-host cells of various species was not established, it was shown that protoplasts of both host and nonhost plants are sensitive to sirodesmin. The sensitivity decreases during culture, and after cell division this phytotoxin was differentially toxic to cells of host plants. Culture filtrates from *P. lingam* were reported to have a differential effect on cell culture lines of *B. napus* providing a mean to select resistant lines and possibly resistant regenerated plants (Sacristan 1982).

3.4.2 Protocol

Bioassay with Protoplasts. Protoplasts of *B. napus* are isolated from leaf material by a procedure detailed by Wetter and Constabel (1982) or as described by Glimelius (1984). The protoplasts are suspended at 2.5×10^4/ml in multi-well dishes (24 wells, 0.5 ml/well) in protoplast culture medium (e.g., Glimelius et al. 1986) containing plant hormones (e.g., 4.5 µM 24D, 2.2 µM benzylamino purin). Other protoplasts can be isolated and culture by procedures that are appropriate for each case. Sirodesmin PL is kept as a methanol stock solution that is diluted with water according to the required final concentration (1–5 µM). The phytotoxin is added to the protoplast suspension; shortly afterwards the viability of the protoplasts is evaluated microscopically by adding Evans Blue (0.5% dissolved in 0.4 M sorbitol). The ability of protoplasts to form cell walls after culture in the presence of sirodesmin can be evaluated by the fluorescence of these walls when stained with 0.1% calcofluor white (American Cyanamid) in 0.4 M sorbitol.

Bioassay of Cultured Cells. For assaying protoplast-derived cell cultures, the protoplasts are first trapped in 0.4% agarose beads containing the same culture medium as mentioned above and the beads are inserted in 3.5-cm petri dishes also containing the same medium. This enables the addition of the sirodesmin PL at any given stage of the culture. After exposure to the phytotoxin for specific time periods, the latter can be removed and the beads "washed" with a salt solution (e.g., CPW) after which fresh culture medium (without phytotoxin) is added. The number of cell colonies is counted after a given period of further culture (e.g., total 20 days from plating).

3.5 Bioassay of *Helminthosporium maydis* Race T Toxin with Maize Protoplasts

3.5.1 General Considerations and Background

The following assay represents a case where the phytotoxin is purified and is specifically affecting only plants that are susceptible to the pathogen from which the phytotoxin was isolated. Moreover, the phytotoxin *Helminthosporium maydis* race T (*Hm* T) targets to mitochondria that are susceptible to it. Cells (and protoplasts) harboring the latter type of mitochondria can be used in bioassays. There is ample literature on: (1) differential sensitivity of maize plants having different "cytoplasm" (i.e., mitochondria); (2) on the utilization of *Hm* T to select *H. maydis* resistant plants from scutellar calli; (3) on the possible involvement of *Hm* T in suppressing mitochondrial metabolism (e.g., Pelcher et al. 1975; Gengenbach et al. 1977; Earle 1978, 1982, 1983; Walton and Earle 1984; Holden and Sze 1989). The mitochondria of cytoplasmic male sterile, type T (CMS-T) maize plants contain a gene, T-*urf* 13, that was implicated with both male sterility and sensitivity to *H. maydis* (see Holden and Sze 1989; and review by Breiman and Galun 1990). Although direct procedures to assess the effects of *Hm* T on specific mitochondrial activities are available (Holden and Sze 1989), the bioassay of this phytotoxin by maize protoplasts is rather sensitive and specific. The following protocol is based mainly on studies of Earle and her collaborators.

3.5.2 Protocol

The Hm T *Phytotoxin*. This phytotoxin is insoluble in water but can be solubilized in either dimethyl sulfoxide or ethanol. The concentrations of Hm T that will drastically affect sensitive protoplasts are rather low. It is therefore possible to add the minute volumes of stock-solution directly to the cultured protoplasts (keeping the final percentage of solvent in the culture below 1%).

Protoplasts. Young maize leaves are surface sterilized as detailed previously. They are then either shredded or abrased and incubated in petri dishes containing maceration fluid. This fluid may consist of 2% Cellulysin in 0.5 M sorbitol and 10 mM $CaCl_2$, or of a mixture of macerating enzymes dissolved in the appropriate osmoticum (e.g., CPW as noted in Sect. 3.3.2). The duration of incubation is dependent on the enzyme-composition of the maceration fluid, but should not be longer than about 3–5 h. The macerated leaves are filtered through a nylon screen (80 μm) and the protoplasts are then centrifugated at a low speed (~100 *g*). The precipitated protoplasts are resuspended with a wash solution (e.g., 0.5 M sorbitol, 10 mM $CaCl_2$, 5 mM MES, pH 5.8) and centrifuged again. Precipitation and washing should be repeated one or two more times. The precipitated protoplasts are resuspended at a density of 1–2 × 10^5 ml^{-1} in a wash solution (or in a culture medium, such as MS) and aliquots are placed into petri dishes or multiwell plates. When 35-mm plastic petri dishes are used, 1 ml of suspension is placed in each dish.

Exposure to Hm *T and Assays.* The *Hm* T stock solution is mixed into the petri dishes or wells that contain the protoplast suspensions, at various volumes, to reach a final concentration of 0–50 mg ml^{-1}. The protoplasts are observed periodically with an inverted microscope. Evaluation of toxicity can be performed by counting the damaged protoplasts in small samples placed in a hemocytometer (with a depth of 0.2 rather than the common 0.1 mm). Otherwise the vitality can be determined by staining (e.g., FDA, see Sect. 3.3.2 above) or by other means noted in the protocols described above.

References

Aviv D, Galun E (1985) An in vitro procedure to assign pigment mutations in *Nicotiana* to either the chloroplast or the nucleus. J Hered 76:135–136

Aviv D, Galun E (1988) Transfer of cytoplasmic organelles from an oligomycin resistant *Nicotiana* cell suspension to tobacco protoplasts yielding oligomycin resistant cybrid plants. Mol Gen Genet 215:128–133

Avni A (1990) The molecular genetic basis for tentoxin sensitivity in higher plants' chloroplasts. PhD Thesis, The Weizmann Institute of Science, Rehovot, Israel

Bajaj YPS (1989) Plant protoplasts and genetic engineering, vol I. Springer, Berlin Heidelberg New York, pp 444

Breiman A, Barash I (1981) Partial characterization of phytotoxic compounds from culture filtrates of *Phytophthora citrophthora*. Phytopathol Z 102:1–9

Breiman A, Galun E (1981) Plant protoplasts as tool in quantitative assays of phytotoxic compounds from culture filtrates of *Phytophthora citrophthora*. Physiol Plant Pathol 19:181–191

Breiman A, Galun E (1990) Nuclear-mitochondrial interrelation in angiosperm – a review article. Plant Sci 71:3–19

Daub ME (1982) Cercosporin, a photosensitizing toxin from *Cercospora* species. Phytopathology 72:270–274

Daub ME (1984) A cell culture approach for the development of disease resistance: studies on the phytotoxin cercosporin. Hortscience 19:382–387

Daub ME (1986) Tissue culture and the selection of resistance to pathogens. Annu Rev Phytopathol 24:159–186

Daub ME (1987) The fungal photosensitizer cercosporin and its role in plant disease. In: Heitz JR, Downum KR (eds) Light-activated pesticides (ACS Symp Ser 339) Am Chem Soc, pp 271–280

Daub ME, Briggs SP (1983) Changes in tobacco cell membrane composition and structure caused by cercosporin. Plant Physiol 71:763–766

Daub ME, Hangarter RP (1983) Light-induced production of singlet oxygen and superoxide by the fungal toxin, cercosporin. Plant Physiol 73:855–857

Durbin RD (1981) Applications. In: Durbin RD (ed) Toxins in plant disease. Academic Press, New York, pp 445–505

Durbin RD, Graniti A (1989) Possible applications of phytotoxins. In: Phytotoxins and plant pathogenesis, NATO ASI series, vol H27. Springer, Berlin Heidelberg New York, pp 337–355

Earle ED (1978) Phytotoxin studies with plant cells and protoplasts. In: Thorp TA (ed) Frontiers of plant tissue culture 1978. Proc 4th IAPTC Congress. Univ. of Calgary, Calgary, pp 363–372

Earle ED (1982) Cytoplasmic specific effects of *Helminthosporium maydis* race T toxin on corn protoplasts and mitochondria. In: Earle ED, Demarly Y (eds) Variability in plant regenerated from tissue culture. Praeger Scientific, New York, pp 351–367

Earle ED (1983) The use of phytotoxins for in vitro selection of disease-resistant plants. In: Cell and tissue culture techniques for cereal crop improvement. Science Press, Beijing; Gordon and Breach, New York, pp 265–278

Evans DA, Sharp WR, Ammirato PV, Yamada Y (1983) Handbook of plant cell culture, vol 1. Macmillan, New York, pp 970

Ferezou JP, Riche C, Quesneau-Thierry A, Pascard-Billy C, Barbier M, Bousquet JF, Roudart G (1977) Structures de deux toxines isolées des cultures due champignon *Phoma lingam* Tode: la sirodesmine PL et al. desacetyl sirodesmine P.L. Nouv J Chem 1:327–334

Galun E (1982) Screening for chloroplast composition in hybrids and cybrids resulting from protoplast fusion in angiosperms. In: Edelman M, Hallick RB, Chua NH (eds) Methods in chloroplast molecular biology. Elsevier, Amsterdam, pp 139–148

Gengenbach BG, Rines HW (1986) Use of phytotoxins in selection of disease resistant mutants in tissue culture. Iowa State J Res 60:449–476

Gengenbach BG, Green CE, Donovan CM (1977) Inheritance of selected pathotoxin resistance in maize plant regenerated from cell cultures. Proc Natl Acad Sci USA 74:5113–5117

Glimelius K (1984) High growth rate and regeneration capacity of hypocotyl protoplasts in some Brassicaeae. Physiol Plant 61:38–44

Glimelius K, Djupsjobacka M, Fellner-Feldegg H (1986) Selection and enrichment of plant protoplasts heterokaryons of Brassiceae by flow sorting. Plant Sci 45:133–141

Hammerschlag FA (1984) Optical evidence for an effect of culture filtrates of *Zanthomonas campestris* p.b. *pruni* on peach mesophyll cell membranes. Plant Sci Lett 34:295–304

Hildebrandt AC, Riker AJ, Duggar BM (1946) The influence of the composition of the medium on growth in vitro of excised tobacco and sunflower tissue culture. Am J Bot 33:591–597

Holden MJ, Sze H (1989) Effects of *Helminthosporium maydis* race T toxin on electron transport in susceptible corn mitochondria and prevention of toxin actions by dicyclohexylcarbodiimide. Plant Physiol 91:1296–1303

Maliga P, Breznovits S, Marton L (1973) Streptomycin resistant plants from callus of haploid tobacco. Nature New Biol 224:29–30

Patil SS, Tam LQ, Sakai WS (1972) Mode of action of the toxin from *Pseudomonas phaseolicola*. I. Toxin specificity, chlorosis and ornithin accumulation. Plant Physiol 49:803–807

Pelcher LE, Kao KN, Gamborg OL, Yoder OX, Gracen VE (1975) Effect of *Helminthosporium maydis* race T toxin on protoplasts of resistant and susceptible corn (*Zea mays*). Can J Bot 53:427–431

Perl A, Aviv D, Galun E (1991) Protoplast fusion mediated transfer of oligomycin resistance from *Nicotiana sylvestris* to *Solanum tuberosum* by intergeneric cybridization. Mol Gen Genet 225:11–16

Pierik RLM (1987) In vitro culture of higher plants. Martinus Nijhoff Publ, Dordrecht, 344 pp

Reinert J, Bajaj YPS (1977) Plant cell tissue and organ culture. Springer, Berlin Heideberg New York

Sacristan MD (1982) Resistance responses to *Phoma lingam* of plants regenerated from selected cell and embryogenic cultures of haploid *Brassica napus*. Theor Appl Genet 61:193–200

Sjödin C, Glimelius K (1989) Differences in response to the toxin sirodesmin PL produced by *Phoma lingam* (Tode ex. Fr.) Desm. on protoplasts, cell aggregates and intact plants of resistant and susceptible *Brassica* accessions. Theor Appl Genet 77:76–80

Sjödin C, Nyhammar T, Glimelius K (1988) Effects of toxic metabolites produced by *Phoma lingam* (Tode ex Fr.) Desm. on protoplasts, cells and plants of host and non-hosts of the pathogen. Physiol Mol Plant Pathol 32:301–312

Steele JA, Uchytill TF, Durbin RD, Bhatnagar P, Rich DH (1976) Chloroplast coupling factor 1: a species specific receptor for tentoxin. Proc Natl Acad Sci USA 73: 2245–2248

Street HE (1977) Plant tissue and cell culture, 2nd edn. Blackwell, Oxford 503 pp

Strobel GA (1977) Bacterial phytotoxins. Annu Rev Microbiol 31:205–224

Takebe I, Otsuki Y (1969) Infection of tobacco mesophyll protoplasts by tobacco mosaic virus. Proc Natl Acad Sci USA 64:843–848

Takebe L, Otsuki Y, Aoki S (1968) Isolation of tobacco mesophyll cells in intact and active state. Plant Cell Physiol 9:115–124

Turner NX, Graniti A (1969) Fusicoccin: a fungal toxin that open stomata. Nature 223:1070–1071

Ullrich CI, Novacky AJ (1991) Electrical membrane properties of leaves, roots and single root cap cells susceptible *Avena sativa*: effect of victorin C. Plant Physiol 95:675–681

Vardi A, Galun E (1989) Isolation and culture of *Citrus* protoplasts. In: Bajaj YSP (ed) Plant protoplasts and genetic engineering I. Springer, Berlin Heidelberg New York, pp 147–159

Vardi A, Epstein E, Breiman A (1986) Is the *Phytophthora citrophthora* culture filtrate a reliable tool for the in vitro selection of resistant *Citrus* variants? Theor Appl Genet 72:569–574

Vasil IK (1984) Cell culture and somatic cell genetics in plants vol I. Academic Press, Orlando, 825 pp

Walton JD, Earle ED (1984) Isolation and bioassay of fungal phytotoxins. In: Vasil IK (ed) Cell culture and somatic cell genetic, vol I. Academic Press, Orlando, pp 598–607

Wetter LR, Constabel F (1982) Plant cell culture methods. Natl Res Counc Can, Saskatoon, 146 pp

Widholm, JM (1972) The use of fluorescein diacetate and phenosafranin for determining viability of cultured plant cells. Stain Technol 47:188–193

Yoder OC (1980) Toxins in pathogenesis. Annu Rev Phytopathol 18:103–129

Yoder OC (1981) Assay. In: Durbin RD (ed) Toxins in plant disease. Academic Press, New York, pp 45–78

Zelcer A, Galun E (1976) Culture of newly isolated tobacco protoplasts – precursors incorpation into RNA and DNA. Plant Sci Lett 12:305–315

Determination of Host-Selective Toxins

K. Kohmoto

1 Introduction

Since Hutchinson's first toxigenic hypothesis of fungal plant disease, proposed in 1913, there have been many reports of phytotoxins isolated from culture filtrates of pathogens. Some toxins could be related to symptom development, and others were found to have no positive relation to the pathogenicity of the producing fungus. Over many years, the evaluation of toxin in pathogenes is fluctuated from optimistic excitement to disappointment. The concept of host-specific or host-selective toxins (HST) launched in 1964 by Pringle and Scheffer became a historical turning point, and subsequent toxin studies in plant pathology developed accordingly. Many discoveries, suggestions and prospects concerning the development of the HST concept have been reviewed (Scheffer and Briggs 1981; Nishimura and Kohmoto 1983a,b; Scheffer 1983, 1989; Scheffer and Livingston 1984).

More than 14 examples of HST-involving diseases have been reported so far (Table 1). Almost all HSTs were isolated, and their structures determined, during the last decade (Scheffer and Livingston 1984; Kohmoto and Otani 1991), and in addition, some were confirmed by total synthesis. These toxins were biologically active at concentrations as low as ng/ml. Their roles in pathogenesis and modes of action could also be investigated in depth by quantitative use of chemically defined toxin preparations (Nishimura 1987; Kohmoto and Durbin 1989; Kohmoto et al. 1989; Otani et al. 1991). After recent progress in toxin research, the hypothetical HST scheme began to materialize as the realistic one. The mechanism for determination of HST-mediated host specificity involves at least three basic processes: (1) prior to penetration, fungal spores release a specificity factor (HST) as a signal of the parasite; (2) the shot signal molecules selectively bind to putative receptor sites in host cells; and (3) signal transduction gives rise to conditioning the cells toward an accessible state for penetration by immobilizing the host's defense reaction before cell death. Even in the so-called perthophytes, a virulent spore can penetrate and establish its initial colonization without requiring host-cell death, in an HST-exposed cell at early stages of pathogenesis (Yoder and Scheffer 1969; Comstock and Scheffer 1973; Kohmoto et al. 1989).

To date, there are diverse and increasing demands to determine such HSTs, not only in pathological and ecological studies of host-parasite interactions, but also in various fields of modern agri-biotechnology.

Table 1. A list of host-selective toxins

Producing pathogen (Previous name)	Short-hand designation of toxin	Chemical nature of toxin
Alternaria alternata Apple pathotype (*A. mali*)	AM-toxin I, II, III	Cyclic depsipeptide
Alternaria alternata Japanese pear pathotype (*A. kikuchiana*)	AK-toxin I, II	Ester of amino acid and epoxydecatrienoic acid
Alternaria alternata Rough lemon pathotype (*A. citri*)	ACR-toxin I	Polyhydroxyalkenyl dihydropyrone
Alternaria alternata Strawberry pathotype	AF-toxin I, II, III	Ester of amino acid and epoxydecatrienoic acid
Alternaria alternata Tangerine pathotype (*A. citri*)	ACT-toxin Ib, IIb[a]	Ester of amino acid and epoxydecatrienoic acid
Alternaria alternata Tobacco pathotype (*A. longipes*)	AT-toxin	Not determined
Alternaria alternata Tomato pathotype (*A. alternata* f.sp. *lycopersici*)	AAL-toxin TA, TB (AL-toxin I, II)	Ester of propanetricar-boxylic acid and amino-dimethylheptadecapentol
Bipolaris sacchari (*Helminthosporium sacchari*)	HS-toxin A, B, C	Sesquiterpene glucoside
Cochliobolus carbonum race 1 (*Helminthosporium carbonum* race 1)	HC-toxin I, II, III	Cyclic tetrapeptide
Cochliobolus heterostrophus race T (*Helminthosporium maydis* race T)	HMT-toxin band 1, 2	Linear polyketols
Cochliobolus victoriae (*Helminthosporium victoriae*)	HV-toxin C (Victorin C)	Unusual oligopeptide containing chlorine atoms
Periconia circinata	PC-toxin (Peritoxin A, B)[b]	Unusual peptide containing chlorine atoms[b]
Phyllosticta maydis	PM-toxin	Linear polyketols

[a] ACTG-toxins were isolated from the same pathogen and elucidated in structure by Kono et al. (1986), but their involvement in early pathogenesis was unclear.
[b] Presented by Dr. V. Macko at a U.S.-Japan Cooperative Science Seminar held in Honolulu in 1990.

In plant pathology there are many questions about HSTs necessitating quantitative consideration. For example, how much HST does a single virulent spore produce during early germination at the infection site? What is the minimum amount of HST required for the induction of accessibility in a host cell? At what concentration does HST affect host defense systems? How is HST metabolically changed in plants? How does HST play a role in parasitic fitness in fields?

In recent years, HSTs and related phytotoxins with pathological significance have attracted enthusiastic attention from crop breeders and plant genetic engineers, mainly because they want to use toxins as screening agents or surrogates of pathogens for mass selection, gene analysis, and characterization for disease-resistant genotypes. However, even if culture filtrates were used, as occasionally observed, the projects, since they were conducted without a thorough pathological evaluation of the toxin in culture filtrates, were found eventually unsuccessful. Thus, it is strongly recommended in such experiments to employ toxin preparations that have been pathologically evaluated and chemically defined. Hence, the methods for sensitive qualitative and quantitative determination of toxins are still the basis of these conventional and biotechnology-oriented researches. The problems and prospects concerning the use of phytotoxins in breeding have been critically reviewed by Wenzel (1985), Daub (1986) and Gengenbach and Rines (1986).

2 Definition of Host-Selective Toxin

Originally, HSTs could hardly be determined by chemical methods, and bioassay based on their biological activity on host plants was the only means to characterize them as significant pathogen products. Toxin researchers must remember Yoder's notion in 1981 that a biological definition is the only definition until the chemical structure has been confirmed. Even now, in practical studies it is still reasonable and persuasive to fulfill the following four requirements, stated by Scheffer and his colleagues, for a candidate toxin to be called an HST:

1. Correlation between toxin productivity and pathogenicity of the pathogen. All isolates of the pathogen that produce the toxin are pathogenic to the specific host. All isolates that fail to produce the toxin lose pathogenicity.
2. Correlation between toxin sensitivity and susceptibility of plants to the pathogen. Plants that are susceptible to the pathogen are sensitive to the toxin it produced.
3. Toxin must be released on spore germination at the site of penetration in advance (Nishimura and Scheffer 1965).
4. Toxin must confer host cells accessible to hyphal invasion, and induce the initial colonization of the pathogen only in the host cells (Yoder and Scheffer 1969).

Currently, genetic and/or molecular biological analyses must without doubt be the most crucial way of evaluating the primary and inevitable involvement of a toxin in disease development; but it may be difficult to complete the analysis if the pathosystem is totally new or poorly understood. The second way is the careful examination of whether the toxin in question can satisfy the conditions mentioned above. To evaluate toxins by only a single criterion may often be misleading. In addition, there might be some obstacles to overcome, and discrepancies may be found among the data obtained. It is necessary to provide evidence as to why the conditions are not satisfied. Some explanations are given in review articles (Yoder 1980, 1981).

3 Bioassay of Host-Selective Toxins

For the bioassay of HSTs, there are too many diverse methods to list all of those reported to date. Pathologically meaningful studies of toxins greatly depend on bioassays, as mentioned above. The choice of a suitable bioassay for HST obviously depends on the purpose of the particular research system under investigation: we must first consider "what kind of information is needed?" Further criteria for the relevant choice are sensitivity, selectivity, simplicity to manipulate, ability to quantify, accuracy and reproducibility of assay methods. The intention here is not to describe a comprehensive collection of bioassay procedures, but to introduce some fundamental principles and reliable methods of bioassays, and to make comments on interpreting data that might be obtained.

3.1 Leaf Assay

A bioassay procedure based on toxin-induced leaf necrosis is one of the most popular assays employed for various types of experiments; for example, detecting HSTs in culture filtrates of pathogens, estimating amounts of toxins in spore germination fluids or sampling solutions, investigating the toxin sensitivity spectra of plants, and so on.

In general, young leaves of the same age are harvested shortly before use. Sometimes, to adequately detect a slight but significant difference between treatments, small pieces cut from a leaf can be employed to minimize accidental variation in toxin response due to different leaves. Leaves and leaf tissues are slightly scratched on the surface with a needle or the sharp edge of a glass capillary, and a small drop of toxin-containing solution is applied immediately to each wound. A longer time lag between wounding and toxin application might cause a reduction in sensitivity to toxin. The leaves are incubated in a moist chamber. Necrotic areas developed around the application sites are measured after a given incubation period, in many cases, 24 or 48 h or more. Usually, plant response is recorded by measuring necrotic leaf area or by reading the maximum dilution end-point for necrosis. For measurement of the necrotic area, a computer-assisted planometer or an image analyzer is efficiently used if available. A section paper method is also employed: the necrotic area is traced on fine section paper and calculated by counting the number of sections or by weighing cut paper pieces corresponding to the necrotic leaf area. This method is a time-consuming and less accurate procedure.

Many environmental factors occasionally influence the necrosis development in the toxin-treated leaf tissues: for example, in apple leaves treated with AM-toxin, light conditions are very critical for necrosis development during the early incubation period (Tabira et al. 1989). Light at 570–680 nm suppresses the AM-toxin-induced necrosis of apple and Japanese pear leaves. The inhibition by light is evident at an intensity as low as $10\,\mu W/cm^2$. The protective effect of light appears to be specific to toxin species, because activities of AK-toxin and other nonspecific toxins are not affected by light exposure. Similar photoprotection from necrosis is observed in rough lemon leaves treated with ACR-toxin and Tms-cytoplasm maize leaves treated with HMT-toxin (Kohmoto et al. 1989). On the other hand, there is a contrasting phenomenon: Bhullar et al. (1975) reported that when cut ends of maize leaves bound or unbound with aluminum foil were immersed in HMT-toxin for 72 h in light (2500 ft-c), the unbound parts became chlorotic, while the banded parts remained unchanged. Experimental conditions for incubation might explain the different light effects. Another example of light-activated toxicity is described for a photosensitizing and nonspecific toxin, cercosporin, which generates singlet oxygen and superoxide, and hence exerts its toxic action when exposed to light (Daub 1982; Daub and Hangarter 1983). Whenever an unknown sample is assayed, one should remember that light conditions during incubation may induce inhibiting or promoting effects or no effect on necrosis formation of leaves.

Incubation temperatures also greatly affects toxin-induced necrosis. Reliable bioassays for comparative studies should be done at a standardized temperature.

Toxin sensitivity varies with the leaf age of test plants: young leaves of Japanese pear and apple have a higher sensitivity to AK- and AM-toxin, respectively, than old ones, while young leaves of tobacco are less sensitive to AT-toxin than old ones (Nishimura 1987). Strawberry leaves have almost the same sensitivity to AF-toxin, regardless of age.

Growth conditions also affect the toxin sensitivity of the plants. Air temperature, soil characteristics, fertilizer, and shading are factors which critically determine the sensitivity. In combinations of AK-toxin and Japanese pear, AM-toxin and apple, HS-toxin and sugarcane, HV-toxin and oat, and PC-toxin and sorghum, leaves of plants grown in a greenhouse, where air temperature rises over 35 °C in the summer season, become temporarily toxin-insensitive. To recover the normal sensitivity to toxin, leaves have to be kept at room temperature (20 °C) for a day or so. Hot greenhouses can obviously hinder toxin research.

Determining the plant responses to toxin is sometimes difficult, but essential. Necrosis is the final response of plant cells to toxin; the analysis of the dose-response relations by means of necrotic leaf area is fundamental for the elucidation of physiological and pathological roles of a toxin. Necrotic tissues, however, may vary in color, depending on plant species and genotypes. They are often accompanied by water congestion, suggesting that a toxin may, directly or indirectly, affect plasma membrane function.

To determine the specific activity (toxin activity per unit weight) and the amount of toxin in extracts or fractions in its purification process, an assay of the dilution end point causing necrosis is more accurate and useful than the necrotic area assay mentioned above.

3.2 Shoot Assay (Cutting Assay)

Cut shoots with leaves are sometimes employed as a tester, where the cut end is dipped in sample solutions and allowed to transpire. However, if crude preparations are assayed, the results may give a false information, because many coexisting compounds in the solution are apt to affect delicate transpiration. Furthermore, incubation conditions that stimulate rapid transpiration may induce a quick desiccation of toxin-treated tissues before manifesting other toxigenic syndromes such as water congestion and browning. When vein necrosis occurs on gramineous leaves, so-called runner lesions are produced and their length is related to toxin concentration.

3.3 Root Growth Assay (Seed Assay)

The root growth assay was first applied for measuring relative toxicity of HV-toxin (Meehan and Murphy 1947; Litzenberger 1949), and is one of the most popular bioassays. Oat seeds are germinated on moist paper and uniform seedlings are selected and incubated in petri dishes containing test solutions. The longest root on each seedling is measured 72 h after incubation at 22 °C, and the dilution which gives 50% inhibition of root growth is determined. Luke and Wheeler (1955) expressed relative activity of toxin in a test solution as "units" and "milliunits" of toxicity: a milliunit is defined to be the amount of toxin in 1 ml that reduces root growth of susceptible seedlings by 50%. This unit concept was satisfactorily applied to the comparative estimation of toxin in various sample solutions during purification and other processes, but caused some difficulties in determining the dilution end point. Consequently, the maximum dilution that gives complete inhibition of root growth became the universal criterion for the toxicity (Pringle and Braun 1957).

This bioassay have significant advantages. Whenever seeds of near-isogenic lines with good quality are available, the genotype-specific toxin action can be easily determined in terms of differential responses between resistant and susceptible genotypes. For example, HMT-toxin was satisfactorily used to determine the proportion of Tms and normal cytoplasm corn in any blended seed lot by seedling bioassay (Lim et al. 1971). The disadvantage of this process is that seeds of crop plants that are propagated by vegetative organs may respond differently to a toxin, because they lack genetic uniformity.

It should also be noted that even a single species of HST may have dual, host-selective and nonselective, effects on plants: e.g., AAL-toxin, causing Alternaria stem canker in tomato, exerts its toxic action on root growth of both the susceptible and resistant genotypes of tomato, whereas leaf necrosis by the toxin occurs only on the susceptible genotype (Umezawa et al. 1988; Witsenboer et al. 1988).

3.4 Pollen Assay

An assay based on the inhibition of pollen germination was first applied to HMT-toxin (Laughnan and Gabay 1973). Toxin preparations are incorporated into a modified agar medium supplemented with maize leaf extract, on which maize pollen is incubated and examined for germination. This procedure does not require aseptic conditions, because the assay can be completed within 2 or 3 h and the medium employed is unfavorable for the growth of other microbes. The simplicity and quickness are significant merits of the pollen assay. However, an analysis of the data may require knowledge of the genetic background of the disease (or toxin) resistance of plants from which pollen was collected.

3.5 Isolated Cell Assay

Isolated intact cells and protoplasts are often utilized to analyze physiological effects of toxin, mode of toxin action, and toxin receptor. This method can be advantageous for advancing HST studies, because some uncertainties that arise from the use of whole seedlings or plant tissues, such as variability of toxin penetration, thus producing different responses by variously damaged cells, can be eliminated. Populations of isolated single cells and protoplasts are easily subjected to statistical analysis and interpretations.

Hawes and Wheeler (1982) first employed root cap cells of oats and examined toxin-induced cell death in relation to temperature and plasmolysis. Their protocol for isolation is as follows: to minimize microbial contamination, all stages of cell isolation are conducted under aseptic conditions. Deglumed oat seeds are immersed in 50 ml of 0.5% sodium hypochlorite in a 200-ml funnel plugged with a serum stopper. A 10-mm square of vinyl mesh screen placed in the neck of the funnel allows the passage of surface debris, but prevents the loss of seeds. Air is introduced through a hypodermic needle inserted into the stopper, and the seeds are vigorously agitated for 10 min.

The sodium hypochlorite solution is then replaced with water and the seeds are again vigorously agitated. After 5–10 min, the seeds are rinsed and fresh water is added. This process is repeated until the water is free of trichomes and surface debris. The seeds are allowed then to germinate for 36 h at 25 °C in sterile petri dishes.

The germinated seedlings are soaked for 5 min in 50 ml of water in the funnel apparatus described above, and then a slow stream of air is introduced and seedlings are gently agitated for 5 min. The stopper is removed from the funnel, the immersion water is collected and rinsed over the seedlings five times. The cell suspension thus obtained is centrifuged at 5 × g for 1 min and the supernatant discarded. The pellet, consisting of isolated root cap cells, is resuspended in water, and adjusted to an appropriate concentration using a hemacytometer. According to their experiments, each root yielded 1000–2000 cells. About 50% are singly separated, while the rest are in chains or monolayer clumps of two to ten cells.

Cell viability can be assessed by vital staining with fluorescein diacetate (FDA) or Evans blue, and by the ability to plasmolyse in a hypertonic solution after experimental treatments. Incubation temperature may affect toxin sensitivity. Oat cells have a critical temperature at around 12 °C (Briggs et al. 1984). They are insensitive to HV-toxin when incubated at 10 °C but not at 15 °C (Hawes and Wheeler 1982). A similar technique, using isolated corn root cap cells, is successfully applied to the quantitative assay of the HMT-toxin (Hawes 1983a).

Kodama et al. (1991) have recently reported that viability of cultured, single cells can quantitatively be determined by a colorimetric assay using 3-(4,5-dimethylthiazol-2-yl)-2,5-diphenyl-tetrazolium bromide (MTT). The standardized protocol is as follows: AL-toxin-treated or untreated tomato cell suspension (1×10^4 cells) in a 0.45-ml medium is placed in a well of 24-well trays. MTT solution (0.05 ml, 500 μg/ml in 0.05 M sodium phosphate buffer at pH 7.5) is added to each well, and incubated for 6 h at 37 °C. Cell suspensions are transferred to test tubes and centrifuged at 600 × g for 5 min to collect the cells. To extract the formazan produced, 3 ml isopropanol is added to each tube, throughly mixed, and cell debris are removed by centrifugation at 600 × g for 5 min. Absorbance at 570 nm of the extracts is determined.

The MTT method is not only a rapid and quantitative assay for HST effects, but also allows a large number of samples to be handled. In addition to AL-toxin, AF- , AK- , AM- and ACR-toxins are satisfactorily assayed. The viability of cell preparations that are ambiguously judged by the staining with FDA or Evans blue can be distinctly determined by the MTT method.

3.6 Protoplast Assay

Protoplasts can be isolated from various plant tissues and calli. They are also satisfactorily used for bioassay. Since the general and comprehensive methods will be described by Galun and Breiman in another chapter of this Volume, I will only briefly mention some of protoplast assays for HSTs.

1. Hawes' procedure (1983b): Root cap protoplasts are isolated from inner root-cap cells of oat seedlings grown under aseptic conditions for 36 h, as described in the preceding section. Root tip segments (about 1 mm) are preplasmolysed for 20 min in 0.6 M sorbitol before enzyme treatment. The segments are incubated in a cell-wall degrading enzyme solution consisting of 5% Cellulysin (Calbiochem-Behring), 0.5% potassium dextran sulfate and 0.6 M sorbitol in 0.1 M 2-(*N*-morpholino)-ethanesulfonic acid buffer, pH 5.5. After incubation for 2 h at 25 °C, the suspension is shaken gently to free surface cells but leave the main body of the root cap intact. The root tips are discarded, and cells and protoplasts are harvested as follows: the cell-protoplast mixture is centrifuged for 1 min at 90 g, washed twice in 0.6 M sorbitol, resuspended in sorbitol containing 0.01% carbenicillin and cephaloridine, and the concentration of cells and protoplasts is determined with a hemacytometer.
2. Briggs' procedure (1984): Mesophyll protoplasts are isolated from primary leaves of oats, grown in vermiculite for 1 week under fluorescent light at 22 °C by peeling away the lower epidermis and floating the leaves (peeled surface facing down) on a solution containing 0.5% Cellulysin (Calbiochem-Behring) and 0.6 M sorbitol, adjusted to pH 5.6 with KOH. The preparation is incubated at 30 °C for 2 h in the dark, then is swirled gently to release protoplasts. The protoplasts are filtered through a layer of Miracloth and the suspension is centrifuged at 40 g for 10 min. The supernatant is removed and the pellet is washed by centrifugation in a suspension medium containing 0.6 M sorbitol and 10 mM $CaCl_2$, adjusted to pH 5.7 with KOH. It requires about 3 h to prepare the protoplasts. Protoplasts are used at a concentration of about 10^6/ml. (As will be referred to later, the toxin dose-response relation varies with protoplast concentration employed.) They can be observed with a microscope in dark-field and in dark-field with UV light (for observations of fluorescence). They are stained by adding 5 µl of 0.5% FDA in acetone to 0.5 ml of the sample and incubated for 5 min. The sample is viewed first in dark-field and then with UV light so that the same protoplasts can be scored for viability. Viable protoplasts fluoresce pale green in UV light, whereas dead protoplasts emit a dim, blood-red fluorescence. Briggs et al. (1984) showed that the time to detect the toxin effect appeared to depend on the activity of HV-toxin preparation used. When susceptible protoplasts were exposed to a high concentration (1.4 µg/ml) of toxin, all were killed in 60 min, while at a low concentration (1.4 ng/ml), only 84% were killed in 19 h.

Gendloff et al. (1987) improved Briggs' assay method: Protoplast suspensions are adjusted by dilution to a density of 5×10^4/ml, and 0.2 ml (10^4 protoplasts) is placed in each well (flat-bottomed) of polystyrene microtiter plates (Dynatech). HV-toxin is dissolved in the osmoticum containing 0.6 M sorbitol and 10 mM $CaCl_2$; 50 µl is then added to each microtiter well. Plates are covered with parafilm and incubated without agitation for 3 h at 35 °C in the dark. Viable protoplasts are determined by FDA staining (2.5 µl of a 0.5% solution). According to their data, survival rates were equal in wells containing 2×10^3 to 2×10^4 protoplasts; however, survival of protoplasts increased when concentration was higher than 2×10^4 per well. Toxin at 3.0 ng/ml killed 50% of the protoplasts and 10 ng/ml killed 90% in 3 h. Toxin sensitivity of the protoplast was lower than that in the root assay, although the sensitivity increased as exposure time increased longer than 3 h.

The new version of protoplast assay has many advantages over previously employed electrolyte leakage and root growth assays for HV-toxin. (1) The reduction in assay time is so significant that results can be obtained in 6 h including protoplast preparation. (2) The assay can be performed with very small volumes of reaction solution containing toxin. (3) Possible uncertainties or complications arising from the cell wall and intercellular space are eliminated. However, the improved protoplast assay is still ten times less sensitive to HV-toxin than the root growth assay.

A precaution noted by Briggs et al. (1984) is that toxin may subtly affect protoplast appearance under the microscope; even dead protoplasts determined by FDA staining had a seemingly normal appearance. Moreover, there was a 20–40 min delay before toxin-induced death of protoplasts was evident (Gendloff et al. 1987). An additional precaution that should be taken is that light may exert opposite effects on the toxin sensitivity of protoplasts (MacRae and Yoder 1988): e.g., protoplasts from etiolated Tms-cytoplasm maize leaves are insensitive to the toxin, whereas protoplasts from green leaves are highly sensitive, even though their mitochondria are always sensitive to HMT-toxin. When etiolated leaves are exposed to light for 24 h, protoplasts prepared from them become totally sensitive to HMT-toxin. Conversely, protoplasts from green tissue become insensitive to HMT-toxin when exposed to light for more than 6 h. Thus, procedures for handling assay plants should be standardized as to light conditions.

3.7 Electrolyte Leakage Assay

An assay based on toxin-induced electrolyte leakage from tissues is the most popular and simple physiological assay employed for HST studies. This method was originally suggested by Wheeler and Black (1963), and developed by Damann et al. (1974) in the United States for HV-toxin and by Otani et al. (1977) in Japan for AK-toxin. Damann's procedure is described in the following.

Tissue samples (0.1–2 g) from the first true leaf of 6- to 21-day-old oat plants are enclosed in cheesecloth, vacuum-infiltrated for 10 min, rinsed throughly in distilled water, placed in beakers containing a given volume of

HV-toxin solution or distilled water, and incubated for 1–5 h in a reciprocal shaker at 100–120 strokes/min. Conductance of the ambient solution is measured at intervals with a conductivity meter, using a dip-type electrode (K = 1) (Damann et al. 1974).

Otani's procedure: five leaf disks, 1 cm in diameter, are cut from freshly harvested young pear leaves with a leaf puncher. The disks are vacuum-infiltrated for 20 min with AK-toxin solution or deionized water as a control. After rinsing with deionized water, the leaf tissues are placed in flasks containing 10 ml deionized water and incubated in a reciprocal shaker at 27 °C. At intervals, the reaching solutions are removed and 10 ml fresh water is placed in each flask. Conductance in the solutions is measured with a conductivity meter. Content of each cation in leaching solutions is measured by an atomic absorption spectrophotometer or an inductively coupled plasma spectrometer.

These bioassay procedures can be modified depending on plant species, organs, tissues, and experimental aims, if necessary.

The profile of ion leakage with time after toxin treatment often suggests possible target organelles and the mode of toxin action. Linear plots with time immediately after toxin exposure suggest a dysfunction of the plasma membrane due to the toxin. If the plot line depicted with time has an inflexion point where the rate of leakage increases, toxin may indirectly affect plasma membranes through initial damage to the target organelles. It is also possible to hypothesize that an early, subtle modification of membrane function by toxin develops, with time lag, into the severe destruction of membranes. A parabolic line suggests that induced electrolyte leakage is an indirect effect of the toxin, or a recovery mechanism like a self-repair system in membranes could possibly be working after initial slight modification of the membranes by the toxin.

Aging of cut leaves occasionally causes an unusual decline in toxin sensitivity: sorghum tissues that are freshly excised and aged for 3 h before exposure to PC-toxin lose about 50% as much electrolytes as do freshly cut, toxin-treated tissues. Leaves held for 24 h prior to exposure have little or no toxin-induced leakage (Bronson and Scheffer 1977). The decline can be prevented when cut ends of leaves are kept in 10^{-4} M kinetin solution during aging. However, no decline in toxin sensitivity is observed in the combination of oats and HV-toxin.

3.8 CO_2-Fixation Assay

A simple, quantitative method based on toxin-induced inhibition of CO_2 fixation by leaf tissues in the dark was satisfactorily utilized to determine HMT-toxin (Bhullar et al. 1975). This method has routinely been employed for toxin research, although the primary target of the toxin is undoubtedly the mitochondria of Tms cytoplasm maize. Their protocol for dark CO_2 fixation is as follows: leaf disks (5 mm in diameter) from young maize plants are cut and floated on water. Disks (10–12) are randomly transferred to steel planchets (25 × 7 mm, usually 22 to 24 per assay), which contain 2.8 ml of test solution each, in a 150 × 25 mm plastic petri dish modified to serve as a $^{14}CO_2$ tagging chamber. The leaf disks are preincubated, in the presence of toxin, under 1300 ft-c of fluorescent light for 6 h at 26 °C before permitting fixation of $^{14}CO_2$ in the dark. A planchet containing 1 ml of 3 N H_2SO_4 then is placed in the center of the chamber. After sealing with vacuum grease, the chamber is covered with aluminum foil, and $^{14}CO_2$ is released by injecting 0.5 µmol of $NaH^{14}CO_3$ into the acid through a serum stopper. After correcting the volumes of planchets and liquid, the amount of CO_2 in the air space of the chamber is 3.5 µmol (specific radioactivity approximately 1×10^7 dpm/µmol). The air inside the chamber is mixed quickly with the help of a large syringe. After 2 h of exposure to $^{14}CO_2$ in the dark, the chamber cover is removed under a fume hood. After allowing the disks to remain for 15 min in the dark, four to ten disks are transferred to a scintillation counting vial containing 0.5 ml of digestion mixture, prepared by mixing 60% perchloric acid and 30% H_2O_2 in 1:2 ratio. The vials are capped and incubated for 1.5 h at 70 °C. After cooling, 10 ml of scintillation fluid [for example, Biofluor (NEN)] are added. Scintillation fluid can be prepared by mixing 15 g of Omnifluor (NEN), 1 l of toluene, and 1 l of Triton X-100. Radioactivity is determined by a liquid scintillation spectrometer.

In carrying out dark CO_2-fixation assays, it was observed occasionally that fixed CO_2 was released in the first few minutes after disks were collected. To minimize these effects, Bhullar et al. (1975) recommend that all replicates of a dilution series should routinely be removed in a planned sequence rather than randomly. In addition, nontoxin-treated control disks should be harvested at the start, near the middle, and at the end of a dilution series. A 6-h exposure to light prior to the period of dark CO_2 fixation increases sensitivity of dark fixation to toxin and can minimize errors due to prior light conditions affecting toxin sensitivity of leaves, and a diurnal increase in the dark CO_2-fixation rate.

Tabira et al. (1989) have assayed AM-toxin by a similar method based on inhibition of photosynthetic CO_2 fixation in apple leaves. The protocol: disks (7 mm in diameter) from young leaves of susceptible and resistant apples are vacuum-infiltrated in AM-toxin solution or water for 30 min. (To protect fragile leaf tissues from damage, the recovery of pressure must be very carefully manipulated, to be as slow as possible.) The disks are incubated in the light or the dark for 4 h at 26 °C. The leaf disks are transferred into a transparent $^{14}CO_2$-tagging chamber. A small petri dish containing 1 ml of 2 N HCl is then placed in the center of the chamber. $^{14}CO_2$ is released by injecting 20 µl of $NaH^{14}CO_3$ (0.1 mCi/mmol, 50 µCi/ml) into the acid through a silicon stopper by using a microsyringe. The leaf disks are illuminated with a photoreflecting lamp at 9500 µW/cm^2 through a water phase (3 cm thick) containing 1% $CuSO_4$ for 15 min at room temperature. Unfixed $^{14}CO_2$ is then trapped with 3 N KOH. The leaf disks are treated as mentioned above, and radioactivity is measured by a liquid scintillation spectrometer.

3.9 Amino Acid Incorporation Assay

Gardner et al. (1985) tried to assay HST from the rough lemon pathotype of *Alternaria alternata* by inhibiting the incorporation of proline into protein in leaf tissues, and devised a sensitive and quantitative method for toxin assay. The protocol is summarized as follows:

Ten leaf disks (0.6 cm in diameter) from young citrus leaves are incubated, with or without toxin, in scintillation vials with constant rapid shaking in 0.95 ml of 0.025 M MES-Tris buffer (pH 6.5). After 1–4 h, 0.05 ml of 0.4 mM ^{14}C-proline (100–500 dpm/nmol) is added and the vials are incubated another 2–3 h. Disks are briefly washed with buffer, then soaked for 12 h in 15 ml of 95% EtOH and rinsed with 5 ml of the same solvent.

Usually 10–30 nmol of proline was incorporated (Gardner et al. 1985). ACRL-toxin I, the major HST of the rough lemon pathotype, showed 50% inhibition (ED_{50}) at 18–30 ng/ml on susceptible rough lemon. The inhibition was specifically observed on susceptible cultivars and an intermediate one. On the other hand, ACTG-toxin A of the tangerine pathotype had ED_{50} at 0.5 μg/ml on susceptible cv. Dancy tangerine, and, in addition, substantially affected resistant cultivars (Gardner et al. 1986).

3.10 Infection Induction Assay

In recent years, many lines of evidence supported the concept that the key role of HSTs in initial colonization is not to kill host cells prior to penetration but to predispose the host cells beneath germinating spores to accept penetration (Yoder and Scheffer 1969; Comstock and Scheffer 1973; Otani et al. 1985; Kohmoto et al. 1989; Kohmoto and Durbin 1989). From this viewpoint, a sophisticated bioassay system was developed to detect the induction of fungal penetration to plant tissues by toxin or a host recognition factor apparently nontoxic.

A saprophytic *Alternaria alternata* that is not pathogenic to plants is basically agressive, namely, it has the ability to penetrate even thin plastic film as well as plant substrates, and is employed as a challenging test organism. Aviruent strains of the particular fungus under investigation can also be used as a test organism. Spores collected are thoroughly washed in deionized water with two or three cycles of centrifugation (500 × g, for 3–5 min). Washed spores are suspended in diluted toxin solutions to be tested and deionized water as a control. Spore concentration is adjusted to 5×10^5 spores/ml, using Thoma's hemacytometer. The spore suspensions are sprayed on the lower surface of freshly detached host leaves, and the leaves placed on

Table 2. Effect of AK-toxins I and II on infection by avirulent *Alternaria alternata* of leaves from susceptible pear cv. Nijisseiki (Otani et al. 1985)

A. alternata	Treatment		Infection hypha formation[a] (%)	Lesion formation[a] (spots/cm^2 leaf)
Avirulent spore (O-94)	AK-toxin I	10^{-7} M	37.7	26.5
		10^{-8}	20.7	5.2
		10^{-9}	12.2	0.0
	AK-toxin II	10^{-6} M	38.4	12.1
		10^{-7}	15.3	0.3
		10^{-8}	9.0	0.0
	Water		13.8	0.0
Virulent spore (K-103)	Water		29.5	44.2

[a] Observed at 24 h after spore inoculation (5×10^5 spores/ml).

sponge mats in a moist chamber and incubated for 24 h at 28 °C. Results are interpreted on the basis of two different criteria. One criterion is the formation rate of infection hyphae, and the other is the number of necrotic lesions. (1) The leaves inoculated with spores are immersed in 0.1% aniline blue (cotton blue) in lactophenol for 2 h, and then boiled in a lactophenol-ethanol solution until the chlorophyll is removed. After rinsing with fresh lactophenol, infection hyphae formed in the leaves are microscopically observed (Table 2). A penetration peg of avirulent conidium is capable of forming a conspicuous invading hypha in a susceptible pear leaf tissue when inoculated with a small amount of AK-toxin. (2) Visible necrotic spots on the leaves are easily scored, as shown in Table 2. Avirulent spores can invade susceptible tissues like virulent spores, provided they are suspended in a dilute HST solution and inoculated.

The concentration of the toxin solutions should be such that the toxin does not cause any necrosis under these conditions: it is generally recommended to employ about tenfold the minimum concentration of the toxin in question determined by leaf necrosis assay, so that a significant infection induction would be observable without developing toxigenic necrosis in this assay.

This method is a key bioassay for evaluating whether a candidate fungal toxin is involved in very early host-parasite interactions and plays a role in host-specificity or host-recognition factors.

4 HPLC Analysis of Host-Selective Toxin

Simple and rapid procedures for the quantitative determination of HSTs have been requested in order to determine toxins in germinating fluids, culture filtrates of the pathogen, diseased host tissues, and other materials. However, conventional procedures were time-consuming and found to be unsatisfactory in sensitivity, resolution, rapidity, and manipulation. Recent application of high performance liquid chromatography or high pressure liquid chromatography (HPLC) has made extensive progress in the analysis of chemistry, biology, and pathology of HST. Here, I would like to describe several examples of HPLC application for HST analysis, because various modifications in the choice of columns and mobile phase solvents in HPLC will help develop an efficient assay for other HSTs. In this respect, excellent books on the basic theory and further practice of HPLC (Engelhardt 1986; Fallon et al. 1987) are expedient for reference, and the information provided there will help to devise appropriate conditions for different systems.

4.1 AK-Toxin and AF-Toxin

The Japanese pear pathotype of *Alternaria alternata* causing a destructive black-spot disease of the Japanese pear, produces HSTs, AK-toxin I and II, which are esters of *N*-acetyl-β-methyl(or demethyl)phenylalanine and ω-epoxydecatrienoic acid (Nakashima et al. 1982; Otani et al. 1985). These toxins in spore-germinating fluids were quantitatively determined by HPLC (Hayashi et al. 1990). The procedure employed is described as follows.

Spores of the pathogen are washed three times with deionized water by centrifuging at 600 × g for 5 min. The 100-ml spore suspension at a concentration of 5×10^5 spores/ml is spread uniformly onto paper towels (or spread out as a thin layer in flat plastic containers) and incubated in a moist chamber for appropriate periods at 26 °C. After incubation, the rate of spore germination is examined under a microscope, and the spore-germinated fluid is collected by squeezing the towels and filtered through filter paper to remove spores. The filtrate is adjusted to pH 5.5 with 1 M KH_2PO_4 and extracted twice with diethyl ether. Under these conditions, more than 95% of AK-toxins are extracted at the ether phase. After removal of ethyl ether, the residue is dissolved in 1 ml methanol. Aliquots (1–20 μl) of this sample solution are loaded onto a reverse phase Fine-pak SIL-C_{18} HPLC column (4.6 × 250 mm, Japan Spectroscopic Co.) and chromatographed with a model BIP-1 instrument (Japan Spectroscopic Co.). The mobile phase is a mixture of methanol/acetic acid/water (60:1:39, v/v/v), the flow rate is adjusted at 1 ml/min, and absorbance is monitored at 290 nm. The retention time (Rt) for AK-toxin I is 17.5 min, and II is 13.5 min, nevertheless they are very similar in structure. The minimum amount of toxins detectable by this protocol is approximately 10 ng. The regression line for the relationship between amounts of toxin and their peak height is linear in the range 10–200 ng. The amount of each toxin released from one germinated spore can be calculated as follows:

$$\text{Amount of toxin released on germination} = \frac{\text{Total amount of toxin in the sample (pg)}}{(5 \times 10^5 \times 10^2 \text{ spores}) \times \text{rate of spore germination}}$$

Using this procedure, Hayashi et al. (1990) demonstrated that one spore of the Japanese pear pathotype produced about 0.02 pg AK-toxin I during the first 6 h of germination, and a trace amount of AK-toxin II after 24-h incubation. Similarly, one spore of the strawberry pathotype of *Alternaria alternata* released about 0.08 pg AF-toxin I (Rt, 15.5 min) during the first 6 h of germination. Retention time for AF-toxin II is 12 min and for III, 20.5 min. No toxins were detected in the dormant spores, however.

4.2 ACT-Toxin

Kohmoto et al. have recently isolated, employing HPLC, the host-specific ACT-toxin Ib and Ic from the tangerine pathotype of *Alternaria alternata* causing brown spot disease in the Dancy tangerine, Emperor mandarin, and Marcott tangor (unpubl. data). The toxins were also found to be esters of *N*-polyhydroxyacylvaline and ω-epoxydecatrienoic acid. The procedure is given below.

Since culture filtrates generally contain various metabolites, medium components and other impurities, cleanup procedures are very imporant for successful isolation. Culture filtrate of the pathogen is mixed with Amberlite XAD-2 at pH 5.5 to adsorb toxins. Toxins are eluted with methanol from the adsorbed Amberlite resin, and the eluate is concentrated to the residual water phase from which toxins are extracted three times with ethyl acetate at pH 5.5 (1 M KH_2PO_4). The ethyl acetate extract is applied to reverse phase HPLC on Develosil ODS-5 (20 × 250 mm, Nomura Chemical Co.), using a gradient solvent system composed of acetonitrile/water/acetic acid (25:75:1 to 60:40:1, v/v/v) at a flow rate of 5 ml/min for 40 min. Analytical and preparative HPLC are conducted with a Hitachi HPLC system equipped with an L-3000 photodiode array detector, L-5000 LC controller, L-6000 pump, and D-2500 chromatointegrator. Sample solutions should be filtered through a 0.2-μm Milipore filter to remove insoluble materials prior to injection. Several peaks are detected by UV absorbance at 320 nm. One peak fraction (Rt, 27.5 to 30.5 min) contains much ACT-toxin Ib and some ACT-toxin Ic. This fraction is rechromatographed by HPLC, using a Develosil ODS column (20 × 250 mm) and a gradient solvent system composed of methanol/water/acetic acid (60:40:1 to 100:0:1, v/v/v) at a flow rate of 5 ml/min for 40 min. One major peak (Rt, 21.7 to 23.3 min) is detected by UV absorbance at 290 nm; this fraction gives a pale brown syrup of ACT-toxin Ib after evaporation.

In our experiment, yield was about 22 mg from 60 l of culture fluid. ACT-toxin Ib induces vein necrosis at 10 ng/ml and a rapid increase in electrolyte leakage from susceptible leaves of the Emperor mandarin, but does not affect resistant citrus at 100 μg/ml. ACT-toxin Ic, a photoinducible but still phytotoxic isomer, would be detected at an earlier retention time if it occurs.

We further simplified this procedure for quantitative analysis of ACT-toxin Ib, the major HST of the tangerine pathotype: culture filtrate or spore-germinating fluid is extracted with ethyl acetate at pH 5.5 adjusted with 1 M KH_2PO_4, and then chromatographed by HPLC, by using a Develosil ODS-5 (4.6 × 250 mm) and a gradient solvent system composed of acetonitrile/water/acetic acid (30:70:1 to 60:40:1, v/v/v) at a flow rate of 1 ml/min for 30 min. ACT-toxin Ib has a retention time of 17.55 min, while toxin Ic has a retention time of 16.03 min. One spore was found to release 0.01 pg of ACT-toxin Ib within 1 h of incubation. The amount rapidly increased with time: 0.035 pg per spore in 12 h, 0.21 pg per spore in 18 h and 0.25 pg per spore in 24 h. ACT-toxin Ic was detected only after 24 h.

4.3 AM-Toxin

Host-selective AM-toxin I, a cyclic depsipeptide from the apple pathotype of *Alternaria alternata* causing leaf blotch of apple, can also be quantified by HPLC. Ten ml of germination fluid is subjected to a Waters Sep-Pak C_{18} cartridge, washed with deionized water and eluted with ethanol. The ethanol eluate is chromatographed by HPLC, using a Develosil ODS-5 (4.6 × 250 mm) with a Waters Guard-Pak (μBondapak C_{18}) as precolumn and a mixture of ethanol/water (1:1, v/v) at a rate of 0.6 ml/min. AM-toxin I (Rt, 9.44 min) and II (Rt, 11.72 min) are monitored by UV absorbance at 224 nm.

4.4 AAL-Toxin

Alternaria alternata f.sp. *lycopersici*, the pathogen of Alternaria stem canker in tomato, is known to produce host-selective AAL-toxins consisting of 4 esters (at C-13 and C-14) of propane-tricarboxylic acid and 19 carbon amino pentols (Bottini et al. 1981). Each component is designated as TA-1, TA-2, TB-1, and TB-2.

Siler and Gilchrist (1982) employed maleic acid as a reversible derivatizing reagent for amino groups in the toxins, because the native toxins lack significant absorbance at wavelengths above 210 nm. Ten-μl aliquots of 1 M maleic anhydride in dioxane, prepared immediately before use, are added to solutions of semi-purified toxins in 0.1 M sodium carbonate (pH 9.2) at room temperature. The progress of the reaction can be monitored by loss of ninhydrin positive reaction through spot testing on a silica gel TLC plate. The pH should be maintained above 9 by the dropwise addition of 0.1 M sodium hydroxide, and would be more than 90% complete in less than 5 min. At least a 10:1 molar excess of maleic anhydride to toxin is used to ensure complete derivatization. The maleylation is specific for amino groups under the conditions used. Before HPLC analysis, maleylated toxin solutions are adjusted to pH 6–7 by the addition of hydrochloric acid.

HPLC analysis is performed on a Varian 5060 HPLC equipped with a VISTA CDS 401 data processor, by using a 2.6 × 250 mm column packed with 5 μC-18 bonded silica (Alltech Associates) and a 60 min gradient from 50:50 to 70:30 methanol/0.1 M potassium acetate buffer at a flow rate of 1 ml/min. Elution of maleylated toxins is followed by UV absorbance at 250 nm. Retention time of TA-1 is 23.2 min; TA-2, 25.7 min; TB-1, 41.7 min; and TB-2, 43.4 min under these conditions.

Quantitative estimates of AAL-toxins in leaves of naturally infected and spore-inoculated tomato plants by Siler and Gilchrist (1983) indicate that the toxin concentrations in diseased leaves were comparable with those required to induce necrosis in detached leaves (approximately 10 ng/ml).

4.5 HV-Toxin (Victorin)

HV-toxin (victorin C) is the most predominant host-selective toxin from *Cochliobolus victoriae*, causing victoria blight in oats. To extensively purify and isolate victorin C from culture filtrates of the pathogen, reverse phase HPLC was efficiently used by Walton and Earle (1984), Macko et al. (1985), Wolpert et al. (1985), and Mayama et al. (1986).

For a simple determination procedure for HV-toxin, Mayama's protocol is summarized as follows: 5–10 ml of the pathogen's culture fluid adjusted to pH 4 are centrifuged and passed through a Waters Sep-Pak cartridge that had been previously washed with 5 ml of acetonitrile containing 0.075% trifluoroacetic acid (TFA) and with 5 ml of 0.075% TFA water solution, and the cartridge is washed by 5 ml of water and then 2 ml of 16% acetonitrile plus 0.075% TFA. These washes are discarded since they contain little toxicity. Finally, the toxic activity is eluted with 4 ml of 50% acetonitrile plus 0.075% TFA. The eluted toxic fraction is dried at 45 °C and redissolved in 0.1 ml of HPLC solvent. Generally, 1 ml equivalent of culture fluid is injected per run. Samples are chromatographed on a Waters Radial-Pak C18 column (on a Waters RCM 1000 module) and with 22% acetonitrile in water containing 0.075% TFA at a flow rate of 2 ml/min. Peaks are detected by absorbance at 254 nm. The major toxin (victorin D according to Mayama's criterion; this probably corresponds to victorin C by Wolpert's nomenclature) is eluted around at Rt 14.5 min.

Gradient reverse phase HPLC by Walton and Earle (1984) showed good resolution of the major toxin. A Waters HPLC system with gradient programmer and a Waters μBondapak C18 column (3.9 × 30 mm) are used. Aliquots of alumina column eluants of culture filtrates are injected and eluted with a linear gradient of 0% B (100% A) to 60% B in 60 min, at 1.5 ml/min, where solvent A is 0.05% TFA in glass-distilled water and solvent B is 0.05% TFA in acetonitrile. The eluant can be monitored by absorbance at 254 nm. The toxin is detected in a single peak fraction with a retention time of 30–32 min.

Most recently, polyclonal antibodies against victorin C were prepared with victorin-BSA conjugates as an alternative measure for determining the toxin (Akimitsu et al. 1992). The antibodies were highly specific, could detect as little as 10 pg/ml of victorin in an indirect ELISA, and used to demonstrate victorin binding sites in oat leaf tissues by Western blotting.

4.6 HC-Toxin

Rasmussen and Scheffer (1988) devised a new and simple purification method for host-selective HC-toxins from *Helminthosporium carbonum* race 1. Their protocol and precautions are as follows:

Cultures of the pathogen are filtered and concentrated to 0.1 volume under reduced pressure. An equal volume of methanol is added to the concentrated filtrate and the solutions are stored overnight at −20 °C; the resulting precipitate is removed by filtration. After removal of methanol under reduced pressure, the aqueous solution is extracted three times with methylene chloride. Steps to this point should be completed as quickly as possible to avoid hydrolysis of the epoxide of HC-toxins. The methylene chloride extracts are concentrated and placed on a column (4.1 × 35 cm) of Sephadex LH-20 previously equilibrated with methanol. The column is developed with methanol, and fractions containing selective toxic activity are identified by bioassay. The toxin-containing fractions are combined and concentrated under reduced pressure, and subjected to flash chromatography. The diameter of the silica gel (Whatman LPS-2) column is determined by size (dry weight) of the sample, but the length of the packed silica bed should be about 15 cm regardless of column diameter. The column was developed with hexane/methylene chloride/acetone (1:1:1, v/v/v; solvent A) and 50-ml fractions are collected. Each fraction is tested for epoxide, by TLC (Merck silica gel-60, 0.25 mm, developed with acetone/methylene chloride, 1:1, v/v), and by spraying with an epoxide indicator, 2% (w/v) 4-(*p*-nitrobenzyl)-pyridine (NBP) in acetone and by subsequent spraying with 10% (v/v) tetraethylene-pentamine in acetone after heating the TLC plate for 5 min at 100 °C; a positive test is indicated by a blue spot at an R_f value of 0.55, which is shown to be HC-toxin I. After toxin I is eluted in solvent A (first fraction), epoxide-containing, minor host-selective toxins are eluted with 1000 ml of methylene chloride/acetone (1:1, v/v, solvent B). HC-toxin I is crystallized from diethylether solutions.

In preparation of HPLC, the solvent B eluate is concentrated, dissolved in water (25 ml), and placed on a flash chromatography column (2.2 × 15 cm) of octadecyl C18 (40 μm, J.T. Baker Chemical Co.). Some nontoxic materials are removed by passing 100 ml of water through the column. Minor toxins are eluted with 30% ethanol in water (100 ml) and dried under reduced pressure. The residue is dissolved in water (200 μl or more). Aliquots of the preparation are loaded onto a Waters μBondapak C_{18} HPLC column (7.8 × 300 mm) on a Varian 5000 instrument. The initial solvent of 7% ethanol in water is changed linearly over a 30-min period to 20% ethanol; the concentration is then held constant for the next 15 min. The flow rate is at 2 ml/min, and absorbance is monitored at 215 nm. A peak fraction at about a retention time of 25 min contains HC-toxin II and III, and a peak fraction at about a retention time of 33 min indicates the presence of toxin IV. The former fraction is again chromatographed on a 4 × 250 mm column of Whatman Partisil 5 with an isocratic mixture of hexane/ethanol, 95:5, v/v, at a rate of 3 ml/min. HC-toxin II had a retention time of approximately 22 min, and III approximately 34 min, under these conditions.

4.7 HS-Toxin

Helminthosporium sacchari, the cause of eye spot disease in sugarcane, produces three host-selective toxins (HS-toxin) that are isomeric forms of oligosaccaride-sesquiterpene, and their various, nontoxic analogs or homologs as well. Macko et al. (1981) purified HS-toxins from culture filtrates of the pathogen. Toxins were adsorbed on a column of Ambersorb XE 348 and then eluted with a linear gradient of water to acetone. Toxic fractions were subjected to silica-gel 60 column chromatography with 80% acetonitrile in water or with chloroform/methanol/water, 60:39:1, v/v/v. Toxins were purified by droplet countercurrent chromatography in descending mode with the two phases of a mixture of *n*-butanol/acetone/water, 33:10:50, at a flow rate of 0.1 ml/min. Final purification was achieved by HPLC on a column of μ-Bondapak C_{18} 7.8 × 300 mm with 22% acetonitrile in water at a flow rate of 2 ml/min. Toxin I, II and III had a retention time of 10, 14, and 18 min, respectively.

For cleanup of samples, Livingston and Scheffer (1984) employed flash chromatography: 30 g of Whatman LPS-2 is poured as a slurry in acetonitrile into a column giving a bed with dimensions of 2.2 × 19 cm. A methanol solution of the sample is mixed with 1.5 g of LPS-2 and gently stirred while the solvent evaporates. The final powder is applied to the top of the column bed, and the sample is eluted rapidly under pressure with stepwise gradient solutions consisting of 100 ml volumes of a acetonitrile/water in the following ratios: 100:0; 97.5:2.5; 95:5; 90:10; 85:15; and 80:20. Fractions (9 ml) are collected and monitored for toxin by TLC.

Toxins and analogs isolated from culture filtrates are then chromatographed on a Varian 5000 HPLC equipped with a Waters μBondapak C_{18} column (7.8 × 300 mm) and acetonitrile/water mixtures as follows: for toxins, 20:80; for 3-gal analogs, 19:81; for the 2-gal analogs with the 1,1 gal arrangement, 20:80; and for the 2-gal analogs with the 0,2 or 2,0 arrangement, 26:74. The flow rate is 2 ml/min. Monitoring is done by absorbance at 215 nm.

Acknowledgments. I would like to express my thanks to Profs. S. Ouchi and H. Otani for critical reading the manuscript and helpful suggestions. This work was supported in part by grants from the Ministry of Education, Science and Culture, Japan (Nos. 63440010, 02044099, and 023004019).

References

Akimitsu K, Hart LP, Walton JD, Hollingsworth R (1992) Covalent binding sites of victorin in oat leaf tissue detected by anti-victorin polyclonal antibodies. Plant Physiol 98:121–126

Bhullar BS, Daly JM, Rehfeld DW (1975) Inhibition of dark CO_2 fixation and photosynthesis in leaf discs of corn susceptible to the host-specific toxin produced by *Helminthosporium maydis*, race T. Plant Physiol 56:1–7

Bottini AT, Bowen JR, Gilchrist DG (1981) Phytotoxins. II. Characterization of a phytotoxic fraction from *Alternaria alternata* f. sp. *lycopersici*. Tetrahedron Lett 22:2723–2726

Briggs SP, Scheffer RP, Haug AR (1984) Selective toxin of *Helminthosporium victoriae*: thermal relationships in effects on oat tissues and protoplasts. Phytopathology 74:768–773

Bronson CR, Scheffer RP (1977) Heat- and aging-induced tolerance of sorghum and oat tissues to host-selective toxins. Phytopathology 67:1232–1238

Comstock JC, Scheffer RP (1973) Role of host-selective toxin in colonization of corn leaves by *Helminthosporium carbonum*. Phytopathology 63:24–29

Damann KE Jr, Gardner JM, Scheffer RP (1974) An assay for *Helminthosporium victoriae* toxin based on induced leakage of electrolytes from oat tissues. Phytopathology 64:652–654

Daub ME (1982) Cercosporin, a photosensitizing toxin from *Cercospora* species. Phytopathology 72:370–374

Daub ME (1986) Tissue culture and the selection of resistance to pathogens. Annu Rev Phytopathol 24:159–186

Daub ME, Hangarter RP (1983) Light-induced production of singlet oxygen and superoxide by the fungal toxin, cercosporin. Plant Physiol 73:855–857

Engelhardt H (ed) (1986) Practice of high performance liquid chromatography. Springer, Berlin Heidelberg New York

Fallon A, Booth RFG, Bell LD (1987) Applications of HPLC in biochemistry. Elsevier, Amsterdam

Gardner JM, Kono Y, Tatum JH, Suzuki Y, Takeuchi S (1985) Plant pathotoxins from *Alternaria citri*: the major toxin specific for rough lemon plants. Phytochemistry 24:2861–2867

Gardner JM, Kono Y, Chandler JL (1986) Bioassay and host-selectivity of *Alternaria citri* toxins affecting rough lemon and mandarins. Physiol Molec Plant Pathol 29:293–304

Gendloff EH, Scheffer RP, Somerville SC (1987) An improved bioassay for victorin based on the use of oat protoplasts. Physiol Molec Plant Pathol 31:421–427

Gengenbach BG, Rines HW (1986) Use of phytotoxins in selection of disease resistant mutants in tissue culture. Iowa State J Res 60:449–476

Hawes MC (1983a) Technique for using isolated root cap cells in a simple, quantitative assay for the pathotoxin produced by *Helminthosporium maydis* race T. Phytopathology 73:1184–1187

Hawes MC (1983b) Sensitivity of isolated oat root cap cells and protoplasts to victorin. Physiol Plant Pathol 22:65–76

Hawes MC, Wheeler H (1982) Factors affecting victorin-induced root cap cell death: temperature and plasmolysis. Physiol Plant Pathol 20:137–144

Hayashi N, Tanabe K, Tsuge T, Nishimura S, Kohmoto K, Otani H (1991) Determination of host-selective toxin production during the spore germination of *Alternaria alternata* pathogens with high performance liquid chromatography. Phytopathology 80:1088–1091

Kodama M, Yoshida T, Otani H, Kohmoto K, Nishimura S (1991) Effect of AL-toxin on viability of cultured tomato cells determined by MTT-colorimetric assay. Ann Phytopathol Soc Jpn 57:663–670

Kohmoto K, Durbin RD (eds) (1989) Host-specific toxins: recognition and specificity factors in plant disease. Tottori Univ Press, Tottori
Kohmoto K, Otani H (1991) Host recognition by toxigenic plant pathogens. Experientia (Basel) 47:755–764
Kohmoto K, Otani H, Kodama M, Nishimura S (1989) Host-recognition: can accessibility to fungal invasion be induced by host-specific toxins without necessitating necrotic cell death? In: Graniti A, Durbin RD, Ballio A (eds) Phytotoxins and plant pathogenesis. Springer, Berlin Heidelberg New York, pp 249–265
Kohmoto K, Itoh Y, Shimomura N, Kondoh Y, Otani H, Kodama M, Nishimura S, Nakatsuka S Isolation and biological activities of two host-specific toxins from the tangerine pathotype of *Alternaria alternata*. Phytopathology (submitted)
Kono Y, Gardner JM, Suzuki Y, Takeuchi S (1986) Studies on host-selective toxins produced by a pathotype of *Alternaria citri* causing brown spot disease of mandarins. Agric Biol Chem 50:1597–1606
Laughnan JR, Gabay SJ (1973) Reaction of germinating maize pollen to *Helminthosporium maydis* pathotoxins. Crop Sci 13:681–684
Lim SM, Hooker AL, Smith DR (1971) Use of *Helminthosporium maydis* race T pathotoxin to determine disease reaction of germinating corn seed. Agron J 63:712–713
Litzenberger SC (1949) Nature of susceptibility to *Helminthosporium victoriae* and resistance to *Puccinia caronata* in Victoria oats. Phytopathology 39:300–318
Livingston RS, Scheffer RP (1984) Selective toxins and analogs produced by *Helminthosporium sacchari*. Plant Physiol 76:96–102
Luke HH, Wheeler H (1955) Toxin production by *Helminthosporium victoriae*. Phytopathology 45:453–458
Macko V, Goodfriend K, Wachs T, Renwick JAA, Acklin W, Arigoni D (1981) Characterization of the host-specific toxins produced by *Helminthosporium sacchari*, the causal organism of eyespot disease of sugarcane. Experientia (Basel) 37:923–924
Macko V, Wolpert TJ, Acklin W, Jaun B, Seibl J, Meili J, Arigoni D (1985) Characterization of victorin C, the major host-selective toxin from *Cochliobolus victoriae*: structure of degradation products. Experientia (Basel) 41:1366–1370
MacRae WD, Yoder OC (1988) Light has opposite effects on sensitivity of maize protoplasts to T-toxin from *Cochliobolus heterostrophus*. Physiol Molec Plant Pathol 32:293–300
Mayama S, Tani T, Ueno T, Midland SL, Sims JJ, Keen NT (1986) The purification of victorin and its phytoalexin elicitor activity in oat leaves. Physiol Molec Plant Pathol 29:1–18
Meehan F, Murphy HC (1947) Differential phytotoxicity of metabolic by-products of *Helminthosporium victoriae*. Science 106:270–271
Nakashima T, Ueno T, Fukami H (1982) Structure elucidation of AK-toxins, host-specific phytotoxic metabolites produced by *Alternaria kikuchiana* Tanaka. Tetrahedron Lett 23:4469–4472
Nishimura S (1987) Recent development of host-specific toxin research in Japan and its agricultural use. In: Nishimura S, Vance CP, Doke N (eds) Molecular determinants of plant diseases. Japan Sci Soc Press, Springer, Berlin Heidelberg New York, pp 11–26
Nishimura S, Kohmoto K (1983a) Host-specific toxins and chemical structures from *Alternaria* species. Annu Rev Phytopathol 21:87–116
Nishimura S, Kohmoto K (1983b) Roles of toxins in pathogenesis. In: Daly JM, Deverall BJ (eds) Toxins and plant pathogenesis. Acad Press, Sydney, pp 137–157
Nishimura S, Nakatsuka S (1989) Trends in host-selective toxin research in Japan. In: Kohmoto K, Durbin RD (eds) Host-specific toxins: recognition and specificity factors in plant disease. Tottori Univ Press, Tottori, pp 19–31
Nishimura S, Scheffer RP (1965) Interactions between *Helminthosporium victoriae* spores and oat tissue. Phytopathology 55:629–634
Otani H, Morikawa M, Nishimura S, Kohmoto K (1977) Nature of specific susceptibility to *Alternaria kikuchiana* in Nijisseiki cultivar among Japanese pears (8). Ann Phytopathol Soc Jpn 43:562–568

Otani H, Kohmoto K, Nishimura S, Nakashima T, Ueno T, Fukami H (1985) Biological activities of AK-toxins I and II, host-specific toxins from *Alternaria alternata* Japanese pear pathotype. Ann Phytopathol Soc Jpn 51:285–293
Otani H, Kohmoto K, Kodama M, Nishimura S (1991) Role of host-specific toxins in the pathogenesis of *Alternaria alternata*. In: Patil S, Ouchi S, Mills D, Vance C (eds) Molecular strategies of pathogens and host plants. Springer, Berlin Heidelberg New York, pp 139–149
Pringle RB, Braun AC (1957) The isolation of the toxin of *Helminthosporium victoriae*. Phytopathology 47:369–371
Pringle RB, Scheffer RP (1964) Host-specific plant toxins. Annu Rev Plant Pathol 2:133–156
Rasmussen JB, Scheffer RP (1988) Isolation and biological activities of four selective toxins from *Helminthosporium carbonum*. Plant Physiol 86:187–191
Scheffer RP (1983) Toxins as chemical determinants of plant disease. In: Daly JM, Deverall BJ (eds) Toxins and plant pathogenesis. Acad Press, Sydney, pp 1–40
Scheffer RP (1989) Host-specific toxins in phytopathology: origins and evolution of the concept. In: Kohmoto K, Durbin RD (eds) Host-specific toxins: recognition and specificity factors in plant disease. Tottori Univ Press, Tottori, pp 11–17
Scheffer RP, Briggs SP (1981) Introduction: a perspective of toxin studies in plant pathology. In: Durbin RD (ed) Toxins in plant disease. Acad Press, New York, pp 1–20
Scheffer RP, Livingston RS (1984) Host-selective toxins and their role in plant diseases. Science 223:17–21
Siler DJ, Gilchrist DG (1982) Determination of host-selective phytotoxins from *Alternaria alternata* f. sp. *lycopersici* as their maleyl derivatives by high-performance liquid chromatography. J Chromatogr 238:167–173
Siler DJ, Gilchrist DG (1983) Properties of host specific toxins produced by *Alternaria alternata* f. sp. *lycopersici* in culture and in tomato plants. Physiol Plant Pathol 23:265–274
Tabira H, Otani H, Shimomura N, Kodama M, Kohmoto K, Nishimura S (1989) Light-induced insensitivity of apple and Japanese pear leaves to AM-toxin from *Alternaria alternata* apple pathotype, Ann Phytopathol Soc Jpn 55:567–578
Umezawa T, Yamamoto M, Lee SS, Tsuge T, Kohmoto K, Nishimura S (1988) Two different phases in phytotoxicity of AL- and AF-toxins. Abstracts of Papers of 5th International Congress of Plant Pathology, Kyoto, 233 pp
Walton JD, Earle ED (1984) Characterization of the host-specific phytotoxin victorin by high-pressure liquid chromatography. Plant Sci Lett 34:231–238
Wenzel G (1985) Strategies in unconventional breeding for disease resistance. Annu Rev Phytopathol 23:149–172
Wheeler H, Black HS (1963) Effects of *Helminthosporium victoriae* and victorin upon permeability. Am J Bot 50:686–693
Witsenboer HMA, Van Schaik CE, Bino RJ, Loffer HJM, Nijkamp HJJ, Hille J (1988) Effects of *Alternaria alternata* f. sp. *lycopersici* toxins at different levels of tomato plant cell development. Plant Sci 56:253–260
Wolpert TJ, Macko V, Acklin W, Jaun B, Seibl J, Meili J, Arigoni D (1985) Structure of victorin C, the major host-selective toxin from *Cochliobolus victoriae*. Experientia (Basel) 41:1524–1529
Yoder OC (1980) Toxins in pathogenesis. Annu Rev Phytopathol 18:103–129
Yoder OC (1981) Assay. In: Durbin RD (ed) Toxins in plant disease. Acad Press, New York, pp 45–78
Yoder OC, Scheffer RP (1969) Role of toxin in early interactions of *Helminthosporium victoriae* with susceptible and resistant oat tissues. Phytopathology 59:1954–1959

Evaluating the Phototoxicity and Photogenotoxicity of Plant Secondary Compounds

R.R. Aucoin, E. Schneider, and J.T. Arnason

1 Introduction

It is well known that the toxicity of many plant secondary metabolites is greatly enhanced in the presence of sunlight or UVA (320–400 nm). There are at least eight major biosynthetic classes of photo-activated chemicals which are distributed in many distinct families of higher plants (Table 1). Representative structures from each class are given in Fig. 1. It should be noted that some groups such as the polyines are quite large and include more than 500 known molecular structures which are distributed among the more than 10000 species of the largest plant family, the Asteraceae (reviewed in Towers 1984; Arnason et al. 1990). These phototoxins have been shown to be toxic to a wide variety of organisms including bacteria, fungi, viruses, insects, and mammals (Towers 1984). Although they have no known function in either plant physiology or metabolism, extensive studies in recent years have demonstrated their role as a defensive mechanism against attack by plant pathogens or insect herbivores.

The mode of action of these phototoxins may be used to divide these compounds into two major categories: the photogenotoxins and the photooxidants. In the former group the phototoxins intercalate with DNA and, upon UV-irradiation, react with pyrimidine bases to form mono- or bifunctional adducts (Song and Tapley 1979). This mode of action is characteristic of the furanocoumarins such as 8-methoxypsoralen, the furanoquinolines, the furochromenes, and the beta-carboline alkaloids, although these compounds may also cause oxidations. The naturally occurring photooxidants include many polyines and their thiophene derivatives, and extended quinones such as hypericin. Their toxicity is primarily manifested through the production of activated oxygen species such as singlet oxygen and superoxide radicals (Bakker et al. 1979; Arnason et al. 1981; Knox and Dodge 1985) and the subsequent damage to important biomolecules such as lipids and proteins, leading to membrane lysis (MacRae et al. 1980; Larson and Berenbaum 1988). With most phototoxins examined to date there exists both light-dependent and light-independent toxicity.

An emerging phototoxin technology has developed in recent years because light-activated phytochemicals have many medical and agricultural applications. Psoralen, for example, is a long-established component in the photochemotherapeutic treatment of psoriasis and other skin conditions and

Table 1. Phototoxins from higher plants

Biosynthetic class	Precursor	Examples[a]	Occurrence (families)	Principal mode of action[b]
Furanocoumarins	Shikimic acid	Psoralen Angelicin	Rutaceae Apiaceae Fabaceae Moraceae Solanaceae Pittosporaceae Thymeleaceae Orchidaceae Asteraceae	DNA monoadducts and cross-linking
Extended quinones	Acetate	Hypericin	Hypericaceae	Singlet oxygen, free radicals
Polyines	Oleic acid	PHT[c] Alpha-T[c]	Asteraceae Apiaceae Campanulaceae Pittosporaceae Olacaceae Euphorbicaceae Valerianaceae Annonaceae Opiliceae Sapindaceae Araliaceae	Singlet oxygen, free radicals
Furanochromones	Acetate	Khellin Visnagin	Fabaceae	DNA monoadducts
Furanoquinolines	Tryptophan	Dictamnine	Rutaceae	Photoadducts
Isoquinolines	Tyrosine, Phenylalanine	Berberine Sanguinarine	Annonaceae Papaveraceae Berberidaceae Juglandaceae Magnoliaceae Menispermaceae Ranununculaceae Rubiaciae Rutaceae	Singlet oxygen, free radicals, photogeno-toxicity
Beta-carbolines	Tryptophan	Harmane	Simaroubaceae Rutaceae Cyperaceae Fabaceae Polygonaceae Rubiaceae Sapindaceae Passifloraceae Zygophyllaceae Solanaceae	Singlet oxygen, photoadducts
Isoflavonones	Shikimic acid	Pisatin	Fabaceae	Free radicals

[a] See also Fig. 1.
[b] For review, see Towers (1984), Knox and Dodge (1985).
[c] PHT, phenylheptatryne; alpha-T, alpha-terthienyl.

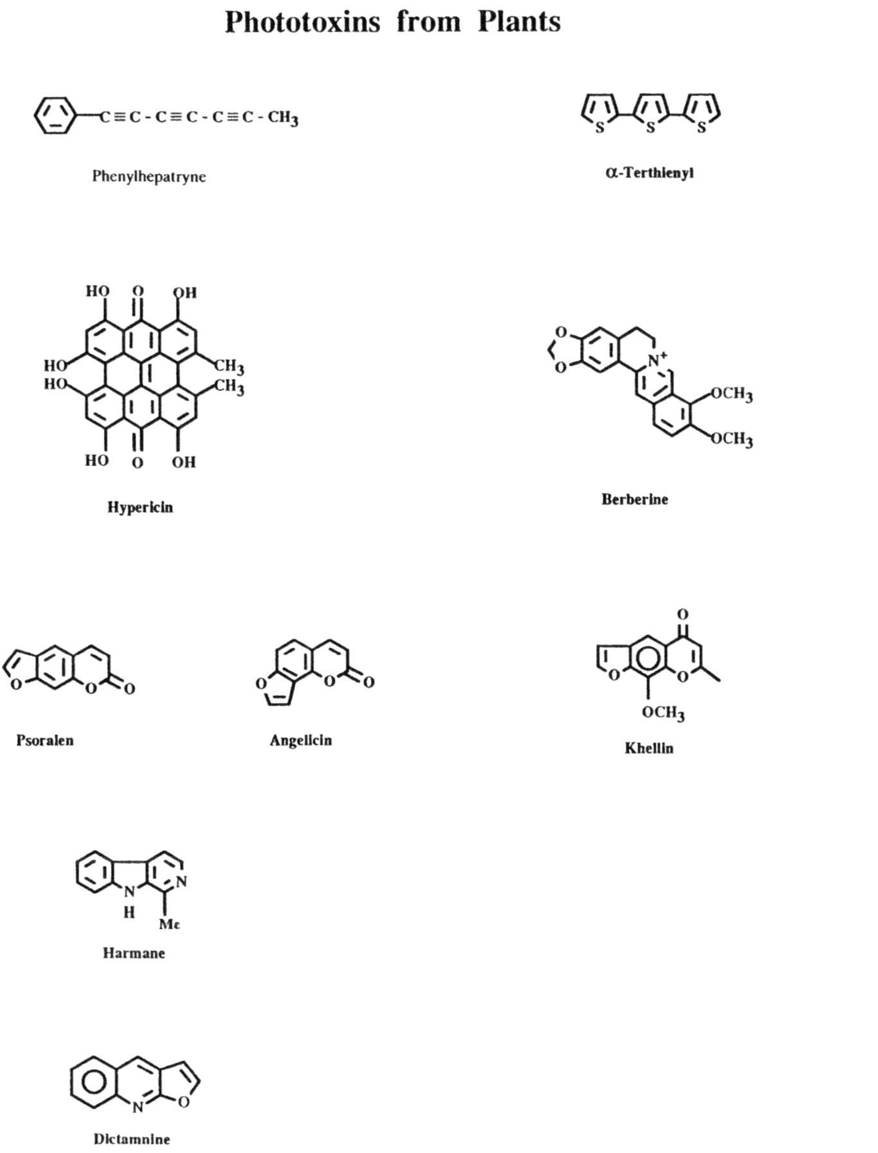

Fig. 1. Representative chemical structures of phototoxins. Phenylheptatryne and alpha-terthienyl are polyines. Hypericin is an extended quinone, berberine is an isoquinoline alkaloid, psoralen and angellicin are furanocoumarins, and khellin is a furanochromone. Harmane is a beta-carboline alkaloid and dictamnine is a furoquinoline alkaloid

is now being used for treatment of T-cell lymphoma (Edelson et al. 1987). Many furocoumarins have strong antiviral properties (see Hudson 1989) while other phototoxins may have potential as insect control agents. The polyine alpha-terthienyl, for example, is a potent mosquito larvicide which has been successful in advanced field trials (Arnason et al. 1981, 1989; Philogene et al. 1985). These light-activated plant phototoxins are also of ecological significance, particularly with respect to insect-plant relationships where they provide one of the best examples of stepwise coevolution. The importance of these secondary phytochemicals has led to the development of both qualitative and quantitative bioassays to assess the toxicity and phototoxicity of these compounds. In this Chapter we outline the bioassays which have been used successfully to evaluate phototoxicity and photogenotoxicity, and their specific applications.

2 Bioassays for Phototoxicity

2.1 Light Sources and Filters

Most secondary phytochemicals which have been observed to be light-activated absorb maximally in the UVA range (320–400 nm). A notable exception is the extended quinone pigment hypericin which absorbs maximally in the visible range (590–600 nm). Bioassays for phototoxicity therefore depend mainly on UV light sources such as blacklight-blue (BLB) lamps which emit a narrow band of radiation at about 350 nm (e.g., Westinghouse F20T12). The UV sources used, as well as the duration of exposure of test organisms, varies greatly among bioassays used to assess phototoxicity. We have found that a bank of 4 × 20 W BLB lamps providing 5 W/m^2 at a height of 20 cm provides sufficient UVA for most applications. Photographic filters can be used to reduce the light intensity or to narrow the band received by the test materials. Full-spectrum, full-intensity solar-simulating conditions can be achieved with a 1000-W xenon arc lamp. Solar-simulating fluorescent lamps (e.g., Durotest Vitalites) provide full-spectrum irradiation at less than solar intensities.

2.2 Yeast Bioassay

The yeast bioassay provides a simple and rapid method of screening plant tissues or extracts for phototoxicity (Daniels 1965; Camm et al. 1975; Kavli and Volden 1984). It is particularly useful where only small amounts of extracted chemicals are available. Sterile agar (Sabouraud dextrose broth) in petri dishes is first coated with a thin lawn of yeast solution (*Candida albicans*). Plant extracts (in suitable solvents) are then applied to filter paper discs (6 mm diameter) and, after drying, placed onto the agar plates which are then covered and sealed with parafilm to prevent contamination. Alternatively, small weighed pieces of plant tissues may be used. Each plate typically contains three or four sample discs and one control disc (solvent only). After 0.5–4 h under UVA light at room temperature (with or without UV filters) the plates are transferred to an incubator for an additional 12–44 h at room temperature under nonphotosensitizing light conditions. Toxicity of the plant extracts or plant specimens are then determined by the presence of a zone of inhibition (clearing) extending outward from the discs. Extracts or plant materials which cause a clearing in the dark control groups are considered to have antibiotic activity, while those which result in a clearing in the UV-treated groups are considered to be phototoxic.

2.3 Bacterial Bioassays

A number of bacterial bioassays have been used which are based on the yeast bioassay. These include the use of *Bacillus subtilis*, *Staphylococcus albus*, *Pseudomonas fluorescens*, and *Escherchia coli* (Wat et al. 1979, 1980; McKenna and Towers 1981; Abeysekera et al. 1983; Ashwood-Smith et al. 1983, 1986; Proksch et al. 1983; Tuveson et al. 1986; Marchant and Cooper 1987; Towers and Hudson 1987). In the photobiological assay of Ashwood-Smith et al. (1983), plant extracts or test substances are first chromatographed on TLC plates, and then small discs are cut from the plates and placed onto solidified agar in petri dishes. The plates are preincubated for a period of 2–30 min which allows time for a small amount of the test substance to diffuse into the agar. The indicator microorganisms, in a soft agar solution (10 g of bactotryptone, 9 g bactoagar, and 5 g KCl in 1000 ml water) are then overlayed onto each plate. After the soft agar has set, plates are irradiated with near-UV for 1 h, followed by 12–24 h in an incubator under normal light conditions.

Ashwood-Smith et al. (1986) have furthur developed these bacterial bioassays to examine the photosensitivity of DNA to furanocoumarins. This was done using the DNA-repair-deficient mutant *E. coli* B_{s-1} (rec^+, exr^-, hrc^-) which is extremely sensitive to UV light (Igali et al. 1970). Although qualitative in nature, this extremely sensitive bioassay allowed for the detection of furanocoumarins as low as 10^{-8} g/g wet weight in plant tissue.

Tuveson et al. (1986) have developed an assay to study the inactivation of *E. coli* strains with differing sensitivities to near- and far-UV. They used four *E. coli* strains which carried all the possible combinations of genes controlling sensitivity to oxidative (photodynamic) damage (NUV; *nur* versus nur^+) and excision proficiency (FUV; *uvrA6* versus $uvrA^+$). Using these test organisms it was possible to distinguish between those phototoxins whose effects are due to oxygen-independent photoadduct formation (e.g., many furanocoumarins) and those which cause photodynamic (oxygen-dependent) damage (e.g., thiophenes). Cell suspensions of these bacteria were combined with presumed phototoxins and then inactivated using broad-spectrum NUV. Fluence-response curves were then determined and the inactivation kinetics used to assess the phototoxicity and confirm the presumed mode of action of the various phototoxins.

2.4 Mosquito Larvae Bioassay

The mosquito larval bioassay has been successfully used to examine the relative phototoxicity of a wide range of naturally occurring and synthetic polyacetylenes and thiophene derivatives (Arnason et al. 1981), as well as to examine the effects of berberine on mosquito larval development. (Philogene et al. 1984a). It is based on the World Health Organization's established tests for the susceptibility and resistance of mosquito larvae to insecticides (World Health Organization 1963).

With new substances, a pretest is conducted to establish a suitable range of concentrations for optimum bioactivity. Ten early fourth instar larvae are placed in 100 ml of dechlorinated water after which the test compound (in 1 ml acetone) is added to obtain a final concentration of 0, 10, 100, or 1000 ppb. Assays are performed in duplicate and include control groups (UV filter) and groups treated with UV light for 4 h followed by 20 h of nonphotosensitizing light conditions. After the pretest, a range of six to eight geometric concentrations of the test compounds are selected (e.g., 2, 4, 8, 16, 32 ppb). Twenty larvae are added to 250 ml dechlorinated water in 500-ml jars and 1 ml of acetone containing the test compounds placed on the water. After 10 min, the groups are treated with or without UV as before. Assays are performed in triplicate. Probit analyses are then used to determine 24- and 48-h LC50's, LC90's, and fiducial limits.

2.5 Brine Shrimp Bioassay

The brine shrimp assay provides a convenient, general assay for examining the toxicity of new bioactive compounds extracted from plants (Meyer et al. 1982; Alkofahi et al. 1989) and has recently been used to determine the relative phototoxicity (and structure-activity relationships) of a large number of thiophenes (R. Marles et al. 1990, in prep.). Eggs of the brine shrimp (*Artemia salina*) are hatched (24–48 h) in a brine solution and then transferred to vials at a final concentration of about ten larvae (nauplii) per 5 ml of brine containing the appropriate plant extract concentrations. As with the mosquito bioassay, a logarithmic series of concentrations are first tested to arrive at an approximate LD50, followed by a geometric series of dilutions to obtain dose-response data. After 4 h under UVA, vials are left at ambient room conditions for an additional 20 h. The number of dead and moribund larvae are counted at the bottom of the vials with the aid of a dissecting microscope. To kill the remaining larvae, 1 ml of methanol is then added to each vial and the total number of dead larvae counted after 1 h. This procedure avoids the tedious problem of counting out a precise number of organisms for each vial. Probit analysis provides the relative toxicities of the compounds tested. The procedure is simple enough to be used in the field and has been used to assess the phototoxicity of medicinal plant extracts in Borneo (Leaman and Arnason 1990, in prep.).

3 Photogenotoxicity

Because the target of most furanocoumarins appears to be at the level of DNA, bioassays have been developed which are specific for this aspect of phototoxicity. This is in part due to the current and potential cosmetic and medical uses that exist for these compounds (Towers 1979). These assays include examinations of bacterial frameshift mutations such as those induced by the fuorquinoline alkaloid dictamnine (Ashwood-Smith et al. 1982), the inactivation of DNA and RNA viruses by beta-carboline alkaloids such as harmine (Hudson et al. 1986), and the photo-induction of DNA cross-linking such as that observed using the furanochromone visnagin (Altamirano-Dimas et al. 1985). One of the more commonly used photogenotoxicity bioassays involves the use of Chinese hamster ovary (CHO) cells. These cells are phototoxin-sensitive and can be used to study lethal effects, mitotic inhibition, and chromosomal aberrations (Ashwood-Smith et al. 1982; Abeysekera et al. 1983; Towers and Abramowski 1983; Philogene et al. 1984b).

The following method has been used to examine the UV-induced chromosomal aberrations caused by the isoquinoline alkaloid berberine (Abeysekera et al. 1983; Philogene et al. 1984b): CHO cells were cultured in Eagles minimum essential medium (MEM) supplemented with 10% fetal calf serum (FCS), antibiotics (streptomycin sulfate 29.6 μg/ml, penicillin 125 μg/ml, kanamycin 100 μg/ml, fungizone 2.5 μg/ml), and 7.5% sodium bicarbonate (1 μg/ml). Cells of stock cultures were grown in 240-ml plastic culture flasks at 37 °C in water and resuspended in the fresh medium. For seeding, the suspension was diluted to to an approximate density of 70000 cells/ml. An aliquot (2 ml) of this dilution was seeded onto coverslips in plastic dishes (35 × 10 mm) and kept in MEM with 10% FCS at 37 °C for 2 days (60–80% confluency). Test compounds were dissolved in 95% ethanol and diluted in MEM with 2.5% FCS. Dilutions were made and 1 ml added to each petri dish after removing the culture medium. Tests were carried out in duplicate, one series irradiated and the other series kept in the dark (30 min). After irradiation, the coverslips were washed two times with MEM, and fresh medium with 10% FCS added to the dishes. Samples were incubated for 16 h. Four hours prior to harvesting, cells were pretreated with 0.2 ml colchicine (0.01%) in 2.5% MEM. Sodium citrate (1%) was used for 20 min during harvesting. Air-dried coverslips were then stained with 2% acid orecin, mounted, and 100 metaphase plates analyzed for chromosome breaks and exchanges.

4 Plant Pest Bioassays

4.1 Phytophagous Insects

The activities of many phototoxins in irradiated and unirradiated trials have been evaluated using phytophagous insects, in particular lepidopteran larvae of economic importance. A thorough examination of phototoxicity in these larvae is often difficult, however, because the tests require both a large amount of pure secondary compound and that the insects be carried through their whole life cycle (Arnason et al. 1990). Some of the parameters which have been used to phototoxicity in larvae include larval mortality, mean development time, time to pupation, and nutritional indices such as those used by Waldbauer (1968).

The furanocoumarins have been the most extensively studied in larval bioassays, in part because of their commercial availability. Berenbaum (1978) was the first to demonstrate the UV-enhanced toxicity of the furanocoumarin 8-methoxypsoralen (xanthotoxin) and its effects on larval survivalship and development. Over the last decade, a number of studies have explored the phototoxicity of a variety of secondary phytochemicals to lepidopteran larvae. These include seven polyacetylenes and thiophenes (Champagne et al. 1984, 1986; Downum et al. 1984), the flavonoid quercitin (Ahmad et al. 1987), the beta-carboline alkaloid harmane (Lee and Berenbaum 1989), and the extended quinone hypericin (Downum et al. 1984; Samuels and Knox 1989; Sandberg and Berenbaum 1989).

To bioassay phototoxins, purified compounds are usually incorporated into artificial diets such as the wheat germ-based diet which is commonly used for the tobacco hornworm *Manduca sexta*. Final concentrations should reflect the expected concentrations found in plant tissues and fresh diet should be made often to avoid degradation of test compounds. Groups of 24 neonate larvae are placed individually in multiwelled trays with each well containing a diet cube. These are covered with UV-transparent polyethylene film or appropriate filters. Four groups are used for each treatment consisting of dark controls with and without test compounds and light-treated groups with and without compounds. UV treatment may consist of a bank of 8 × 40 W Durotest Vitalites augmented with a bank of 4 × 20 W BLB lamps. Larvae are typically weighed every 2–4 days to examine growth rates and, if required for nutritional studies, diet remaining and frass weighed as well. Statistical analysis usually involves one-way ANOVA followed by Duncans multiple range test on mean weight data (parametric) and the Kruskall-Wallace test on mean development times (nonparametric).

4.2 Fungal Pathogens

Most fungal pathogens are not able to become established because of host plant defense mechanisms which are chemical in nature (Swain 1977). The invasion of plant tissues by hyphae may induce an accumulation of antifungal substances (see, for example, De Wit and Kodde 1981). The evaluation of plant compounds and phototoxins and their role in the phytoalexin response has been based on the development of an antifungal bioassay. In assays with the plant pathogens *Fusarium culmorum* and *F. oxysporum* (Bourque et al. 1985; Kourany et al. 1988) a number of polyacetylenes and thiophenes were shown to have inhibitory effects towards conidial germination and mycelial growth.

In these assays small amounts of test compounds were dissolved in ethanol and aliquots placed in the cavities of microscope well slides. After the solvent had evaporated, the dried deposits were covered with 100 µl of a spore suspension containing 5×10^4 spores/ml in Fries medium. The slides were placed in covered petri dishes lined with moistened filter paper and incubated for 18 h at 25 °C in the dark, or in the presence of 4 W m^{-2} near-UV. Percentage germination was determined by examining more than 100 spores per sample by phase contrast microscopy. All treatments were duplicated and each experiment carried out twice.

To examine the effects of plant compounds on mycelial growth, Erlenmyer flasks (25 ml), each containing 10 ml of sterilized Fries medium, were innoculated with 100 µl of a 5×10^9 spore suspension. Test compounds were added to flasks and cultures incubated at 25 °C with continuous shaking at 150 rpm in the dark or in the presence of near-UV. Following a 72-h incubation, the fungi were collected on Whatman filter paper using a vacuum filtration apparatus, dried at 75 °C for 25 h and weighed.

5 Summary/Conclusions

Bioassays for phototoxicity of naturally occurring and synthetic substances have applications in medicine, biology, and agriculture. Many of the tests described in this Chapter are rapid and simple to use and some may be used in either field or laboratory situations. Qualitative tests such as the yeast bioassay may provide rapid results as to the phototoxicity of plant tissues or extracted compounds. The brine shrimp and mosquito larval bioassays have provided more quantitative data on the relative toxicity of secondary metabolites. Other tests, such as the bacterial bioassays, are very useful in examining photogenotoxicity and may be used to examine the specific mode of action of secondary phytochemicals. Important contributions in understanding the compatible-incompatible reactions between host plants and fungal pathogens have come from the use of fungal bioassays. The molecular basis for these reactions will be elucidated as the results from research at the DNA and RNA levels becomes available.

References

Abeysekera BF, Abramowski Z, Towers GHN (1983) Genotoxicity of the natural furochromones, khellin and visnagin and the identification of a khellin-thymine photoadduct. Photochem Photobiol 38:311–315

Ahmad S, Pritsos CA, Bowen SM, Kirkland KE, Blomquist GJ, Pardini RS (1987) Activities of enzymes that detoxify superoxide anion and related toxic oxyradicals in *Trichoplusia ni*. Arch Insect Physiol Biochem 6:85–96

Alkofahi A, Rupprecht JKJ, Anderson JE, McLaughlin JL, Mikolajczak KL, Scott BA (1989) Search for new pesticides from higher plants. In: Arnason JT, Philogene BJR, Morand P (eds) Insecticides of plant origin. ACS Symp Ser 387, Am Chem Soc, Washington, DC, pp 25–43

Altamirano-Dimas M, Hudson JB, Abramowski Z, Towers GHN (1985) The photoinduction of DNA crosslinks by the furochromone visnagin. Photobiochem Photobiophys 10:121–130

Arnason JT, Swain T, Wat CK, Graham EA, Partington S, Towers GHN (1981) Mosquito larvicidal activity of polyacetylenes from species in the asteraceae. Biochem Syst Ecol 9:63–68

Arnason JT, Philogene BJR, Morand P, Imrie K, Iyengar S, Duval F, Soucy-Breau C, Scaino J, Werstiuk NH, Hasspieler B, Downe AER (1989) Naturally occurring and synthetic thiophenes as photoactivated insecticides. In: Arnason JT, Philogene BJR, Morand P (eds) Insecticides of plant origin. ACS Symp Ser 387, Am Chem Soc, Washington, DC, pp 164–172

Arnason JT, Philogene BJR, Towers GHN (1990) Phototoxins in plant-insect interactions. In: Rosenthal G, Berenbaum MR (eds) Herbivores, 2nd edn. Academic Press (in press)

Ashwood-Smith MJ, Towers GHN, Abramowski Z, Poulton GA, Liu M (1982) Photobiological studies with dictamnine, a furoquinoline alkaloid. Mutat Res 102:401–412

Ashwood-Smith MJ, Poulton GA, Ceska O, Liu M, Furniss E (1983) An ultrasensitive bioassay for the detection of furocoumarins and other photosensitizing molecules. Photochem Photobiol 38:113–118

Ashwood-Smith MJ, Ceska O, Chaudary SK, Warrington PJ, Woodcock P (1986) Detection of furocoumarins in plants and plant products with an ultrasensitive biological photoassay employing a DNA-repair deficient bacterium. J Chem Ecol 12:915–932

Bakker J, Gommers EJ, Nieuwenhuis I, Wynberg H (1979) Photoactivation of the nematicidal compound alpha-terthienyl from roots of marigolds. J Biol Chem 254:1841–1844

Berenbaum M (1978) Toxicity of a furanocoumarin to armyworms: a case of biosynthetic escape from insect herbivores. Science 201:532–534

Bourque G, Arnason JT, Madhosingh C, Orr W (1985) The photosensitization of the plant pathogen *Fusarium culmorum* by phenylheptatryne from *Bidens pilosa*. Can J Bot 63:899–902

Camm EL, Towers GHN, Mitchell JC (1975) UV-mediated antibiotic activity of some compositae species. Phytochemistry 14:2007–2011

Champagne DE, Arnason JT, Philogene BJR, Campbell G, McLaughlin D (1984) Photosensitization and feeding deterrence of *Euxoa messoria* by alpha-terthienyl, a naturally occurring thiophene from the asteraceae. Experientia 40:577–578

Champagne DE, Arnason JT, Philogene BJR, Morand P, Lam J (1986) Light-mediated allelochemical effects of naturally occurring polyacetylenes and thiophenes from asteraceae on herbivorous insects. J Chem Ecol 12:835–858

Daniels F (1965) A simple microbiological method for demonstrating phototoxic compounds. J Invest Dermatol 44:259

De Wit PJGM, Kodde E (1981) Induction of polyacetylene phytoalexins in *Lycopersicon esculentum* after innoculation with *Cladosporium fulvum*. Physiol Plant Pathol 18:143–148

Downum KR, Rosenthal GA, Towers GHN (1984) Phototoxicity of the plant metabolite alpha-terthienyl to larvae of *Manduca sexta*. Pestic Biochem Physiol 22:104–109

Edelson R, Berger C, Gasparro F (1987) Treatment of cutaneous T-cell lymphoma by extracorporeal photochemotherapy: preliminary results. N Engl J Med 316:297–303

Hudson JB (1989) Antiviral properties of photosensitizers used in dermatology: recent advances and alternative interpretations. Photodermatology 6:155–165

Hudson JB, Graham EA, Fong R, Hudson L, Towers GHV (1986) Furthur studies on the antiviral activity of harmine, a photoactive beta-carboline alkaloid. Photochem Photobiol 44:483–487

Igali S, Bridges BA, Ashwood-Smith MJ, Scott BR (1970) Mutagenesis in *E. coli*: photosensitization to near UV by 8-methoxypsoralen. Mutat Res 9:21–30

Kavli G, Volden G (1984) The *Candida* test for phototoxicity. Photodermatology 1:204

Knox JP, Dodge AD (1985) Isolation and activity of the photodynamic pigment hypericin. Plant Cell Environ 8:19–25

Kourany E, Arnason JT, Schneider E (1988) Accumulation of phototoxic thiophenes in *Tagetes erecta* (Asteraceae) elicited by *Fusarium oxysporum*. Physiol Mol Plant Pathol 33:287–297

Larson RA, Berenbaum M (1988) Environmental phototoxicity. Environ Sci Technol 22:354–360

Lee KW, Berenbaum M (1989) Action of antioxidant enzymes and cytochrome P-450 monooxygenases in the cabbage looper in response to plant phototoxins. Arch Insect Physiol Biochem 10:151–162

MacRae WD, Irwin DAJ, Bisalputra T, Towers GHN (1980) Membrane lesions in human erythrocytes induced by the naturally occurring compounds alpha-terthienyl and phenylheptatryne. Photobiochem Photobiophys 1:309–318

Marchant YY, Cooper GK (1987) Structure and function relationships in polyacetylene photoactivity. In: Hietz J, Downum KR (eds) Light-activated pesticides. ACS Symposium Series 339, Am Chem Soc, Washington, DC, pp 241–253

Marles RJ, Arnason JT, Compadre L, Compadre C, Mehta B, Morand P, Redmond RW, Scaiano JC (1990) Quantitative structure activity relationship analysis of natural products: Phototoxic thiophenes. Rec Adv Phytochem 25:371–396

McKenna DJ, Towers GHN (1981) Ultra-violet mediated cytotoxic activity of beta-carboline alkaloids. Phytochemistry 20:1001–1004

Meyer BN, Ferrigni NR, Putnam JE, Jacobsen JB, Nichols DE, McLaughlin JL (1982) Brine shrimp: a convenient general bioassay for active plant constituents. Planta Med 45:31–34

Philogene BJR, Arnason JT, Duval F (1984a) Effect of khellin and 8-methoxypsoralen on the development and survival of *Aedes atropalpus*. Can Entomol 17:1153–1157

Philogene BJR, Arnason JT, Towers GHN, Abramowski Z, Campos F, Champagne D, McLaughlin D (1984b) Berberine: a naturally occurring phototoxic alkaloid. J Chem Ecol 10:115–123

Philogene BJR, Arnason JT, Berg CW, Duval F, Champagne D, Taylor RG, Leitch LC, Morand P (1985) Synthesis and evaluation of the naturally occurring phototoxin alpha-terthienyl as a control agent for larvae *Aedes tropalpus*, *Aedes intrudens* (Diptera: Culicidae) and *Simulium verecundum* (Diptera: Simulidae). J Econ Entomol 78:121–126

Proksch P, Proksch M, Towers GHN, Rodriguez E (1983) Phototoxic and insecticidal activities of chromenes and benzofurans from *Encelia*. J Nat Prod 46:331–334

Samuels R, Knox P (1989) Insecticidal activity of hypericin towards *Manduca sexta* larvae. J Chem Ecol 15:855–862

Sandberg SL, Berenbaum MR (1989) Leaf-tying by tortricid larvae as an adaptation for feeding on phototoxic *Hypericum perforatum*. J Chem Ecol 15:875–885

Song PS, Tapley KJ (1979) Photochemistry and photobiology of psoralens. Photochem Photobiol 29:1177–1197
Swain T (1977) Secondary compounds as protective agents. Annŭ Rev Plant Physiol 284:479–536
Towers GHN (1979) Photosensitizers in plants and their photodynamic action (a review). Prog Phytochem 6:183–202
Towers GHN (1984) Interactions of light with phytochemicals in some natural and novel systems. Can J Bot 62:2900–2911
Towers GHN, Abramowski Z (1983) UV-mediated genotoxicity of furanoquinoline and certain tryptophan-derived alkaloids. J Nat Prod 46:576–581
Towers GHN, Hudson JB (1987) Potentially useful antimicrobial and antiviral phototoxins from plants. Photochem Photobiol 46:61–66
Tuveson RW, Berenbaum MR, Heininger EE (1986) Inactivation and mutagenesis by phototoxins using *Escherichia coli* strains differing in sensitivity to near- and far-ultraviolet light. J Chem Ecol 12:933–947
Waldbauer GP (1968) The consumption and utilization of food by insects. Adv Insect Physiol 5:229–288
Wat CK, Biswas RK, Graham EA, Bohm L, Towers GHN, Waygood ER (1979) Ultraviolet-mediated cytotoxic activity of phenylheptatryne from *Bidens alba*. J Nat Prod 42:103–111
Wat CK, MacRae D, Yamamoto E, Towers GHN, Lam J (1980) Phototoxic effects of naturally occurring polyacetylenes and alpha-terthienyl on human erythrocytes. Photochem Photobiol 32:167–172
World Health Organization (1963) Insecticide resistance and vector control. WHO 13th Rep No 265

Toxic Extracellular Enzymes

H.M. KALISZ and M.E. KALISZ

1 Introduction

This Chapter describes the methods used for the assay, identification and determination of "toxic" action of extracellular enzymes of microbial (mainly fungal) and higher plant origin involved in plant pathogenesis. Methods used for the general characterization of an enzyme, such as its catalytic properties, molecular weight, amino acid composition, protein sequence analysis, have been omitted. A detailed account of the theoretical and practical aspects of enzyme characterization may be found in most biochemical textbooks (e.g., Segel 1975; Cornish-Bowden 1979; Page 1984; Darbre 1986; Franks 1988).

The term "toxic enzyme" is taken to mean an enzyme capable of directly attacking specific structures of the host or pathogen. Thus, microbial enzymes, such as pectinases, ligninases and cellulases, and plant enzymes, such as chitinase and β-1,3-glucanase, are classified as toxic. Discussion of carnivorous plant hydrolases involved in insect digestion (Clancy and Coffey 1977; Heslop-Harrison Y. and Heslop-Harrison J. 1981), as well as plant enzymes involved in the synthesis of toxins or metabolites, such as phenolics and phytoalexins, in response to pathogen attack (Stoessl 1983; Hahlbrock and Scheel 1989) is outside the scope of this Chapter.

The term "extracellular" refers to enzymes which cross the cytoplasmic membrane (Glenn 1976). Secretion of eukaryotic enzymes involves the transport of the proteins from the rough endoplasmic reticulum via the Golgi complex and secretory vesicles to the plasma membrane. In bacteria, the cytoplasmic membrane accomplishes export of proteins (Priest 1984; Akazawa and Hara-Nishimura 1985). Synthesis of extracellular enzymes is usually controlled by induction, derepression, catabolite repression or end-product inhibition (Glenn 1976).

A systematic classification of enzymes has been adopted on the recommendation of the International Union of Biochemistry (1984). Consequently, enzymes are divided into six major classes as follows: (1) oxidoreductases; (2) transferases; (3) hydrolases; (4) lyases; (5) isomerases; (6) ligases, according to the general nature of reactions they catalyze. The six classes are further subdivided into subclasses, to specify the type of reaction more fully and to indicate the chemical nature of the reactants.

Each enzyme is assigned a recommended name (e.g., chitinase), a systematic name, e.g., poly(1,4-[N-acetyl-β-D-glucosaminide])glucanohydrolase,

Table 1. Plant enzymes implicated in disease resistance

Enzyme	Source	References
Chitinase and β-1,3-Glucanase	Tobacco	Kauffmann et al. (1987); Legrand et al. (1987)
	Tomato	Joosten and De Wit (1989)
	Potato	Kombrik et al. (1988)
	Cucumber	Boller and Métraux (1988)
	Muskmelon	Roby and Esquerre-Tugaye (1987)
	Bean	Vögeli et al. (1988)
	Pea	Mauch et al. (1988b)
	Forest trees	Wargo (1975)
Proteases	Various	Boller (1986)
Cellulases	Most plants	Sexton and Roberts (1982)
α-Mannosidase	Most plants	Dey and Campillo (1984); Mellor et al. (1984)
β-N-Acetyl-glucosaminidase	Various	Dey and Campillo (1984)
α-Glucosidase	Tobacco	Edreva and Georgieva (1980)
β-Glucosidase	Tobacco	Edreva and Georgieva (1980)
Trehalase	Various	Veluthambi et al. (1981)
Polysaccharases	Legumes	Solheim and Fjellheim (1984)
Ribonuclease	Various	Sacher et al. (1982)
Lipases	Potato	Racusen (1984)

which identifies the reaction catalyzed (e.g., random hydrolysis of N-acetyl-β-D-glucosaminide 1,4-β-linkages in chitin and chitodextrins), and a classification number (e.g., EC 3.2.1.14), which enables accurate and unambiguous identification of an enzyme.

2 Sources of Enzymes

2.1 Plant Enzymes

Plants, like animals, use post-infectional biochemical responses to resist diseases (Bell 1981; Lamb et al. 1989). These include the modification of cell walls, production of lignins, tannins and melanins (Bell 1981), phytoalexins (Bell 1981; Bailey and Mansfield 1982; Darvill and Albersheim 1984), lectins (Etzler 1985; Broekaert et al. 1989), inhibitors of enzymes and viruses (Abu-Goukh et al. 1983; Ryan 1984), and lytic enzymes (Bell 1981; Boller 1987).

A number of plant enzymes have been implicated in disease resistance (Table 1). However, only chitinase and β-1,3-glucanase have been conclusively demonstrated to play an important role in the prevention of microbial attack (Schlumbaum et al. 1986; Mauch et al. 1988b; Benhamou et al. 1989, 1990; Broekaert et al. 1989; Coté et al. 1989).

2.2 Microbial Enzymes

Many pathogenic and non-pathogenic microorganisms produce an array of enzymes capable of attacking plant cell components. Their secretion, therefore, does not enable these organisms to be plant pathogenic. However, the ability to produce such enzymes is an essential feature of paramount importance for the pathogenicity of some microorganisms. The exact role of these enzymes in pathogenicity is still unclear, although the pectic enzymes (Collmer and Keen 1986) and cutinase (Kolattukudy 1985) have been conclusively shown to play an important role in plant pathogenesis. Other enzymes implicated in plant pathogenesis include the proteinases, cellulases, hemicellulases and lignin peroxidases.

3 Purification

3.1 Isolation and Initial Fractionation

The first step in the purification of a protein is the preparation of an extract containing the protein in a soluble form. Microbial extracellular enzymes are located external to the cell wall and thus can be isolated without the need for cell disruption. On the other hand, most plant extracellular enzymes are located inside the cell wall. Their isolation, therefore, requires the disruption of cells. A number of chemical, physical and mechanical methods are available for cell disruption (Suelter 1985; Scopes 1987; Harris and Angal 1989). Grinding with a pestle and mortar is normally used for the isolation of plant chitinases and β-1,3-glucanases.

Cell extracts usually have to be treated immediately after disruption to remove contaminants, such as nucleic acids and proteinases, and to prevent enzyme inactivation by proteolysis, acidification or oxidation (Suelter 1985; Janson and Rydén 1989). A particular problem with many plants is the presence of phenolic compounds which upon homogenization are rapidly converted by phenol oxidases to quinones, which often cause enzyme inactivation. Quinone formation can be significantly reduced by the addition of copper-chelating agents, such as diethyldithiocarbamate or thiourea, or thiols, such as β-mercaptoethanol or cysteine. The reversible binding of phenols to proteins may be prevented either by using an extraction buffer at or above pH 8, or by adding phenol-complexing polymers, such as polyvinylpyrrolidone, polyethylene glycols or anion-exchange resins.

Clarification or initial fractionation of protein solutions is achieved by methods such as centrifugation, cross-flow microfiltration, ultrafiltration or precipitation. The most commonly used precipitants are ammonium sulphate, sodium sulphate, polyethylene glycol, ethanol and acetone. Organic solvents usually cause partial denaturation of proteins and are not recommended as first choice precipitants. Precipitation with ammonium sulphate and polyethylene glycol often stabilizes the proteins and is a convenient and useful purification step (Suelter 1985; Scopes 1987).

An alternative method of fractionation is aquenous two-phase extraction, whereby separation is based on the partition of proteins between two water-rich phases, which are composed of water and two polymers, usually polyethylene glycol and dextran (Albertsson 1986; Walter et al. 1985). The partition of proteins between the phases depends on the concentration and molecular weight of the polymers, salt and buffer composition, pH and temperature.

3.2 Liquid Chromatography

A large number of methods, based on differences in molecular size, charge and structure, and activity, are available for the purification of proteins (Table 2; Heftmann 1983; Scopes 1987; Janson and Rydén 1989; Harris and Angal 1989). It is advisable to plan a purification scheme before commencing the actual purification procedure. The sequence of chromatographic steps should be designed to avoid or at least minimize buffer changes and concentration steps. In addition, the smallest possible number of purification steps should be used to reduce yield loss. In practice one should start with methods, such as ion exchange or hydrophobic interaction chromatography, which enable the application of large sample volumes and have the highest capacities. The final purification step should be used to remove possible aggegates or degradation products and to prepare the purified protein for its use or storage. Gel filtration is most suitable for this purpose. Concentration is often required after this step and is usually achieved by ultrafiltration.

Protein elution from a chromatography column is usually monitored by ultraviolet absorption at 280 nm. Enzyme activity may also be detected online (Künnecke et al. 1989).

The degree of purity of an enzyme is established by measurement of protein content and enzyme activity of the sample, and determination of its specific activity (units/mg protein).

3.3 Electrophoretic Analysis Methods

Electrophoresis has become an almost mandatory analytical procedure for the assessment of protein purity. It is also used to monitor the progress of a purification procedure and to identify fractions containing the protein of interest. In addition, electrophoresis can be used to characterize purified proteins (Andrews 1986; Dunn 1986; Hames and Rickwood 1989).

Polyacrylamide gel electrophoresis (PAGE), due to its high resolution capacity, is the method of choice for protein separation. Electrophoresis

Table 2. Liquid chromatography methods available for protein purification

Chromatography method	Separation criteria	Important parameters
Gel filtration	Molecular size	Column length
Ion exchange	Charge	pH, ionic strength
Hydroxylapatite	Charge, adsorption	Ionic strength
Hydrophobic interaction	Hydrophobicity	Polarity, ionic strength
Reversed phase	Hydrophobicity	Polarity, ionic strength
Affinity	Activity, biospecific interaction	Ligand, eluent

separates proteins on the basis of their size, net charge or relative hydrophobicity. PAGE under native conditions is often used to analyze proteins in their native state. It is also used to determine the molecular weight of a protein using gels containing a polyacrylamide concentration gradient. A widely used procedure is PAGE in the presence of sodium dodecyl sulphate (SDS), which resolves proteins on the basis of the molecular weight of their constituent polypeptides. Isoelectric focusing and titration curves separate proteins according to their relative net charge over a pH spectrum. Two-dimensional gel electrophoresis can be used to separate complex protein mixtures according to charge and size by isoelectric focusing and SDS-PAGE, respectively.

Proteins are normally visualized by Coomassie Brilliant Blue R-250 (Weber and Osborn 1969; Wong et al. 1985) or silver staining (Switzer et al. 1979; Heukeshoven and Dernick 1988). Silver staining is up to 100 times more sensitive than Coomassie staining, with a detection limit of less than 1 ng protein. In addition, specific detection methods, such as for carbohydrates (Gander 1984; Munoz et al. 1988; Doerner and White 1990), lipids (Cutting 1984), enzyme activity (Sect. 4.3) and antigenic properties (Beisiegel 1986) may be used.

3.4 Western, or Protein, Blotting

Identification can be made easier, faster and more sensitive by transferring the proteins electrophoretically from the gel to a membrane, such as nitrocellulose or polyvinyldifluoride (PVDF), by a technique referred to as Western, or protein, blotting (Gershoni 1988; Prieur and Russo-Marie 1988). The membranes can be stained for protein (Rohringer and Holden 1985; Glenney 1986; Root and Reisler 1989), glycoprotein (Eriksson S. et al. 1989; Heimgartner et al. 1989) and activity (Sect. 4.3), as well as radiographically (Burnette 1981; Howe and Hershey 1981) and immunologically (Howe and Hershey 1981; Glenney 1986).

4 Analytical Methods

4.1 Determination of Protein Concentration

4.1.1 Ultraviolet Absorption

This is the method of choice when using a pure protein with a known extinction coefficient, as it is rapid, sensitive (sensitivity range at 280 nm = 100–3000 μg) and nondestructive. However, due to large variations in the contents of the tyrosine and tryptophan residues, the method is inaccurate unless the extinction coefficient of the protein is known. In addition, crude extracts contain many substances, such as nucleic acids, which absorb strongly in the ultraviolet region and may lead to erroneous protein estimations.

An alternative and more sensitive method which is less affected by the protein amino acid content is absorption in the far-ultraviolet (Suelter 1985).

4.1.2 Biuret Method

This method is used little due to its insensitivity (sensitivity range = 1–10 mg protein). However, it provides an accurate estimate of protein concentration with little variation in colour intensity from protein to protein. The copper reagent reacts with peptide bonds to form a purple-coloured complex with an absorption maximum at 540–560 nm. Ammonium ions and lipids interfere with the assay.

4.1.3 Lowry Method

The most widely used procedure for protein determination is the Folin-Lowry method (Lowry et al. 1951). The method is based on the copper-catalyzed oxidation of peptides and proteins by phosphomolybdic/phosphotungstic acid (Folin phenol reagent) to form a strong blue colour (Legler et al. 1985) which is usually measured at 660 nm. However, the reaction is affected by a diverse range of substances (Peterson 1979). Consequently, a large number of modified procedures have been developed to overcome this interference (Peterson 1979; Rodríguez-Vico et al. 1989). The sensitivity range is 2–30 μg at 660 nm and 1–20 μg at 750 nm.

4.1.4 Dye-Binding Method

The method of Bradford (1976) is based on the differential colour change of a dye (Coomassie Brilliant Blue G-250) in response to protein concentration.

The method is rapid, sensitive (1–15 μg protein) and simple, but shows large variation in response to different proteins, and is affected by an array of common laboratory chemicals (Friedenauer and Berlet 1989).

4.1.5 Bicinchoninic Acid (BCA) Method

This method, like the Lowry assay, relies on the formation of a coloured complex between peptide bonds and copper atoms in an alkaline environment (Smith et al. 1985). However, unlike the Lowry method, the BCA assay uses the sodium salt of BCA, a highly specific chromophore for Cu^{1+}. The intense purple colour has an absorbance maximum at 562 nm. The method is much simpler to perform, more sensitive (can detect 0.5 μg protein) and more tolerant to interference than the Lowry assay. Interference by chemicals, such as glucose and β-mercaptoethanol, may be circumvented by treating the sample with iodoacetamide (Hill and Straka 1988) or by selectively precipitating the protein (Brown et al. 1989).

4.2 Measurement of Enzyme Activity

The amount of enzyme in a sample is usually expressed in terms of the rate at which it catalyzes a reaction under optimal conditions of substrate concentration, effectors, pH, buffer, ionic strength and temperature. Most enzyme reactions decrease in velocity with time. Thus, only the initial rate of reaction, which is proportional to enzyme concentration, is a true measure of enzyme activity.

Two types of units, International Unit (IU) and katal, are commonly used to express the amount of enzyme in a sample. The IU, recommended by the International Union of Biochemistry (1984), is defined as the amount of enzyme that transforms 1 μmol substrate min^{-1}; the katal is the amount of enzyme that catalyzes the transformation of 1 mmol substrate s^{-1}.

Continuous and end point assays may be employed to follow enzyme reactions. In both cases, either the substrate or product of reaction may be measured. For simple reactions product formation methods tend to be used, as they are more sensitive. However, for multistage reactions, in which the intermediates accumulate, only substrate utilization provides the true rate of reaction.

In an end-point assay, an enzyme reaction is allowed to proceed for a fixed period of time before being stopped, mainly by acidification or rapid heating, and the amount of product formed or substrate remaining is measured by enzymatic, chemical or radiochemical methods.

In a continuous assay, changes in the concentration of substrate or product are measured continuously. Most of the reactions must be coupled to a second enzyme assay [usually the NAD(P) oxidoreductase system] where one of the products of the first reaction is acted upon by the coupling enzyme in the reaction mixture (Suelter 1985; Scopes 1987).

4.2.1 Reducing Sugar Analysis

A number of methods are available for reducing sugar analysis (Rexová-Benková and Markovic 1976). However, only a few, such as the dinitrosalicylic acid (DNS; Miller et al. 1960) and the Somogyi-Nelson (Sprio 1966) methods are commonly used.

4.2.1.1 Somogyi-Nelson Method

Sugar solution (5–200 µg) is mixed with 2 ml Somogyi reagent and made up to 4 ml with distilled water. The mixture is boiled for 15 min, cooled, mixed with 2 ml Nelson reagent and 4 ml distilled water. The absorbance is read at 520 nm and the glucose concentration is estimated from a standard graph.

4.2.1.2 Dinitrosalicylic Acid (NDS) Method

Sugar solution (0.1–3.0 mg) in 1 ml is mixed with 3 ml DNS reagent, placed in a boiling water bath for 5 min, then cooled. The absorbance is read at 540 nm and the glucose concentration is estimated from a standard graph.

4.2.2 Measurement of β-1,3-Glucanase Activity

β-1,3-Glucanase (EC 3.2.1.39) exists in at least two isoforms (Mauch et al. 1988a), with a molecular weight of ca. 34 kDa, an isoelectric point of ca. 9.2 and a pH optimum of pH 5.

β-1,3-Glucanase activity can be determined by measuring the release of reducing sugars from laminarin, pre-treated with sodium borohydride to eliminate the high background of reducing sugars (Mauch et al. 1984). β-1,3-Glucanase is incubated for 20 min at 37 °C with 1 mg reduced laminarin in 20 mM sodium acetate buffer, pH 5, in a final volume of 0.5 ml. One milliunit (mU) is defined as the amount of enzyme which liberates 1 nmol min^{-1}.

4.2.3 Measurement of Chitinase Activity

Chitinase (EC 3.2.1.14), along with β-1,3-glucanase, is a potent inhibitor of microbial growth causing lysis of fungal and bacterial cell walls (Boller et al. 1983; Mauch et al. 1988b; Broekaert et al. 1989). Chitinase, a monomer of

30 kDa with optimal activity at pH 4.2–4.7, exists in at least four isoforms: two acidic and two basic (Legrand et al. 1987; Mauch et al. 1988a).

4.2.3.1 *Radiometric Assay for Chitinase*

This assay is based on the formation of soluble oligosaccharides from tritium-labelled chitin (Cabib 1988). The assay mixture contains 0.5 mg [^{3}H]-chitin suspension (6 kBq mg^{-1}), 5 µl 1 M phosphate buffer, pH 6.5, and chitinase in a final volume of 100 µl. The reaction is stopped after 30–120 min at 37 °C by the addition of 0.2 ml 10% trichloroacetic acid and the suspension is either filtered through glass-fibre filters or centrifuged (1000 g, 5 min). The radioactivity in the filtrate or supernatant is measured and chitinase activity is calculated using N-acetylglucosamine as standard (Boller et al. 1983).

4.2.3.2 *Viscosimetric Assay for Chitinase*

The assay is based on the measurement of the reduction in the viscosity of a glycol chitin or 6-O-hydroxypropylchitin solution (Ohtakara 1988). Chitinase is incubated with 0.3% (w/v) glycol chitin in 50 mM phosphate buffer, pH 6.5, and the flow time of the reaction mixture is measured at different time intervals. One unit is defined as the amount of enzyme which reduces the specific viscosity of the assay mixture to 50% in 1 min.

4.2.3.3 *Colorimetric Assay for Chitinase*

The assay is based on the determination of monomeric N-acetylglucosamine with p-dimethyl-aminobenzaldehyde (DMAB) following incubation with insoluble chitin (Boller and Mauch 1988). The procedure is as follows: 0.4 ml chitinase solution in sodium acetate buffer, pH 4.5, is incubated with 0.1 ml colloidal chitin at 37 °C in a shaking water bath. The reaction is stopped after 2 h by centrifugation (1000 g, 3 min); 0.3 ml of the supernatant is incubated for 1 h with 20 µl desalted snail gut enzyme (30 mg ml^{-1}) and 30 µl 1 M phosphate buffer, pH 7.1, to hydrolyze the chitin oligomers. After 1 h, 70 µl 1 M borate buffer, pH 9.8, is added, the mixture is incubated in a boiling water bath for 3 min, then rapidly cooled. DMAB (2 ml) is added to the mixture and the extinction at 585 nm is measured after 20 min. Chitinase activity is calculated using N-acetylglucosamine as standard.

4.2.4 Measurement of Proteinase Activity

Many prokaryotic and eukaryotic organisms, including higher plants and plant pathogens produce extracellular proteinases (North 1982; Kalisz 1988).

Most extracellular proteinases have wide substrate specificities and are involved mainly in the degradation of exogenous proteins to peptides and amino acids before cellular uptake (Payne 1980). Proteinases are classified by their catalytic mechanism into four groups; serine (EC 3.4.21), cysteine (EC 3.4.22), aspartic (EC 3.4.23) and metalloproteinases (EC 3.4.24).

Proteinase activity is usually assayed using either chromogenic (Rinderknecht et al. 1968; Chavira et al. 1984) or fluorogenic (Evans and Ridella 1984; Ostoa-Saloma et al. 1989) indicators conjugated to substrates, or radiolabelled substrates (Jensen 1984).

4.2.5 Measurement of Pectic Enzyme Activity

Pectic enzymes attack pectic polymers, causing plant tissue maceration, cell lysis and modification of cell wall structure, which enables subsequent depolymerization of other plant components (Collmer and Keen 1986; Collmer 1987). Pectic enzymes are classified into two main groups: pectinesterases (EC 3.1.11) and depolymerases. Depolymerases catalyze cleavage either by hydrolysis (polygalacturonase, EC 3.2.1.15; exo-poly-α-D-galacturonosidase, EC 3.2.1.82) or by β-elimination (pectate lyase, EC 4.2.2.2; pectin lyase, EC 4.2.2.10; exopolygalacturonate lyase, EC 4.2.2.9). The enzymes exist in multiple isoforms and exhibit differences in their physical and catalytic properties (Rombouts and Pilnik 1980; Ried and Collmer 1986).

4.2.5.1 Pectinesterase Assay

Pectinesterase activity is usually measured at pH 7.2 and 30 °C by continuously recorded titration of liberated carboxyl groups, using $2\,mg\,ml^{-1}$ high methyl pectin in 10 mM phosphate buffer as substrate (Rexová-Benková and Markovic 1976).

4.2.5.2 Polygalacturonase Assay

Polygalacturonase is usually assayed at 37 °C by measuring the rate of increase of reducing groups formed on release of oligogalacturonates from 0.5% (w/v) polygalacturonate in acetate buffer, pH 4.5 (Collmer et al. 1988).

The decrease in viscosity of the substrate solution may also be measured. This is a very sensitive assay with a 50% decrease in viscosity of a pectate solution to the hydrolysis of only a few percent of the glycosidic bonds (Rombouts and Pilnik 1980). However, this percentage may vary for different enzymes due to differences in their action pattern.

4.2.5.3 *Pectate Lyase and Exopolygalacturonate Lyase Assay*

Pectate and exopolygalacturonate lyases are assayed by measuring the increase in absorbance at 235 nm of 4,5-unsaturated reaction products, released from 0.5% (w/v) polygalacturonic acid in 50 mM acetate buffer, pH 4.5, in the presence of calcium ions. The molar extinction coefficient of the unsaturated product at 235 nm is 4600 M^{-1} cm^{-1} (Collmer et al. 1988).

4.2.5.4 *Pectin Lyase Assay*

Pectin lyase activity is assayed as above using highly esterified pectins as substrate and in the absence of calcium. However, most pectin lyases are markedly stimulated by calcium and other cations, the stimulation being dependent on pH and degree of substrate esterification. The molar extinction coefficient of unsaturated oligogalacturonates is 5550 M^{-1} cm^{-1} (Collmer et al. 1988).

4.2.6 Measurement of Cutinase Activity

Cutinase catalyzes the hydrolysis of cutin, which is composed of saturated and unsaturated fatty acids (Kolattukudy 1985). Cutinase, a glycoprotein containing 4–6% carbohydrate, is a monomer of molecular weight of 25 kDa, with optimum activity at pH 10.

4.2.6.1 *Radiometric Assay for Cutinase*

The reaction mixture contains 0.4 ml of a labelled cutin suspension (4 mg) in 50 mM glycine-NaOH buffer, pH 10, and 0.1 ml enzyme (1 µg) (Kolattukudy et al. 1981). The mixture is incubated at 30 °C for 10 min with shaking in a water bath, the reaction is stopped by the addition of 0.2 ml 1 M HCl, and the mixture is filtered through a glass wool plug. Soluble material non-covalently bound to cutin is removed with 2 × 0.5 ml acetone. Aliquots (0.5 ml) of the combined filtrate are mixed with 10 ml scintillation fluid and the radioactivity counted.

4.2.6.2 *Colorimetric Assay for Cutinase*

The reaction mixture contains 1.6 ml 0.1 M phosphate buffer, pH 8, 0.2 ml Triton X-100 [0.4% (w/v)], 0.2 ml enzyme solution and 0.1 ml p-nitrophenyl butyrate. Initial linear rates of activity are measured spectrophotometrically at 405 nm. Assays are performed at pH 8 to minimize nonenzymatic hydrolysis of the substrate (Kolattukudy et al. 1981).

4.2.7 Measurement of Cellulase Activity

Most cellulolytic organisms contain a complex cellulase enzyme system which consists of three components: (1) endo-1,4-β-glucanase or carboxymethylcellulase (EC 3.2.1.4); (2) exo-1,4-β-glucanase or cellobiohydrolase (EC 3.2.1.91); and (3) β-glucosidase or cellobiase (EC 3.2.1.21; Wood and McCrae 1979). A large number of assays have been developed in an attempt to establish the mechanism of cellulose breakdown (Canevascini and Gattlen 1981; Wood and Kellogg 1988). Standard procedures for the measurement of cellulase activity have been recommended by the Commission on Biotechnology (Ghose 1984).

Fungal cellulases exist in multiple isoforms (Goksoyr and Eriksen 1980) which differ in their isoelectric (pI 3.4–5.8) and molecular (40–50 kDa) properties. Cellulases are glycoproteins with optimal activity at pH 4.8 and 50 °C.

4.2.7.1 Total Cellulase Activity

The recommended assay (Ghose 1984) is based on the estimation of 2 mg glucose from a 50 mg sample of filter paper. The reaction mixture is incubated at 50 °C for 60 min and the glucose concentration measured by the DNS method. The concentration of enzyme releasing 2 mg of glucose is calculated and expressed as filter paper units of activity (FPU).

4.2.7.2 Endo-1,4-β-glucanase Assay

Carboxymethylcellulose (CMC; 1% (w/v) in 50 mM Tris buffer, pH 7.5) is widely used for the measurement of endo-1,4-β-glucanase. Endoglucanase activity can be determined by measuring the increase in reducing sugars or the decrease in viscosity. The two methods are not alternatives as cleavage at one or two sites remote from the end of the CMC chain would produce large changes in viscosity without greatly affecting the reducing power of the solution. Conversely, enzymes cleaving terminal sites of the CMC chain would effect little change in viscosity, but a relatively large change in reducing power. Hence, both methods may have to be used to prevent erroneous conclusions.

4.2.7.3 Exo-1,4-β-glucanase (Exocellobiohydrolase) Assay

Exo-1,4-β-glucanase activity can be measured using Avicel. However, as some endo-1,4-β-glucanases hydrolyse Avicel to a limited extent it is advisable to measure activity on CMC, a specific substrate for endo-1,4-β-glucanase, and Avicel. Low activity on CMC but high activity on Avicel indicates the presence of an exo-1,4-β-glucanase. Conversely, high activity on both substrates is indicative of complete cellulases.

4.2.7.4 β-Glucosidase or Cellobiase Assay

The recommended cellobiase assay (Ghose 1984) is based on the conversion of cellobiose to glucose. The assay mixture is incubated at 50 °C for 30 min, and the reaction is terminated by boiling for 5 min. The amount of glucose released is determined using a glucose oxidase assay kit. Activity is expressed as the amount of enzyme releasing 1 mg of glucose.

Cellobiase activity may also be measured directly, using 100 μM cellobiose, 25 μM dichlorophenol-indophenol (DCPIP) and 10–50 μg enzyme in 20 mM phosphate buffer, pH 6. Activity is followed at 600 nm; the rate of decrease in absorbance, due to DCPIP reduction, being proportional to enzyme activity. The molar extinction coefficient of oxidized DCPIP at $600\,nm = 11.8\,mM^{-1}\,cm^{-1}$ (Renganathan et al. 1990).

Alternatively, β-1,3-glucosidase is incubated at 50 °C for 30 min in 25 mM citrate buffer, pH 4.8, with 1 mM p-nitrophenyl-β-D-glucopyranoside as substrate. The reaction stopped by the addition of 3 ml 0.1 M NaOH, and the absorbance measured at 420 nm. One unit is defined as the amount of enzyme liberating 1 mg of p-nitrophenol.

4.2.8 Measurement of Xylanase Activity

Complete degradation of branched acetyl xylan requires the concerted action of several enzymes; the major enzymes being β-1,4-D-xylosidase (EC 3.2.1.37) and endo-1,4-β-xylanase (EC 3.2.1.8). The enzymes differ considerably in their physical and catalytic properties (Eriksson et al. 1990).

β-Xylosidase activity is measured spectrophotometrically at 420 nm using 20 mM p-nitrophenyl-β-D-xylopyranoside in phosphate buffer, pH 6.5, as substrate.

Endo-xylanase is assayed at 50 °C by incubating the enzyme with 0.5–1.0% (w/v) xylan in acetate buffer, pH 5, and measuring the amount of reducing sugars released.

4.2.9 Measurement of Ligninase (Lignin Peroxidase) Activity

Lignin degradation involves a complex array of enzymes, the key enzyme being a H_2O_2-requiring, haem-containing glycoprotein, ligninase (lignin peroxidase, EC 1.11.1.7). Ligninase exists in multiple isoforms, which differ in their isoelectric points (pI 3.7–4.7), molecular weights (40–50 kDa), carbohydrate content (12–19%), substrate specificity and stability (Eriksson K.-E. et al. 1990; Glumoff et al. 1990). Optimal activity is observed at pH 2–3.

Ligninase activity is assayed at 37 °C by measuring the increase in absorbance at 310 nm due to the oxidation of veratryl alcohol to veratraldehyde. The reaction mixture contains 2 mM veratryl alcohol, 0.4 mM H_2O_2, 50 mM tartaric acid, and an appropriate dilution of ligninase to give an absorbance change of 0.2 min^{-1}. The molar extinction coefficient of veratraldehyde at 310 nm is 9300 $M^{-1} cm^{-1}$ (Tien and Kirk 1988).

4.3 Activity Stain for Enzymes Resolved by Electrophoresis

Detection of enzymes separated by PAGE is a very useful technique for the identification and characterization of the individual components of an enzyme.

4.3.1 Substrate Incorporation in Electrophoretic Gel

This procedure involves the incorporation of the substrate in the gel prior to electrophoresis. Endo-1,4-β-glucanase activity has been detected using 0.1% (w/v) CMC as substrate. Following electrophoresis the gels are washed in buffer, incubated for 30 min at a suitable temperature, then immersed in DNS reagent and incubated at 100 °C for 15 min. Activity is located by the appearance of dark brown bands (Saddler and Khan 1981). Alternatively, activity is visualised with Congo red after 2 h incubation at 37 °C (Schwarz et al. 1987).

β-1,3-Glucanase activity may be detected by incorporating 0.15% (w/v) laminarin in polyacrylamide gels. Following electrophoresis the gels are incubated at 37 °C for 45 min in 50 mM sodium acetate buffer, pH 5, then stained with Aniline Blue and activity zones are observed by ultraviolet transillumination (Coté et al. 1989).

4.3.2 Substrate Incorporation in Overlay Gel

Following electrophoresis, gels are overlaid with a thin film of agarose containing the substrate and incubated under appropriate conditions. This method has been used for the detection of individual components of the cellulase and xylanase complexes (MacKenzie and Williams 1984; Biely et al. 1985), pectinases (Ried and Collmer 1985), proteinases (Westergaard et al. 1980) and chitinases (Trudel and Asselin 1989).

4.3.3 Immersion of Gels in Buffered Substrate Solution

Following electrophoresis, gels are immersed and incubated in a buffered solution containing the substrate. This technique has been used for the detection of individual components of the cellulase complex (Coughlan 1988; McHale et al. 1989).

4.4 Demonstration of Toxic Action of Enzymes

A selection of the methods used to demonstrate the involvement of β-1,3-glucanase and chitinase in plant disease resistance and cutinase and pectinase in plant pathogenesis are described. However, similar procedures have been, and may be, used to determine the roles of many other enzymes in plant pathogenesis.

4.4.1 Demonstration of Antimicrobial Activity of Plant Enzymes

4.4.1.1 Induction of Plant Enzymes

The activity of a variety of plant enzymes increases strongly in response to pathogen attack. However, as microbial enzymes may also contribute to the newly appearing activities, it is imperative to demonstrate that the increased enzyme activity in plant cells originates from the host and is not part of the pathogen's armour.

Inoculation of healthy plant tissues with heat-killed pathogens, pathogen cell walls or elicitors, demonstrated that the increased β-1,3-glucanase and chitinase activities were of plant origin. The two glycosidases were induced coordinately and with very similar kinetics in pea pods inoculated with either conidia of the pathogen, *Fusarium* sp., or the elicitor, chitosan (25 µg per pod; Mauch et al. 1984). Immunoblotting following SDS-PAGE and transfer to nitrocellulose was used to confirm the induction of chitinase and β-1,3-glucanase (Vögeli et al. 1988; Joosten and De Wit 1989).

Further information about enzyme expression was obtained by in vitro translation of RNA isolated from infected plants (Roby and Esquerre-Tugaye 1987; Vögeli et al. 1988). Total RNA was isolated from infected plant leaves by homogenization and re-extraction with phenol:chloroform (1:1, v/v). Polyadenylated RNA was isolated by oligo-(dT)-cellulose chromatography. In vitro translation was performed in the presence of ^{35}S-methionine in a rabbit reticulocyte lysate translation system. The translation products were then immunoprecipitated using antibodies raised against the purified enzymes and the immunoprecipitates analyzed by SDS-PAGE.

4.4.1.2 Enzymatic Degradation of Isolated Cell Walls

Boller et al. (1983) and Mauch et al. (1988b) incubated chitinase and β-1,3-glucanase at pH 5.5 for 2 h at 37 °C in a shaking water bath with a suspension of isolated cell walls (1.2 mg) of the pathogen, *Fusarium solani* f. sp. *phaseoli*. The reaction was stopped by centrifugation (2000 g, 5 min). The release of reducing sugars and of chitin oligosaccharides (after post-incubation with snail gut enzyme) was indicative of enzyme activity against fungal walls.

4.4.1.3 Inhibition of Microbial Growth by Plant Enzymes

Schlumbaum et al. (1986) demonstrated a concentration-dependent inhibition of fungal growth by purified bean chitinase by examining the growth of *Trichoderma viride* on malt extract agar in the presence of various protein preparations. *T. viride* was grown on petri dishes containing 2% malt extract agar at the bottom and 1.5% malt extract agar at the top. One day after inoculation, when the colony diameter was 3–4 cm, wells were punched into the 1.5% agar layer 1 cm from the growing front and were filled with 20 μl of an enzyme preparation. The cultures were incubated for a further 8–12 h and their growth around the wells examined. An inhibition zone was observed around wells containing either extract from infected bean leaves or purified chitinase. This antifungal activity was blocked by the addition of antibodies raised against purified bean chitinase. Extract from uninfected leaves and boiled chitinase had no effect on fungal growth.

Broekaert et al. (1988, 1989) incubated a fungal spore suspension (200 μl at 2×15^5 spores ml^{-1}) in a glucose-peptone-salts medium at 22 °C in penicillin flasks until the hyphae of the germlings attained an average length of 30 μm. Test solutions (50 μl), containing different concentrations of tobacco, wheat or thorn-apple chitinases were then added to the germlings and the flasks were reincubated at 22 °C until the control germlings (50 μl sterile water added) attained a length of 500 μm. Chitinases inhibited the growth of the fungal hyphae by causing lysis of the fungal wall; the antifungal activity being blocked by immunoprecipitation with antiserum to the chitinase. Moreover, chitinase significantly inhibited spore germination; the inhibitory potency being dependent on the source of chitinase and on the fungal species. A freshly prepared spore suspension (200 μl at 4×10^5 spores ml^{-1}) in a glucose-peptone-salts medium was incubated with 50 μl chitinase at 22 °C in a moist, dark chamber. Spore germination rate was determined microscopically by counting the number of germinated spores (of a total number of 200 spores per culture) at timed intervals.

Mauch et al. (1988b) demonstrated coordinated antifungal activity of chitinase and β-1,3-glucanase. Fungal spores (1×10^6 spores) were plated out on nutrient agar and were incubated for up to 24 h at room temperature to allow spore germination and initial vegetative growth. Sterile filter paper discs (4 mm diameter) were then placed on the agar surface and 40 μl of a test solution was applied to the discs. The plates were then incubated for up to 72 h and the effect on hyphal growth examined.

4.4.1.4 Subcellular Localization of Plant Enzymes

Benhamou et al. (1989, 1990) used antibodies raised against the two glycosidases to study the subcellular localization of each enzyme in fungus-infected plant tissues. Roots of tomato seedlings at the three-leaf stage (sensitive and resistant strains) were inoculated with a suspension of fungal microconidia (2×10^6 spores ml^{-1}). Control plants were treated with distilled water. Root samples (1 mm^3) were collected every 24 h after inoculation, fixed by immersion in 3% (v/v) glutaraldehyde with 0.1 M sodium cacodylate buffer, pH 7.2, for 2 h at 4 °C, then dehydrated in a graded ethanol series and embedded in Epon 812. Ultrathin tissue sections were collected on formvar-coated nickel grids and processed for immunogold labelling. The sections were floated on a drop of phosphate buffered saline (PBS), pH 7.2, containing 0.25% (w/v) ovalbumin for 5 min, transferred to a drop of normal goat serum for 60 min at room temperature, and incubated for 2 h at 37 °C with the primary antibody appropriately diluted in PBS-ovalbumin. Grids were rinsed with Tris-HCl, pH 8.2, containing 0.5 M NaCl and 1% (w/v) bovine serum albumin and incubated on a drop of colloidal gold (10 nm)-conjugated goat antiserum to rabbit immunoglobulins for 30–60 min at room temperature. The grids were then washed with PBS, rinsed with distilled water, contrasted with uranyl acetate and lead citrate, and examined electron microscopically. In both types of plants, enzyme deposition was correlated with pathogen distribution; the enzyme accumulating earlier in resistant plants than in susceptible ones.

4.4.2 Demonstration of Importance of Microbial Enzymes in Plant Pathogenesis

4.4.2.1 Induction of Microbial Enzymes

Jones et al. (1972) examined the secretion of several polysaccharide-degrading enzymes by *Fusarium oxysporum* f. sp. *lycopersici*. Unbuffered liquid medium (9 ml) containing 1% (w/v) isolated tomato cell walls was inoculated with 1 ml of a fungal spore suspension (1×10^6 spores) and the cultures were incubated on a rotary shaker (180 rpm) at 30 °C. Flasks were removed at 24 h intervals, insoluble material was separated from the medium by filtration through glass fibre filter paper and the filtrate was tested for enzyme activity.

Lin and Kolattukudy (1978), and Woloshuk and Kolattukudy (1986) used a similar approach to demonstrate cutinase appearance in a time-dependent manner on incubation of *Fusarium solani* f. sp. *pisi* spores in a medium containing cutin monomers or powdered cutin. Antibodies raised against the purified enzyme were used to confirm that the induced enzyme was cutinase. Moreover, addition of cycloheximide (an inhibitor of protein synthesis; $0.5–5.0\,\mu g\,ml^{-1}$) to spore suspensions 30 min prior to the addition of cutin powder demonstrated that induction of the enzyme involved de novo synthesis.

Involvement of pretranslational control in the induction of cutinase and pectate lyase synthesis was demonstrated by measuring the levels of RNA specific for the enzyme (Woloshuk and Kolattukudy 1986; Dean and Timberlake 1989), and by in vitro translation and cDNA cloning of polyadenylated RNA (Flurkey and Kolattukudy 1981; Solidary et al. 1984) following enzyme induction.

4.4.2.2 Induction of Disease Symptoms

Dickman et al. (1982) treated *Mycosphaerella* spp., a fungal parasite which is normally unable to penetrate intact papaya fruits, with a cutinase isolated from another species, then placed it on the surface of intact papaya fruits. This enabled the parasite to infect the fruit; host penetration being inhibited on addition of anti-cutinase antibodies.

A correlation between virulence and pathogenicity has also been demonstrated for the cutinases and pectinases (Kolattukudy 1985; Collmer and Keen 1986).

4.4.2.3 Prevention of Infection by Inhibition of Microbial Enzymes

Maiti and Kolattukudy (1979) used specific antibodies raised against the cutinase from *Fusarium solani* f. sp. *pisi*, and diisopropylfluorophosphate (DFP), a potent inhibitor of the enzyme, to prevent infection of the pea epicotyl by the pathogen. Conidial suspensions of the fungus were mixed with antibody or DFP, and 5 μl droplets of the suspension were placed on the surface of 1.2-cm sections of epicotyl from 6-day-old etiolated pea seedlings. The epicotyl sections were incubated for 72 h at 22 °C in the dark on moist filter paper in petri dishes. In a control experiment, the epicotyls were breached with a pin. The percentage of epicotyl segments infected, epitomized by the development of dark spots, was calculated. Inoculated areas were examined microscopically to check for germination and growth of the conidia. Inhibition of cutinase activity prevented fungal entry into intact plants.

4.4.2.4 *Use of Mutants*

Pectate lyase-deficient mutants of *Erwinia* spp. have been shown to be unable to cause maceration of plant tissues (Chatterjee and Starr 1977). In addition, an out mutation affecting pectate lyase (and cellulase) export but not synthesis brought about an inability to cause pathogenicity (Andro et al. 1984).

Conversely, insertion of a cutinase gene into the parasitic fungus, *Mycosphaerella* spp. enabled it to infect intact papaya fruits (Dickman et al. 1989). A construct containing the cutinase coding region and extensive portions of the 5′- and 3′-flanking regions from the *Fusarium* genome was transferred into *Mycosphaerella* spp. Transformants with cutinase activity were used in the infection experiments. The ability of the cutinase-positive mutants to invade intact fruits was reversed by treatment with cutinase inhibitors or antibodies.

The induction of pathogenicity in non-pathogenic microorganisms has also been demonstrated for the pectinases. *Escherichia coli* clones producing pectinases have been shown to be able to macerate plant tissues (Keen et al. 1984; Collmer et al. 1985).

4.4.2.5 *Subcellular Localization of Microbial Enzymes*

Shaykh et al. (1977) demonstrated the presence of cutinase in the region of fungal attack with ferritin-conjugated antibodies prepared against purified cutinase. A fungal spore suspension was placed on the intact surface of pea stem sections and the progress of penetration of the germinating spores was monitored by scanning electron microscopy. At the time of penetration, the area was treated with the antibodies and transmission electron microscopy was used to demonstrate the association of ferritin with the plant cuticle.

References

Abu-Goukh AA, Greve LC, Labavitch JM (1983) Purification and partial characterisation of "Bartlett" pear fruit polygalacturonase inhibitors. Physiol Plant Pathol 23:111–122

Akazawa T, Hara-Nishimura I (1985) Topographic aspects of biosynthesis, extracellular secretion and intracellular storage of proteins in plant cells. Annu Rev Plant Physiol 36:441–472

Albertsson PA (1986) Partition of cell particles and macromolecules, 3rd edn. J Wiley & Sons, New York

Andrews AT (1986) Electrophoresis – theory, techniques, and biochemical and clinical applications. Clarendon, Oxford

Andro T, Chambost J-P, Kotoujansky A, Cattaneo J, Bertheau Y, Barras F, Van Gijsegem F, Coleno A (1984) Mutants of *Erwinia chrysanthemi* defective in secretion of pectinase and cellulase. J Bacteriol 160:1199–1203

Bailey JA, Mansfield JW (eds) (1982) Phytoalexins. J Wiley & Sons, New York

Beisiegel U (1986) Protein blotting. Electrophoresis 7:1–18
Bell AA (1981) Biochemical mechanisms of disease resistance. Annu Rev Plant Physiol 32:21–81
Benhamou N, Grenier J, Asselin A, Legrand M (1989) Immunogold localization of β-1,3-glucanases in two plants infected by vascular wilt fungi. Plant Cell 1:1209–1221
Benhamou N, Joosten MHAJ, De Wit PJGM (1990) Subcellular localization of chitinase and of its potential substrate in tomato root tissues infected by *Fusarium oxysporum* f. sp. *radicis-lycopersici*. Plant Physiol 92:1108–1120
Biely P, Markovic O, Mislovicová D (1985) Sensitive detection of endo-1,4-β-glucanases and endo-1,4-β-xylanases in gels. Anal Biochem 144:147–151
Boller T (1986) Role of proteolytic enzymes in interactions of plants with other organisms. In: Dalling M (ed) Plant proteolytic enzymes. CRC Press, Boca Raton, Florida
Boller T (1987) Hydrolytic enzymes in plant disease resistance. In: Kosuge T, Nester EW (eds) Plant-microbe interactions, molecular and genetic perspectives, vol 2. Macmillan, New York, pp 385–414
Boller T, Mauch F (1988) Colorimetric assay for chitinase. Methods Enzymol 161:430–435
Boller T, Métraux JP (1988) Extracellular localization of chitinase in cucumber. Physiol Mol Plant Pathol 33:11–16
Boller T, Gehri A, Mauch F, Vögeli U (1983) Chitinase in bean leaves: induction by ethylene, purification, properties, and possible function. Planta 157:22–31
Bradford MM (1976) A rapid and sensitive method for the quantitation of microgram quantities of protein utilizing the principle of protein-dye binding. Anal Biochem 72:248–254
Broekaert WF, Van Parijs J, Allen AK, Peumans WJ (1988) Comparison of some molecular, enzymatic and antifungal properties of chitinases from thorn-apple, tobacco and wheat. Physiol Mol Plant Pathol 33:319–331
Broekaert WF, Van Parijs J, Leyns F, Joos H, Peumans WJ (1989) A chitin-binding lectin from stinging nettle rhizomes with antifungal properties. Science 245:1100–1102
Brown RE, Jarvis KL, Hyland KJ (1989) Protein measurement using bicinchoninic acid: elimination of interfering substances. Anal Biochem 180:136–139
Burnette WN (1981) "Western blotting": electrophoretic transfer of proteins from sodium dodecyl sulfate – polyacrylamide gels to unmodified nitrocellulose and radiographic detection with antibody and radioiodicated protein A. Anal Biochem 112:195–203
Cabib E (1988) Assay for chitinase using tritiated chitin. Methods Enzymol 161:424–426
Canevascini G, Gattlen C (1981) A comparative investigation of various cellulase assay procedures. Biotechnol Bioeng 23:1573–1590
Chatterjee AK, Starr MP (1977) Donor strains of the soft-rot bacterium *Erwinia chrysanthemi* and conjugational transfer of the pectolytic capacity. J Bacteriol 132:862–862
Chavira R, Burnett TJ, Hageman JH (1984) Assaying proteinases with Azocoll. Anal Biochem 136:446–450
Clancy FG, Coffey D (1977) Acid phosphatase and protease release by the insectivorous plant *Drosera rotundifolia*. Can J Bot 55:480–488
Collmer A (1987) Pectic enzymes and bacterial invasion of plants. In: Kosuge T, Nester EW (eds) Plant-microbe interactions, molecular and genetic perspectives, vol 2. Macmillan, New York, pp 253–284
Collmer A, Keen NT (1986) The role of pectic enzymes in plant pathogenesis. Annu Rev Phytopathol 24:383–409
Collmer A, Schoedel C, Roeder DL, Ried JL, Rissler JF (1985) Molecular cloning in *Escherichia coli* of *Erwinia chrysanthemi* genes encoding multiple forms of pectate lyase. J Bacteriol 161:913–920
Collmer A, Ried JL, Mount MS (1988) Assay methods for pectic enzymes. Methods Enzymol 161:329–335
Cornish-Bowden A (1979) Fundamentals of enzyme kinetics. Butterworths, London
Coté F, Letarte J, Grenier J, Trudel J, Asselin A (1989) Detection of β-1,3-glucanase activity after native polyacrylamide gel electrophoresis: application to tobacco pathogenesis-related proteins. Electrophoresis 10:527–529

Coughlan MP (1988) Staining techniques for the detection of the individual components of cellulolytic enzyme systems. Methods Enzymol 160:135–144
Cutting JA (1984) Gel protein stains: phosphoproteins. Methods Enzymol 104:451–455
Darbre A (ed) (1986) Practical protein chemistry – a handbook. Wiley & Sons, New York
Darvill AG, Albersheim P (1984) Phytoalexins and their elicitors – a defence against microbial infection in plants. Annu Rev Plant Physiol 35:243–275
Dean RA, Timberlake WE (1989) Regulation of the *Aspergillus nidulans* pectate lyase gene (pelA). Plant Cell 1:275–284
Dey PM, Campillo ED (1984) Biochemistry of the multiple forms of glycosidases in plants. Adv Enzymol Relat Areas Mol Biol 56:141–249
Dickman MB, Patil SS, Kolattukudy PE (1982) Purification and characterization of an extracellular cutinolytic enzyme from *Colletotrichum gloeosporioides* on *Carica papaya*. Physiol Plant Pathol 29:333–347
Dickman MB, Podila GK, Kolattukudy PE (1989) Insertion of cutinase gene into a wound pathogen enables it to infect intact host. Nature 342:446–448
Doerner KC, White BA (1990) Detection of glycoproteins separated by nondenaturing polyacrylamide gel electrophoresis using the periodic acid-Schiff stain. Anal Biochem 187:147–150
Dunn MJ (ed) (1986) Gel electrophoresis of proteins. Wright, Bristol
Edreva AM, Georgieva ID (1980) Biochemical and histochemical investigations of α- and β-glucosidase activity in an infectious disease, a physiological disorder and in senescence of tobacco leaves. Physiol Plant Pathol 17:237–243
Eriksson K-E, Blanchette RA, Ander P (1990) Microbial and enzymatic degradation of wood and wood components. Springer, Berlin Heidelberg New York
Eriksson S, Bhikhabhai R, Hammar L (1989) Analysis of glycoproteins – lectin binding of HIV glycoproteins. PhastSystem Application File No. 301, Pharmacia LKB Biotechnology, Uppsala, Sweden
Etzler ME (1985) Plant lectins: molecular and biological aspects. Annu Rev Plant Physiol 36:209–234
Evans CH, Ridella JD (1984) An evaluation of fluorometric proteinase assays which employ fluorescamine. Anal Biochem 142:411–420
Flurkey WH, Kolattukudy PE (1981) In vitro translation of cutinase in mRNA: evidence for a precursor form of an extracellular fungal enzyme. Arch Biochem Biophys 212: 154–161
Franks F (ed) (1988) Characterisation of proteins. Human Press, Clifton, New Jersey
Friedenauer S, Berlet HH (1989) Sensitivity and variability of the Bradford protein assay in the presence of detergents. Anal Biochem 178:263–268
Gander JE (1984) Gel protein stains: glycoproteins. Methods Enzymol 104:447–451
Gershoni JM (1988) Protein blotting: a manual. Methods Biochem Anal 33:1–58
Ghose TK (ed) (1984) Measurement of cellulase activities. Commission on Biotechnology, International Union of Pure and Applied Chemistry, Biochemical Engineering Research Centre, Indian Institute of Technology, New Delhi, India
Glenn AR (1976) Production of extracellular proteins by bacteria. Annu Rev Microbiol 30:41–62
Glenney J (1986) Antibody probing of Western blots which have been stained with indian ink. Anal Biochem 156:315–319
Glumoff T, Harvey PJ, Molinari S, Goble M, Frank G, Palmer JM, Smit JDG, Leisola MSA (1990) Lignin peroxidase from *Phanerochaete chrysosporium*. Molecular and kinetic characterization of isozymes. Eur J Biochem 187:515–520
Goksoyr J, Eriksen J (1980) Cellulases. In: Rose AH (ed) Microbial enzymes and bioconversions, Economic microbiology, vol 5. Academic Press, New York, pp 283–330
Hahlbrock K, Scheel D (1989) Physiology and molecular biology of phenylpropanoid metabolism. Annu Rev Plant Physiol Plant Mol Biol 40:347–369
Hames BD, Rickwood D (eds) (1989) Gel electrophoresis of proteins – a practical approach, 2nd edn. IRL Press, Oxford

Harris ELV, Angal S (eds) (1989) Protein purification methods – a practical approach. IRL Press, Oxford
Heftmann E (ed) (1983) Chromatography. Elsevier, Amsterdam
Heimgartner U, Kozulíc B, Mosbach K (1989) Polyacrylic polyhydrazides as reagents for detection of glycoproteins. Anal Biochem 181:182–189
Heslop-Harrison Y, Heslop-Harrison J (1981) The digestive glands of *Pinguicula*: structure and cytochemistry. Ann Bot 47:293–319
Heukeshoven J, Dernick R (1988) Improved silver staining procedure for fast staining in PhastSystem development unit. I. Staining of sodium dodecyl sulfate gels. Electophoresis 9:28–32
Hill HD, Straka JG (1988) Protein determination using bicinchoninic acid in the presence of sulfhydryl reagents. Anal Biochem 170:203–208
Howe JG, Hershey JWB (1981) A sensitive immunoblotting method for measuring protein synthesis initiation factor levels in lysates of *Escherichia coli*. J Biol Chem 256:12836–12839
International Union of Biochemistry (1984) Enzyme nomenclature. Academic Press, London
Janson J-C, Rydén L (eds) (1989) Protein purification – principles, high resolution methods, and application. VCH Publishers, New York
Jensen HS (1984) Adsorbed soluble [^{3}H]elastin as substrate for proteinase activity. A new microassay technique. Biochem J 218:645–648
Jones TM, Anderson AJ, Albersheim P (1972) Host-pathogen interactions. IV. Studies on the polysaccharide-degrading enzymes secreted by *Fusarium oxysporum* f. sp. *lycopersici*. Physiol Plant Pathol 2:153–166
Joosten MHAJ, De Wit PJGM (1989) Identification of several pathogenesis-related proteins in tomato leaves inoculated with *Cladosporium fulvum* (syn *Fulvia fulva*) as 1,3-β-glucanases and chitinases. Plant Physiol 89:945–951
Kalisz HM (1988) Microbial proteinases. Adv Biochem Eng Biotechnol 36:1–65
Kauffmann S, Legrand M, Geoffroy P, Fritig B (1987) Biological function of pathogenesis-related proteins: four PR proteins of tobacco leaves have 1,3-β-glucanase activity. EMBO J 6:3209–3212
Keen NT, Dahlbeck D, Staskawicz B, Belser W (1984) Molecular cloning of pectate lyase genes from *Erwinia chrysanthemi* EC16 and their high level expression in *Escherichia coli*. J Bacteriol 159:825–831
Kollatukudy PE (1985) Enzymatic penetration of the plant cuticle by fungal pathogens. Annu Rev Phytopathol 23:223–250
Kollatukudy PE, Purdy RE, Maiti IB (1981) Cutinases from fungi and pollen. Methods Enzymol 71:652–664
Kombrik E, Schröder M, Hahlbrock K (1988) Several pathogenesis-related proteins in potato are β-1,3-glucanases and chitinases. Proc Natl Acad Sci USA 85:782–786
Künnecke W, Kalisz HM, Schmid RD (1989) Flow injection zymography – a novel procedure for the on-line detection of enzyme activity. Anal Lett 22:1471–1484
Lamb CJ, Lawton MA, Dron M, Dixon RA (1989) Signals and transduction mechanisms for activation of plant defences against microbial attack. Cell 56:215–224
Legler G, Müller-Platz CM, Mentges-Hettkamp M, Pflieger G, Jülich E (1985) On the chemical basis of the Lowry protein determination. Anal Biochem 150:278–287
Legrand M, Kauffmann S, Geoffroy P, Fritig B (1987) Biological function of pathogenesis-related proteins: four tobacco pathogenesis-related proteins are chitinases. Proc Natl Acad Sci USA 84:6750–6754
Lin TS, Kollatukudy PE (1978) Induction of a biopolyester hydrolase (cutinase) by low levels of cutin monomers in *Fusarium solani* f. sp. *pisi*. J Bacteriol 133:942–951
Lowry OH, Rosebrough NJ, Farr AL, Randall RJ (1951) Protein measurement with the Folin phenol reagent. J Biol Chem 193:265–275
MacKenzie CR, Williams RE (1984) Detection of cellulase and xylanase activity in isoelectric-focused gels using agar substrate gels supported on plastic film. Can J Microbiol 30:1522–1525

Maiti IB, Kollatukudy PE (1979) Prevention of fungal infection of plants by specific inhibition of cutinase. Science 205:507–508
Mauch F, Hadwiger LA, Boller T (1984) Ethylene: symptom, not signal for the induction of chitinase and β-1,3-glucanase in pea pods by pathogens and elicitors. Plant Physiol 76:607–611
Mauch F, Hadwiger LA, Boller T (1988a) Antifungal hydrolases in pea tissue. I. Purification and characterisation of two chitinases and two β-1,3-glucanases differentially regulated during development and in response to fungal infection. Plant Physiol 87:325–333
Mauch F, Mauch-Mani B, Boller T (1988b) Antifungal hydrolases in pea tissue. II. Inhibition of fungal growth by combinations of chitinase and β-1,3-glucanase. Plant Physiol 88:936–942
McHale AP, Hackett TJ, McHale LM (1989) Specific zymogram staining procedure for the exocellobiohydrolase components produced by *Talaromyces emersonii* CBS 814.70. Enzyme Microb Technol 11:17–20
Mellor RB, Mörschel E, Werner D (1984) Legume root response to symbiotic infection: enzymes of the peribacteroid space. Z Naturforsch 39c:123–125
Miller GL, Blum R, Glennon WE, Burton AL (1960) Measurement of carboxymethylcellulase activity. Anal Biochem 1:127–132
Munoz G, Marshall S, Cabrera M, Horvat A (1988) Enhanced detection of glycoproteins in polyacrylamide gels. Anal Biochem 170:491–494
Nakamura K, Tanaka T, Kuwahara A, Takeo K (1985) Microassay for proteins on nitrocellulose filter using dye-staining procedure. Anal Biochem 148:311–319
North MJ (1982) Comparative biochemistry of the proteinases of eucaryotic microorganisms. Microbiol Rev 46:308–340
Ohtakara A (1988) Viscosimetric assay for chitinase. Methods Enzymol 161:426–430
Ostoa-Saloma P, Ramírez J, Perez-Montfort R (1989) Measurement of casein digestion by a fluorometric method. Anal Biochem 176:239–243
Page MI (ed) (1984) The chemistry of enzyme action. Elsevier, Amsterdam
Payne JW (ed) (1980) Microorganisms and nitrogen sources. J Wiley & Sons, London
Peterson GL (1979) Review of the Folin phenol protein quantitation method of Lowry, Rosebrough, Farr and Randall. Anal Biochem 100:201–220
Priest FG (1984) Extracellular enzymes. Aspects of microbiology ser 9. Van Nostrand Reinhold (UK) Co Ltd, Berkshire, England
Prieur B, Russo-Marie F (1988) An automated western blot analysis using the PhastSystem. Anal Biochem 172:338–343
Racusen D (1984) Lipid acyl hydrolase of patatin. Can J Bot 62:1640–1644
Renganathan V, Usha SN, Lindenburg F (1990) Cellobiose-oxidizing enzymes from the lignocellulose-degrading basidomycete *Phyanerochaete chrysosporium*: interaction with microcrystalline cellulose. Appl Microbiol Biotechnol 32:609–613
Rexová-Benková L, Markovic O (1976) Pectic enzymes. Adv Carbohydr Chem Biochem 33:323–385
Ried JL, Collmer A (1985) Activity stain for rapid characterisation of pectic enzymes in isoelectric focusing and sodium dodecyl sulfate-polyacrylamide gels. Appl Environ Microbiol 50:615–622
Ried JL, Collmer A (1986) Comparison of pectic enzymes produced by *Erwinia chrysanthemi*, *Erwinia carotovora*, subsp. *carotovora*, and *Erwinia carotovora*, subsp. *atroseptica*. Appl Environ Microbiol 52:305–310
Rinderknecht H, Geokas MC, Silverman P, Haverback BJ (1968) A new sensitive method for the determination of proteolytic activity. Clin Chim Acta 21:197–203
Roby D, Esquerre-Tugaye M-T (1987) Induction of chitinases and of translatable mRNA for these enzymes in melon plants infected with *Colletotrichum lagenarium*. Plant Sci 52:175–185
Rodríguez-Vico F, Martínez-Cayuela M, García-Pergrín E, Ramírez H (1989) A procedure for eliminating interferences in the Lowry method of protein determination. Anal Biochem 183:275–278

Rohringer R, Holden DW (1985) Protein blotting: detection of proteins with colloidal gold, and of glycoproteins and lectins with biotin-conjugated and enzyme probes. Anal Biochem 144:118–127

Rombouts FM, Pilnik W (1980) Pectic enzymes. In: Rose AH (ed) Microbial enzymes and bioconversions. Economic microbiology, vol 5. Academic Press, New York, pp 227–282

Root DD, Reisler E (1989) Copper iodide staining of protein blots on nitrocellulose membranes. Anal Biochem 181:250–253

Ryan CA (1984) Systemic responses to wounding. In: Kosuge T, Nester EW (eds) Plant-microbe interactions, vol 1. Macmillan, New York, pp 307–320

Sacher JA, Tseng J, Williams R, Cabello A (1982) Wound-induced RNAse activity in sweet potato. Plant Physiol 69:1060–1065

Saddler JN, Khan AW (1981) Cellulolytic enzyme system of *Acetivibrio cellulolyticus*. Can J Microbiol 27:288–294

Schlumbaum A, Mauch F, Vögeli U, Boller T (1986) Plant chitinases are potent inhibitors of fungal growth. Nature 324:365–367

Schwarz WH, Bronnenmeier K, Gräbnitz F, Staudenbauer WL (1987) Activity staining of cellulases in polyacrylamide gels containing mixed linkage β-glucans. Anal Biochem 164:72–77

Scopes RK (1987) Protein purification – principles and practise, 2nd edn. Springer Berlin Heidelberg New York

Segel IH (1975) Enzyme kinetics. Wiley & Sons, New York

Sexton R, Roberts JA (1982) Cell biology of abscission. Annu Rev Plant Physiol 33: 133–162

Shaykh M, Soliday C, Kolattukudy PE (1977) Proof for the production of cutinase by *Fusarium solani* f. *pisi* during penetration into its host, *Pisum sativum*. Plant Physiol 60:170–172

Smith PK, Krohn RI, Hermanson GT, Mallia AK, Gartner FH, Provenzano MD, Fujimoto EK, Goeke NM, Olson BJ, Klenk DC (1985) Measurement of protein using bicinchoninic acid. Anal Biochem 150:76–85

Solheim B, Fjellheim KE (1984) Rhizobial polysaccharide-degrading enzymes from roots of legumes. Physiol Plant 62:11–17

Soliday CL, Flurkey WH, Okita TW, Kolattukudy PE (1984) Cloning and structure determination of cDNA for cutinase, an enzyme involved in fungal penetration of plants. Proc Natl Acad Sci USA 81:3939–3943

Spiro RG (1966) Analysis of sugars found in glycoproteins. Methods Enzymol 8:3–26

Stoessl A (1983) Secondary plant metabolites in preinfectional and postinfectional resistance. In: Bailey JA, Deverall BJ (eds) The dynamics of host defence. Academic Press, New York, pp 71–122

Suelter CH (1985) A practical guide to enzymology. J Wiley & Sons, New York

Switzer RC, Merril CR, Shifrin S (1979) A highly sensitive silver stain for detecting proteins and peptides in polyacrylamide gels. Anal Biochem 98:231–237

Tien M, Kirk TK (1988) Lignin peroxidase of *Phanerochaete chrysosporium*. Methods Enzymol 161:238–249

Trudel J, Asselin A (1989) Detection of chitinase activity after polyacrylamide gel electrophoresis. Anal Biochem 178:362–366

Veluthambi K, Mahadevan S, Maheshwari R (1981) Trehalose toxicity in *Cuscuta reflexa*. Plant Physiol 68:1369–1374

Vögeli U, Meins F, Boller T (1988) Co-ordinated regulation of chitinase and β-1,3-glucanase in bean leaves. Planta 174:364–372

Walter H, Brooks DE, Fisher D (eds) (1985) Partitioning in aqueous two-phase systems: theory, methods, uses, and applications in biotechnology. Academic Press, London

Wargo PM (1975) Lysis of the cell wall of *Armillaria mellea* by enzymes from forest trees. Physiol Plant Pathol 5:99–105

Weber K, Osborn M (1969) The reliability of molecular weight determinations by dodecyl sulfate-polyacrylamide gel electrophoresis. J Biol Chem 244:4406–4412

Westergaard JL, Hackbarth C, Treuhaft MW, Robterts RC (1980) Detection of proteinases in electrophoretograms of complex mixtures. J Immunol Methods 34:167–175

Woloshuk CP, Kollatukudy PE (1986) Mechanism by which contact with plant cuticle triggers cutinase gene expression in the spores of *Fusarium solani* f. sp. *pisi*. Proc Natl Acad Sci USA 83:1704–1708

Wong P, Barbeau A, Roses AD (1985) A method to quantitate Coomassie blue-stained proteins in cylindrical polyacrylamide gels. Anal Biochem 150:288–293

Wood TM, McCrae SI (1979) Synergism between enzymes involved in the solubilisation of native cellulose. Adv Chem Ser 181:181–209

Wood WA, Kellogg ST (eds) (1988) Methods in enzymology, vol 160. Academic Press, New York

Analysis of Toxic Extracellular Polysaccharides

K. Wydra and K. Rudolph

1 Introduction

Extracellular polysaccharides are derived from plant cell walls or extruded by secretory vesicles. Pectic substances such as arabinogalactans, arabinans and galactans belong to the class of extracellular polysaccharides, as do glycoconjugates, such as arabinogalactan proteins (AGPs), some hydroxyproline-rich glycoproteins (HRGPs) and some lectins.

Although the biological function is not clearly understood for some of these extracellular polysaccharides, "toxic" properties have been described. In this Chapter (Sect. 2) "toxicity" is defined in terms of phytopathological aspects, and thus describes the role of extracellular polysaccharides as being part of the defence response of the plant in interaction with pathogens (Rudolph et al. 1989).

Since the extracellular polysaccharides form a group of polymers with a great diversity in properties and structure, procedures for the analysis have to be selected from a wide range of methods. Methods for extraction and solubilization, as well as purification and characterization, have to be selected according to the properties of the polymers to be analyzed (Dey

Table 1. Carbohydrate polymers of plant cell walls (Aspinall 1980)

General category	Structural classification
Cellulose	β-1,4-D-Glucan
Pectic substances	Galacturonans and rhamnogalacturonans
	Arabinans, galactans and arabinogalactans I[a]
Hemicelluloses	Xylans [including arabinoxylans and (4-O-methyl) glucuronoxylans]
	Glucomannans and galactoglucomanans
	Xyloglucans
	β-D-Glucans (1,3 and 1,4)
Other polysaccharides	β-1,3-linked D-glucans (callose)
	Arabinogalactans II[a]
	Glucuronomannans
Glycoproteins	

[a] Arabinogalactans of type I are characterized by essentially linear β-1,4-D-galactan chains, whereas those of type II contain highly branched interior chains with 1,3- and 1,6-intergalactose linkages.

1990). Thus, the properties influencing extraction and purification techniques (Sect. 3) will be discussed before describing the methods in detail.

In the first part of specific methods (Sect. 4), those for extraction and solubilization for each group of glycoproteins and polysaccharides (lectins, HRGPs, AGPs and pectic polysaccharides) are presented. In the second part (Sect. 5) "Purification and characterization", the lectin techniques, affinity chromatography and precipitation techniques are described in detail. Physicochemical methods for structural analysis are briefly mentioned and references for further studies are given.

In a third part (Sect. 6), cytochemical techniques, classical and modern electron microscopical methods, which elucidate the host-pathogen relationship, are described. Using specific staining methods based on enzyme-, lectin- or antibody-gold complexes, the molecular nature of the interaction between cell wall polymers of the plant and surface polymers of the pathogen can be elucidated.

Abbreviations

AGP	arabinogalactan protein
api*f*	apiofuranosyl
ara*f*	arabinofuranosyl
ara*p*	arabinopyranosyl
asn	asparagine
BSL	*Bandeira simplicifolia* lectin
CM	carboxymethyl
^{13}C-n.m.r.	nuclear magnetic resonance spectroscopy
CNBr	cyanogen bromid
DEAE	diethylaminoethyl
EAL	*Erythrina arborescens* lectin
EDTA	ethylenedinitrolotetraacetic acid tetrasodium salt
EIL	*Erythrina indica* lectin
EPS	extracellular polysaccharides
f.a.b.	fast atom bombardment
fuc*p*	fucopyranosyl
gal	galactose
galA	galacturonic acid
g.l.c.	gas liquid chromatography
glcNAc	N-acetyl-glucosamine
HPLAC	High Performance Liquid Affinity Chromatography
HR	hypersensitive reaction
HRGP	hydroxyproline-rich glycoprotein
hyp	hydroxyproline
IEC	ion exchange chromatography
isodityr	isodityrosine
lys	lysine

man	mannose
m.s.	mass spectrometry
PATAg	periodic acid thiocarbohydrazide silver proteinate
PAW	PhOH-HOAc-H_2O
PBS	phosphate buffered saline
PHA	phytohemagglutinin (*Phaseolus vulgaris*)
RCA	*Ricinus communis* agglutinin
rha	rhamnose
SBA	soybean agglutinin
ser	serine
tyr	tyrosine
WGA	wheat germ agglutinin
xyl*p*	xylopyranosyl

2 General Description of Plant Cell Wall Polysaccharides

2.1 Classification

The plant cell wall is composed of molecules with extraordinary structural complexity. Polysaccharides and glycoproteins form the main part of the dry cell wall and are present in a ratio of approximately 90% polysaccharide to 10% protein (McNeil et al. 1984). Growing plant cells are surrounded by the primary cell wall, while mature plant cells are surrounded by the additional secondary cell wall.

Both the primary and secondary wall contain the same type of polysaccharides, mainly cellulose, hemicellulose and pectic substances (Table 1), besides lignin, suberin and protein (Cassab and Varner 1988). Pectic substances are primarily a group of closely associated polysaccharides from the primary cell wall and intercellular regions of the higher plant. The mannose-rich polysaccharides [mannans, glucomannans and galactomannans (Cartier et al. 1988)], the xylans and the noncarbohydrate-polymer lignin are mainly secondary wall components. Galactoglucomannans are one of the major components of cell walls of woody tissues of gymnosperms and angiosperms, and have also been isolated from primary cell walls of suspension-cultured tobacco cells (Eda et al. 1985). Xyloglucans from a variety of sources have been characterized (Eda and Kato 1978); glucuronoxylans (Eda et al. 1977) and arabinoxyloglucans have been isolated from tobacco leaves (Siddiqui and Rosa 1985).

2.2 Localization

Many of these polysaccharides occur extracellularly: (1) arabinogalactans and arabinogalactan proteins (Aspinall 1973; Ghanekar and Pérombelon

1980; Fincher et al. 1983), which are primarily localized in the extracellular matrix and in gums and exudates (Showalter and Varner 1989), and released in response to wounding and infection (Stuart and Varner 1980); (2) cell wall-derived pectic material (Brown and Mansfield 1988), such as galacturonans (McNeil et al. 1984); and (3) pectic fragments, as well as glycoproteins (Yamaoka and Sato 1977), especially hydroxyproline-rich glycoproteins (Benhamou et al. 1990a,b).

2.3 Function (Toxicity)

The primary role of the cell-wall polysaccharides is to stabilize the plant cell and to provide a physical and chemical barrier between the cell and its environment. Structural variations of polysaccharides, although belonging to the same general family, influence their physical properties and hence their biological functions. Due to their enormous variability and the peripheral location, the cell-wall macromolecules (pectins, cellulose, xyloglucans and glycoproteins) are thought to be involved in a number of ways in several reaction chains during the resistance response, including specific recognition, degradation and defence events (Rigas and Osgood 1955; Sequeira 1978; Anderson 1982; Keen and Holliday 1982; Hahn et al. 1989).

Extracellular polysaccharides and glycoproteins may be directly toxic to cells or organisms. They may also initiate plant responses mediated by surface-localized complementary mechanisms, such as accumulation of phytoalexins (Hahn et al. 1981; Nothnagel et al. 1983) and/or phenols (Carrasco et al. 1978), production of proteinase inhibitors (Bishop et al. 1984; Walker-Simmons et al. 1984), lignification (Robertsen 1986) and necrosis (Mussell 1973; Cervone et al. 1987).

2.3.1 Direct Toxicity

2.3.1.1 Toxicity to Pathogens

Examples for direct toxicity are reported for lectins (Table 2). The high-molecular weight toxins ricin (*Ricinus communis*) and abrin (*Abrus precatorius*; Callow 1977; Olsner and Pihl 1978; Kauss 1981; Gabius et al. 1988) are toxic to animals. Phytohemagglutinin (PHA, *Phaseolus vulgaris*) shows mitogenic activity against animal cells (Sharon and Lis 1972; Albersheim and Anderson-Prouty 1975; Sharon 1977). The presence of toxic constituents in plant organs may be regarded as providing some form of protection against animal attack, e.g., the larvae of the bruchid beetle are killed by hemagglutinin found in black bean (*Phaseolus vulgaris*; Janzen et al. 1976).

Toxicity to fungi is manifested by inhibition of spore germination, mycelial and germ-tube growth, which may be accompanied by agglutination

Table 2. Effects of lectins on animal cells (Callow 1977)

1. Agglutination of erythrocytes and other cells
2. Preferential agglutination of tumour cells (Burger and Goldberg 1967; Inbar and Sachs 1969)
3. Induction of mitosis (Nowell 1960)
4. Insulin-like effects on fat cells (Cuatrecasas and Tell 1973)
5. Induction of platelet release reaction (Tollefson et al. 1974)
6. Induction of insulin release from pancreatic cells (Lockhart-Ewart et al. 1973)
7. Inhibition of phagocytosis by granulocytes (Berlin 1972)
8. Inhibition of fertilization of ovum by sperm (Nicolson 1974)
9. Inhibition of protein synthesis (Olsner et al. 1974)

(After Kornfeld et al. 1974)

of fungal spores (Kojima and Uritani 1974; Sharon 1977), e.g., the effects of wheat germ agglutinin on *Trichoderma viride* (Mirelman et al. 1975; Weinhold and Hancock 1980). Numerous studies report toxic root exudates (e.g., exudates from Chinese cabbage against *Pythium butleri*; Tripathi and Grover 1978), which suggest these compounds may be involved in plant resistance to pathogens. In most cases the root exudates consist of non-carbohydrate components such as phenolic compounds and amino acids. Toxic substances have also been obtained from leaf washings (Weinhold and Hancock 1980).

2.3.1.2 Toxicity to Plant Cells: Hypersensitive Reaction

Directly toxic compounds associated with or released from plant cell walls are thought to be involved in the hypersensitive reaction (HR) of plants against pathogens (Brown and Mansfield 1988; Doares et al. 1989). Glycoproteins (Brown and Kimmins 1973), toxic fragments from the primary cell wall and pectic polysaccharides (Yamazaki et al. 1983) have been found to be related to a quick plant cell death, thus preventing the spread of the pathogen. Growing evidence from several studies supports the concept that pectic components of the plant cell wall, which are not directly toxic to the pathogen, accumulate after the infection and act as elicitors. Thus, these substances initiate metabolic changes in plants, which lead to necrotization (resistance reaction; Cervone et al. 1989; Kopp et al. 1989; Benhamou et al. 1990; Tepper and Anderson 1990).

2.3.2 Indirect Toxicity: Role in Resistance Response

Extracellular or cell wall-associated polymers are obviously involved in the recognition as well as in the subsequent defence response of plants towards pathogens or other unspecific stress factors (El-Banoby and Rudolph

1980, 1981; Wydra and Rudolph 1990). Two groups of compounds can be differentiated:

1. Glycoproteins: cell surface or secreted lectins and agglutinins, hydroxyproline-rich glycoproteins (HRGPs), arabinogalactan proteins (AGPs) and "extensins";
2. Polysaccharides: arabinogalactans and other pectic polysaccharides and gums.

Specific recognition may be evoked by the interaction of complementary macromolecules present at the surfaces of the two organisms, which results in a host response preventing multiplication of the potential pathogen.

However, it is known that an unspecific accumulation of cell wall-associated glycoproteins, e.g., extensins (Monro et al. 1976; Cassab and Varner 1988), and of gum exudates as a response to stress and wounding, including pathogen-infection, prevents spread of infection. Moisture loss is inhibited by plugging the site of injury with pectic substances.

The structural variability of complex carbohydrates is greater than that of proteins or nucleic acids, on a weight basis (Keen and Holliday 1982). Therefore, elucidating the structure of extracellular cell wall polymers with regard to their potential for specific or nonspecific interactions with macromolecules or small ligands is an exciting challenge.

3 Characterization of Polysaccharides

Soluble, extracellular polysaccharides can be obtained without the necessity of using chemical extraction for release from insoluble cell wall material. Often polysaccharides or fragments from cell walls are isolated from suspension-cultured cells or the intercellular space of plant tissue (Terry et al. 1981; Terry and Jones 1981; Yamazaki et al. 1983; McNeil et al. 1984; Cartier et al. 1988; Suzuki et al. 1990). However, even if mild isolation and fractionation procedures are applied, structural modifications may occur during isolation. Thus, saponification of ester-groups, base-catalyzed degradations by β-elimination from methyl galacturonate residues in pectins, cleavage of covalent interpolymeric linkages in the cell wall and the O-glycosidic linkage of glycans to protein under alkaline conditions and of glycosidic bonds during extraction under acidic conditions have been reported (Aspinall 1980).

3.1 Glycoproteins

Glycoproteins occur as surface-located components of the plant cell wall. This group includes the classical lectins, agglutinins, hydroxyproline-rich glycoproteins (HRGPs), covalently bound wall glycoproteins (extensins) and arabinogalactan-proteins. These compounds may play a decisive role in highly specific interactions, e.g., between rhizobia and legumes (Albersheim and Wolpert 1977; Bauer et al. 1977), in the attachment of the crown gall bacteria (*Agrobacterium tumefaciens*) to its host (Lippincott and Lippincott 1977), as bacteria-agglutinating agents in the intercellular space (Huang et al. 1975; Mew et al. 1984), and as cell wall-associated HRGPs accumulating in response to pathogen attack (Lamport and Catt 1981).

Glycoproteins cannot be isolated by a uniform procedure, due to variability in amount and properties of carbohydrate moieties (Montreuil et al. 1986). The type of extraction and purification to be used should be chosen according to the purpose of the investigation. In order to avoid irreversible modification of the molecular structure, mild conditions for extraction and fractionation should be used, especially when studying the biological activity or physico-chemical properties such as viscosity (Burdon and van Knippenberg 1985). Since cell wall glycoproteins are always isolated under degradative conditions (Fincher et al. 1983), the in vivo structures of these glycoproteins are not fully elucidated.

Many glycoproteins are freely soluble in dilute salt solutions, while extraction and solubilization of cell wall- or membrane-bound glycoproteins require more drastic methods (Callow 1975; Burdon and Knippenberg 1985). Glycoproteins which are associated with the membrane surface are solubilized by varying parameters such as pH and ionic strength. More firmly associated glycoproteins require the use of detergents, strong chaotropes or organic solvents for solubilization. The latter can cause denaturation and loss of function associated with the peptide chain, but may be used when primary structures are to be analyzed (Burdon and Knippenberg 1985).

3.1.1 Lectins

The majority of lectins are easily soluble and generally extracted from plant tissues (seeds, leaves, roots, tubers, fruits or bark) by homogenizing the tissue in physiological saline or dilute buffers (Fincher et al. 1983).

Less soluble lectins are associated with membranes or occur as cell wall subfractions in certain plants (Callow 1977). When extracted in salt solution, portions of wall-associated lectins could be eluted, indicating that lectins are not always bound to the wall via carbohydrate binding sites (Kauss 1981). An extracellular appearance of lectins is documented, e.g., in the rhizobium-clover symbiosis (Dazzo 1977).

Lectins are glycoproteins or proteins which are able to bind specific saccharides and thus can attach to membrane glycoproteins via the projecting sugar residues and cause cells to agglutinate (Sequeira 1978).

Lectins have multiple binding sites for sugars; binding can be inhibited by adding a suitable sugar (hapten; Sharon 1977). Some lectins do not interact with monosaccharides, but only with complex oligosaccharides, and for some lectins the hapten is not yet known. Until now the amino-acid sequences of ca. 10 lectins have been reported [e.g., Concanavalin A, peanut lectin, soja lectin (Gabius et al. 1988)].

Many, if not all lectins are composed of subunits and undergo association-dissociation reactions, which result in different molecular weights (Sharon and Lis 1972). Therefore, these multiple molecular forms of lectins, isolectins, differ in their electrophoretic mobilities. The isolectins may be the product of closely related genes and may originally have had differences in the carbohydrate side chain, or they may be formed prior to, or during isolation, as a result of side chain modifications, such as hydrolysis of the amide group of glutamine or asparagine in the protein. The properties of *Phaseolus vulgaris* isolectins are described by Leavitt et al. (1977).

Purification from plant extracts can be carried out by the conventional techniques of protein chemistry such as salt fractionation and chromatography on ion exchangers, other types of adsorbents or affinity chromatography.

3.1.2 Hydroxyproline-Rich Glycoproteins (HRGPs)

3.1.2.1 Localization and Function

HRGPs were discovered 30 years ago by Lamport and Northcote (1960). They are major structural proteins of the plant cell wall (Bell et al. 1971; Lamport and Catt 1981; Showalter and Varner 1989; Wilson and Fry 1986) and have been located either intra- or extracellularly in many different sources (Anderson et al. 1977; Esquerré-Tugayé and Lamport 1979), and recently also in monocotyledones (maize; Kieliszewski and Lamport 1987). In cytochemical studies HRGPs have been detected in healthy tissue: intracellularly in the Golgi apparatus, and in the primary and secondary cell walls (Benhamou et al. 1990b). In response to infection by a wide variety of microorganisms (Clarke et al. 1981; Esquerré-Tugayé et al. 1979; Hammerschmidt et al. 1984; Mazau and Esquerré-Tugayé 1986), HRGPs are accumulated in paramural bodies (Chamberland et al. 1989), papillae (Mazau et al. 1987a), in wall appositions, as vesicles between plasmalemma and the plant cell wall (Bishop and Cooper 1983), as wall-like excrescences formed along cell walls, and as depositions between the pathogen and the host cell wall in the intercellular space (Benhamou et al. 1990a,b). It has been suggested that the accumulation of these glycoproteins is part of a general defence reaction against pathogenic attack (Esquerré-Tugayé and Lamport 1979; Benhamou et al. 1990c). Accumulation of HRGPs, extensins and arabinogalactans has also been observed in plant tissue in response to wounding, infection, oligosaccharide elicitors or ethylene (Chrispeels et al. 1974; Stuart and Varner 1980; Varner and Lin 1989).

In a recent study, based on a molecular approach, the accumulation of HRGPs in response to infection could be confirmed. Using a cloned carrot HRGP-gene (Delaat and van Loon 1983) the accumulation of HRGP in melon leaves, 3 and 4 days after inoculation with *Colletotrichum lindemuthianum*, could be correlated with the accumulation of HRGP m-RNA. The accumulation of HRGP m-RNAs after infection was reported in leaves and stems which were directly affected by the disease and also in tissue distant from the infection site (Benhamou et al. 1990b).

Cytochemical studies on HRGPs using lectin-gold and immuno-gold staining techniques are dealt with in Sections 6.3 and 6.4.

The exact role of HRGPs in defence response is not clearly established; however, the possibility that they act as molecular barriers to pathogen invasion (Cassab and Varner 1987), and provide matrixes for the deposition of lignin and/or act as agglutinins (Melon and Helgeson 1982; Wydra and Rudolph 1989; Wydra 1991) has often been emphasized.

3.1.2.2 Structure

Monosaccharide and Amino Acid Constituents. Common structural features of HRGPs are a linear molecule composed of a carbohydrate moiety with arabinose and galactose in variable amounts; rhamnose, galactosamine and galacturonic acid may occur (Hori and Sato 1977; Hori 1978; Mazau et al. 1987b; Chamberland et al. 1989), and a protein moiety with a high content of hydroxyproline (Hyp) and the β-hydroxy-amino acids serine and threonine and either glycine or serine (Ser) (Vretblad 1976; Touzé and Esquerré-Tugayé 1980; Cassab and Varner 1988).

Hydroxyproline is usually glycosylated by one to four arabinosides (Mazau and Esquerré-Tugayé 1986). In the protein moiety, repeated peptides with the unique sequence Ser-$(Hyp)_4$ were described in HRGPs from different plants including melon (Esquerré-Tugayé and Lamport 1979); carrot (Chen and Varner 1985); tomato (Smith et al. 1986); and bean (Corbin et al. 1987). The similarity between different HRGPs is demonstrated by the fact that antibodies raised against purified HRGPs from melon (*Cucumis melo* L. "Cantaloup Charentais"; Mazau et al. 1988) were found to cross-react specifically with purified tomato HRGPs and HRGPs from tobacco cells (Benhamou et al. 1990a).

Carbohydrate and Protein Linkage. Serine and Hyp are involved in the carbohydrate-protein linkage. The glycosyl-serine (or -threonine) linkage (α-D-gal*p* Ser) is usually labile in alkaline conditions (Neuberger et al. 1972; Lamport et al. 1973; Cho and Chrispeels 1976; Hardy et al. 1988), while glycosidic linkages to Hyp and hydroxylysine (β-L-ara*f*-Hyp) are quite stable under alkaline conditions (Lamport 1977).

The differing lability of the linkages in alcali has been used in isolating Hyp glycosides from glycoproteins (Lamport and Miller 1971) and proteoglycans (Fincher et al. 1974; McNamara and Stone 1978). Arabinofuranosyl and rhamnopyranosyl linkages are acid labile (BeMiller 1967) and therefore easily hydrolyzed.

The structural properties described above have to be considered when selecting the extraction method.

The biosynthesis of HRGPs has not been studied in detail, but several studies provided evidence that the Golgi apparatus is involved in the processing and transport of secretory HRGPs (Gardiner and Chrispeels 1975; Chrispeels 1976).

3.1.2.3 Solubility

Not only does the intra- and extracellular distribution of HRGPs change during development, as demonstrated in soybean seed extract (Cassab and Varner 1988), and after infection, but also their solubility changes. Cooper and Varner (1983) and Smith et al. (1984) observed a slow insolubilization of extensins in the cell wall due to covalent binding. Since several different HRGPs may be present in the same tissue, the preparation often contains a continuous spectrum of closely related molecular species, as well as other chemically related distinct macromolecules (Clarke et al. 1979). As numerous different procedures for extracting HRGPs have been described, a "standard" method does not exist. Some methods for extraction and the first steps of purification of soluble and insoluble HRGPs are briefly presented. For further studies, references are given.

3.1.3 Extensins

Clarke et al. (1979) distinguish between the two arabinose- and galactose-containing glycoproteins, "extensins" and arabinogalactan proteins.

Insoluble, high base HRGPs firmly bound to the cell wall are called extensins (McNeil et al. 1984). Extensins are characteristically associated with cell wall fractions of higher and lower plants (Lamport 1965; Talmadge et al. 1973; Catt et al. 1976; Clarke et al. 1979; Cassab and Varner 1988; Wagner and Jordan 1988), are mostly insoluble, have a lower alanine content than AGPs, only short oligosaccharide chains (Pope 1977; Fincher et al. 1983) and are distinguished from the Solanaceae lectins by the absence of cysteine (Allen et al. 1978).

Extensins could be extracted from cell walls by treatment with mildly acidified $NaClO_2$ (O'Neill and Selvendran 1980; Fry 1982). Solubilization of extension with $NaClO_2$ and results obtained from the characterization of extensins of tomato (Epstein and Lamport 1984) and several other plants (Fry 1982, 1986) suggest a covalent link by the amino acid isodityrosine formed between two tyrosine residues from different extensin molecules (Cassab and Varner 1988).

3.1.4 Arabinogalactan Proteins (AGPs)

Arabinogalactan-proteins, another class of HRGPs, have been isolated from seeds of every taxonomic group of flowering plants, as well as from leaf extracts and callus-culture filtrates (Gleeson and Clarke 1979; Gleeson et al. 1979; Akiyama and Kato 1981). AGPs are major components of gum exudates. Some of the Hyp-rich AGPs were found to show agglutinating activity and can be classified as lectins (Allen and Neuberger 1973; Jermyn and Yeow 1975).

AGPs are distinguished from extensin by a protein moiety of mostly 2–10% (extensin: 50%) which is acidic and rich in Hyp, Ser, Ala and Gly (Fincher et al. 1983) [extensin: basic, rich in Hyp, Ser, Lys, Tyr, Isodityr (Stuart and Varner 1980)]. The carbohydrate chains of AGPs are characterized by a relatively high degree of polymerization and are generally composed of gal, ara*f*, ara*p*, rha, man, galA, and/or glcA (Fincher et al. 1983). In contrast, extensins are polymerized by a degree of 1 to 4 in the carbohydrate moiety and composed only of ara*f* and *gal* (Lamport and Miller 1971; Muray and Northcote 1978; O'Neill and Selvendran 1980).

Another difference, relevant for extraction conditions, is the high solubility of AGPs, in contrast to that of extensins, which rapidly become covalently attached within the cell wall (Pope 1977).

The arabinogalactans and AGPs have been grouped into three structural types by Aspinall (Aspinall 1970, 1973): Type I, the arabino-4-galactans which are linear polymers of galacto-pyranose residues in β-1,4-linkage, more or less substituted by arabinose; Type II, the arabino-3,6-galactans; and Type III, polysaccharides with arabinogalactan side chains.

The general structural feature of this family is a branched β-galactopyranose framework possessing predominantly 1,3-linkages with varying amounts of 1,-6-linked gal*p* oligosaccharide substituents. Short side chains on the linear backbone may be substituted by ara*f* and ara*p* residues and other less abundant monosaccharides such as glycosyluronic acids, which often occur in terminal positions (Fincher et al. 1983).

3.2 Hemicelluloses and Pectic Polysaccharides

Deposition of polysaccharides in plant tissues is known as a response to infection by pathogens, and to environmental stress, such as high (Smith and McCully 1977) or low temperature, drought and wounding by chemical or mechanical means, or by ultrasound (Currier and Webster 1964). Accumulation in response to unspecific stress may induce disease resistance (Stermer and Hammerschmidt 1985). These deposites originating from the plant are difficult to isolate and analyze, and their composition is largely unknown. A characterization is possible using cytochemical methods, such as specific staining reagents, lectin-based or immunological techniques.

The following polysaccharides will be treated in this Chapter:

1. Hemicelluloses: xyloglucan, xylan and arabinoxalan;
2. Neutral pectic polysaccharides: arabinan, galactan, arabinogalactan protein;
3. Acidic pectic polysaccharides: homogalacturonan, rhamnogalacturonan I and II.

3.2.1 Sources and Localization

The primary cell wall of higher plants consists of a framework of cellulose microfibrils embedded in a matrix of hemicelluloses and pectic polysaccharides (Talmadge et al. 1973). Immunocytochemical studies showed that the polysaccharides are not distributed evenly in the primary cell wall, the middle lamella or intercellular layers and corners between cells (Moore et al. 1986).

The cell wall matrix polysaccharides are synthesized in the Golgi apparatus, as demonstrated for the main pectic polysaccharide rhamnogalacturonan I and the hemicellulose xyloglucan (Moore and Staehelin 1988) and for arabinose-containing cell wall materials and their precursors (Owens and Northcote 1982; Kawasaki 1987).

Acidic pectic polysaccharides have been localized in the middle lamella, in cell corners (Ruel and Joseleau 1984) and as amorphous deposits accumulating on the inner and outer surface of the cell wall, as well as amorphous aggregates in the intercellular space associated with bacterial colonies during the incompatible interaction (Brown et al. 1990). The location of insoluble pectins, which may account for 10–50% of the total pectins (Selvendran 1985) and which appear to be cross-linked to other cell wall polymers, is not yet elucidated.

3.2.1.1 Xyloglucan

Xyloglucan, a hemicellulose, accounts for about 20% of the dry weight of the primary walls of dicots and 2% for monocots. It has been isolated from sycamore (York et al. 1984); common bean (Wilder and Albersheim 1973); tobacco (Akiyama and Kato 1982); rose (Barnoud et al. 1977); and soybean cells (Hayashi et al. 1980).

In the cell wall xyloglucan is highly hydrogen-bound with cellulose microfibrils. Xyloglucans are nearly exclusively located in the cellulose-microfibril region of the cell walls of organized tissue and have been demonstrated in the forming cell plate (Hayashi et al. 1987; Moore and Staehelin 1988).

In suspension-cultured sycamore cells, xyloglucan is found throughout the wall and the middle lamella (Moore et al. 1986). It seems to form a layer on the surface of cristalline regions of the microfibrils and to penetrate into the amorphous regions, which mainly consist of pectic polysaccharides (Varner and Lin 1989).

A xyloglucan highly substituted with residues of arabinofuranose and galactopyranose was isolated from cell walls of *Solanum tuberosum* (Ring and Selvendran 1981). During auxin-induced cell wall growth, release of xyloglucan fragments has been observed (Hayashi et al. 1984).

3.2.1.2 Xylan and Arabinoxylan

Xylan is the principle hemicellulose of graminaceous monocots (McNeil et al. 1978). As demonstrated by xylanase-gold-complexes, xylan is concentrated in the primary walls of parenchyme cells of *Arundo donax*, but absent from the middle lamella (Ruel and Joseleau 1984).

Arabinoxylan has been detected in the extracellular polysaccharides of suspension-cultures of sycamore (Keegstra et al. 1973); sugarcane (Takeuchi and Komamine 1980); wheat (Burke et al. 1974); tobacco (Akiyama and Kato 1982); and periwinkle. The constituent of the extracellular polysaccharide of suspension cultures are supposed to correspond to those of the matrix polysaccharides in the cell wall.

Another hemicellulosic polysaccharide, the acidic glucuronoarabinoxylan, was isolated from sycamore cells (Darvill et al. 1980b).

3.2.1.3 Arabinans and Galactans

The non-cellulosic neutral polysaccharides, arabinans and galactans, are supposed to be part of the amorphous materials of the primary cell wall, in which the skeletal cellulose microfibrils are embedded (Kawasaki 1987). According to Moore and Staehelin (1988), it is not yet determined if arabinans and galactans are present in the matrix around the microfibrils or if they are located exclusively in the middle lamella and the corner junctions, similar to rhamnogalacturonans I.

3.2.1.4 Arabinogalactan and Arabinogalactan Protein

Arabino-4-galactans, classified as Type I by Aspinall (1973) are minor components of plants and were isolated from pectic complexes of seeds, bulbs, leaves and coniferous compression wood. The arabino-3,6-galactans classified as Type II were isolated from coniferous woods, gums, saps and exudates of angiosperms and from seeds, leaves, rods, fruits and tissue cultures (Aspinall 1970, 1973). Arabinogalactans and arabinogalactan proteins are primarily localized in the extracellular matrix and in gums and exudates (Showalter and Varner 1989). They are partly associated with the cell walls, freely soluble in water and are thought to play a role in cell-cell recognition.

3.2.1.5 Homogalacturonan

The α-1,4-linked homogalacturonans have been isolated from the EPS of suspension-cultured sycamore (York et al. 1985) and tobacco cells (Yamaoka and Sato 1977). Homogalacturonans are thought to occur in primary cell walls, although they may be covalently attached to other cell wall polymers (Dabrowski et al. 1982).

In infected plants, polygalacturonic acid containing molecules, possibly derived from homogalacturonans, were detected in papillae, wall appositions, intercellular spaces and accumulated in uninvaded phloem cells neighbouring infected ones (Benhamou et al. 1990c).

3.2.1.6 Rhamnogalacturonan I

Rhamnogalacturonan I, the major pectic polysaccharide, is located exclusively in the middle lamella, in the corner junctions and along air spaces (Moore et al. 1986; Moore and Staehelin 1988).

Polysaccharides containing a backbone of alternating α-linked 2-rhamnosyl and 4-linked D-galactosyluronic acid residues are suggested to be present in most, if not all higher plant cell walls (McNeil et al. 1984).

3.2.1.7 Rhamnogalacturonan II

Rhamnogalacturonan II has been identified in the primary cell walls of dicotyledons such as pea, pinto bean and tomato seedlings, as well as in the cell walls of oat (Darvill et al. 1978). Rhamnogalacturonan II is covalently attached to other components of the primary cell wall (Varner and Lin 1989), perhaps through homogalacturonans (McNeil et al. 1984).

3.2.2 Structure

3.2.2.1 Xyloglucan

Xyloglucan has a backbone of β-4-linked D-glucosyl residues with D-xylosyl side chains α-linked to O-6 of some of the glucosyl residues. Some of the xylosyl residues are substituted at O-2 with D-galactosyl or L-fucosyl-α-D-galactosyl (McNeil et al. 1984; Thomas et al. 1987).

3.2.2.2 Xylan and Arabinoxylan

All known xylans consist of a backbone of 4- and 2,4-linked xylosyl residues (McNeil et al. 1984). Different xylans have different arrays and combinations of side chains, which include L-arabinofuranosyl, D-galactosyl-β-5-L-arabinofuranosyl, D-xylosyl-β-2-L-arabinofuranosyl and D-glucuronosyl moieties. Terminal arabinosyl residues are the most common side chain of graminaceous monocot xylans and were also found in sycamore glucuronoarabinoxylan (Darvill et al. 1980b; Thomas et al. 1987). Feruloyl esters on the arabinosyl side chains are hypothesized to form diferulate-polysaccharide cross-links. This coupling of polysaccharides may have effects on the resistance of the cell wall to enzymic digestion (Friend 1976). Xylan fragments containing specific side chains may have biological activities (McNeil et al. 1984).

3.2.2.3 Arabinan and Galactan

Arabinans are principally constituted of 5-linked α-L-arabinofuranosyl residues, some of which gave α-L-arabinofuranosyl residues attached to O-2 and/or O-3.

Galactans are pure β-4-linked galactans, some may have 6-linked galactosyl residues in addition to 4-linked galactosyl residues (McNeil et al. 1984; Varner and Lin 1989).

3.2.2.4 Arabinogalactan and Arabinogalactan Protein

The most common type of arabinogalactans consists of terminal-, 3-, 6-, and 3,6-linked galactosyl residues and 3- or 5-linked arabinofuranosyl residues.

In some samples, glucuronic acid is found, often in terminal positions (Fincher et al. 1983). The other type of arabinogalactans are usually composed of 4- and 3,4-linked galactosyl residues more or less substituted by 5-linked or terminally linked arabinofuranosyl residues (Aspinall 1981).

3.2.2.5 Homogalacturonan

Homogalacturonans are consisting mainly of α-4-linked galacturonic acid residues. The frequence and pattern of substitution with other sugar residues is not known (Varner and Lin 1989). Homogalacturonan fragments can be produced by partial acid hydrolysis or by treatment with an endo-α-1,4-polygalacturonic acid lyase, resulting in oligogalacturonides from 10 to 13 residues long (Nothnagel et al. 1983; Davis et al. 1982).

3.2.2.6 Rhamnogalacturonan I

Rhamnogalacturonan I (RG I) is a polysaccharide or a family of closely related polysaccharides with a molecular weight of approximately 200 000. It consists of a backbone of an alternating sequence of 2-linked L-rhamnosyl and 4-linked D-galacturonic acid residues, with about one-half of the rhamnosyl residues glycosylated at O-4.

At least 30 different kinds of side-chains with an average length of 7 glycosyl residues, arabinosyl, galactosyl and small amounts of fucosyl and galacturonic acid residues are found (McNeil et al. 1984).

Since rhamnogalacturonan I is only present in the intercellular layer, it can be a source of fragments with signal function (Moore et al. 1986).

3.2.2.7 Rhamnogalacturonan II

Rhamnogalacturonan II (RG II) is unrelated in structure to RG I, but it is, as RG I, a structurally extremely complex molecule. RG II is a polymer of approximately 60 glycosyl residues and is covalently attached to other components of the wall through α-4-linked galacturonic acid residues. RG II contains many unusual sugars, such as 2-0-methylfucosyl, 2-0-methylxylosyl, apiosyl (a branched pentosyl residue), 3-C-carboxy-5-deoxy-L-xylosyl (aceric acid), 3-deoxy-manno-octulosonic acid (KDO) (York et al. 1985a).

3.2.3 Function (Toxicity)

Toxicity of polysaccharides has been demonstrated for intact molecules and for fragments. The latter have received increasing importance in the recent relevant literature.

3.2.3.1 *Intact Molecule*

The increase in the proportion of pectic polysaccharides and depositions of "gums" or "gels" in xylem vessels has been demonstrated by VanderMolen et al. (1977).

The function of these depositions may be to restrict spread of a pathogen by encapsulation (Mendgen 1978), to agglutinate the pathogen (Wydra and Rudolph 1990) or to plug or seal a wound, thus restricting interchange of ions, metabolites and toxins between injured and healthy host cells or between pathogen and host cells.

Arabinogalactans are characterized by a generally adhesive nature, a water-holding capacity and a capability to associated with other macromolecules. Therefore, it has been assumed that arabinogalactans function by way of preventing moisture loss and interacting specifically with carbohydrate binding molecules or lectins (Clarke et al. 1979). However, arabinogalactans and AGPs are also carbohydrate-binding macromolecules, since they bind specifically with β-glucosyl substituted dyes, such as Yariv artificial antigen (Jermyn and Yeow 1975; Anderson et al. 1977; Clarke et al. 1978).

After bacterial infection, pectic material increased on the outer surface of plant cell walls. These pectic polysaccharides were also demonstrated between bacterial cells in the intercellular space (Brown et al. 1990; O'Connell et al. 1990), suggesting an interaction with acidic extracellular polysaccharides from the bacteria. Therefore, bacteria-agglutinating properties of the pectic substances seem to play a role in the plant's defence response (Brown and Mansfield 1990; Wydra 1991).

3.2.3.2 *Fragments*

Plant cell wall fragments have been demonstrated to play a role in the defence reactions of the plant. The fragments are released if the plant cell walls are incubated with extracellular components of a fungal pathogen, with microbial, cell-wall degrading enzymes (polygalacturonase, pectate lyase, xylanase) or after treatment with hot water or partial acidic hydrolysis (Hahn et al. 1981). The fragments possess elicitor activity, acting as a signal to initiate plant defence reactions, such as accumulation of phenolics, phytoalexins or stress metabolites such as casbene in castor beans (*Ricinus communis* L.; West et al. 1985). Further examples for induction of phytoalexins by polygalakturonase and/or pectic fragments are given in Table 3.

Pectic fragments released from wounded cell walls by the activity of a α-1,4-endopolygalacturonase have been shown to regulate the expression of proteinase inhibitor genes in plants (Ryan et al. 1985) and to induce endo-polygalacturonase inhibiting proteins (PGIP; Cervone et al. 1989b).

Table 3. Phytoalexin-producing plants in which polygalacturonase and/or pectic fragments have been shown to serve as elicitors (West et al. 1985)

Phytoalexin	Plant	Family	Reference
Phaseollin	*Phaseolus vulgaris*	Leguminosae	Wickham
Pisatin	*Pisum sativum*	Leguminosae	Hadwiger
Medicarpin	*Trifolium repens*	Leguminosae	Gustine
Glyceollins	*Glycine max*	Leguminosae	Hahn et al. (1981)
			Davis et al. (1982)
Casbene	*Ricinus communis*	Euphorbiaceae	Bruce and West (1982)
			Lee and West (1981)
Ipomeamarone	*Ipomea batatas*	Convolvulaceae	Sato et al. (1982)
Not identified	*Beta vulgaris*	Chenopodiaceae	Vasil'eva et al. (1982)

Proteinase inhibitors are considered to be part of the natural defence of plants towards attacking pests (Ryan 1978). PGIPs are suggested to regulate the activities of pectic enzymes and thus the fragmentation of cell wall polysaccharides. Fragments released by acid treatment or by cell wall degrading enzymes can kill plant cells and are thought to be a trigger for hypersensitive cell death in response to a pathogenic attack.

The possible mechanisms involved are treated in detail for three aspects:

Elicitor-Active Fragments. Heat-stable, elicitor-active fragments have been isolated from endopolygalacturonic acid-lyase-treated soybean cell walls, citrus pectin and citrus sodium polypectate. The structure of the oligogalacturonides play a decisive role for their activity. α-1,4-linked, deca -and undeca-galacturonides, that contained 4,5-unsaturated galactosyluronic acid at the non-reducing termini showed the highest specific elicitor activity (Nothnagel et al. 1983; Davis et al. 1984, 1986a). In a preparation of oligomers from castor bean, a minimum degree of polymerization of nine units appeared to be required for elicitor activity; trideca-α-1,4-D-galacturonide was the most active of the oligomers tested. Since methylesterification of the carboxyl groups diminished the elicitor activity of the oligomers, it was suggested that, besides other structural features, a polyanionic character of the oligomer is important (Jin and West 1984). This was supported by Davis et al. (1986b), who demonstrated that carboxyl groups of the galacturonic acid residues were required for elicitor activity.

Elicitor activity of the decagalacturonide-rich fraction was inhibited by fractions rich in oligogalacturonides with degrees of polymerization from 3 to 8 residues, and those with degrees of polymerization greater than 12 residues (Davis et al. 1986b). The larger oligogalacturonides form gels, that may trap some of the decagalacturonides, thereby decreasing elicitor activity. Gel formation in interactions between pectic polymers and extracellular components from a pathogen may play a role in the compatible interaction between host and pathogen (Wydra and Rudolph 1990; Wydra

1991). Pectic fragments synergistically acting with fungal components such as galactoglucomannans or glucans have been observed to stimulate the accumulation of phenolic defence-related products. This indicates a regulative function of plant and fungal factors in the expression of defence reactions (Tepper and Anderson 1990). The interaction of two types of elicitors suggests that the pectic elicitors may act as a general signal of wall damage, which serves to modulate the plant's response to other types of signals reflecting the source of wall damage, e.g., microbial attack, mechanical injury, etc. (West et al. 1985). Also, galactose residues have been observed to accumulate in vesicles or paramural bodies in response to pathogenic attack (Benhamou et al. 1988).

Pectic Fragments Inducing Synthesis of Enzyme Inhibitors. In response to wounding of leaves mechanically or by insect feeding, the accumulation of proteinase inhibitors has been demonstrated in tomato, potato and other plants (Ryan 1980). Proteinase inhibitors were induced by pectic oligomers which were fragmented from cell walls by cellular α-1,4-endopolygalacturonases interacting with cell walls as a result of wounding (Ryan et al. 1985). The signal for wound-induced proteinase inhibitor synthesis is transported throughout the plant within 2 h following wounding (Ryan 1978).

Ryan et al. (1985) suggested that small oligosaccharides are the systemic signals that activate proteinase inhibitor genes, whereas larger fragments are signals of localized response. As one source exhibiting PIIF (proteinase inhibitor-inducing factor)-active pectic fragments, a rhamnogalacturonan I was purified from sycamore cell walls (Ryan et al. 1981). The location of RG I in the intercellular layer and around the air spaces seems ideal for signalling mechanical damage to surrounding cells (Moore et al. 1986).

Polygalacturonase-inhibiting proteins (PGIP) have been isolated from a variety of dicotyledonous plants, including bean (Cervone et al. 1987); peas (Hoffmann and Turner 1982); oranges (Barmore and Nguyen 1985); and alfalfa (Degrà et al. 1988). PGIPs inhibit fungal endopolygalacturonases, which depolymerize polygalacturonic acid to elicitor-inactive oligogalacturonides with a degree of polymerization lower than nine. In the presence of PGIP, the rate of production of elicitor-active oligogalacturonide with a degree of polymerization higher than nine increases. These fragments are signals that activate plant defence responses (Cervone et al. 1989a). PGIP is suggested to regulate the physiological activities of pectic enzymes (Cervone et al. 1989b). In the presence of PGIP, endopolygalacturonase lost its macerating activity, but its killing activity was maintained (De Lorenzo et al. 1987). Killing of plant cells is part of the defence response of plants in the hypersensitive reaction.

Cell Wall Fragments That Kill Plant Cells. Specific polysaccharide-degrading enzymes associated with the hypersensitive reaction of plant cells to microbial attack release heat-stable molecules from the plant cell wall, supposingly carbohydrates, that kill the plant cell. The enzymes pectin lyase, pectin methylesterase and xylanase act synergistically in releasing the killing activity. The cell wall fragments are supposed not to originate from homogalacturonan, oligogalacturonides, RG I or RG II. Xylanase was shown to form reducing xylose from larchwood xylan (Bucheli et al. 1990). A xylanase isolated from *Trichoderma viride* culture fluid released heat-stable killing activity and was thus stable to kill rice cells (Ishii 1988). Xylanase also seems to be one of the factors involved in the release of cell wall fragments capable of killing maize cells (Bucheli et al. 1990).

Killing fragments could be solubilized by partial acid hydrolysis from sycamore, maize and rice cell walls (Yamazaki et al. 1983; Doares et al. 1989). A monosaccharide-rich fraction of endo-β-1,4-glucanase which digested xyloglucan has been shown to effectively inhibit the elongation of pea stem segments (York et al. 1984). In this fraction, the neutral sugars glucose, xylose, galactose, fucose and arabinose were detected.

Hahne and Lörz (1988) reported the release of phytotoxic heat-stable factors from cell walls of cultured plant cells after treatment with pectinases and hemicellulases.

Enzymes adapted to the constituents of the primary cell walls of monocotyledons, such as arabinase, arabinosidase, xylanase, galactanase, endo-β-1,4-glucanase and polygalacturonase are produced by cereal pathogens. Release of arabino-xylan fragments, arabinose and xylose from cell walls was detected in connection with wall degradation, but a directly toxic effect was not yet reported.

3.2.4 Physicochemical Properties

Structural polydispersity in preparations of pectic polysaccharides is high, between preparations from different plants as well as in fractions from single preparations, due to differences in nature, proportions and linkage of peripheral substituents (Turner and Lifner 1975; Fincher et al. 1983).

The degree of branching, the degree of esterification of D-galacturonic acid residues, attachment of O-acetyl and insertion of L-rhamnose residues in the galacturonan side chains influence the physical properties, and hence the biological functions of pectins (Aspinall 1982a). In contrast to purified proteins, in which cell molecules are identical in structure and molecular weight, for polysaccharides, even for those which are chemically and physically homogeneous, variations occur from molecule to molecule. Variations may be continuous in respect of all parameters, such as molecular size, proportions of sugar constituents and particular linkage types; in this case, separation into discrete molecular species is impossible. Pectic polysaccharides are always found in association with one another. Other polysaccharides may be homogenous in chemical composition, but physically (i.e., molecular size, viscosity) heterogeneous. According to Aspinall (1982a), conclusive evidence for an existing homogeneity is unattainable.

Physicochemical properties of arabinogalactans and AGPs, such as viscosity, adhesive nature, the overall shape and behaviour in solution and their possibility to interact with other macromolecules, are determined by (1) the proportion of regions with highly branched structure and regions of regular carbohydrate sequence; (2) by the extent of substitution of the proteins of the AGPs by polysaccharide chains; and (3) the shape flexibility and nature of the individual substituent chains.

3.2.5 Principles for Extraction and Isolation

Depending on the great variation in chemical and physical properties, a wide range of isolation procedures can be used. They are based on different principles (Pazur 1986):

1. Solubility – using techniques such as fractional precipitation, fractional dissolution and distribution between immiscible solvents;
2. Ultracentrifugation, including density gradient centrifugation;
3. Membrane ultrafiltration;
4. Chromatography – adsorption, ion exchange and partition chromatography; and
5. Gel filtration.

For any method used, the possibility of structural modification during isolation must be considered. Saponification of ester groups, base-catalyzed degradation by β-elimination from methyl galacturonate residues in pectins and acid degradation by cleavage of glycosidic bonds may occur (Aspinall 1982b).

3.2.6 Extraction Media

Polysaccharides can be extracted using only water at various temperatures; acidic polysaccharides can be isolated with reagents that effect cation exchange; and pectins are solubilized by complexing tightly coordinated bivalent metal ions with chelating solvents such as disodium ethylene dinitrilotetraacetate, ammonium oxalate or sodium hexamataphosphate (Aspinall 1982b).

RG I and RG II are released from the primary cell walls of suspension-cultured sycamore cells by enzymatic digestion with a fungal endopolygalacturonase (English et al. 1972). The hemicelluloses xyloglucan and glucuronoarabinoxylan are extracted with aqueous NaOH or KOH from purified cell walls after removal of pectic material by enzymatic or chemical extraction (York et al. 1985b).

3.2.7 Fractionation

Fractionation based on selective precipitation of polysaccharides can be carried out using:

1. The noncovalent binding of polysaccharides with lectins or
2. Formation of salts or complexes of pectic acid with neutral cupric acetate (Aspinall et al. 1969; Cartier et al. 1988), or cationic detergents such as cetyltrimethylammonium bromide (Cetavlon) and cetylpyridinium bromide (Scott 1965).

Insoluble complexes are formed by some polysaccharides with dyes and make a selective fractionation possible, e.g., precipitation of arabino-3,6-galactans with artificial carbohydrate antigens prepared by coupling diazotized 4-aminophenyl β-D-glucopyranoside with phloroglucinol (Yariv antigen; Yariv et al. 1962; Hirst et al. 1965). A frequently used initial fractionation of water extracts from plant tissue or tissue culture is treatment to saturation with $(NH_4)_2SO_4$, which does not usually precipitate arabinogalactans, due to their high solubility (Fincher and Stone 1974; Mau et al. 1982; Wydra 1991).

Anion exchange chromatography is the most useful technique for initial fractionation of extracellular polysaccharides. Before applying to the column, pectic polysaccharides are treated with NaOH and then digested with endopolygalacturonase. Pectic polysaccharides are separated by means of a combination of ion-exchange and gel-permeation chromatography. This procedure removes many contaminants, including protein and neutral polymers (York et al. 1985b).

Xyloglucan, as a neutral polysaccharide, elutes with the void volume of the column; glucuronoarabinoxylans are isolated by ion-exchange chromatography and subsequent gel permeation chromatography on a BioGel P-100 and Agarose A 1.5-m column. Arabinogalactans and arabinogalactan proteins can be isolated from extracellular polysaccharides of suspension cultures by combinations of anion-exchange, cation-exchange, cellulose and gel-filtration chromatography, as well as extraction and precipitation procedures.

3.2.8 Structural Analysis

3.2.8.1 Homogeneity

Constant chemical composition can be demonstrated by hydrolysis for individual sugars, functional group analysis or analysis for particular classes of sugars, e.g., hexuronic acids and by spectroscopic examination, especially by nuclear magnetic resonance spectroscopy. Physical properties, which are determined, are optical rotation and solution properties such as viscosity. Charge heterogeneity can be shown by electrophoresis (Aspinall 1982b).

3.2.8.2 Molecular Structure

For the elucidation of the complete molecular structure of a polysaccharide the following structural features have to be characterized:

1. The types of monosaccharide residues established by paper chromatography, thin layer chromatography, or HPLC after hydrolysis of the polysaccharide (Lan et al. 1985; Pazur 1986; Hardy et al. 1988); by gas chromatography of the trimethylsilyl methylglycoside, methylester derivatives (Chaplin 1982; Davis et al. 1986a); or by ion-exchange chromatography, e.g., borate-anion exchange chromatography (Lee 1972; Mikami and Ishida 1983); or cation exchange chromatography complexed with various metal ions (Thomas and Lorel 1976), which may be followed by pulsed amperometric detection (Hardy et al. 1988).

2. The D or L configuration of the residues, determined by measurements on the monosaccharides liberated by acid or enzyme hydrolysis of the polysaccharide, e.g., measurement of specific rotation, circular dichroism (c.d.) absorptions or enzymic susceptibilities, butanolysis of the residues and GC-analysis of the trimethylsilylated derivatives (Gerwig et al. 1978, 1979); the configuration can be established by a comparison between the values of the unknown and that of the known compounds. Detailed instructions are given by Pazur (1986).

3. Number of monosaccharides per molecule – the degree of polymerization can be evaluated by FPLC of oligosaccharides (York et al. 1985b), e.g., on a Mono Q anion exchange column for establishing the size distribution of oligogalacturonides (Cervone et al. 1989a). The molecular weight of the oligosaccharides can be determined by fast atom bombardment mass spectroscopy (f.a.b.-m.s.) (Barber et al. 1981; Nothnagel et al. 1983; Davis et al. 1986b).

4. The positions of glycosidic linkages between monosaccharide residues. The methods used are: complete methylation of the polysaccharide; hydrolysis of the methylated product to a mixture of partially methylated monosaccharides; reduction of the methylated monosaccharides; acetylation of the reduced products; and finally the analysis of the partially methylated alditol acetates by gas liquid chromatography (g.l.c.) and mass spectrometry (m.s.). From the positions of the free hydroxyl groups, the positions of the glycosidic linkages between monosaccharide units can be deduced. A description of the procedure is given in Valent et al. (1980), McNeil et al. (1982), and Pazur (1986).

5. Ring structure of monosaccharides. A frequently used method for determination of the ring structure of the component monosaccharides is based on converting the glycosyl residues of the complex carbohydrate being analyzed into residues whose ring form can readily be ascertained. This includes methylation of the complex carbohydrate; partial hydrolysis; reduction and ethylation of the partially methylated oligosaccharides; and preparation of alkylated alditol acetates. The partially methylated, partially ethylated alditol acetates are identified by g.l.c. and m.s. (Darvill et al. 1980a).

6. Sequence of monosaccharides and the α or β configuration of the glycosidic bonds. In addition to the data obtained by chromatographic and mass spectra data, further methods are necessary to determine the anomeric configuration of the bonds and the sequence of residues in the polysaccharide.

Nuclear magnetic resonance spectroscopy, ^{13}C-n.m.r, provides information on anomeric configuration and location of the glycosidic linkages, as well as the monosaccharide composition and the conformation of the polysaccharide. After fractionation of complex mixtures of peralkylated oligosaccharides by reversed-phase, high-pressure liquid chromatography, the identification of disaccharides, almost all trisaccharides, and in some cases tetrasaccharides by g.l.c. and m.s. is possible. Thus, the oligosaccharide sequences may be determined and, with the glycosyl-linkage composition of the intact complex carbohydrate, can be used to determine the glycosyl sequence of the carbohydrate (Valent et al. 1980).

For determination of the anomeric configuration of the glycosyl residues, per-O-alkylated glycosyl-alditols are analyzed by ^{1}H-NMR spectroscopy (Bebault et al. 1978).

Enzyme hydrolysis can be carried out with two types of enzymes. The exo- and endohydrolases can provide valuable information obout the structure and residue sequences of the fragments. Some enzymes have a specificity for a particular type of linkage.

Degradation reactions are used to obtain structural information. Alkaline degradation only provides a limited amount of information due to the complex nature of reactions, however, β-elimination has been useful in structural studies on polysaccharides, especially on those containing uronic acid residues (Pazur 1986).

3.2.8.3 Immunological Methods

Immunological methods such as precipitation reactions, agar diffusion assays and enzyme linked immunoassays using antibodies specific for the carbohydrate residues of the antigen, can be used for different structural characterizations. The number of polysaccharides and the purity of these polysaccharides can be determined. Polysaccharides with similar structural features can be identified using cross-reactions in precipitin formation, and using hapten inhibition studies, the nature of the combining sites can be elucidated (Pazur 1982). In this context, the techniques of affinity chromatography can be used to isolate, purify and characterize polysaccharides (Sect. 5.1; Pazur 1986). Galactose-binding lectins are used for the isolation of AGPs (Gleeson and Clarke 1979; Gleeson et al. 1979), e.g., lectin from the clam *Tridacna maxima*, peanut agglutinin and mouse IgA myeloma (Baldo et al. 1978; Gleeson et al. 1978), *Ricinus communis* lectin (Wagner and Jordan 1988) and lectin from *Abrus precatorius* (Olsner et al. 1974). All these lectins recognize galactose in the ligand, but their affinity for galactose oligosaccharides and substituted galactose residues is different.

4 Methods for Extraction and Solubilization

4.1 Glycoproteins

4.1.1 Lectins

4.1.1.1 Soluble lectins

Method 1. (Shet and Madaiah 1988)

1. Plant material (seeds, leaves, roots etc.) is homogenized with 0.15 M NaCl in a Waring blender.
2. The homogenate is filtered through cheese cloth and centrifuged at 4 °C.
3. $(NH_4)_2SO_4$ is added to the supernatant until saturation of 30–70% is reached.
4. The precipitate is dissolved and dialyzed extensively against 0.01 M Na-P-buffer pH 7.2, containing 0.15 M NaCl (PBS) in the cold.

The retentate is centrifuged, and the clear supernatant used for further purification.

Variations of this method for the extraction of soluble lectins are described in the following publications:

lectin extraction from potato (*Solanum tuberosum*) (Allen et al. 1978, Ghanekar and Pérombelon 1980); from tomato (*Lycopersicon esculentum*) (Kilpatrick 1980, Nachbar et al. 1980); from thorn apple (*Datura stramonium*) (Desai et al. 1981); from *Phaseolus vulgaris* (phytohemagglutinin) (Felsted et al. 1975, Osborn et al. 1983); from apple (*Malus silvestris* X. *ranetka*) (Romeiro et al. 1981); from soybean

(*Glycine max*) (Bhuvaneswari et al. 1977, Fett and Sequeira 1980); from jack bean (*Canavalia ensiformis*); from wheat (*Triticum aestivum*) (Allen et al. 1973); and from *Wistaria sinensis* (Ahmed and Chatterjee 1988).

Method 2. (Vretblad 1976) The plant material is defatted before extracting the lectins.

1. Soybeans are ground in a Sorvall Omni-mixer and treated with hexane for 2 h under stirring. The solvent is removed by filtration through sinter glass and the meal dried overnight in a fume hood.
   The dry meal is extracted with 0.9% saline for 1 h at room temperature under vigorous stirring. The precipitate from a 50–80% ammonium sulphate precipitation is redissolved, dialyzed and applied to an affinity column.
2. Wheat germ is treated with hexane and dried as above. The dry meal is extracted with 0.05 M sodium-acetate buffer pH 4.5 for 1 h at room temperature under stirring. Ammonium sulphate is added to 40% saturation. The precipitate is redissolved, dialyzed and applied to an affinity column.

4.1.1.2 Bound Lectins

Method 3. Extraction with detergent (Kauss and Glaser 1974; Kauss and Bowles 1976):

plant material is homogenized and the walls subsequently extracted firstly with 0.5 M P-buffer pH 7.1 and then with 0.1 M EDTA/0.06% Triton-X-100. Thus, a buffer-soluble and a detergent-soluble part of wall-associated lectin can be extracted.

Method 4. Romeiro et al. (1981) compared different solutions for extracting an agglutination factor from apple:

1. The solution used by Stokes et al. (1968) for separation of mitochondria: 2 mM sodium metabisulfite in 0.1 M K-phosphate, pH 7;
2. 0.85% NaCl in water;
3. The following solution was used by Marinkovich (1964) and others (Sequeira and Graham 1977) for extraction of potato lectin; by Woods et al. (1979) for extraction of corn seeds; and by Mellon and Helgeson (1982) for extraction of tobacco callus: seeds of *Malus sylvestris* were frozen, immersed in liquid N_2 and ground to a fine powder. The powder was extracted in a blender at high speed for 5 min with a solution of 3 vol of absolute ethanol mixed with 1 vol of 1.25 N HCl. One gram of insoluble PVP (polyvinyl pyrrolidone) was added to each 10 ml of this

extract. The suspension was filtered through cheese cloth, the extraction repeated.

The filtrates were centrifuged (16000 g, for 20 min, at 4 °C) and the supernatant passed successively through Millipore filters (5-, 1.2-, 0.45-μm pore size). The filtrate was mixed with 2 vol of acetone at 4 °C and stored overnight. The supernatant liquid was discarded, and the white precipitate was suspended in 0.1 M sodium acetate buffer (pH 3.6). One ml buffer was added per gram of seeds extracted. The suspension was centrifuged at 27000 g, for 20 min, at 4 °C. The supernatant liquid was retained, and the pellet was resuspended in the same amount of buffer and centrifuged again. The pellet was discarded, and the two supernatant liquid fractions were dialyzed against 400 vol of 0.1 M sodium acetate buffer (pH 3.6), at 5 °C with continuous stirring. The dialyzed material, termed "crude agglutinating factor" was stored at −20 °C.

This procedure is quite harsh and would denature most proteins (Mellon and Helgeson 1982).

4.1.2 HRGPs

4.1.2.1 Soluble HRGPs

Method 1. Soluble, hyp-rich glycoproteins were extracted from apple fruit tissue by Knee (1975) in sequence in the presence of detergent or the absence of chloride ions with the media:

1. a) Acetone
   b) 5 mM phosphate
   c) 1 M KCl
2. a) 0.2 M Tris phosphate
   0.01 M $Et_2N \cdot CSSNa$ pH 8
   b) 1 M KCl pH 7.5
   or
3. a) 0.2 M Tris phosphate
   0.01 M $Et_2N \cdot CSSNa$ pH 8
   b) 0.2 M Tris base pH 10

followed by ultrafiltration, de-esterification after addition of NaCl to 0.1 M and ion exchange chromatography on DEAE cellulose.

Method 2. A water-soluble HRGP was extracted from potato tubers in a 0.1 M sodium acetate buffer, pH 3.8, using 2 mM sodium metabisulphite as an antioxidant, at 0–4 °C. It was described as a lectin with agglutinating activity (Allen and Neuberger 1973).

Purification of the lectin was achieved by ammonium sulphate precipitation, subsequent column chromatography of the dialyzed and centrifuged precipitate on DEAE cellulose, CM-cellulose, G 100 Sephadex and SP-Sephadex C-50, from which the lectin eluted as a single peak.

Further examples are as follows:

Other soluble HRGPs with lectin properties, rich in Hyp, Ser but also half-cysteine as opposed to the cell wall glycoproteins, were isolated from tomato (*Lycopersicon esculentum*; Nachbar et al. 1980) and thorn apple (*Datura stramonium*; Desai et al. 1981).

Often the soluble HRGPs e.g., the salt-soluble HRGP from carrot (Stuart and Varner 1980) are not sufficiently characterized to distinguish them from AGPs (Fincher and Stone 1983).

4.1.2.2 Cell Wall-Bound HRGPs (Extensins)

Cell wall HRGPs from runner beans were solubilized by sodium chlorite-acetic acid treatment (Selvendran 1975; Selvendran et al. 1975; O'Neill and Selvendran 1980). By this method partial oxidation or hydrolysis may occur or some amino acids may be either destroyed or modified:

Method 3. After sequential treatment of the wall tissue with 1% sodium deoxycholate and PhOH-HOAc-H_2O (PAW) 2:1:1, cell wall material from runner beans was extracted twice with 2% hexametaphosphate at 95 °C for 2 h. After treatment of the residual material with sodium chlorite-HOAc at 70 °C in a water bath, holocellulose was extracted with 1 and 4 N KOH for 2 h, to remove hemicellulose. The residue was washed with distilled H_2O four times, and finally with absolute ethanol and diethylether. The supernatant obtained after chlorite-ethanol and diethylether. The supernatant obtained after chlorite-HOAc treatment was treated with EtOH, containing 1% HOAc, the resulting precipitate ("crude wall protein") was extracted with PhOH-H_2O and glycoproteins isolated from the H_2O and PhOH layer. The glycoprotein from the H_2O layer contained most of the HRGPs.

Further examples: Mellon and Helgeson (1982) isolated a HRGP from tobacco (*Nicotiana tabacum* L.) callus tissue cultures by the acidic-ethanol extraction procedure of Marinkovich (1964) (see also Sect. 4.1.1.2). Purification by IEC on CM cellulose resulted in a preparation with agglutinating activity against avirulent strains of the bacterial pathogen *Pseudomonas solanacearum*.

Method 4. HRGPs have been isolated by Esquerré-Tugayé et al. (1979) from the wall fraction of moss callus cells (homogenization medium 50 mM Tris buffer, pH 8, 1 mM EDTA, 0.1 M $MgCl_2$) by sequential extraction, using:

1. Boiling detergent (1 mM EDTA, 100 °C, 2 h) to remove pectic substances, or chaotropic salt (guanidine thiocyanate, 6 M, at room temperature) to solubilize glycoprotein; or
2. Two successive extractions with 10% NaOH at 4 and 24 °C to remove hemicellulose.

The wall-bound HRGP was extractable with these media. The results indicated that some of the glycoproteins were not linked covalently together or to other cell wall constituents.

4.1.3 AGPs

Method 1. The effect of the different extraction conditions (variation of pH and temperature) showed differences in amino acid composition and total protein content of β-lectins, isolated from leaves of *Hedera helix* (Gleeson and Jermyn 1979). Ethanol-insoluble material was extracted with 0.01 M sodium phosphate buffer pH 7 with NaCl at 100 or 0 °C. The β-lectins were precipitated from the extract with β-glycosyl Yariv artificial antigen (these AGPs are called β-lectins) (Yariv et al. 1967; Jermyn and Yeow 1975). The supernatant, after centrifugation of the 0 °C-fraction, was treated for 45 min at 100 °C with a pH adjusted either to pH 4 or 7; the β-lectin again precipitated with Yariv antigen.

Method 2. AGPs from wheat endosperm were extracted with water, 3 min at 90 °C (a pretreatment with hot 80% ethanol may be useful), followed by amylase digestion and a subsequent 90 °C water treatment for 1 h. This was followed by treatment with hot 80% ethanol, then a water extraction at 40 °C for 30 min (Fincher and Stone 1974).

Other extraction media are (1) 0.15 M NaCl; (2) 0.05 M Tris HCl, 0.15 M NaCl, 0.001 M $CaCl_2$, pH 7.4 at 4 °C (Gleeson et al. 1979; Gleeson and Clarke 1979); or (3) 0.1 M Na-P-buffer in 1% NaCl, pH 7.0 at 100 °C for 3 min or 0 °C for 1 h (Gleeson and Jermyn 1979); 50 mM maleate-Tris buffer pH 6.5, 5 mM KCl, 2 mM $Na_2S_2O_5$, 1 mM $MgCl_2$, 1 mM EDTA, 0.26 M sucrose (Van Holst et al. 1981).

AGPs are easily obtained from filtrates of suspension culture medium (Hawes and Adams 1972; Akiyama and Kato 1981, 1982a).

4.2 Polysaccharides

Method 1. Slusarenko et al. (1983) isolated a bacteria-agglutinating pectic polysaccharide from different plant species and cultivars by the following method:

Two- to 3-week-old plants were placed in the dark for 24 h to destarch the leaves; tulip leaves were extracted immediately after harvesting. Leaves were comminuted in 300 cm^3 absolute ethanol for each 100 g fresh weight of tissue. The slurry was filtered through a double layer of muslin and resuspended in 300 cm^3 acetone. Filtration and acetone washing was repeated three times and the final residue was air dried. Aliquots of 10 g of the acetone powder were suspended in 200 cm^3 of 0.1 M citrate-oxalate buffer, pH 4.0, and heated in a boiling water bath for 15 min (Jarvis et al. 1978). The slurry was cooled and centrifuged at 16000 g for 10 min. Two volumes of absolute ethanol were added to the supernatant to give a gelatinous precipitate corresponding to the acidic pectic polysaccharide fraction of Barrett and Northcote (1965). The precipitate was collected and dried for ca. 12 h at 50 °C. A further 2 vol of absolute ethanol were added to the residual supernatant to give the neutral pectic polysaccharide fraction (Barrett and Northcote 1965), as a powdery precipitate which was also collected and dried at 50 °C. Dried precipitates were ground and stored in desiccator.

Method 2. Jasalavich and Anderson (1981) isolated an agglutinin for saprophytic pseudomonads from legume tissues by water extraction, followed by DEAE and CM gelfiltration. The filtrate was precipitated with 95% ethanol at 4 °C, the precipitate resuspended in sterile, distilled water, octanol was added, the mixture centrifuged and the supernatant dialyzed against water. Further purification was achieved by an affinity procedure based on the absorption of the agglutinin from the extracts onto *Pseudomonas putida* cells in the presence of Mg^{2+}, followed by its release in the absence of Mg^{2+} ions. The water-extracted agglutinin had properties in common with the plant's "β-lectins" (arabinogalactans).

Method 3. Stevenson et al. (1986) purified and analyzed xyloglucan, arabinoxylan; arabinogalactans; a rhamnogalacturonan-II-like polysaccharide; and a polygalacturonic acid-rich polysaccharide in the extracellular polysaccharide fraction of suspension-cultures of sycamore cells. The filtrate of the culture medium was concentrated with a Pellicon Cassette System (Millipore Corp.) with a 10000-mol.-wt. cut-off cassette (PTGC) after which absolute ethanol was added to a final concentration of 70%. The precipitate was centrifuged, redissolved in K-phosphate (10 mM, pH 7) and dialyzed against buffer and distilled H_2O.

After deesterification with NaOH (pH 12, 2 °C, 2 h), neutralization with acetic acid and dialysis against sodium acetate buffer (50 mM, pH 5.2), the solutions were treated with endopolygalacturonase. Glycosyl and glycosyl-linkage analyses are described in Stevenson et al. (1986).

5 Purification and Characterization: Lectin Techniques

Classical purification steps are: salt (ammonium sulphate: Allen and Neuberger 1973; Fincher and Stone 1983; Young et al. 1984; Shet and Madaiah 1988; Cu-salts: Cartier et al. 1988, aluminium salts) or solvent precipitation (Hawes and Adams 1972), differential centrifugation (Menon and Hartmann 1988), density gradient ultracentrifugation (Fincher and Stone 1974), deproteinization in sodium tetraborate buffer pH 9.8 with 20 vol PhOH (Akiyama and Kato 1981), complexing with Yariv antigen (Gleeson and Jermyn 1979; Gleeson and Clarke 1979) and fractionation by chromatographic methods (IEC and gel filtration, Akiyama and Kato 1981). For purification of glycoconjugates and studying their structure, lectin techniques are very suitable.

In cytochemistry, carbohydrates on the surface of the plant cell wall or in the intercellular space can be localized and quantified by binding of conjugated lectins. Lectin-affinity chromatography may be used for isolation, fractionation and structural analysis of glycoconjugates. Glycoconjugates can be detected in extracts from plant material by staining gels with labelled lectins or by precipitation methods.

5.1 Affinity Chromatography

5.1.1 Principle

The principle of affinity chromatography is depicted in Fig. 1.

A mixture of macromolecules and contaminants is applied onto the column, but only molecules with specific affinities for the ligands are retained or retarded, others are eluted. The adsorbed molecules can be eluted by altering the composition of the solvent (e.g., adding sugar hapten or increasing the ionic strength of the buffer). In affinity chromatography the selection of matrix, ligand and spacer arm may influence the results.

Affinity chromatography thus exploits the unique specificity inherent in a ligand-biomacromolecule interaction. In biospecific affinity chromatography, the immobilized ligand is a biomolecule (e.g., concanavalin A or nicotinamide adenine dinucleotide); in pseudoaffinity chromatography, the ligand is a non-biological molecule (e.g., dyes or metals).

Affinity chromatography in general results in an enrichment of classes of heterogeneous glycoproteins, with similar binding sites recognized by the immobilized lectin receptor. However, an adsorbent can also be constructed that is specific for the cell, macromolecule or oligosaccharide to be purified (Lowe 1979).

Whether or not homogeneous preparations can be obtained by lectin affinity chromatography depends upon the frequency of a specific carbo-

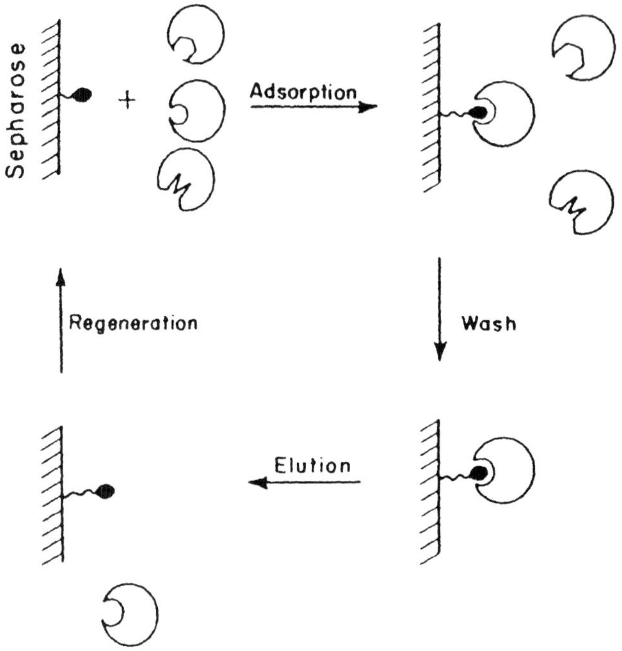

Fig. 1. The principle of affinity chromatography (Wilchek and Givol 1971)

hydrate determinant in the crude extract, or on the degree of previous purification. Even relatively impure plant extracts may be employed, resulting in a rapid and facile separation of the desired molecule from inhibitors and contaminants. Such a "single step" procedure has been reported, e.g., the lectin from winged bean tubers, purified on cross-linked guar gum (Shet and Madaiah 1988) and a lectin from *Gladiolus*-style mucilage (Gleeson et al. 1979).

The "haptens" with the most potency towards lectins in most cases are complex oligosaccharides which may be several 1000-fold more efficient than monosaccharides (Fig. 2; Kornfeld and Ferres 1975; Debray et al. 1981, 1983).

5.1.2 Support Materials

The most important properties of the support material are a high specificity and high binding capacity, along with a good chemical stability. Additionally, an ideal matrix should exhibit a very low general surface charge, a good permeability, no hydrophobic binding sites, excellent mechanical rigidity, good recovery and reproducibility.

Table 4. Particulate support materials for affinity chromatography (Narayanan and Crane 1990)

Trade name	Supplier	Support material
Low to medium pressure		
Trisacryl	IBF, France	Totally synthetic
Ultrogel	IBF, France	Polyacrylamide/agarose
Sepharose	Pharmacia LKB, Sweden	Agarose
Sephadex	Pharmacia LKB, Sweden	Crosslinked dextran
Sephacryl	Pharmacia LKB, Sweden	Polyacrylamide/dextran
Macrosorb	Sterling Organic, UK	Various organic powders
Eupergit C	Roehm Pharma, FRG	Polyacrylamide
Affi-gel	Biorad, USA	Crosslinked polyacrylamide
Matrex Cellufine	Amicon, USA	Crosslinked cellulose
CPG	Pierce, USA	Controlled pore glass
Bakerbond Wide-Pore (Prepscale)	J.T. Baker Inc., USA	Polymer-clad silica
Synthetic		
Dynospheres	Dyno Particles, Norway	Synthetic polymer
Separon H 1000	Tessek, Czechoslovakia	Synthetic polymer
TSK PW	Tosoh Inc., Japan	Polymer
Fractogel		
TSK/Toyopearl	E. Merck, FRG	Vinyl polymer
High pressure		
Hypersil WP300	Shandon, UK	Silica
Lichrospher	E. Merck, FRG	Silica
Ultraspere	Beckman, USA	Silica
Spheron	Waters, USA	Silica
Superose	Pharmacia LKB, Sweden	Crosslinked agarose
Zorbax	E.I. Dupont, USA	Silica

Agarose, polyacrylamide, cellulose and porous glass beads have been most widely used as support materials (Pazur 1981). Limiting properties of agarose include low flow rate, low resolution, susceptibility to microbial degradation, low binding capacity (binding capacity of ConA Sepharose is 10–16 mg ConA $\times$ ml^{-1} drained gel), solubility in hot water and non-aqueous solvents, as well as partial retention of adsorbed molecules. Newly developed support materials are cross-linked agarose (reported to withstand higher back-pressure than agarose), polyacrylamide, synthetic polymers and inorganic materials like silica (Table 4).

The agarose derivatives Sepharose 2B and 4B have been widely used, as have polyacrylic hydrazide-Sepharose (Lotan et al. 1977) and the polyacrylamide derivative Affi-Gel (Davey et al. 1976; Nilson and Waxdal 1976).

Silica, with its small particle size and its rigidity, is used in High Performance Liquid Affinity Chromatography (HPLAC), first introduced by

Concanavalin A (Con A)	GlcNAc(β1-2)Man(α1-3) GlcNAc(β1-2)Man(α1-6) → Man(β1-4)-R_1or R_2[a]
	Man(α1-2)Man(α1-2)Man(α1-3) Man(α1-3), Man(α1-6) → Man(α1-6) → Man(β1-4)-R_1
	Gal(β1-4)GlcNAc(β1-4), GlcNAc(β1-2) → Man(α1-3) GlcNAc(β1-4) Man(α1-3), Man(α1-6) → Man(α1-6) → Man(β1-4)-R_1
Lens culinaris (LCA)	GlcNAc(β1-2)Man(α1-3) GlcNAc(β1-2)Man(α1-6) → Man(β1-4)-R_2
	Gal(β1-4)GlcNAc(β1-2)Man(α1-3) Gal(β1-4)GlcNAc(β1-2)Man(α1-6) Gal(β1-4)GlcNAc(β1-6) → Man(β1-4)-R_2
Ricinus communis agglutinin[b] (RCA_I)	Gal(β1-4)GlcNAc(β1-4), Gal(β1-4)GlcNAc(β1-2) → Man(α1-3) Gal(β1-4)GlcNAc(β1-2), Gal(β1-4)GlcNAc(β1-6) → Man(α1-6) → Man(β1-4)-R_1 or R_2
Leukoagglutinating lectin from *Phaseolus vulgaris* (L_4-PHA)	Gal(β1-4)GlcNAc(β1-2)Man(α1-3) Gal(β1-4)GlcNAc(β1-2), Gal(β1-4)GlcNAc(β1-6) → Man(α1-6) → Man(β1-4)-R or R_2
	Gal(β1-4)GlcNAc(β1-4), Gal(β1-4)GlcNAc(β1-2) → Man(α1-3) Gal(β1-4)GlcNAc(β1-2), Gal(β1-4)GlcNAc(β1-6) → Man(α1-6) → Man(β1-4)-R_1 or R_2
Wheat germ agglutinin (WGA)[b]	NeuAc(α2-6)Gal(β1-4)GlcNAc(β1-2)Man(α1-3) GlcNAc(β1-4) NeuAc(α2-6)Gal(β1-4)GlcNAc(β1-2)Man(α1-6) → Man(α1-4)-R_1

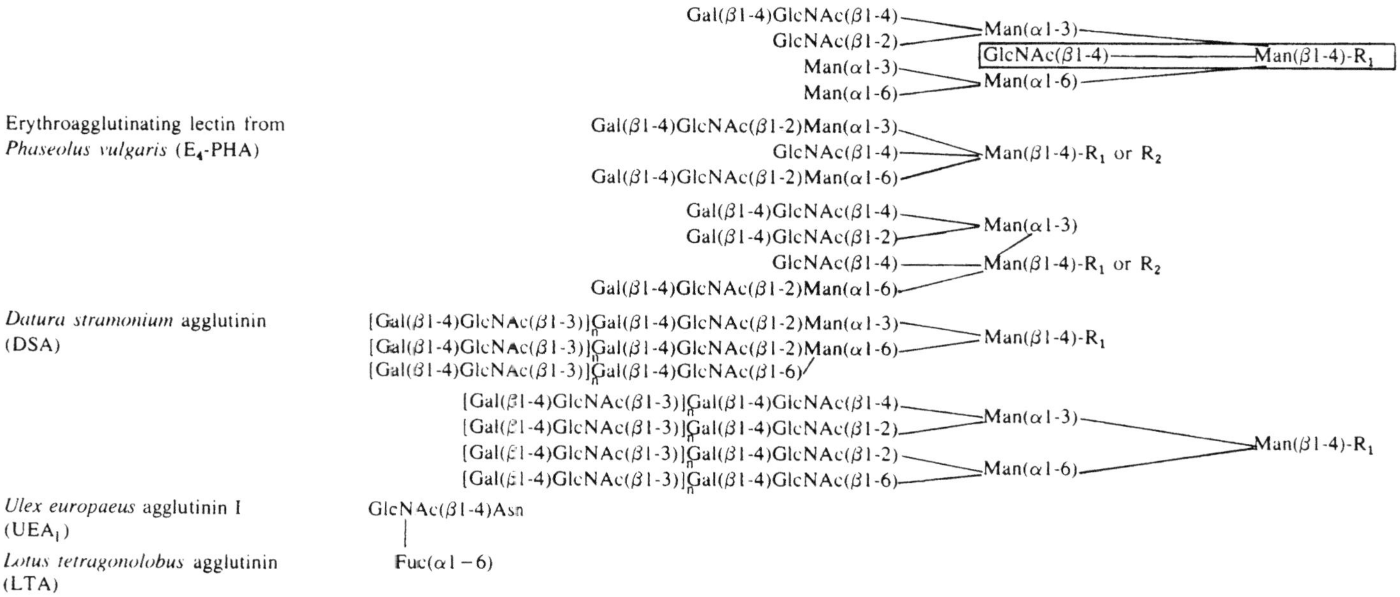

[a] R_1: GlcNAc(β1-4)GlcNAc(β1-N) Asn; R_2: GlcNAc(β1-4)[(Fuc(α1-6)]GlcNAc(β1-N)Asn;
[b] Sequences in boxes are the minimal oligosaccharide structure necessary for lectin recognition.

Fig. 2. Specificity of commonly used lectins towards oligosaccharide sequences belonging to N-glycosyl proteins (Montreuil et al. 1986)

Ohlson et al. in 1978 (Ohlson et al. 1978). Disadvantages of silica are nonspecific adsorption due to residual charge on the surface and solubility at aljaline pH (Narayanan and Crane 1990).

The pore size and the surface area are important factors for each affinity ligand-macromolecule system. To increase the accessibility of the coupled lectin, a spacer arm is used between the support and the ligand. The structural parameters of the ligand and the choice of the spacer strongly influence quality and quantity of the purified lectins.

Gabius (in press) compared 13 different types of linkage and spacer for their efficiency to purify the galactose-binding proteins from two different sources.

5.1.3 Coupling Reaction

For the coupling of a support material with a ligand, an activated form of the support reacts with a derivative of the ligand.

The most frequently used technique for the coupling of lectins to the supporting medium is the activation of agarose beads with cyanogen bromide (Cuatrecasas 1970) followed by the coupling to lectin (Davey et al. 1976; Procedure 1).

The use of glutaraldehyde for the coupling of lectins to polyacrylic hydrazide-Sepharose has been described by Lotan et al. (1977; Procedure 2). A wide range of procedures for coupling proteins covalently to insoluble supporting media is given by Lowe (1979).

Procedure 1. CNBr activation of Sepharose 4B and coupling to lectins according to March et al. (1974):

1. Wash 50 ml of wet Sepharose 4B beads (Pharmacia) on a sintered glass filter with 1 l of chilled distilled water and transfer into a beaker cooled in an ice-bath.
2. Mix 100 ml of ice-cold 2 M K_2CO_3 solution with the wet filtered Sepharose beads, under gentle magnetic stirring. Add 6.25 ml of a freshly prepared CNBr solution in acetonitrile ($2\,g\,ml^{-1}$) and maintain at 0 °C for 2 min in a well-ventilated hood.
3. Quickly pour the entire reaction mixture into a sintered glass filter funnel and wash thoroughly with 3 l of ice-cold distilled water. Transfer the wet activated Sepharose into a 200-ml measuring cylinder and add the ice-cold solution of the lectin to be coupled (0.25 g of lectin dissolved in 0.5 M NaCl + 0.2 M $NaHCO_3$ + 0.2 M haptenic sugar, pH 8.5). Stopper the measuring cylinder and gently stir for 24 h at 4 °C. Filter the reaction mixture through a sintered glass filter funnel and successively wash the coupled gel thoroughly with 1 l each of ice-cold distilled water, 0.1 M $NaHCO_3$ and water.

4. Block the unreacted iminocarbonate groups by suspending the preparation in 100 ml of 1 M glycine solution, under gentle stirring, for 3 h at room temperature.
5. Estimate the approximate amount of uncoupled lectin by measuring the absorbance at 280 nm (protein content) of the reaction mixture and recover the immobilized lectin by filtration, after a 24-h coupling period.
6. Wash the glycine-blocked gel thoroughly with cold distilled water and then with the equilibration buffer containing 0.02% sodium azide. Store the immobilized lectins at 4 °C in this equilibration buffer until required.

Procedure 2. Coupling of lectins to polyacrylic hydrazido-Sepharose with glutaraldehyde (Montreuil et al. 1986).

1. To washed polyacrylhydrazido-Sepharose (Miles), add 3 vol of 10% (w/v) glutaraldehyde solution, under gentle stirring, for 4 h at 4 °C.
2. Dissolve the lectin to be coupled in 0.1 M $NaHCO_3$ + 0.9 M NaCl + 0.1 M haptenic monosaccharide (4–5 mg of lectin ml^{-1} of coupling buffer) and add to the derivatized gel (4–5 mg of lectin ml^{-1} of settled gel beads). Stir the mixture slowly overnight.
3. Wash the gel with phosphate-buffered saline (PBS) and add $NaBH_4$ dissolved in a gel volume of PBS (0.5 mg $NaBH_4$ ml^{-1} of settled gel), for 3 h at 4 °C. After washing with PBS containing 0.02% sodium azide, store the immobilized lectin at 4 °C.

Agarose may also be activated by reaction with divinyl sulphone, followed by the coupling of the vinyl group with appropriate ligands (Fornstedt and Porath 1975; Gabius and Engelhardt 1987; Hajto et al. 1989).

Thus activated, support materials couple to the hemiacetal hydroxyl group of the carbohydrate by the vinyl group. The affinity adsorbents used for this method contain glycosyl units (Pazur 1981).

Another activation-procedure used for the activation of cellulose, glycogen, guaran and some types of porous glass is the 1,1′ carbonylbis (imidazole) method (Bethell et al. 1979).

Montreuil et al. (1986) propose the use of polyacrylic-hydrazido-agarose to avoid non-specific (hydrophobic or ionic) adsorption of glycoproteins on the affinity adsorbents, resulting in a stable, non leaking, uncharged and hydrophilic absorbent. Non-specific adsorption has been observed in some cases with CNBr activated agarose. Non-specific ionic interactions can be reduced using buffers with a moderate ionic strength (0.1–1 M NaCl) while the affinity chromatography is carried out.

Cyanogen bromide-, divinylsulphone- and tresylchloride-activated matrices can rapidly be prepared and coupled, even after the activated matrix has been packed into a column.

Several activation procedures and coupling reactions are given in Table 5 and have been reviewed by Pazur (1981).

Table 5. Activating procedures in affinity chromatography (Narayanan and Crane 1990)

Activating agent	Reagent toxicity	Activation time (h)	Ligand coupling time (h)	pH of coupling	Type of linkage	Stability of complex	Non-specific interactions
Glutaraldehyde	Moderate	1–8	6–16	6.5–8.5	Michael's adduct, Schiff's base	Good	–
CNBr	High	0.2–0.4	2–4 h at 25 °C, overnight at 4 °C	8–10.0	Isourea, imido, carbamate, *N*-substituted carbamate	Unstable at pH <5 or >10	Cationic
Bisoxiranes	Moderate	5–18	15–48	8.5–12.0	Alkylamide, ether, thioether	Excellent	–
DVS	High	0.5–2.0	Rapid	8–10.0	Michael's adduct	Unstable at high pH	–
CDI	Moderate	0.2–0.4	Overnight to 6 days	9–9.5	*N*-substituted carbamate	Unstable at pH >10	–
Periodate	Non-toxic	14–20	Overnight	7.5–8.5	Alkylamine	Good	–
Trichloro-S-triazine	High	0.5–2.0	4–16	7.5–9.0	Triazinyl	Good	π-π Type aromatic
Tresyl chloride	Moderate	0.5–0.8	Rapid	7.5–10.5	Alkylamine	Good	–
Diazonium	Moderate	–	0.5–1.0	6–8	Azo	Moderate	π-π Type aromatic
Hydrazine	High	1–3	3–16	7–9	Amide	Excellent	–
NHS	–	2–4	Rapid	6.5–8.0	Amide	hydrolysis, pH-sensitive	–
4-Nitrophenyl chloroformate	–	0.5	Rapid to 2 days	–	Mixed anhydride, carbamate/urethane	Hydrolysis, pH-sensitive	–

5.1.4 Spacer Molecules

Steric hindrance effects can be reduced using a spacer group between the ligand and the support (Cuatrecasas 1972). Long chain hydrocarbons containing reactive groups on the terminal carbon atom are used as spacer residues. They are first attached to the activated matrix and then coupled to the ligand, or the spacer is first attached to the ligand and then coupled to the support.

Spacer arms of different length and nature are used:

1. Hydrophobic compounds (e.g., 1,6 diaminohexane and 6-aminohexanoic acid covalently attached to CNBr-activated Sepharose);
2. Hydrophilic spacer arms, (prepared by for example coupling 1,3 diamino-2-propanol to CNBr-activated agarose, bromoacetylating and addition of further 1,3 diamino-2-propanol);
3. Multivalent macromolecular molecules, e.g., coupling polypeptides or proteins as poly-(L-lysine) to CNBr-activated agarose through the ε-amino groups of the lysyl residues; or
4. Charge-free spacer molecules, e.g., coupling adipic acid dihydrazide, succinic acid dihydrazide and polyglutamic acid hydrazide or polyacrylic acid hydrazide to CNBr-activated agarose (Lowe 1979).

In order to couple ligands to spacer arms, or directly to CNBr-activated agarose, two types of reactive groups may be introduced into the support matrix: (1) nucleophilic groups: amino, thiol and hydroxyl groups; and (2) electrophilic groups: cyclic imidocarbonates, oxirane groups, isocyanates, isothiocyanates, diazonium salts, phenols, aromatic amines, anhydrides and activated carbonyls, such as acid chlorides and those activated by carbodiimides. This procedure allows affinity chromatography to be applied to almost all ligands.

Examples for coupling of ligands containing amino, carboxyl, aldehyde or ketone, hydroxyl or thiol groups or aromatic functions are given by Lowe (1979).

As already described, nonspecific interactions may arise from the matrix, the spacer arm, the ligand or the coupling agent (e.g., CNBr).

Significant and continuous leakage of the ligand from CNBr-activated supports has been reported (Tesser et al. 1974). The rate of leakage is dependent on pH values (Hauzer et al. 1979), temperature and flow rate (Lowe 1979).

The coupling conditions may influence the properties of the ligand, e.g., a multi-point attachment of ConA leads to deformation of the molecule and changes its chromatographic properties (Davey et al. 1976).

5.1.5 Procedure

Affinity chromatography of glycoproteins on immobilized lectins generally results in three fractions:

1. Non-reactive compounds elute with the void volume of the column with the equilibration buffer,
2. Weakly reactive compounds elute with some retardation with the equilibration buffer,
3. Strongly reactive compounds dissociate from the ligand by addition of the specific sugar at a concentration between 0.1 and 0.5 M or by non specific elution with buffers of high ionic strength (≥1 M NaCl).

A few examples using different chromatographic conditions are briefly described (the method for ConA-affinity chromatography is described by Kogel et al., this Vol.).

Method 1. Three lectin fractions were obtained from PBS-extracts of seeds of the legume *Cytisus scoparius* (Scotch broom) by means of affinity chromatography on an N-acetyl-D-galactosamine medium (Young et al. 1984).

- 50 mg of p-aminophenyl α-acetamido-2-deoxy-α-D galactopyranoside (Calbiochem Behring Corp.) were coupled to an N-hydroxysuccimide-activated agarose gel (Affinity-Gel 10, BioRad Laboratories Canada Ltd.), type described by Cuatrecasas and Parikh (1972) in 0.1 M $NaHCO_3$ at 4 °C overnight.
- the dialysed extract was applied to the column containing the affinity medium, which was then washed with PBS
- three fractions of lectins were obtained by sequential elution with (a) 10 mM D-galactose, (b) 200 mM D-galactose and (c) 50 mM N-acetyl-D-galactosamine, each in PBS.

Method 2. Ahmed and Chatterjee (1988) used a column of starch conjugated with 2-acetoamido-2-deoxy-D-galactose. An agglutinin from *Wistaria sinensis* was fractionated by 40–70% saturation with $(NH_4)_2SO_4$. The column was equilibrated with 0.15 M NaCl and eluted with 0.05 M glycine-HCl buffer (pH 4). The agglutinin appeared as a single peak. The lectin was homogeneous in polyacrylamide disc gel electrophoresis (Ahmed and Chatterjee 1988).

Method 3. A lectin from the tubers of the winged bean (*Psophocarpus tetragonolobus*) was purified in a "single-step" procedure by affinity chromatography on cross linked guar gum (Shet and Madaiah 1988). Guar gum is a galactomannan composed of a chain of β-1,4-linked D-mannopyranosyl residues possessing side stubs of single α-D-galactopyranosyl residues linked α-1,6 to ca. one-half of the β-D-mannosyl residues.

The lectin from winged bean could be purified ten-fold.

The column was irrigated with PBS. The fraction precipitating between 30 and 70% ammonium sulphate saturation was dialyzed against PBS and applied to the column. Desorption of the lectin was achieved using 0.1 M galactose in PBS.

Method 4. An acidic arabinogalactan from *Viscum album* "berries" was isolated by DEAE-Trisacryl M column chromatography and by affinity chromatography on Sephacryl S1000-bound *Ricinus communis* lectin (Wagner and Jordan 1988).

Terminal galactopyranosyl residues in the arabinogalactan provided the binding site for the lectin.

Sephacryl S-1000 (Pharmacia) agarose was CNBr activated. The RCA was dissolved in H_2O containing 1% $NaHCO_3$, added to the agarose and stirred for 2 days at 7 °C, then washed with H_2O and 0.05 M Na-Pi pH 7, 0.15 M NaCl, which was also used as the elutant. The retained arabinogalactan was eluted with 0.2 M lactose in the buffer.

Method 5. An arabinogalactan protein from the style canal of *Gladiolus* was isolated by affinity chromatography with tridacnin (a lectin binding β-galactosyl residues from the clam *Tridacna maxima*) in a one-step procedure (Gleeson and Clarke 1979; Gleeson et al. 1979).

Tridacnin was coupled by 0.1 M $NaHCO_3$, pH 8, containing 0.5 M NaCl to washed CNBr-activated Sepharose 4B. After 2 h at room temperature, any remaining active groups in the gel were blocked by incubating the gel in 0.1 M Tris, pH 8.1, overnight at 1 °C. As the lectin binding is calcium-ion-dependent, after washing, the gel was equilibrated with 0.15 M NaCl containing 0.01 M $CaCl_2$. The crude style extract, dissolved in the equilibrating buffer was loaded on the column and the adsorbed material was specifically eluted by washing the column with 0.15 M NaCl in the absence of calcium ions.

Further Methods. Purification of macromolecules by affinity chromatography is produced by:

1. The purification of phytohemagglutinin lectins from red kidney beans (*Phaseolus vulgaris*) on porcine thyroglobulin Sepharose. The elution was accomplished with 0.05 M glycine-HCl (pH 3.0) containing 0.5 M NaCl (Felsted et al. 1975; Leavitt et al. 1977).
2. The isolation of a lectin from *Amaranthus leucocarpus* by affinity chromatography using a blood group "A" stroma (erythrocyte membrane residues) column. Elution can be carried out with distilled water and/or 0.2 M glycine (pH 2.5) or with N-acetyl galactosamine solutions. The *A.*

leucocarpus binding site on the red cell surface is not yet determined (Callow 1977).

3. Isolectins of *Phaseolus vulgaris* were purified by Pusztai and Palmer (1977), Pusztai and Stewart (1978) on a fetuin-Sepharose 4B column.
4. Lectins from soybean and wheat germ could be purified from Epoxy-activated Sepharose 6B coupled with N-acetyl-D-galactosamine or N-acetyl-glucosamine, respectively. The adsorbed material was eluted with saline containing lactose or acetate buffer pH 4.5 with N-acetyl-D-glucosamine (gradient 0–100 ng ml^{-1}), respectively (Vretblad 1976).
5. Further examples: Soybean agglutinin was purified by affinity chromatography on a column of Sepharose-N-ε-aminocaproyl-β-galactosylamine (Gordon et al. 1972); wheat germ agglutinin on a column of Sepharose-2-acetamido-N-ε-aminocaproyl-2-deoxy-β-glucopyranosylamine (Lotan et al. 1973a); *Lotus tetragonolobus* agglutinin and *Ulex europeus* agglutinin on a column of Sepharose-N-ε-amino-caproyl-β-L-fucopysamosylamine (Blumberg et al. 1972). *Ricinus* agglutinins were eluted from a Sepharose-N-ε-aminocaproyl-β-galactopyranosylamine column with 0.1 M galactose (Gordon et al. 1972).

5.2 Fractionation and Characterization of Oligosaccharides

Another frequent application of affinity chromatography is the structural analysis of binding sites of lectins, especially in fractionation and characterization of their oligosaccharide structures.

5.2.1 Preparation

Oligosaccharides can be obtained from glycoproteins by chemical or enzymatic cleavage. Frequently used methods to release oligosaccharides from glycoproteins or glycopeptides are cited in the references given below and a more detailed description of these methods is given in Burdon and Knippenberg (1985) and Chaplin and Kennedy (1986). In plant glycoproteins, N-glycosidic linkages occur between β-D-glcNAc-asn (Clarke and Shannon 1976) and O-glycosidic linkages between α-D-gal-Ser (Lamport et al. 1973), which are alkali labile, and gal-Hyp (Fincher et al. 1974) and β-L-ara-Hyp (Allen et al. 1978), which are alkali resistant.

O-glycosidic linkages are mostly cleaved by alkaline β-elimination; N-glycosidic linkages are cleaved in most cases by alkaline hydrolysis, hydrazinolysis, trifluoracetolysis (Burdon and Knippenberg 1985; Montreuil et al. 1986) or enzymes. The latter are the mildest and most specific form of cleaving oligosaccharide units from glycoproteins.

Endoglycosidases specifically hydrolyze internal glycosidic linkages and release a part or all of the carbohydrate moiety. The endoglycosidases are divided into three categories: endo-β-N-acetylglucosaminidases, endo-α-N-acetylgalactoaminidases and endo-β-galactosidases. Methods for cleavage by enzymes are described in detail in Burdon and Knippenberg (1985), Chaplin and Kennedy (1986). The liberated oligosaccharides have to be homogenous before attempting structural analysis.

5.2.2 Classical Fractionation Methods

A fractionation of oligosaccharides can be achieved by gel filtration, paper chromatography or thin-layer chromatography. Oligosaccharides containing phosphate or sulphate groups or uronic acids can be fractionated by methods depending on charge such as paper electrophoresis, ion exchange chromatography and certain HPLC systems (Baenziger and Natowicz 1981).

5.2.3 HPLC

Oligosaccharides can be separated with high speed and good resolution by HPLC, using gel permeation, ion exchange, normal phase or reverse phase chromatography. Instructions for neutral oligosaccharide fractionation by normal phase HPLC and reversed phase HPLC are given by Mellis and Baenziger (1981) and Wells et al. (1982), respectively.

5.2.4 Affinity Chromatography

Generally, non-specific interactions rarely occur and an exact description of the primary structure of glycopeptides and oligosaccharides is only possible if the specificity of the immobilized lectin is well defined.

Once the binding specificities of lectins are well established, as they are for plant lectins such as *Ricinus communis* agglutinins (Green et al. 1987), phytohemagglutinins and *Vicia villosa* agglutinin, lectin affinity chromatography can be utilized to distinguish among a large number of closely related oligosaccharide structures (Green and Baenziger 1987; Moroney et al. 1987).

Investigation of the oligosaccharide specificity of twelve lectins have shown that different lectins are able to recognize different saccharidic sequences on the same glycan structure (Debray et al. 1981). These authors elaborated the chemical structures of saccharide determinants recognized by agglutinins. Other studies also showed that different lectins bind to the same cell-surface glycoproteins (Nilsson and Waxdal 1978). The definition of carbohydrate structures with lectins should therefore be interpreted very cautiously.

Homogenous oligosaccharides of known structure were chromatographed on columns of leukoagglutinating and erythroagglutinating phytohemagglutinins (L-PHA, E-PHA) to characterize the specific binding sites of these lectins (Green and Baenziger 1987). The purified oligosaccharides were obtained after enzymatic release (N-glycanase) and chemical reduction ($NaBH_4$).

Methods for fractionation of glycopeptides and oligosaccharides using lectin affinity chromatography are described by Montreuil et al. (1986).

To fractionate and analyze oligosaccharides, lectin affinity chromatography can be preceded by HPLC.

5.2.5 HPLAC

Recently a technique of lectin-affinity high-performance liquid chromatography (HPLAC) for analyzing and separating oligosaccharides has been established. This method was used in order to define the oligosaccharide specificities of *Ricinus communis* agglutinin I and II, using characteristic elution positions obtained for each oligosaccharide of known structure (Green et al. 1987).

5.3 Precipitation Techniques

The methods used in lectin precipitation technique are derived from the antibody technique of immunoprecipitation, immunodiffusion (Ouchterlony 1967), immunoelectrophoresis (Bøg-Hansen et al. 1977), crossed antigen-antibody electrophoresis (Laurell 1965) and agglutination.

5.3.1 Precipitation with Labelled Lectins

By specific precipitation or staining of gels with labelled lectins, glycoconjugates can be characterized. In immunoprecipitation, labelled glycoproteins form a complex with a lectin and the complex is then precipitated by addition of anti-lectin antibody (Juliano and Li 1978). The lectin-glycoprotein-antibody precipitate is dissolved, the labelled glycoprotein isolated under denaturing conditions, and detected by SDS-gel electrophoresis and autoradiography.

5.3.2 Molecular Basis of Precipitation Reactions

Hydrogen bonds between hydroxyl groups of sugars and the binding sites of the proteins were studied by Quiocho and Vyas (1984) and Goldstein and Poretz (1986) as one of the primary factors that are responsible for the stability and specificity of carbohydrate-lectin interactions.

In hemagglutination-inhibition assays/deoxy-, thio- and fluorogalactoses were used to gain insight into the strength of the hydrogen bonds and the nature of donor/acceptor groups in the galactose-specific lectins from *Erythrina indica* (EIL); *Erythrina arborescens* (EAL); *Ricinus communis* (RCA-I); *Abrus precatorius* (APA); and *Bandeira simplicifolia* (BSL-I; Bhattacharyya and Brewer 1988a).

Studies on the free energy contribution of the hydrogen bonds demonstrated that strong hydrogen bonds occur with C-3 and C-4 hydrogen groups of sugars via charged groups in the proteins (Bhattacharyya and Brewer 1988a).

5.3.3 Precipitation and Inhibition Assays: Establishing Branching Patterns

Further studies on binding and precipitation of the lectins RCA-I, EIL, EAL and *Glycine max* (soybean, SBA) with complex-type oligosaccharides and glycopeptides revealed that the extent of precipitation depended on the valency of the oligosaccharides as well as on their branching patterns (Bhattacharyya et al. 1988). Bi-, tri- and tetraantennary oligosaccharides were found to be bi-, tri- or tetravalent for binding to the lectins. Each arm of the branched chain oligosaccharides could bind individual lectin molecules, which led to cross-linking and precipitation. In the case of two triantennary oligosaccharides with nearly the same affinities, the lectins showed different degrees of sensitivity to their branching patterns. These results were obtained by quantitative precipitation of the lectins in the presence of different oligosaccharides and the inhibition of precipitation. The precipitation assays were performed either with 0.1 M Tris-HCl buffer, pH 7.2 containing 0.9 M KCl, 1 mM $MnCl_2$ and 1 mM $CaCl_2$; or with PBS as described in Battacharyya et al. (1987a,b).

Broad similarity in the carbohydrate binding properties of RCA-I, EIL, EAL and APA were indicated by the results of inhibition of hemagglutination by the five lectins with complex-type oligosaccharides.

The binding and precipitation activities of complex-type carbohydrates can also be mimicked by cluster glycosides, as reported for the interaction of EIL and RCA-I with synthetic cluster glycosides (Battacharyya and Brewer 1988b).

With quantitative precipitation and precipitation-inhibition assays, and the use of immobilized lectin affinity chromatography on phytohemagglutinin agarose (L-or-E-PHA), the specificities of L-PHA and E-PHA or oligosaccharides with different branching patterns were established (Hammarstrom et al. 1982; Green and Baenziger 1987).

Quantitative precipition-inhibition assays with RCA I and 31 complex carbohydrates revealed that hydrophobic interactions are important for binding (Wu et al. 1988). Precipitation of lectins with proteins is also dependent on the pH and ionic strength of the buffer solution (Freier and

Rüdiger 1987). To characterize glycolipid-lectin interactions, binding of ^{125}I-labelled lectins to glycolipids on thin-layer chromatograms can be used to rapidly analyze glycolipid extracts for specific oligosaccharide structures (Smith 1983; Torres et al. 1988).

5.3.4 Agglutination

Lectins can bind to specific carbohydrate determinants of cell surface glycoproteins or glycolipids. This non-covalent, reversible interaction between lectins, which contain at least two carbohydrate-binding sites per molecule, and carbohydrate chains of the cell surface, leads to an agglutination of cells. Using lectins with well-defined specificities, agglutination and its specific inhibition by oligosaccharides with known structure can be used to quantify and determine the structure of carbohydrate units at the cell surface.

To form cross-links between glycoproteins and glycolipid determinants on the cell, the multivalent lectins have to overcome mutual repulsion due to charged groups at the cell surface.

5.3.4.1 Techniques

A widely used agglutination method involves an initial mixing of cells, followed by a period of time in which the cells settle due to gravity. This procedure is carried out in wells of microtitre trays, which are available in different forms and volumes. Aggregated cells stay in suspension or form a carpet of cells covering the bottom of the well, while non-aggregated cells slide down to form a compact button or ring at the centre of the well (Herbert 1978; Wydra 1991). The agglutination is usually graded as ++++ (an even carpet of cells), +++, ++, +, ±, − (a negative button).

With serial dilutions of the lectin, the titre may be recorded as the end point of agglutination or as the reciprocal of this dilution, known as agglutination index.

Microtube slides can be used for the detection of agglutinated cells under the microscope, but the exact estimation of the degree of agglutination is rather difficult. Amongst agglutination tests with phytopathogenic bacteria, the method showing the most distinct differences proved to be the simple evaluation of agglutination in a drop in a petri dish under a dissecting microscope (Wydra 1991). These experiments can reveal information about the surface structure of cells and their potential interaction.

5.3.4.2 Inhibition of Agglutination

The ability of mono-, oligo- or polysaccharides of known specificity to inhibit agglutination of cells by a lectin may indicate that antigenic structures are present on the cell. Agglutination of phytopathogenic bacteria by an agglutinin derived from the host plant cell wall and the inhibition of the agglutination in vitro by bacterial extracellular polysaccharide, may give insight into possible specific interactions occurring in host-pathogen systems (Wydra and Rudolph 1989, 1990; Wydra 1991).

In general, specific binding is distinguished by its reversibility by simple monosaccharides, oligosaccharides or glycoside inhibitors. Inhibitors can be added before or after the lectin and should produce a reversal of lectin binding at concentrations of about 100 mM. If sugar inhibitors, added after the lectin, fail to completely reverse binding, despite their ability to prevent binding when added prior to lectin, internalization of lectins by cells can be assumed (Philips et al. 1974).

5.3.4.3 Factors Affecting Agglutination

The complex process of agglutination is influenced by:

1. The nature of the lectin, its size and valency (Lotan et al. 1973b);
2. The binding sites of the lectins: many lectins interact with more than a single monosaccharide unit, and often the most potent inhibitors are oligosaccharides or glycopeptides (Allen and Neuberger 1973);
3. The number, distribution, mobility and steric factors arising from the structure of the carbohydrate receptors within a carbohydrate unit (Beeley et al. 1983);
4. Non-specific binding by hydrophobic interactions, e.g., hydrophobic sites were detected in Concanavalin A (Edelman and Wang 1978) and lectin from lima bean (*Phaseolus lunatus*; Roberts and Goldstein 1982);
5. Temperature, which may influence the association-dissociation equilibria of lectins (Huet et al. 1974). The rates of association and dissociation generally increase with temperature.

Besides temperature, other experimental factors affect the complex process of agglutination and have often led to varying, occasionally contradictory results (Schnebli 1976). The rate of agglutination may be influenced by the degree and kind of mixing, the molarity and pH of the buffer used in the test, by the tendency of certain bacteria to self-agglutinate and by the pretreatment of the bacteria (e.g., washing of bacteria in 0.001 M NaCl or with EDTA; Wydra 1991).

Purification and isolation of agglutinins may be improved considerably when the specific and unspecific effects on the agglutination reaction are known.

5.3.4.4 Function in Plant-Pathogen Interactions

Studies in vitro on agglutination (El-Banoby and Rudolph 1980) and inhibition between bacteria, bacterial lipopolysaccharides, extracellular polysaccharides (EPS) and agglutinin from plant cell walls, suggest complex interactions between these macromolecules. These interactions are strongly affected by the specific composition of the macromolecules. The chemical composition of the bacterial component in the agglutination reaction is affected by pathogenicity and virulence of the bacterial species/strains and by the culture conditions. In the case of EPS of *Pseudomonas syringae* pv. *phaseolicola*, decisive features were: (1) the ratio of levan to alginate (Wydra 1991); and (2) physical properties, e.g., viscosity and tendency or ability for gel formation with other macromolecules. (3) Differences in composition and biological properties of plant agglutinins according to their origin from resistant or susceptible, healthy or infected plants also affected the agglutination ability.

The examples described for precipitation and agglutination assays may reveal the precise recognition phenomena between carbohydrate structures. A further characterization of the interactions may help to understand the structure-function relationships in cellular recognition phenomena and host-pathogen interactions.

6 Cytochemical Techniques

Electron microscopic studies are a valuable means to demonstrate specific polysaccharides in situ. In conventional electron microscopy, staining was conducted with more or less specific dyes, whereas newly developed methods such as enzyme and lectin-gold staining or staining based on immunological interactions allow more specific studies.

6.1 Classical Staining Methods

In conventional staining methods, uranyl acetate and lead citrate have often been used (Clarke et al. 1981; Leach et al. 1982; Harper et al. 1987). A standard procedure is described in the following method.

6.1.1 Staining with Uranyl Acetate and Lead Citrate

Solutions (Huang 1990):

1. Uranyl acetate staining solution

   1 g of uranyl acetate is added to 100 ml of 50% ethyl alcohol, shaken at 5-min intervals for 30 min, filtered through a 0.45-μm membrane, and stored in an amber bottle at 4 °C.

2. Lead citrate staining solution

Lead nitrate, $Pb(NO_3)_2$	1.33 g
Sodium citrate, $Na_3(C_6H_5O_7) \cdot 2H_2O$	1.76 g
Distilled water	30.00 ml

Shake the ingredients in a 50 ml volumetric flask for 30 min. Add 8 ml of 1 N NaOH, and adjust the volume to 50 ml with distilled water. Adjust the pH to 12 ± 0.1 with NaOH if necessary. Filter through a 0.45-μm membrane. Store at 4°C.

Staining: The grids prepared for transmission electron microscopy (Huang 1990) are transferred to perforated BEEM capsules and stained with uranyl acetate staining solution for 40 min, washed three times with distilled water (5 min each), post-stained with lead citrate staining solution for 15 min and finally washed three times with distilled water (5 min each). Be sure the grids are completely submerged in staining solution or water throughout the operation.

6.1.2 Staining with Polysaccharide-Specific Dyes

Certain classes of polysaccharides can be differentiated cytochemically, e.g.: (1) ruthenium red strongly stains not only acidic polysaccharides such as pectins, but also the highly branched acidic polymers present in many forms of bacterial extracellular polysaccharides (Leach et al. 1982; Ebrahim-Nesbat and Slusarenko 1983; Harper et al. 1987; Brown and Mansfield 1988). (2) The iron adsorption method results in a blue staining with pectins (Fry and Milholland 1990). (3) Phloroglucenol is used for staining of gums (Fry and Milholland 1990); (4) toluidine blue for acidic polysaccharides (Roebuck et al. 1978); (5) methylene blue to differentially stain cellulose (blue), lignin (green) and pectin (violet) (Cline and Neely 1983; Lawson et al. 1989; Bretschneider et al. 1989). (6) PATAg (periodic acid thiocarbohydrazide silver proteinate), a more specific dye, predominantly reacts with 1,4-linked molecules, notably cellulose and pectic materials.

6.1.3 Staining of 1,4-Linked Molecules with PATAg

Brown and Mansfield (1988) have described the following procedure: inoculated leaves are fixed with 4% formaldehyde in 0.05 M sodium cacodylate buffer (pH 7.2) at 4 °C for 16 h, washed in buffer, dehydrated in a graded acetone series, embedded in Epon-Araldite, and sectioned. Free-floating sections are stained using the periodic acid thiocarbohydrazide silver proteinate procedure (PATAg) which visualizes polysaccharides with vicial glycols (Roland 1978). Treated sections are collected on uncoated 300-mesh copper grids and examined without further staining. The PATAg procedure was also applied to tissue fixed in glutaraldehyde and osmium tetraoxide, but the authors found more intense staining after treatment with formaldehyde. Consequently, this fixative is used for all quantitative studies. Controls may be run as suggested by Roland (1978), using hydrogen peroxide in place of periodic acid; staining is greatly reduced by the peroxide treatment (Brown and Mansfield 1988).

6.2 Enzyme-Gold Staining

Enzyme-gold complexes have been used for the ultrastructural localization of hemicelluloses in the plant cell wall. Thus, gold-conjugated endo-β-1-4 xylanase stains specifically xylan, β-1-4 mannanase visualizes the pectic polymer glucomannan in the cell wall (Ruel and Joseleau 1984), and β-1-4 exoglucanase and β-glucosidase are used to localize cellulose and β-glucosides, respectively (Chamberland et al. 1989). The procedure of Ruel and Joseleau (1984) is briefly described below.

Preparation of Colloidal Gold. The smallest colloidal gold particles were obtained by reducing chloroauric acid ($HAuCl_4$) with white phosphorus as proposed by Faulk and Taylor (1971) and modified in 1977 by Horisberger and Rosset. A 1% solution of chloroauric acid (3 ml) is added to distilled water (240 ml) in a carefully cleaned flask and the solution is neutralized with 0.2 N K_2CO_3. An ether solution consisting of 4 parts of diethyl ether and 1 part of phosphorus-saturated ether is prepared a few hours before use. The mixture is shaken and allowed to stand for 15 min at room temperature. The solution is then heated under reflux for 5 min at which time a wine-red colour develops. Under these conditions the diameter of the gold granules may vary from 4–9 µm.

The preparation of gold complexes needs some preliminary conditions:

1. Enzyme solutions must be freed from salts or remaining ampholines when prepared by electrofocussing.
2. Determination of the minimal amount of enzyme necessary for full stabilization of the colloidal gold. The technique described by Horisberger et al. (1975) may be used. A constant volume of colloidal gold (2.5 ml) is mixed with 50 µl of serial dilutions of each enzyme. After 5 min, 0.25 ml of 10% NaCl solution is added and flocculation is judged visually (when flocculation occurs, the red colour of the enzyme-gold complex initially turns violet, then blue).
3. Formation of the enzyme-gold complexes. Xylanase-gold complex: 3 ml of the xylanase solution diluted 36 times (initial activity 0.75×10^{-6} reducing equivalent released per milliliter of substrate solution) are added to 15 ml of the colloidal gold. The mixture is shaken and allowed

to stand 5 min at room temperature. It is then centrifuged at 54 000 g. The enzyme-gold complex forms a light dark red sediment, which is carefully collected and resuspended in water without adding polyethylene glycol. The activity of the complex is then tested before use by measuring the reducing sugars.

Use of Enzyme-Gold Complexes on Ultrathin Sections. All the samples used are fixed in aldehydic mixtures (glutaraldehyde 3.5% or Karnovsky's medium (1965)) and embedded in glycol methacrylate (GMA) according to Leduc and Bernhard (1967) or in 50/50 methyl-butyl methacrylate mixture with 1% of benzoyl peroxide added as a catalyst.

Ultrathin sections (silver or white) are collected on plastic rings and floated on a drop of the enzyme-gold complex in the dark.

A contact period of 15–30 min is generally sufficient to obtain a good marking of the walls. The rings are thoroughly rinsed with distilled water before being collected on carbon-coated grids and investigated without further staining in an electron microscope.

Controls: sections are incubated with enzyme-gold complexes prepared with preheated enzymes.

6.3 Lectin-Gold Staining

Lectin-gold complexes enable specific staining of polysaccharides and glycoconjugates by exploiting the highly specific interactions between lectins and carbohydrate containing polymers.

Lectins used for the specific indentification of sugar residues are given in Table 6.

Lectin gold complexes have been used to localize and characterize the residues in paramural bodies. D-Galactose, N-acetyl-galactosamine, β-glucosides and galacturonic acid-containing molecules were part of the paramural bodies (Benhamou et al. 1988; Chamberland et al. 1989). The following method describes lectin-gold staining of tomato root samples (Chamberland et al. 1989):

Tissue Processing. Infected and healthy tomato root samples are fixed with 3% glutaraldehyde in 0.1 M sodium cacodylate, pH 7.2, rinsed with the same buffer, and post-fixed with 1% osmium tetroxide in sodium cacodylate buffer. They are then dehydrated in a graded series of ethanol, and embedded in Epon 812. Ultrathin sections are collected on nickel grids and processed for cytochemical labelling.

Preparation of Protein-Gold Complexes. Colloidal gold solutions with particles averaging 15 nm in diameter can be prepared according to Frens (1973). For each protein (Table 6) the gold suspension is adjusted to their

Table 6. List of lectins (proteins) and enzymes, and conditions for complex formation (Chamberland et al. 1989)

Lectin (protein)	Source	Simple sugar specificity	Minimal amount of protein needed to stabilize 10 ml colloidal gold (μg)	pH of colloidal gold for conjugation	pH of pellet resuspension and sections incubation
Ricinus communis agglutinin (RcA_1)	Caster oil plant	D-galactose	60	8.0	6.0
Helix pomatia agglutinin (HpA)	Roman snail	α-*N*-acctyl-D-Galactosamine	100	7.4	7.4
Concanavalin A (ConA)	*Canavalia ensiformis* (jack bean)	α-D-mannose α-Methyl mannoside α-Glocose	250	8.0	7.4
Ulex europaeus agglutinin I (UcA_1)	Gorse	α-L-Fucose	250	6.3	7.4
Wheat germ agglutinin (WGA)	*Triticum vulgare*	*N*-acctyl-D-glucosamine	–	–	–
Ovomucoid	Egg white	WGA	20	4.8	6.0
Aplysia gonad lectin (AGL)	*Aplysia depilans*	Galacturonic acids	50	9.5	8.0
β-(1→4) exoglucanase	*Trichoderma harzianum*	β-(1→4) Glucans	100	9.0	6.0
β-Glucosidase (E.C. 3.2.1.21)	Almonds	β-D-Glucosides	40	9.3	5.2–6.0

isoelectric point, using 0.2 M K_2CO_3. Protein-gold complexes are prepared according to Bendayan (1984). Minimal amounts of the proteins (Table 5) are dissolved in 100 µl of distilled water to which 10 ml of gold suspension at the appropriate pH are added. After centrifugation at 28 000 g for 1 h at 4 °C, red pellets are obtained and recovered in 0.5 ml of phosphate-buffered saline (PBS) containing 0.02% of polyethylene glycol (PEG) 20 000; the pH of the PBS-PEG buffers are previously adjusted to the pH corresponding to the pH optimum for the activity of each protein. Ovomucoid is used as an indirect label for wheat germ agglutinin (WGA) on the basis of WGA affinity for the ovomucoid sugar groups (Roth 1983). The ovomucoid was complexed to colloidal gold at pH 4.8.

Cytochemical Labelling. Sections of samples taken 4–5 days after inoculation are first floated on a drop of PBS-PEG at the appropriate pH (Table 5), then incubated in a drop of enzyme or lectin-gold complex for 30 min at room temperature in a moist chamber. The sections are then washed with PBS at pH 7.4, and thoroughly rinsed with distilled water. For the indirect labelling with WGA, sections are first incubated in a drop of WGA (50 µg $\times$ ml^{-1} in PBS) for 30 min at room temperature, then washed with PBS and incubated in a drop of diluted ovomucoid-gold complex (1:30 in PBS-PEG, pH 6) for 30 min before washing with PBS and distilled water.

Sections are contrasted with uranyl acetate and lead citrate before examination in an electron microscope.

Cytochemical Controls. Specificity of the labelling is assessed by means of the following control tests: (1) incubation with the gold-complexed protein to which its corresponding sugar molecule or substrate has been previously added; (2) incubation with WGA, previously adsorbed with N-acetylchitotriose, followed by incubation with the ovomucoid-gold complex (indirect labelling), (3) incubation with bovine serum albumin (BSA)-gold complex; (4) incubation with stabilized or non-stabilized gold suspension.

Thus, it was found that polygalacturonic acid-rich fragments were released from the cell wall and that polygalacturonic acid containing molecules accumulated at strategic sites such as wall appositions and in intercellular spaces (Benhamou et al. 1990c). These findings provided new insights into the implication of pectic substances in the plant defense reaction and supported the hypothesis that they may function as signaling substances capable of mediating defensive molecular responses of the host. Since these reactions finally lead to necrotization, the compound eliciting this response may be classified as toxic.

6.4 Immunological Staining Techniques

Antibodies raised against purified plant cell wall polymers can most specifically localize specific target molecules.

With antiserum against HRGPs, coupled to fluorescein isothiocyanate (immuno-fluorescence method) or complexed with gold, HRGPs could be localized in or on cell walls (Brown and Kimmins 1978), in excrescences formed along cell walls, in intercellular spaces of infected tissue (Benhamou et al. 1990a), and at places of contact between the pathogen and the plant cell wall (Benhamou et al. 1990b). Antibodies against rhamnogalacturonan and xyloglucan revealed the Golgi apparatus as the site of synthesis of these cell wall polysaccharides and as the source of secretory vesicles (Moore et al. 1986; Moore and Staehelin 1988). The following method describes the demonstration of agglutinins in tobacco and potato tissue by immunofluorescence (Leach et al. 1982).

6.4.1 Immunofluorescence

For fluorescence microscopy, petiole or leaf tissue from tobacco or potato is frozen in an embedding medium containing 40 mg $\times$ ml^{-1} gum Acacia, 125 mg $\times$ ml^{-1} sucrose and 0.13 mg $\times$ ml^{-1} Thimerosol in distilled water and sectioned with a cryostat. Sections are collected on Millipore membrane filters (0.42-µm pore) and rinsed carefully with PBS in a Millipore filtering apparatus. The sections are flooded with sheep normal immunoglobulin (A_{280nm} = 1.0) in PBS, incubated at room temperature for 20 min, and then rinsed. Purified anti-agglutinin antibodies (A_{280nm} = 0.1) in PBS or normal rabbit IgGs (A_{280nm} = 1.0) are added to the sections. After a 20-min incubation, the sections are rinsed again. FITC-conjugated sheep anti-rabbit immunoglobulins are added to the sections which are then incubated in the dark for 20 min. Following a final rinse, the sections are carefully washed off the filters onto slides coated with sodium silicate (1 part Na_2SiO_3 to 3 parts distilled water) and blotted dry. A drop of mounting buffer (9 parts glycerol to 1 part PBS, pH 7.6) is added to the sections which are then covered with a No. 1 coverslip.

In indirect immunofluorescence, the sections were first stained with monospecific antibodies, and then with secondary fluorescent antibodies. Lectins of *Phaseolus vulgaris* were detected in protein bodies in the cell and in the cytoploasma. Using immuno-gold staining, leg hemoglobulin could be localized in the modules of lupin plants (Vivo et al. 1989). Problems which may arise using the immunological techniques are the ability of plant cell walls to adsorb many antibodies and gold-labelled markers by non-specific means through the charged sugar residues of the cell wall polymers (Craig et al. 1987) and, in the case of immunofluorescence, the autofluorescence of many cell walls. Therefore, before the antibody-gold complex is added, non-specific sites of protein adsorption are saturated with bovine serum albumin (Craig et al. 1987) or the antigenic sites on the cell wall can be masked by 0.56 M sodium metaperiodate (Vivo et al. 1989). For comparison, similarly treated control plants can reveal changes in cell wall composition in response to infection.

The next method describes immuno-gold labelling of HRGPs in tobacco tissue.

6.4.2 Immuno-Gold Labelling

Ultrathin sections of inoculated or healthy tobacco leaves are floated on a drop of phosphate-buffered saline (PBS), pH 7.2, containing 0.2% (w/v) ovalbumin for 5 min and incubated in a drop of normal goat serum diluted in 1:10 in PBS-ovalbumin for 60 min at room temperature. The sections are transferred to a drop of the antiserum raised against melon HRGPs at a 1:200 dilution in PBS-ovalbumin and incubated for 2 h at 37 °C in a moist chamber. After washing with Tris-HCl, pH 8.2, containing 0.5 M NaCl and 1% (w/v) BSA, sections are incubated in a drop of gold-conjugated goat antiserum to rabbit immunoglobulins (GAR-gold antibodies) diluted 1:10 in Tris-HCl, NaCl, BSA for 30 min at room temperature. The sections are finally washed with PBS, pH 7.2, rinsed with distilled water, and stained with uranyl acetate and lead citrate prior to examination in an electron microscope. By this method, Benhamou et al. (1990a) demonstrated the accumulation of HRGPs in response to virus infection.

Immunocytochemical Controls. Specificity of the immuno-gold labelling may be assessed by various controls: (1) incubation with the melon HRGP antiserum which has been previously adsorbed with purified melon HRGPs (approx. 1:1 molar ratio); (2) incubation with the melon HRGP antiserum which has been previously adsorbed with purified tobacco HRGPs; (3) incubation with pre-immune serum instead of anti-HRGP antiserum; (4) incubation with GAR-gold antibodies only, the antiserum step being omitted; (5) successive incubations with antiserum, unlabelled goat antiserum to rabbit immunoglobulins, and finally GAR-gold antibodies.

Acknowledgements. We thank Dr. A.J. Buchala, Fribourg, Switzerland, for critically reading the manuscript and for many valuable suggestions to improve the review.

Thanks are also due to Mrs. H. Rollwage for reliably and carefully typing the manuscript.

References

Adams MF, Ettling BV (1973) Larch arabinogalactan. In: Whistler RL (ed) Industrial gums, 2nd edn. Academic Press. New York, p 415

Ahmed H, Chatterjee BP (1988) *Wistaria sinensis* agglutinin: purification, carbohydrate specificity, and characterisation of the combining site. Carbohydr Res 177:173–183

Akiyama Y, Kato K (1981) An extracellular arabinogalactan-protein from *Nicotiana tabacum*. Phytochem 20(11):2507–2510

Akiyama Y, Kato K (1982a) An arabinoxyloglucan from extracellular polysaccharides of suspension-cultured tobacco cells. Phytochemistry 21:2112

Akiyama Y, Kato K (1982b) Methylation analysis of extracellular polysaccharides from suspenion-cultured cells of *Nicotiana tabacum*. Phytochemistry 21:1325
Albersheim P, Anderson-Prouty AJ (1975) Carbohydrates, proteins, cell surfaces, and the biochemistry of pathogenesis. Annu Rev Plant Physiol 26:31–52
Albersheim P, Wolpert J (1977) Molecular determinants of symbiont-host selectivity between nitrogen-fixing bacteria and plants. In: Solheim B, Raa J (eds) Cell wall biochemistry. Universitetsforlaget, Oslo, pp 373–376
Allen AK, Neuberger A (1973) The purification and properties of the lectin from potato tubers: a hydroxyproline containing glycoprotein. Biochem J 135:307–314
Allen AK, Neuberger A, Sharon N (1973) The purification, composition and specificity of wheat-germ agglutinin. Biochem J 131:155–162
Allen AK, Desai NN, Neuberger A, Creeth JM (1978) Properties of potato lectin and the nature of its glycoprotein linkages. Biochem J 171:665–674
Anderson AJ (1982) Preformed resistance mechanisms. In: Mount MS, Lacy GH (eds) Phytopathogenic prokaryotes, vol 2. Academic Press, New York, pp 119–136
Anderson RL, Clarke AE, Jermyn MA, Knox RB, Stone BA (1977) A carbohydrate-binding arabinogalactan-protein from liquid suspension cultures of endosperm from *Lolium multiflorum*. Aust J Plant Physiol 4:143–158
Aspinall GO (1970) Polysaccharides. Pergamon Press, Oxford
Aspinall GO (1973) Carbohydrate polymers of plant cell walls. In: Loewus F (ed) Biogenesis of plant cell wall polysaccharides. Academic Press, New York, p 95
Aspinall GO (1980) Chemistry of cell wall polysaccharides. In: Preiss J (ed) The biochemistry of plants, vol 3. Academic Press, New York, pp 473–500
Aspinall GO (1981) Constitution of plant cell wall polysaccharides. In: Tanner W, Loewus FA (eds) Plant carbohydrates. II. Extracellular carbohydrates. Springer, Berlin Heidelberg New York, pp 2–8
Aspinall GO (1982a) General introduction. In: Aspinall Go (ed) The polysaccharides, vol 1. Acad Press, New York, pp 1–18
Aspinall GO (1982b) Isolation and fractionation of polysaccharides. In: Aspinall Go (ed) The polysaccharides, vol 1. Acad Press, New York pp 19–34
Aspinall GO, Molloy JA, Craig JWT (1969) Extracellular polysaccharides from suspension-cultured sycamore cells. Can J Biochem 47:1063–1070
Baenziger JU, Natowicz M (1981) Rapid separation of anionic oligosaccharide species by high performance liquid chromatography. Anal Biochem 112:357–361
Baldo BA, Neukom H, Stone BA, Uhlenbruck G (1978) Reaction of some invertebrate and plant agglutinins and a mouse myeloma anti-galactan protein with an arabinogalactan from wheat. Aust J Biol Sci 31:149–160
Barber M, Bordoli RS, Sedgwick RD, Tyler AN (1981) Fast atom bombardment of solids as an ion source in mass spectrometry. Nature 293:270–275
Barmore CR, Nguyen TK (1985) Polygalacturonase inhibition in rind of Valencia orange infected with *Diplodia natalensis*. Phytopathology 75:446–449
Barnoud F, Mollard A, Dutton GGS (1977) Une xyloglucane β 1→4 présente dans le milieu de culture des suspensions cellulaires de *Rosa glauca*. Physiol Veg 15:153
Barrett AJ, Northcote DH (1965) Apple fruit pectic substances. Biochem J 94:617–627
Bauer WD, Talmadge K, Keegstra K, Albersheim P (1973) The structure of plant cell walls. II. The hemicellulose of the walls of suspension-cultured sycamore cells. Plant Physiol 51:174–87
Bauer WD, Bhuvaneswari TV, Pueppke SG, Kettering LF (1977) The involvement of soybean lectin in host plant recognition of the nitrogen-fixing symbiont *Rhizobium japonicum*. In: Solheim B, Raa J (eds) Cell wall biochemistry. Universitetsforlaget, Oslo, pp 377–380
Bebault GM, Dutton GGS, Funnell NA, Mackie KL (1978) Structural investigation of *Klebsiella* serotype K32 polysaccharide. Carbohydr Res 63:183–192
Beeley JG, Altoumah B, Jardine JJ (1983) The interaction between wheat-germ agglutinin and ovomucoid. Biochem Soc Trans 11:103–104
Bell RM, Mavis RD, Osborn J, Vagelos PR (1971) Enzymes of phospholipid metabolism: localization in the cytoplasmic and outer membrane of the cell envelope of *Escherichia coli* and *Salmonella typhimurium*. Biochim Biophys Acta 249:628–635

BeMiller JN (1967) Acid-catalyzed hydrolysis of glycosides. Adv Carbohydr Chem 22:25

Bendayan M (1984) Enzyme-gold electron microscopic cytochemistry: a new affinity approach for the ultrastructural localization of macromolecules. J Elect Micros Tech 1:349–372

Benhamou N, Ouellette GB, Gardiner RB, Day AW (1986) Immunocytochemical localization of antigen-binding sites in the cell surface of two ascomycete fungi using antibodies produced against fimbriae from *Ustilago violacea* and *Rhodotorula rubra*. Can J Microbiol 32:871–883

Benhamou N, Chamberland H, Ouellette GB, Pauzè FJ (1988) Detection of galactose in two fungi causing wilt diseases and in their plant host tissues by means of gold-complexed *Ricinus communis* agglutinin I. Physiol Mol Plant Pathol 32:249–266

Benhamou N, Mazau D, Esquerré-Tugayé MT, Asselin A (1990a) Immunogold localization of hydroxyproline-rich glycoproteins in necrotic tissue of *Nicotiana tabacum* L. cv. Xanthi-nc infected by tobacco mosaic virus. Physiol Mol Plant Pathol 36:129–145

Benhamou N, Mazau D, Esquerré-Tugayé MT (1990b) Immunocytochemical localization of hydroxyproline-rich glycoproteins in tomato root cells infected by *Fusarium oxysproum* f. sp. *radicis-lycopersici*: study of compatible interaction. Phytopathology 80:163–173

Benhamou N, Chamberland H, Pauzé FJ (1990c) Implication of pectic components in cell surface interactions between tomato root cells and *Fusarium oxysporum* f. sp. *radicis-lycopersici*. Plant Physiol 92:995–1003

Berlin RD (1972) Effect of concanavalin A on phagocytosis. Nature 235:44–45

Bethell GS, Ayers JS, Hancock WS, Hearn MTW (1979) A novel method of activation of cross-linked agaroses with 1,1′-carbonyldiimidazole which gives a matrix for affinity chromatography devoid of additional charged groups. J Biol Chem 254:2572–2574

Bhattacharyya L, Brewer CF (1988a) Binding and precipitation of lectins from *Erythrina indica* and *Ricinus communis* (agglutinin I) with synthetic cluster glycosides. Arch Biochem Biophys 262(2):605–608

Bhattacharyya L, Brewer CF (1988b) Lectin-carbohydrate interactions. Studies of the nature of hydrogen bonding between D-galactose and certain D-galactose-specific lectins, and between D-mannose and concanavalin A. Eur J Biochem 176:207–212

Bhattacharyya L, Ceccarini C, Lorenzoni P, Brewer CF (1987a) Concanavalin A in interactions with asparagine-linked glycopeptides. Bivalency of high mannose and bisected hybrid type glycopeptides. J Biol Chem 262:1288–1293

Bhattacharyya L, Haraldsson M, Brewer CF (1987b) Concanavalin A interactions with asparagine-linked glycopeptides. Bivalency of bisected complex type oligosaccharides. J Biol Chem 262:1294–1299

Bhattacharyya L, Haraldsson M, Brewer CF (1988) Precipitation of galactose-specific lectins by complex-type oligosaccharides and glycopeptides: studies with lectins from *Ricinus communis* (agglutinin I), *Erythrina indica*, *Erythrina arborescens*, *Abrus precatorius* (agglutinin), and *Glycine max* (soybean). Biochemistry 27:1034–1041

Bhuvaneswari TV, Pueppke SG, Bauer WD (1977) Role of lectins in plant microorganism interactions. I. Binding of soybean lectin to rhizobia. Plant Physiol 60:486–491

Bishop CD, Cooper RM (1983) An ultrastructural study of root invasion in three vascular wilt diseases. Physiol Plant Pathol 22:15–27

Bishop PD, Pearce G, Bryant J, Ryan CA (1984) Isolation and characterization of the proteinase inhibitor-inducing factor from tomato leaves: identity and activity of poly- and oligo-galacturonide fragments. J Biol Chem 259:13127–13177

Blumberg S, Hildesheim J, Yariv J, Wilson KJ (1972) The use of I-amino-L-fucose bound to sepharose in the isolation of L-fucose-binding proteins. Biochim Biophys Acta 264:171–176

Bøg-Hansen TC, Bjerrum OJ, Brogren CH (1977) Identification and quantification of glycoproteins by affinity electrophoresis. Anal Biochem 81(1):78–87

Brammall RA, Higgins VJ (1988) A histological comparison of fungal colonization in tomato seedlings susceptible and resistant to *Fusarium* crown and root rot disease. Can J Bot 66:915–925

Bretschneider KE, Gonella MP, Robeson DJ (1989) A comparative light and electron microscopical study of compatible and incompatible interactions between *Xanthomonas campestris* pv. *campestris* and cabbage (*Brassica oleracea*). Physiol Mol Plant Pathol 34:285–297

Brown IR, Mansfield JW (1988) An ultrastructural study, including cytochemistry and quantitative analyses, of the interactions between pseudomonads and leaves of *Phaseolus vulgaris* L. Physiol Mol Plant Pathol 33:351–376

Brown IR, Mansfield JW, Mazau D, Rumeau D, Esquerré-Tugayé MT (1990) Immunocytochemical localization of hydroxyproline-rich glycoproteins in bean and melon leaves challenged by bacteria. Proc 7th Int Conf Plant Path Bact, Budapest, Hungary, 1989, pp 51–56

Brown RG, Kimmins WC (1973) Hypersensitive resistance. Isolation and characterization of glycoproteins from plants with localized infections. Can J Bot 51:1917–1926

Brown RG, Kimmins WC (1978) Protein-polysaccharide linkages in glycoproteins from *Phaseolus vulgaris*. Phytochem 17:29–33

Bruce RJ, West CA (1982) Elicitation of casbene synthetase activity in castorbean. Plant Physiol 69:1181–1188

Bucheli P, Doares SH, Albersheim P, Darvill A (1990) A host-pathogen interactions. XXXV. Partial purification and characterization of heat-labile molecules secreted by the rice blast pathogen that solubilize plant cell wall fragments that kill plant cells. Physiol Mol Plant Pathol 36:159–173

Burdon RH, van Knippenberg PH (1985) Isolation and fractionation. In: Work TS (ed) Laboratory techniques in biochemistry and molecular biology, vol 16. Beeley JS (ed): Glycoprotein and proteoglycan techniques. Elsevier, Amsterdam, pp 29–62

Burger MM, Goldberg AR (1967) Identification of a tumor-specific determinant on neoplastic cell surfaces. Proc Natl Acad Sci USA 57:359–366

Burke D, Kaufman P, McNeil M, Albersheim P (1974) The structure of plant cell walls. VI. A survey of the walls of suspension-cultured monocots. Plant Physiol 54:109

Callow JA (1975) Plant lectins. Curr Adv Plant Sci 7:181–193

Callow JA (1977) Recognition, resistance, and role of plant lectins in host-parasite interactions. In: Preston RD, Woolhouse HW (eds) Advances in botanical research. Academic Press, London, pp 1–49

Carrasco A, Boudet AM, Marigo G (1978) Enhanced resistance of tomato plants to *Fusarium* by controlled stimmulation of their natural phenolic production. Physiol Plant Pathol 12:225–232

Cartier N, Chambat G, Joseleau J-P (1988) Cell wall and extracellular galacto-glucomannans from suspension-cultured *Rubus fruticosus* cells. Phytochemistry 27(5):1361–1364

Cassab GI, Varner JE (1987) Immunocytolocalization of extensin in developing soybean seed coats by immunogold-silver staining and by tissue printing on nitrocellulose paper. J Cell Biol 105:2581–2588

Cassab GI, Varner JE (1988) Cell wall proteins. Annu Rev Plant Physiol Plant Mol Biol 39:321–353

Catt JW, Hills GJ, Roberts K (1976) A structural glycoprotein, containing hydroxy-proline, isolated from the cell wall of *Chlamydomonas reinhardtii*. Planta 131: 165–171

Cervone F, De Lorenzo G, Degrà L, Salvi G (1987) Elicitation of necrosis in *Vigna unguiculata* Walp. by homogeneous *Aspergillus niger* endo-polygalacturonase and by α-D-galacturonate oligomers. Plant Physiol 85:626–630

Cervone F, Hahn MG, De Lorenzo G, Darvill A, Albersheim P (1989a) Host-pathogen interactions. XXXIII. A plant protein converts a fungal pathogenesis factor into an elicitor of plant defense responses. Plant Physiol 90:542–548

Cervone F, Lorenzo De, D'Ovidio R, Hahn MG, Ito Y, Darvill AG, Albersheim P (1989b) Phytotoxic effects and phytoalexin-elicitor activity of microbial pectic enzymes. In: Ballio A, Durbin RD, Graniti A (eds) Phytotoxins and plant pathogenesis. NATO ASI Ser 27. Springer, Berlin Heidelberg New York, pp 473–477

Chamberland H, Benhamou N, Ouellette GB, Pauzè FJ (1989) Cytochemical detection of saccharide residues in paramural bodies formed in tomato root cells infected by *Fusarium oxysporum* f. sp. *radicis-lycopersici*. Physiol Mol Plant Pathol 34:131–146

Chaplin MF (1982) A rapid and sensitive method for the analysis of carbohydrate components in glycoproteins using gas-liquid chromatography. Anal Biochem 123:336–341

Chaplin MF, Kennedy JF (1986) Carbohydrate analysis, a practical approach. IRL Press, Oxford

Chen J, Varner JE (1985) Biology and molecular biology of hydroxyproline-rich glycoproteins. In: Strumpf, PK, Conn EE (eds) The biochemistry of plants. Alan R Liss, New York, pp 485–520

Cho YP, Chrispeels MJ (1976) Serine-O-galactosyl linkages in glycopeptides from carrot cell walls. Phytochemistry 15:165–169

Chrispeels MJ (1976) Biosynthesis, intracellular transport and secretion of extracellular macromolecules. Annu Rev Plant Physiol 27:19–38

Chrispeels MJ, Sadava D, Cho YP (1974) Enhancement of extensin biosynthesis in aging discs of carrot storage tissue. J Exp Bot 25:1157–1166

Clarke AE, Gleeson PA, Jermyn MA, Knox RB (1978) Characterization and localization of beta-lectins in lower and higher plants. Aust J Plant Physiol 5(5):707–722

Clarke AE, Anderson RL, Stone BA (1979) Form and function of arabinobalactans and arabinogalactan-proteins. Phytochemistry 18:521–540

Clarke JA, Shannon LM (1976) The isolation and characterisation of the glycopeptides from horseradish peroxidase isoenzyme C. Biochim Biophys Acta 427:428–442

Clarke JA, Lisker N, Lamport DTA, Ellingboe AH (1981) Hydroxyproline enhancement as a primary event in the successful development of *Erysiphe graminis* in wheat. Plant Physiol 67:188–189

Cline MN, Neely D (1983) Wound healing process in geranium cuttings in relationship to basal stem rot caused by *Pythium ultimum*. Plant Dis 67(6):636–638

Cooper, JB, Varner JE (1983) Insolubilization of hydroxyproline-rich cell wall glycoprotein in aerated carrot root slices. Biochem Biophys Res Commun 112:161–167

Corbin DR, Sauer N, Lamb CJ (1987) Differential regulation of a hydroxyproline-rich glycoprotein gene family in wounded and infected plants. Mol Cell Biol 7:4337–4344

Craig S, Moore PJ, Dunahay TG (1987) Immunogold localization of intra- and extracellular proteins and polysaccharides of plant cells. Scanning Electron Microsc 1:1431–1437

Cuatrecasas P (1970) Protein purification by affinity chromatography. Derivatizations of agarose and polyacrylamide beads. J Biol Chem 245:3059–3065

Cuatrecasas P (1972) Affinity chromatography of macromolecules. Adv Enzymol 36:29–89

Cuatrecasas P, Parikh I (1972) Adsorbents for affinity chromatography – use of n-hydroxysuccinimide esters of agarose. Biochemistry 11:2291–2299

Cuatrecasas P, Tell APE (1973) Insulin-like activity of concanavalin A and wheat germ agglutinin-direct interactions with insulin receptors. Proc Natl Acad Sci USA 70:485

Currier HB, Webster DH (1964) Callose formation and subsequent disappearance: Studies in ultrasound stimulation. Plant Physiol 39:843–847

Dabrowski J, Hanfland P, Egge H (1982) Analysis of glycosphingolipids by high-resolution proton nuclear magnetic resonance spectroscopy. Meth Enzymol 83: 69–83

Darvill A, McNeil M, Albersheim P (1978) Structure of plant cell walls. VIII. A new pectic polysaccharide. Plant Physiol 62:418–442

Darvill AG, McNeil M, Albersheim P (1980a) General and facile method for distinguishing 4-linked aldopyranosyl residues from 5-linked aldofuranosyl residues. Carbohydr Res 86:309–315

Darvill AG, McNeil M, Albersheim P (1980b) Structure of plant cell walls. XI. Glucuronoarabinoxylan, a second hemicellulose in the primary cell walls of suspension-cultured sycamore cells. Plant Physiol 66:1135–1139

Davey MW, Sulkowski E, Carter WA (1976) Binding of human fibroblast interferon to concanavalin A-agarose – involvement of carbohydrate recognition and hydrophobic interactions. Biochemistry 15:704–713

Davis KR, Lyon GD, Darvill AG, Albersheim P (1982) A poygalacturonic acid lyase isolated from *Erwinia carotovora* is an elicitor of phytoalexins in soybeans. Plant Physiol 69 (Suppl): 142

Davis KR, Lyon GD, Darvill AG, Albersheim P (1984) Host-pathogen interactions. XXV. Endopolygalacturonic acid lyase from *Erwinia carotovora* elicits phytoalexin accumulation by releasing plant cell wall fragments. Plant Physiol 74:52–60

Davis KR, Darvill AG, Albersheim P, Dell A (1986a) Host-pathogen interactions. XXX. Characterization of elicitors of phytoalexin accumulation in soybean released from soybean cell walls by endopolygalacturonic acid lyase. Z Naturforsch 41c:39–48

Davis KR, Darvill AG, Albersheim P, Dell A (1986b) Host-pathogen interactions. IX. Oligogalacturonides released from sodium polypectate by endopolygalacturonic acid lyase are elicitors of phytoalexins in soybean. Plant Physiol 80:568–577

Dazzo FB (1977) Cross reactive antigens and lectin as determinants of host specificity in the *Rhizobium*-clover symbiosis. In: Solheim B, Raa J (eds) Cell wall biochemistry. Universitetsforlaget, Oslo, pp 389–400

Debray H, Decout D, Strecker G, Spik G, Montreuil J (1981) Specificity of twelve lectins towards oligosaccharides and glycopeptides related to N-glycosylproteins. Eur J Biochem 117:41–55

Debray H, Pierce-Crètel A, Spik G, Montreuil J (1983) Affinity of ten solubilized lectins towards various glycopeptides with N-glycosylamine linkage and related oligosaccharides. In: Bøg-Hansen TC, Spengler GA (eds) Lectins – biology, clinical biochemistry, biochemistry, vol 3. De Gruyter, Berlin, p 335

Degrà L, Salvi G, Mariotti D, De Lorenzo G, Cervone F (1988) A polygalacturonase-inhibiting protein in alfalfa callus cultures. J Plant Physiol 133:364–366

Delaat AMM, van Loon LC (1983) The relationship between stimulated ethylene production and symptom expression in virus-infected tobacco leaves. Physiol Plant Pathol 22:261–273

De Lorenzo G, Salvi G, Degrà L, D'Ovidio R, Cervone F (1987) Induction of extracellular polygalacturonase and its mRNA in the phytopathogenic fungus *Fusarium moniliforme*. J Gen Microbiol 133:3365–3373

Desai NN, Allen AK, Neuberger A (1981) Some properties of the lectin from *Datura stramonium* (thorn apple) and the nature of its glycoprotein linkages. Biochem J 197:345–353

Dey PM, Harborn JB (1990) Oligosaccharides. In: Dey PM, Harborne JB (eds) Methods in plant biochemistry. Carbohydrates, vol 2. Academic Press, London, 657 pp

Dixon RA, Jennings AC, Davies LA, Gerrish C, Murphy DL (1989) Elicitor-active components from French bean hypocotyls. Physiol Mol Plant Pathol 34:99–115

Doares SH, Bucheli P, Albersheim P, Darvill AG (1989) Host-pathogen interactions XXXIV. A heat-labile activity secreted by a fungal phytopathogen releases fragments of plant cell walls that kill plant cells. Mol Plant-Microbe Interact 2(6):346–353

Ebrahim-Nesbat F, Slusarenko AJ (1983) Ultrastructure of the interaction of cells of *Pseudomonas phaseolicola* with cell walls of a resistant and susceptible bean cultivar. Phytopathol Z 108:148–159

Eda S, Kato K (1978) An arabinoxyloglucan from the midrip of the leaves of *Nicotiana tabacum*. Agric Biol Chem 42:351–357

Eda S, Ohnishi A, Kato K (1976) Xylan isolated from stalk of *Nicotiana tabacum*. Agric Biol Chem 40:359–364

Eda S, Watanabe F, Kato K (1977) Studies on chemical structure of tobacco hemicellulose 2. 4-O-Methylglucuronoxylan isolated from midrip of *Nicotiana tabacum*. Agric Biol Chem 41:429–434

Eda S, Akiyama Y, Kato K, Ishizu A, Nakano J (1985) A galaktoglucomannan from cell walls of suspension-cultured tobacco (*Nicotiana tabacum*) cells. Carbohydrate Res 137:173–181

Edelman GM, Wang JL (1978) Binding and functional properties of concanavalin A and its derivatives. III. Interactions with indoleacetic acid and other hydrophobic ligands. J Biol Chem 253:3016–3022
El-Banoby FE, Rudolph K (1980) Purification of extracellular polysaccharides from *Pseudomonas phaseolicola* which induce water-soaking in bean leaves. Physiol Plant Pathol 16:425–437
El-Banoby FE, Rudolph K (1981) Correlation between degree of agglutination of *Pseudomonas phaseolicola* cells by raw lectin preparations from 15 bush bean cultivars and degree of resistance to bacterial infection. Phytopathology 71:872 (Abstr)
English PD, Mglothin A, Keegstra K, Albersheim P (1972) A cell wall-degrading endopolygalacturonase secreted by *Colletotrichum lindemuthianum*. Plant Physiol 49:293
Epstein L, Lamport DTA (1984) An intramolecular linkage involving isodityrosine in extensin. Phytochemistry 23:1241–46
Esquerrè-Tugayè MT, Lamport DTA (1979) Cell surfaces in plant-microorganism interactions. Plant Physiol 64:314–319
Esquerrè-Tugayè MT, Lafitte C, Mazau D, Toppan A, Touzè A (1979) Cell surfaces in plant-microorganism interactions. Evidence for the accumulation of hydroxyproline-rich glycoproteins in the cell wall of diseased plants as a defense mechanism. Plant Physiol 64:320–326
Faulk WP, Taylor GM (1971) An immunocolloid gold for the electron microscope. Immunochemistry 8:1081–1083
Felsted RL, Leavitt RD, Bachur NR (1975) Purification of the phytohemagglutinin family of proteins from red kidney beans (*Phaseolus vulgaris*) by affinity chromatography. Biochem Biophys Acta 405:72–81
Fett WF, Sequeira L (1980) A new bacterial agglutinin from soybean. Plant Physiol 66:847–852
Fincher GB, Stone BA (1974) A water-soluble arabinogalactan-peptide from wheat endosperm. Aust J Biol Sci 27:117–132
Fincher GB, Stone BA (1983) Arabinogalactan-proteins: structure, biosynthesis, and function. Annu Rev Plant Physiol 34:47–70
Fincher GB, Sawyer WH, Stone BA (1974) Chemical and physical properties of an arabinogalactan peptide from wheat endosperm. Biochem J 139:535–545
Fincher GB, Stone BA, Clarke AE (1983) Arabinogalactan-Proteins: structure, biosynthesis and function. Annu Rev Plant Physiol 34:47–70
Fornstedt N, Porath J (1975) Characterization studies on a new lectin found in seeds of *Vicia ervilia*. FEBS Lett 57:187–191
Franz G (1976) The dependence of membrane-bound glucan synthetases on glycoprotein which can act as acceptor molecules. Appl Polymer Symp 28:611
Freier T, Rüdiger H (1987) In vivo binding partners of the *Lens culinaris* lectin. J. Biol Chem Hoppe-Seyler 368:1215–1223
Frens G (1973) Controlled nucleation for the regulation of the particle size in monodisperse gold solutions. Nat Phys Sci 241:20–22
Friend J (1976) Lignification in infected tissue. In: Friend J, Threlfall DR (eds) Biochemical aspects of plant-parasite-relationships. Acad Press, New York, pp 291–303
Fry SC (1982) Isodityrosine, a new cross-linking amino acid from plant cell-wall glycoprotein. Biochem J 204:449–455
Fry SC (1986) Cross-linking of matrix polymers in the growing cell walls of angiosperms. Annu Rev Plant Physiol 37:165–186
Fry SC, Milholland RD (1990) Response of resistant, tolerant, and susceptible grapevine tissues to invasion by the Pierce's disease bacterium, *Xylella fastidiosa*. Phytopathology 80:66–69
Gabius H-J (1991) Comparative analysis of lectin purification by various immobilized affinity ligands. Anal Biochem (in press)
Gabius H-J, Engelhardt R (1987) Interaction of mannose-binding proteins with different types of immobilized ligands. J Chromatogr 391:452–456

Gabius H-J, Rüdiger H, Uhlenbruck G (1988) Lektine. Spektrum der Wissenschaft, Nov 1988, pp 50–60
Gardiner M, Chrispeels MJ (1975) Involvement of the Golgi apparatus in the synthesis and secretion of hydroxyproline-rich cell wall glycoproteins. Plant Physiol 55:536–541
Gerwig GJ, Kamerling JP, Vliegenthart JFG (1978) Determination of the D and L configuration of neutral monosacharides by high-resolution capillary G.L.C. Carbohydr Res 62:349–357
Gerwig GJ, Kamerling JP, Vliegenthart JFG (1979) Determination of the absolute configuration of monosaccharides in complex carbohydrates by capillary G.L.C. Carbohydr Res 77:1–7
Ghanekar A, Pérombelon MCM (1980) Interactions between potato lectin and some phytobacteria in relation to potato tuber decay caused by *Erwinia carotovora*. Phytopath Z 98:137–149
Gleeson PA, Clarke AE (1979) Structural studies on the major component of *Gladiolus* style mucilage, an arabinogalactan-protein. Biochem J 181:607–621
Gleeson PA, Jermyn MA (1979) Alteration of the composition of β-lectins caused by chemical and enzymic attack. Aust J Plant Physiol 6:25–38
Gleeson PA, Clarke AE, Jermyn MA, Knox RB (1978) Arabinogalactan proteins of the female reproductive tissues of *Gladiolus gandavensis*. Proc Aust Biochem Soc 11, Univ Melbourne, Victoria, Australia, p 33
Gleeson PA, Jermyn MA, Clarke AE (1979) Isolation of an arabinogalactan protein by lectin affinity chromatography on Tridacnin-Sepharose 4B. Anal Biochem 92:41–45
Goldstein IJ, Poretz RD (1986) Isolation, physicochemical characterization, and carbohydrate-binding specificity of lectins. In: Liener IE, Sharon N, Goldstein IJ (eds) The lectins. Academic Press, New York, pp 33–247
Gordon JA, Blumberg S, Lis H, Sharon N (1972) Purification of soybean agglutinin by affinity chromatography on sepharose-N-ε-aminocaproyl-β-D-galactopyranosylamine. FEBS Lett 24:193–196
Green ED, Baenziger JU (1987) Oligosaccharide specificities of *Phaseolus vulgaris*. Leukoagglutinating and erythroagglutinating phytohemagglutinins. Interactions with N-glycanase released oligosaccharides. J Biol Chem 262:12018–12029
Green ED, Brodbeck RM, Baenziger JU (1987) Lectin affinity high-performance liquid chromatography. Interactions of N-glycanase-released oligosaccharides with *Ricinus communis* agglutinin I and *Ricinus communis* agglutinin II. J Biol Chem 262(25): 12030–12039
Gross M, Rudolph K (1987) Demonstration of levan and alginate in bean plants (*Phaseolus vulgaris*) infected by *Pseudomonas syringae* pv. *phaseolicola*. J Phytopathol 120:9–19
Hahn MG, Darvill AG, Albersheim P (1981) Host-pathogen interactions. XIX. The endogenous elicitor, a fragment of a plant cell wall polysaccharide that elicits phytoalexin accumulation in soybeans. Plant Physiol 68:1161–1169
Hahn MG, Bucheli P, Cervone F, Doares SH, O'Neill RA, Darvill A, Albersheim P (1989) The roles of cell wall constituents in plant-pathogen interactions. In: Nester E, Kosuge T (eds) Plant microbe interactions, vol 3. McGraw Hill, New York, pp 131–181
Hahne G, Lörz H (1988) Release of phytotoxic factors from plant cell walls during protoplast isolation. J Plant Physiol 132:345–350
Hajto T, Hostanska K, Gabius HJ (1989) Modulatory potency of the β-galactoside-specific lectin from mistletoe extract (iscador) on the host defense system in vivo in rabbits and patients. Cancer Res 49:4803–4808
Hammarström S, Hammarström ML, Sundblad G, Arnarp J, Lonngren J (1982) Mitogenic leukoagglutinin from *Phaseolus vulgaris* binds to a pentasaccharide unit in N-acetyllactosamine-type glycoprotein glycans. Proc Natl Acad Sci USA 79:1611–1615
Hammerschmidt R, Lamport DTA, Muldoon EP (1984) Cell wall hydroxyproline enhancement and lignin deposition as an early event in the resistance of cucumber to *Cladosporium cucumerinum*. Physiol Plant Pathol 24:43–47

Hardy MR, Townsend RR, Lee YC (1988) Monosaccharide analysis of glycoconjugates by anion exchange chromatography with pulsed amperometric detection. Anal Biochem 170:54–62

Harper S, Zewdie N, Brown IR, Mansfield JW (1987) Histological, physiological and genetical studies of the responses of leaves and pods of *Phaseolus vulgaris* to three races of *Pseudomonas syringae* pv. *phaseolicola* and to *Pseudomonas syringae* pv. *coronafaciens*. Physiol Mol Plant Pathol 31:153–172

Hauzer K, Tichá M, Horejsi V, Kocourek J (1979) Studies on lectins. XLIV. The pH dependence of lectin interaction with sugars as determined by affinity electrophoresis. Biochim Biophys Acta 583:103–109

Hawes GB, Adams GA (1972) Extracellular arabinogalactan from suspension-cultured red kidney bean root cells. Phytochemistry 1:1461–1465

Hayashi T, Marsden MPF, Delmer DP (1987) Pea xyloglucan and cellulose. V. Xyloglucan-cellulose interactions in vitro and in vivo. Plant Physiol 83:384–389

Hayashi T, Kato Y, Natsuda K (1980) Xyloglucan from suspension-cultured soybean cells. Plant Cell Physiol 21:1405

Hayashi T, Wong YS, MacLachlan G (1984) Pea xyloglucan and cellulose. II. Hydrolysis by pea endo-1,4-β-glucanases. Plant Physiol 75:605–610

Herbert WJ (1978) Passive haemagglutination with special reference to the tanned cell technique. In: Weir DM (ed) Handbook of experimental immunology, 3rd edn. Blackwell, Oxford, pp 20.1–20.20

Hirst EL, Rees DA, Richardson NG (1965) Seed polysaccharides and their role in germination. A survey of the polysaccharide components of mustard seeds with special reference to the embryos. Biochem J 95:453–458

Hoffmann RM, Turner JG (1982) Partial purification of proteins from pea leaflets that inhibit *Ascochyta pisi* endopolygalacturonase. Physiol Plant Pathol 20:173–187

Hori H (1978) The demonstration of galactosamine in extracellular hydroxyproline-rich macro-molecule purified from the culture media of tobacco cells. Plant Cell Physiol 19:501–595

Hori H, Sato S (1977) Extracellular hydroxyproline-rich glycoproteins of suspension cultured tobacco cells. Phytochemistry 16:1485–1487

Horisberger M, Rosset J, Bauer H (1975) Colloidal gold granules as markers for cell surface receptors in the scanning EM. Experientia 31:1147–1148

Horisberger M, Rosset J (1977) Gold granules, as useful marker for transmission and scanning electron microscopy. J Histochem Cytochem 25:295–305

Huang JS (1990) Encapsulation or agglutination of bacteria. In: Klement Z, Rudolph K, Sands DC (eds) Methods in phytobacteriology. Akademiai Kiado, Budapest, pp 483–486

Huang PJ, Huang JS, Goodman RN (1975) Resistance mechanism of apple shoots to an avirulent strain of *Erwinia amylovora*. Physiol Plant Pathol 6:283–287

Huet C, Longchamp M, Huet M, Bernadec A (1974) Temperature effects on the concanavalin A molecule and on concanavalin A binding. Biochim Biophys Acta 365:28–39

Inbar M, Sachs L (1969) Interaction of the carbohydrate-binding protein concanavalin A with normal and transformed cells. Proc Natl Acad Sci USA 63:1418–1425

Ishii S (1988) Factors influencing protoplast viability of suspension-cultured rice cells during isolation process. Plant Physiol 88:26–29

Janzen DH, Juster HB, Liener IE (1976) Insecticidal action of the phytohemagglutinin in black beans on a bruchid beetle. Science 192:795–796

Jarvis MC, Threlfall DR, Friend J (1978) Alkali-labile bonds in the potato cell wall. Biochem Soc Trans 6:391–392

Jasalavich CA, Anderson AJ (1981) Isolation from legume tissues of an agglutinin of saprophytic pseudomonads. Can J Bot 59:264–271

Jermyn MA, Yeow YM (1975) A class of lectins present in the tissues of seed plants. Aust J Plant Physiol 2:501–531

Johansen DA (1940) Microchemical methods. In: Johansen DA (ed) Plant microtechnique, McGraw-Hill, New York, pp 182–203

Jin DF, West CA (1984) Characteristics of galacturonic acid oligomers as elicitors of casbene synthetase activity in castor bean seedlings. Plant Physiol 74:989–992
Juliano RL, Li G (1978) Glycoproteins of CHO cell membrane – partial fractionation of receptors for concanavalin A and wheat germ agglutinin using a lectin immunoprecipitation technique. Biochemistry 17:678–683
Karnovsky MJ (1965) A formaldehyde glutaraldehyde fixative of high osmolarity for use in electron microscopy. J Cell Biol 27:137 A
Kauss H (1981) 25 lectins and their physiological role in slime molds and in higher plants. In: Tanner W, Loewus FA (eds) Plant carbohydrates II. Springer, Berlin Heidelberg New York, pp 627–657
Kauss H, Glaser Ch (1974) Carbohydrate-binding proteins from plant cell walls and their possible involvement in extension growth. FEBS Lett 45:304–307
Kauss H, Bowles D (1976) Some properties of carbohydrate-binding proteins (lectins) solubilized from cell walls of *Phaseolus aureus*. Planta 130:169–174
Kawasaki S (1987) Synthesis of arabinose-containing cell wall precursors in suspension-cultured tobacco cells. IV. Differential subfractionation and comparison of the acidic precursors. Plant Cell Physiol 28:925–935
Keegstra K, Talmadge KW, Bauer WD, Albersheim P (1973) The structure of plant cell walls. III. A model of the walls of suspension-cultured sycamore cells based on the interconnections of the macromolecular components. Plant Physiol 51:188–196
Keen NT, Holliday MJ (1982) Recognition of bacterial pathogens by plants. In: Mount MS, Lacy GH (eds) Phytopathogenic prokaryotes. Academic Press, New York, pp 179–217
Kieliszewski M, Lamport DTA (1987) Purification and partial characterization of a hydroxyproline-rich glycoprotein in a graminaceous monocot, *Zea mays*. Plant Physiol 85:823–827
Kilpatrick DC (1980) Purification and some properties of a lectin from the fruit juice of the tomato (*Lycopersicon esculentum*). Biochem J 185:269–272
Knee M (1975) Soluble and wall-bound glycoproteins of apple fruit tissue. Phytochemistry 14:2181–2188
Kojima M, Uritani I (1974) The possible involvement of a spore agglutinating factor(s) in various plants in establishing host specificity by various strains of black rot fungus, *Ceratocystis fimbriata*. Plant Cell Physiol 15:733–737
Kopp M, Rouster J, Fritig B, Darvill A, Albersheim P (1989) Host-pathogen interactions. XXXII. A fungal glucan preparation protects Nicotinanae against infection by viruses. Plant Physiol 90:208–216
Kornfeld R, Ferris C (1975) Interaction of immunoglobulin glycopeptides with concanavalin A. J Biol Chem 250:2614
Kornfeld S, Adair WL, Gottlieb C, Kornfeld R (1974) Membrane glycoproteins as plant lectin receptors. In: Lee EYC, Smith EE (eds) Biology and chemistry of eukaryotic cell surfaces. Academic Press, London, pp 291–316
Lamport DTA (1965) The protein component of primary cell walls. Adv Bot Res 2:151–218
Lamport DTA (1977) Structure, biosynthesis and significance of cell wall glycoproteins. Rec Adv Phytochem 11:79–115
Lamport DTA, Catt JW (1981) Glycoproteins and enzymes of the cell wall. In: Tanner W, Loewus FA (eds) Plant carbohydrates II. Springer, Berlin Heidelberg New York, pp 133–165
Lamport DTA, Miller DH (1971) Hydroxyproline arabinosides in the plant kingdom. Plant Physiol 48:454–456
Lamport DTA, Northcote DH (1960) Hydroxyproline in primary cell walls of higher plants. Nature 188:665–666
Lamport DTA, Katona L, Roerig S (1973) Galactosylserine in extensin. Biochem J 133:125–131
Lau JM, McNeil M, Darvill AG, Albersheim P (1985) Structure of the backbone of rhamnogalacturonan I, a pectic polysaccharide in the primary cell walls of plants. Carbohydr Res 137:111–125

Laurell CB (1965) Antigen-antibody crossed electrophoresis. Anal Biochem 10:358–361
Lawson RH, Dienett MM, Civerolo EL (1989) Histopathology of *Xanthomonas campestris* pv. *citri* from Florida and Mexico in wound-inoculated detached leaves of *Citrus aurantifolia*: light and scanning electron microscopy. Phytopathology 79:329–335
Leach JE, Cantrell MA, Sequeira L (1982) A hydroxyproline-rich bacterial agglutinin from potato: its localization by immunofluorescence. Physiol Plant Pathol 21:319–325
Leavitt RD, Felsted RL, Bachur NR (1977) Biological and biochemical properties of *Phaseolus vulgaris* isolectins. J Biol Chem 252(9):2961–2966
Leduc EH, Bernhard W (1967) Recent modifications of the glycol methalcrylate embedding procedure. J Ultrastruct Res 19:196–199
Lee SC, West CA (1981) Polygalacturonase from *Rhizopus stolonifer*, an elicitor of casbene synthetase activity in castor bean (*Ricinus communis* L.) seedlings. Plant Physiol 67:633
Lee YC (1972) Analysis of sugars by automated liquid chromatography. In: Ginsburg V (ed) Methods in enzymology, vol 28. Academic Press, New York, pp 63–73
Lindberg B, Lönngren J, Svensson S (1975) Specific degradation of polysaccharides. Adv Carbohydr Chem Biochem 31:185
Lippincott JA, Lippincott B (1977) Nature and specificity of the bacterium-host attachment in *Agrobacterium* infection. In: Solheim B, Raa J (eds) Cell wall biochemistry. Universitetsforlaget, Oslo, pp 439–451
Lockhart-Ewart RB, Kornfeld S, Kipnis DM (1973) Stimulation of insulin release from isolated rat islets by mushroom phytohemagglutinin in vitro. Clin Res 21:622
Lotan R, Gussin AES, Lis H, Sharon N (1973a) Purification of wheat germ agglutinin by affinity chromatography on a Sepharose-bound N-acetylglucosamine derivative. Biochem Biophys Res Commun 52:656–662
Lotan R, Lis H, Rosenwasser A, Novogrodsky A, Sharon N (1973b) Enhancement of the biological activities of soybean agglutinin by cross-linking with glutaraldehyde. Biochem Biophys Res Commun 55:1347–1355 (+correction 56 (1974) 1100)
Lotan R, Siegelman HW, Lis H, Sharon N (1974) Subunit structure of soybean agglutinin. J Biol Chem 249(4):1219–1224
Lotan R, Beattie G, Hubbell IW, Nicolson GL (1977) Activities of lectins and their immobilized derivatives in detergent solutions. Implication on the use of lectin affinity chromatography for the purification of membrane glycoproteins. Biochemistry 16:1787–1794
Lowe CR (1979) An introduction to affinity chromatography. In: Work TS, Work E (eds) Laboratory techniques in biochemistry and molecular biology, vol 7, part II. North-Holland, Amsterdam
March SC, Parikh I, Cuatrecasas P (1974) A simplified method for cyanogen bromide activation of agarose for affinity chromatography. Anal Biochem 60:149–152
Marinkovich VA (1964) Purification and characterization of the hemagglutinin present in potatoes. J Immunol 93:732–741
Mau SL, Raff JW, Clarke AE (1982) Isolation and partial characterization of components of *Prunus avium* L. styles, including an S-allele-associated antigenic glycoprotein. Planta 156:505–516
Mazau D, Esquerré-Tugayé MT (1986) Hydroxyproline-rich glycoprotein accumulation in the cell walls of plants infected by various pathogens. Physiol Mol Plant Pathol 29:147–157
Mazau D, Rumeau D, Esquerré-Tugayé MT (1987a) Biochemical characterization and immunocytochemical localization of hydroxyproline-rich glycoproteins in infected plants. Plant Physiol (Suppl) 83:658 (Abstr)
Mazau D, Rumeau D, Esquerré-Tugayé MT (1987b) Molecular approaches to understanding cell surface interactions between plants and fungal pathogens. Plant Physiol Biochem 25:337–343
Mazau D, Rumeau D, Esquerré-Tugayé MT (1988) Two different families of hydroxyproline-rich glycoproteins in melon callus. Plant Physiol 86:540–546

Mazzucchi U (1983) Recognition of bacteria by plants. In: Callow JA (ed) Biochemical plant pathology. John Wiley Sons, New York, pp 299–324

McNamara MK, Stone BA (1978) Isolation of a carbohydrate-protein linkage compound from wheat endosperm arabinogalactan-peptide and its chemical synthesis. Abstracts of contributed papers and posters. IXth Int Symp Carbohydr Chem London, 10–14 4 78, IUPAC, Intern. Union of Pure and Applied Chemistry, Oxford, p 43

McNeil M, Darvill AG, Albersheim P (1979) The structural polymers of the primary cell walls of dicots. In: Herz W, Grisebach H, Kirby GW (eds) Progress in the chemistry of organic natural products 37. Springer, Berlin Heidelberg New York, pp 191–249

McNeil M, Darvill AG, Albersheim P (1980) Structure of plant cell walls. X. Rhamnogalacturonan I, a structurally complex pectic polysaccharide in the walls of suspension-cultured sycamore cells. Plant Physiol 66:1128–1134

McNeil M, Darvill AG, Aman P, Franzén L-E, Albersheim P (1982) Structural analysis of complex carbohydrates using high-performance liquid chromatography, gas chromatography, and mass spectrometry. Meth Enzymol 83:3–45

McNeil M, Darvill AG, Fry SC, Albersheim P (1984) Structure and function of the primary cell walls of plants. Annu Rev Biochem 53:625–63

Mellis SJ, Baenziger JU (1981) Separation of neutral oligosaccharides by high-performance liquid chromatography. Anal Biochem 114:276–280

Mellon JE, Helgeson JP (1982) Interaction of hydroxyproline-rich glycoproteins from tobacco callus with potential pathogens. Plant Physiol 70:401–405

Mendgen K (1978) Attachment of bean rust cell wall material to host and non-host plant tissue. Arch Microbiol 119:113–117

Menon MKC, Hartmann E (1988) Secretion of protein-bound hydroxyproline from moss callus cells. J Plant Physiol 132:569–574

Mew TW, Mew TP, Huang JS (1984) Scanning electron microscopy of virulent and avirulent strains of *Xanthomonas campestris* pv. *oryzae* on rice leaves. Phytopathology 74:635–641

Mikami H, Ishida Y (1983) Post-column fluorometric detection of reducing sugars in high-performance liquid chromatography using arginine. Bunseki Kagaku 32:E207–E210

Mirelman D, Galun E, Sharon N, Lotan R (1975) Inhibition of fungal growth by wheat-germ agglutinin. Nature 256:414–416

Monro JA, Bailey RW, Penny D (1976) Hemicellulose fractions and associated protein of lupin hypocotyl cell walls. Phytochemistry 15:175

Montreuil J, Bouquelet S, Debray H, Fournet B, Spik G, Strecker G (1986) Glycoproteins. In: Chaplin MF, Kennedy JF (eds) Carbohydrate analysis, a practical approach. IRL Press, Oxford, pp 143–204

Moore PJ, Staehelin LA (1988) Immunogold localization of the cell wall matrix polysaccharides rhamnogalacturonan-I and xyloglucan during cell expansion and cytokinesis in *Trifolium pratense* L.: implications for secretory pathway. Planta 174:433–445

Moore PJ, Darvill AG, Albersheim P, Staehelin LA (1986) Immunogold localization of xyloglucan and rhamnogalacturonan I in cell walls of suspension-cultured sycamore cells. Plant Physiol 82:787–794

Moroney SE, D'Alarcao LJ, Goldmacher VS, Lambert JM, Blättler WA (1987) Modification of the binding site(s) of lectins by an affinity column carrying an activated galactose-terminated ligand. Biochemistry 26:8390–8398

Muray RHA, Northcote DH (1978) Oligoarabinosides of hydroxyproline isolated from potato lectin. Phytochemistry 17:623–629

Mussel HW (1973) Endopolygalacturonase: evidence for involvement in *Verticillium* wilt of cotton. Phytopathology 63:62–70

Nachbar MS, Oppenheim JD, Thomas JO (1980) Lectins in the U.S. diet. Isolation and characterization of a lectin from the tomato (*Lycopersicon esculentum*). J Biol Chem 255:2056–2061

Narayanan SR, Crane LJ (1990) Affinity chromatography supports: a look at performance requirements. Tibtech 8:1–16
Neuberger A, Gottschalk A, Marshall RD, Spiro RG (1972) Carbohydrate-peptide linkages in glycoproteins and methods for their elucidation. In: Gottschalk A (ed) The glycoproteins, 2nd edn. Elsevier, Amsterdam, p 470
Nicolson GL (1974) The interactions of lectins with animal cell surfaces. Int Rev Cytol 39:89–190
Nilsson SF, Waxdal MJ (1976) Isolation and identification of major concanavalin A binding glycoproteins from murine lymphocytes. Biochemistry 15:2698–2705
Nilsson SF, Waxdal MJ (1978) Isolation and characterization of lectin-binding proteins from murine lymphoid-cells. Biochemistry 17:903–910
Nothnagel EA, McNeil M, Albersheim P, Dell A (1983) Host-pathogen interactions. XXII. A galacturonic acid oligosaccharides from plant cell wall elicits phytoalexins. Plant Physiol 71:916–926
Nowell PC (1960) Phytohemagglutinin: an initiator of mitosis in cultures of normal human leukocytes. Cancer Res 20:462–466
O'Connell RJ, Brown IR, Mansfield JW, Bailey JA, Mazau D, Rumeau D, Esquerré-Tugayé M-T (1990) Immunocytochemical localization of hydroxyproline-rich glycoproteins accumulating in melon and bean at sites of resistance to bacteria and fungi. Mol Plant-Microbe Interact 3:33–40
O'Neill MA, Selvendran RR (1980) Glycoproteins from the cell wall of *Phaseolus coccineus*. Biochem J 187:53–63
Ohlson S, Hansson L, Larsson PO, Mosbach K (1978) High performance liquid affinity chromatography (HPLAC) and its application to the separation of enzymes and antigens. FEBS Lett 93:5–9
Olsner S, Pihl A (1978) Abrin and ricin – two toxic lectins. TIBS 3:7–10
Olsner S, Saltveat E, Pihl A (1974) Isolation and comparison of galactose-binding lectins from *Abrus precatorius* and *Ricinus communis*. J Biol Chem 249:803
Osborn TC, Ausloos KA, Brown JWS, Bliss FA (1983) Bean lectins. III. Evidence for greater complexity in the structural model of *Phaseolus vulgaris* lectin. Plant Sci Lett 31:193–203
Ouchterlony Ö (1964) Gel diffusion techniques. In: Ackroyd JF (ed) Immunological methods. Blackwell, Oxford, pp 55–78
Owens RJ, Northcote DH (1981) The location of arabinosyl: hydroxyproline transferase in the membrane system of potato tissue culture cells. Biochem J 195:661–667
Pazur JH (1981) Affinity chromatography of macromolecular substances on adsorbents bearing carbohydrate ligands. In: Tipson RS, Horton D (eds) Advances in carbohydrate chemistry and biochemistry, vol 39. Academic Press, New York, pp 405–447
Pazur JH (1982) β-D-glucose-1-phosphate. A structural unit and an immunological determinant of a glycan from streptococcal cell walls. J Biol Chem 257:589
Pazur JH (1986) Neutral polysaccharides. In: Chaplin MF, Kennedy JF (eds) Carbohydrate analysis, a practical approach. IRL Press, Oxford, pp 55–96
Philips PG, Furmanski P, Lubin M (1974) Cell surface interactions with concanavalin A. Location of bound radiolabeled lectin. Exp Cell Res 86:301–308
Pope DG (1977) Relationships between hydroxyproline-containing proteins secreted into cell wall and medium by suspension-cultured *Acer pseudoplatanus* cells. Plant Physiol 894:894–900
Pusztai A, Palmer R (1977) Nutritional evaluation of kidney beans (*Phaseolus vulgaris*) – toxic principle. J Sci Food Agric 28:620–623
Pusztai A, Stewart JC (1978) Isolectins of *Phaseolus vulgaris*. Physicochemical studies. Biochim Biophys Acta 536:38–49
Quiocho FA, Vyas NK (1984) Novel stereospecificty of L-arabinose-binding protein. Nature 310:381–386
Ride JP (1983) Cell walls and other structural barriers in defense. In: Callow JA (ed) Biochemical plant pathology. John Wiley & Sons, New York, pp 214–236

Rigas DA, Osgood EE (1955) Purification and properties of the phytohemagglutinin of *Phaseolus vulgaris*. J Biol Chem 212:607
Ring SG, Selvendran RR (1981) An arabinogalactoxyloglucan from the cell wall of *Solanum tuberosum*. Phytochemistry 20:2511–2519
Roberts DD, Goldstein IJ (1982) Hydrophobic binding properties of the lectin from lima beans (*Phaseolus lunatus*). J Biol Chem 257:11274–11277
Robertsen B (1986) Elicitors of the production of lignin-like compounds in cucumber hypocotyls. Physiol Mol Plant Pathol 28:137–148
Roebuck P, Sexton R, Mansfield JW (1978) Ultrastructural observations on the development of the hypersensitive reaction in leaves of *Phaseolus vulgaris* cv. Red Mexican inoculated with *Pseudomonas phaseolicola* (race 1). Physiol Plant Pathol 12:151–157
Roland JC (1978) General preparation and staining of thin sections. In: Hall JL (ed) Electron microscopy and cytochemistry of plant cells. Elsevier, Amsterdam, pp 1–62
Romeiro R, Karr A, Goodman R (1981) Isolation of a factor from apple that agglutinates *Erwinia amylovora*. Plant Physiol 68:772–777
Roth J (1983) Application of lectin-gold complexes for electron microscopic localization of glycoconjugates on thin sections. J Histochem Cytochem 31:987–997
Rudolph K, Gross M, Neugebauer M, Hokawat S, Zachowski A, Wydra K, Klement Z (1989) Extracellular polysaccharides as determinants of leaf spot diseases caused by pseudomonads and xanthomonads. In: Graniti A, Durbin RD, Ballio A (eds) Phytotoxins and plant pathogenesis, vol 27. NATO ASI Ser, Ser H: Cell biology. Springer, Berlin Heidelberg New York, pp 177–218
Ruel K, Joseleau J-P (1984) Use of enzyme-gold complexes for the ultrastructural localization of hemicelluloses in the plant cell wall. Histochemistry 81:573–580
Ryan CA (1978) Proteinase inhibitors in plant leaves – biochemical model for pest induced natural plant protection. TIBS 3:148–150
Ryan CA (1980) Wound-regulated synthesis and vacuolar compartmentation of proteinase inhibitors in plant leaves. Curr Top Cell Regul 17:1
Ryan CA, Bishop P, Pearce G, Darvill AG, McNeil M, Albersheim P (1981) A sycamore cell wall polysaccharide and a chemically related tomato leaf polysaccharide possess similar proteinase inhibitor-inducing activities. Plant Physiol 68:616–618
Ryan CA, Bishop PD, Walker-Simmons M, Brown ME, Graham JS (1985) Pectic fragments regulate the expression of proteinase inhibitor genes in plants. In: Key JL, Kosuge T (eds) Cellular and molecular biology of plant stress, vol 22. Alan R Liss, New York, pp 319–334
Sato K, Uritani I, Saito T (1982) Properties of terpene-inducing factor extracted from adults of the sweet-potato weevil, *Cylas formicarius fabricius* (Coleoptera, Brenthidae). Appl Entomol Zool 17:368
Schachman HK (1957) Ultracentrifugation, diffusion and viscometry. In: Colowick SP, Kaplan NO (eds) Methods in enzymology, vol 4. Academic Press, New York, p 32
Schnebli HB (1976) In: Bittiger H, Schnebli HB (eds) Concanavalin A as a tool. Wiley, London, pp 249–255
Selvendran RR (1975a) Analysis of cell wall material from plant tissues: extraction and purification. Phytochemistry 14:1011–1017
Selvendran RR (1975b) Cell wall glycoproteins and polysaccharides of parenchyma of *Phaseolus coccineus*. Phytochemistry 14:2175–2180
Selvendran RR, Davies MC, Tidder E (1975) Cell wall glycoproteins and polysaccharides of mature runner beans. Phytochemistry 14:2169–2174
Sequeira L (1978) Lectins and their role in host-pathogen specificity. Annu Rev Phytopathol 16:453–481
Sequeira L, Graham TL (1977) Agglutination of avirulent strains of *Pseudomonas solanacearum* by potato lectin. Physiol Plant Pathol 11:43–53
Sharon N (1977) Lectins: these proteins found primarily in plants combine specifically with sugars on cell surfaces and hence bind cells together. They are key tools in the study of the role the cell surface plays in cell behavior. Sci Am 236 (no. 6):108–119
Sharon N, Lis H (1972) Lectins: cell-agglutinating and sugar-specific proteins. Science 177:949–957

Shet MS, Madaiah M (1988) Single step purification of a lectin from winged bean tubers. Phytochemistry 27:1951–1954

Showalter AM, Varner JE (1989) Biology and molecular biology of plant hydroxyproline-rich glycoproteins. In: Stumpf K, Conn EE (eds) The biochemistry of plants: a comprehensive treatise, vol 15, Marcus A (ed). Acad Press, New York, pp 485–520

Siddiqui IR, Rosa N (1985) Structural investigation of the sodium hydroxysoluble polysaccharides of tobacco (*Nicotiana tabacum*): arabinoxyloglucan. Carbohydr Res 138:247–255

Siddiqui IR, Wood P (1974) Structural investigation of oxalate-soluble rapeseed (*Brassica campestris*) polysaccharides. III. An arabinan. Carbohydr Res 36:35–44

Slusarenko AJ, Wood RKS (1981) Differential agglutination of races 1 and 2 of *Pseudomonas phaseolicola* by a fraction from cotyledons of *Phaseolus vulgaris* cv. Red Mexican. Physiol Plant Pathol 18:187–193

Slusarenko AJ, Wood RKS (1983) Agglutination of *Pseudomonas phaseolicola* by pectic polysaccharide from leaves of *Phaseolus vulgaris*. Physiol Plant Pathol 23:217–227

Slusarenko AJ, Epperlein M, Wood RKS (1983) Agglutination of plant pathogenic and certain other bacteria by pectic polysaccharides from various plant species. Phytopath Z 106:337–343

Smith DF (1983) Glycolipid-lectin interactions: detection by direct binding of ^{125}I-lectins to thin layer chromatograms. Biochem Biophys Res Commun 115:360–367

Smith JJ, Muldoon EP, Lamport DTA (1984) Isolation of extensin precursors by direct elution of intact tomato cell suspension cultures. Phytochemistry 23:1233–1239

Smith JJ, Muldoon EP, Willard JJ, Lamport DTA (1986) Tomato extensin precursors P_1 and P_2 are highly periodic structures. Phytochemistry 25:1021–1030

Smith MM, McCully ME (1977) Mild temperature "stress" and callose synthesis. Planta 136:65–70

Stermer BA, Hammerschmidt R (1985) In: Key JL, Kosuge T (eds) Cellular and molecular biology of plant stress, vol 22. Alan R Liss, New York, pp 291–302

Stevenson TT, McNeill M, Darvill AG, Albersheim P (1986) Structure of plant cell walls. XVIII. An analysis of the extracellular polysaccharides of suspension-cultured sycamore cells. Plant Physiol 80:1012–1019

Stokes DM, Anderson JW, Rowan KS (1968) The isolation of mitochondria from potato tuber tissue using sodium metabisulfite for preventing damage by phenolic compounds during extraction. Phytochemistry 7:1509–1512

Stuart DA, Varner JE (1980) Purification and characterization of a salt-extractable hydroxyproline-rich glycoprotein from aerated carrot discs. Plant Physiol 66:787–792

Suzuki K, Amino S, Takeuchi Y, Komamine A (1990) Differences in the composition of the cell walls of two morphologically different lines of suspension-cultured *Catharanthus roseus* cells. Plant Cell Physiol 31:7–14

Takeuchi Y, Komamine A (1980) Turnover of cell wall polysaccharides of *Vinca rosea* suspension culture. Physiol Plant 50:113–118

Talmadge KW, Keegstra K, Bauer WD, Albersheim P (1973) The structure of plant cell walls. I. The macromolecular components of the walls of suspension-cultured sycamore cells with a detailed analysis of the pectic polysaccharides. Plant Physiol 51:158–173

Tepper CS, Anderson AJ (1990) Interactions between pectic fragments and extracellular components from the fungal pathogen *Colletotrichum lindemuthianum*. Physiol Mol Plant Pathol 36:147–158

Terry ME, Jones RL (1981) Soluble cell wall polysaccharides released from pea stems by centrifugation. 1. Effect of auxin. Plant Physiol 68:531–537

Terry ME, Rubinstein B, Jones RL (1981) Soluble cell wall polysaccharides released from pea stems by centrifugation. II. Effect of ethylene. Plant Physiol 68:538–542

Tesser GI, Fisch HU, Schnyzer R (1974) Limitations of affinity chromatography: solvolytic detachment of ligands from polymeric supports. Helv Chim Acta 57:1718

Thomas J, Lorel LH (1976) Chromatographic separation of glucose and mannose on cation-exchange resin. Anal Biochem 73:222–226

Thomas JR, McNeil M, Darvill AG, Albersheim P (1987) Structure of plant cell walls. XIX. Isolation and characterization of wall polysaccharides from suspension-cultured Douglas fir cells. Plant Physiol 83:659–671

Torres BV, McCrumb DK, Smith DF (1988) Glycolipid-lectin interactions: reactivity of lectins from *Helix pomatia*, *Wisteria floribunda*, and *Dolichos biflorus* with glycolipids containing N-acetylgalactosamine. Arch Biochem Biophys 262:1–11

Touzè A, Esquerrè-Tugayè MT (1980) Defence mechanisms of plants against varietal non-specific pathogens. In: Wood RKS (ed) Active defense mechanisms in plants. Plenum Press, New York, pp 103–117

Tripathi NN, Grover RK (1978) Participation of root exudates of susceptible and resistant plants in pathogenesis of diseases caused by *Pythium butleri*. Z. Pflanzenkr Pflanzenschutz 85:15–21

Turner RH, Lifner IE (1975) The use of glutaraldehyde treated eythrocytes for assaying the agglutinating activity of lectins. Anal Biochem 68:651–653

Valent BS, Darvill AG, McNeil M, Robertson BK, Albersheim P (1980) A general and sensitive chemical method for sequencing the glycosyl residues of complex carbohydrates. Carbohydr Res 79:165–192

Van Holst G-J, Klis FM, De Wildt PJM, Hazenberg CAM, Buijs J, Stegwee D (1981) Arabinogalactan protein from a crude cell organelle fraction of *Phaseolus vulgaris* L. Plant Physiol 68:910–913

VanderMolen GE, Beckman CH, Rodehorst E (1977) Vascular gelation: a general response phenomenon following infection. Physiol Plant Pathol 11:95–100

Varner JE, Lin L-S (1989) Plant cell wall architecture. Cell 56:231–239

Vasil'eva KV, Gladikikh TA, Davydova MA, Moldosanova GA, Umralina AR (1982) Phytoalexin induction in sugar beet roots by extracellular endopolygalactrronase from Rhizoctonia aderholdi Kolosh Mikol Fitopatol 16:525–528 (in Russian), Biol. Abstr. 77:54848, 1984

Vivo A, Andreu JM, de la Viña S, de Felipe MR (1989) Leghemoglobin in lupin plants (*Lupinus albus* cv. Multolupa). Plant Physiol 90:452–457

Vretblad P (1976) Purification of lectins by biospecific affinity chromatography. Biochim Biophys Acta 434:169–176

Wagner H, Jordan E (1988) An immunologically active arabinogalactan from *Viscum album* "berries". Phytochemistry 27:2511–2517

Walker-Simmons M, West CA, Hadwiger L, Ryan CA (1984) Comparison of proteinase inhibitor-inducing activities and phytoalexin elicitor activities of a pure fungal endopolygalacturonase, pectic fragments, and chitosans. Plant Physiol 76:833–836

Weinhold AR, Hancock JG (1980) Defense at the perimeter: extruded chemicals. In: Horsfall JG, Cowling EB (eds) Plant disease, vol 5. Acad Press, New York, pp 121–138

Wells GB, Turco SJ, Hanson BA, Lester RL (1982) Resolution of dolichylpyrophosphoryl oligosaccharides by high-pressure liquid chromatography. Meth Enzymol 83:137–139

West CA, Moesta P, Jin DF, Lois AF, Wickham KA (1985) The role of pectic fragments of the plant cell wall in the response to biological stresses. In: Key JL, Kosuge T (eds) Cellular and molecular biology of plant stress, vol 22. Alan R Liss, New York, pp 335–349

Whistler RL, Anisuzzaman KM (1980) Gel permeation chromatography. In: Whistler RL, BeMiller JN (eds) Methods in carbohydrate chemistry, vol 8. Acad Press, New York, p 45

Wilchek M, Givol D (1971) The use of affinity chromatography in protein structure studies. Peptides. North-Holland, Amsterdam, p 203

Wilder BM, Albersheim P (1973) The structure of plant cell walls. IV. A structural comparison of the wall hemicellulose of cell suspension cultures of sycamore (*Acer pseudoplatanus*) and of red kidney bean (*Phaseolus vulgaris*). Plant Physiol 51:889

Wilson LG, Fry JC (1986) Extensin – a major cell wall glycoprotein. Plant Cell Environ 9:239–260

Woods A, Hunter N, Sequeira L, Kelman A (1979) Lectin activity isolated from corn seeds. Plant Physiol 63:134
Wu AM, Sugii S, Gruezo FG, Kabat EA (1988) Immunochemical studies on the N-acetyllactosamine β-(1–6)-linked trisaccharide specificity of *Ricinus communis* agglutinin. Carbohydr Res 178:243–257
Wydra K (1991) Interaktionen zwischen Zellwandpolymeren von Bohnenpflanzen (*Phaseolus vulgaris*) und Zellen von *Pseudomonas syringae* pv. *phaseolicola* sowie deren extrazellulären, polymeren Produkten. PH.D. Thesis, Univ. Göttingen
Wydra K, Rudolph K (1989) An agglutination factor from bush bean leaves against *Pseudomonas syringae* pv. *phaseolicola*. In: Galling G (ed) Proc Braunschweig Symp on Applied plant molecular biology, Braunschweig, Nov 21–23, 1988. Zentralstelle für Weiterbildung der Technischen Universität Braunschweig, Braunschweig (FRG), pp 334–340
Wydra K, Rudolph K (1990) Interactions between cell wall polymers of *Phaseolus vulgaris* and bacterial cell surface polymers of *Pseudomonas syringae* pv. *phaseolicola*. In: Proc 7th Int Conf on Plant pathogenic bacteria, 11–16 June 1989, Budapest, Hungary
Yamaoka T, Sato S (1977) Analysis of extracellular macromolecules of cultured tobacco cells in comparison with cell-wall macromolecules. Bot Mag 90:153–163
Yamazaki N, Fry SC, Darvill AG, Albersheim P (1983) Host-pathogen interactions. XXIV. Fragments isolated from suspension-cultured sycamore cell walls inhibit the ability of the cells to incorporate (^{14}C)leucine into proteins. Plant Physiol 72:864–869
Yariv J, Rapport MM, Graf L (1962) The interaction of glycosides and saccharides with antibody to the corresponding phenylazo glycosides. Biochem J 85:383–388
Yariv L, Lis H, Katchalski E (1967) Precipitation of arabic acid and some seed polysaccharides by glycosylphenylazo dyes. Biochem J 105:1C–2C
York WS, Darvill AG, Albersheim P (1984) Inhibition of 2,4-dichlorophenoxyacetic acid-stimulated elongation of pea stem segments by a xyloglucan oligosaccharide. Plant Physiol 75:295–297
York WS, Darvill AG, McNeil M, Albersheim P (1985a) 3-Deoxy-D-*manno*-2-octulosonic acid (KDO) is a component of rhamnogalacturonan II, a pectic polysaccharide of the primary cell walls of plants. Carbohydr Res 138:109–126
York WS, Darvill AG, McNeil M, Stevenson TT, Albersheim P (1985b) Isolation and characterization of plant cell walls and plant cell wall components. Meth Enzymol 118:3–40
Young NM, Watson DC, Williams RE (1984) Structural differences between two lectins from *Cytisus scoparius*, both specific for D-galactose and N-acetyl-D-galactosamine. Biochem J 222:41–48

Immunological Analysis of Mycotoxins

A.P. Wilkinson, C.M. Ward, and M.R.A. Morgan

1 Introduction

1.1 Definition

Mycotoxins are a chemically diverse group of compounds produced as secondary metabolites of certain fungi that may commonly infest agricultural produce. These compounds can cause illnesses or death in animals or man consuming contaminated commodities. Furthermore, animal products used as human food such as milk, dairy products, meat and eggs, may become contaminated with mycotoxins if contaminated feed is consumed by the producing animal; for reviews see Purchase (1974), Bullerman (1986), Schlatter (1988), Van Egmond (1989a) and Pestka (1988). The chemical structures of some of the mycotoxins of greatest current interest are illustrated in Fig. 1.

1.2 Mycotoxin Production and Distribution

Aspergillus, *Penicillium*, *Fusarium*, *Stachybotrys*, *Trichoderma* and *Alternaria* are examples of fungi that commonly infest crops, although only certain strains are toxigenic (Smith and Moss 1985). Contamination by these fungi can occur at any stage of crop production and after harvest, particularly during storage. Though a diverse range of foods can become contaminated with mycotoxins, the food types most frequently affected are nuts and cereals. The presence of toxigenic fungi on a food crop does not automatically infer the presence of mycotoxins, because whether or not particular toxins are synthesized is determined by prevailing climatic conditions on the macro- and micro-scale, and other factors such as insect or other damage to crops (Patterson 1982; Bullerman 1986; Morgan and Lee 1990). In addition, toxins can be present in the apparent absence of fungal contamination, a situation caused by death or removal of the fungi after toxin biosynthesis has occurred.

An exmaple of the influence of climate on toxin elaboration is provided by aflatoxin, a compound synthesized by toxigenic strains of the fungi *Aspergillus flavus* and *Aspergillus parasiticus* when these fungi experience a temperature between 25 and 40 °C and a relative humidity greater than 80% (Butler 1974). These are climatic conditions typical of tropical and sub-

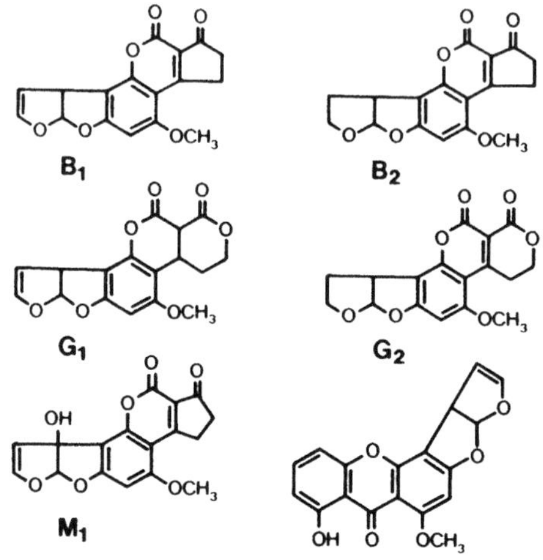

1a

Sterigmatocystin

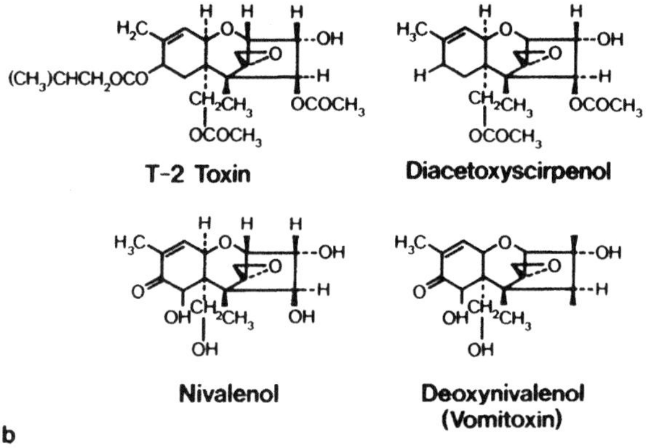

1b

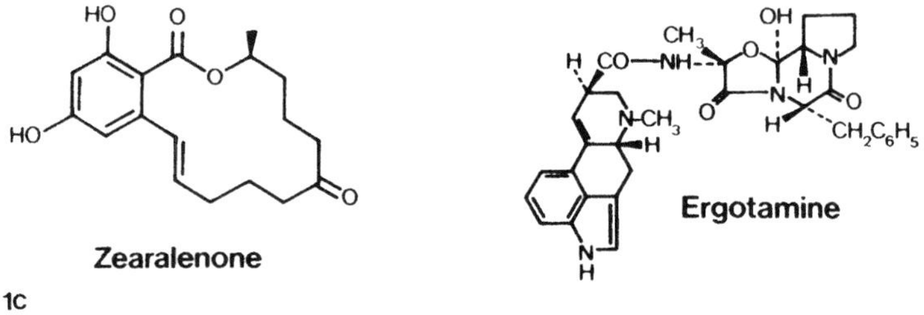

1c

1d

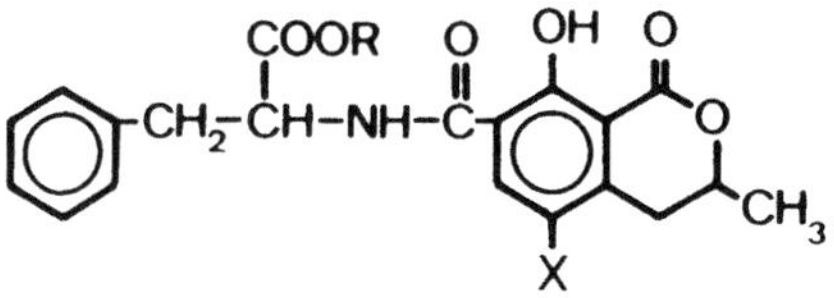

	R	X
Ochratoxin A	H	Cl
Ochratoxin B	H	H

tropical regions of the world (Schmidt and Esser 1985; Denning 1987). On the other hand, fungal strains capable of synthesizing trichothecene mycotoxins such as strains of *Fusarium*, *Trichoderma* and *Stachybotrys*, need the cooler climate often found in temperate regions to form toxin (Ueno 1977).

The numerous factors involved in providing conditions favourable for toxin synthesis means that toxin distribution within a crop can be uneven. For instance, aflatoxin contamination of individual peanut kernels has been shown to range from trace to $1.1\,mg\,kg^{-1}$ (Cucullu et al. 1966) and similar findings have been observed for Brazil nuts, corn and cotton seed (Park and Pohland 1989). This heterogeneity of distribution leads to sampling and analytical problems (Sect. 1.4; Groopman and Donahue 1988).

Control of the various factors necessary for mycotoxin synthesis in the field, especially climate, is difficult, and as mycotoxigenic fungi are fairly ubiquitous in distribution, mycotoxin contamination of agricultural produce in the field may be inevitable. However, once harvested, crops can be stored under conditions that can prevent mycotoxin synthesis. In developing countries storage conditions are often limited by economic factors. These countries also have climates particularly favourable for mycotoxin synthesis, especially aflatoxins, which are very potent toxins.

1.3 Mycotoxin Toxicity

The toxicity of the mycotoxins has been extensively reviewed (Purchase 1974; Newberne and Rogers 1981; Patterson 1982; Schmidt and Esser 1985; Smith and Moss 1985; Neal 1987; Denning 1987; Rousseaux 1988). They exhibit acute, sub-acute and chronic toxicities. The effects are varied because the toxins can interact with almost any organ and physiological system in the body, and the degree of susceptibility of an animal to a particular mycotoxin can be governed by age and sex. Incidents of acute poisonings (mycotoxicoses) of livestock by mycotoxins are common (Purchase 1974), but reports of mycotoxicoses affecting humans are comparatively rare because of differences in suceptibility to the toxins, and because high levels of toxin are often accompanied by visible moulding of foods, which are then normally rejected. However, high toxin levels can be found in produce without visible signs of moulding, as mycotoxins will persist after moulds have disappeared. Contaminated produce may also be processed, for instance a cereal into bread, and when subsequently

Fig. 1. Structures of mycotoxins of current interest **a** Aflatoxins and structurally related sterigmatocystin; aflatoxin M_1 is a mammalian metabolite of aflatoxin B_1. **b** Trichothecenes. **c** Zearalenone and ergotamine. **d** Ochratoxin. Ochratoxin A is the toxic compound found in the presence of greater amounts of non-toxic ochratoxin B

consumed mycotoxicosis may result, as in a recent case in India (Bhat et al. 1989). Other examples of mycotoxicoses that have affected man together with the suspected causative toxin are: aflatoxicosis (aflatoxins); ergotism (ergot alkaloids); Alimentary Toxic Aleukia (trichothecenes); yellow rice disease (citrinin); and Balkan endemic nephropathy (ochratoxin A) (Purchase 1974; Krogh 1977; Bullerman 1986; Van Egmond 1989a).

Generally, acute toxicity is observed at $mg\,kg^{-1}$, parts per million (ppm) levels of exposure, though aflatoxin B_1, an extremely potent toxin, may give rise to acute toxicity at lower concentrations (Denning 1987).

Of great concern for both human and animal health is the effect of repeated sub-acute exposure at $\mu g\ kg^{-1}$, parts per billion (ppb) levels, to mycotoxins because many have now been shown in toxicological experiments to be carcinogenic, mutagenic and teratogenic (Newberne and Rogers 1981; Patterson 1982). One mycotoxin, zearalenone, is oestrogenic in action, especially to pigs. Of the mycotoxins shown to be carcinogenic, aflatoxin B_1 is regarded as the most potent naturally occurring carcinogen in a number of animal species including ducklings, trout and rats (Purchase 1974; Newberne and Rogers 1981; Patterson 1982; Smith and Moss 1985).

1.4 Mycotoxin Regulations, Sampling and Analytical Problems

The biological potency of mycotoxins in trace amounts, particularly of the aflatoxins, makes them a serious concern for farmers, food processors and retailers, and the regulatory authorities. The risk of exposure of a population to mycotoxins is less in technologically developed countries than in developing countries, but prevention of dietary exposure to mycotoxins, especially aflatoxins, will improve the general health of a population (Schmidt and Esser 1985; Groopman and Donahue 1988). To achieve this goal, many countries have formulated statutory and voluntary guidelines governing permissible mycotoxin levels in foods and animal feedstuffs. These regulations have been reviewed by Van Egmond (1989b).

The non-uniform contamination of agricultural produce by mycotoxins means that small parts of a commodity may contain high concentrations of toxin, whilst the bulk will be relatively uncontaminated. This phenomena is analogous to the incidence of contagious diseases which have also been shown to occur in isolated "hot spots" (Cucullu et al. 1966; Whitaker and Wiser 1969; Patterson 1982). To survey a commodity for mycotoxin contamination, statistically derived sampling plans should be employed that take into account the high probability of a trace level of mycotoxin matched with a low probability of high toxin concentration. For reviews see Schuller et al. (1976), Park and Pohland (1989) and Garfield (1989). Only after appropriate samples have been selected can meaningful analysis follow.

Biological matrices in which mycotoxin contamination can occur are chemically complex, comprising large amounts of protein, lipid and carbohydrate in comparison to the trace levels of mycotoxin, often as low as $ng\,kg^{-1}$ (ppt), that need to be quantified. The analytical method chosen must therefore be sensitive and specific to avoid misclassification. The technique should also be simple so that large sample numbers generated in sampling plans can be assayed. Additionally, rapid methodology can provide rapid decisions about acceptance or rejection. Conventional physicochemical analytical techniques such as thin-layer, gas-liquid and high performance liquid (tlc, glc and hplc respectively) chromatographic procedures, as well as mass-spectral analyses, have all been used for analysis of mycotoxins in food (Shotwell 1986). All these methods are capable of resolving relatively simple mixtures of pure mycotoxins at $\mu g\,kg^{-1}$ concentrations, but many components in the complex systems that make up food and other biological matrices can interfere with physicochemical analyses. This necessitates extensive sample cleanup involving time-consuming and costly procedures, before the final analysis can be made. This, together with the sequential operation of many physicochemical techniques, severely limits the number of analyses that can be routinely performed.

Techniques having high sensitivity, high specificity, a capability for high sample throughput and which are relatively cheap in terms of instrumentation and other costs would be of considerable advantage in comparison to conventional techniques. Techniques based on the use of immunological reagents can provide such qualities as a direct consequence of the unique interaction of an antibody for its target molecule. For examples applied to mycotoxins see Lee and Morgan (1988), Morgan (1988), Pestka (1988) and Morgan and Lee (1990).

2 General Immunoassay Principles

Immunological methods of analysis are based on the use of antibodies. Antibodies can be generated that are specific for discrete regions on almost any organic material recognized as "foreign" by the immune system, a property that can be exploited readily in vitro.

Modern immunoanalysis dates from the Nobel Prize-winning work of Yalow and Berson (1959). Immunoassays are designed to quantify analyte by measuring antibody-analyte interactions. The assay first described by Yalow and Berson (1959) involved an unknown quantity of analyte competing with a fixed amount of labelled analyte for a limited number of analyte binding sites provided by the antibody. At equilibrium, antibody can exist in two phases, one bound to analyte and the other analyte-free according to the equation:

$$\underset{\text{(free phase)}}{An + An^* + 2Ab} \rightleftharpoons \underset{\text{(bound phase)}}{AbAn + AbAn^*},$$

where An^* = labelled analyte, Ab = antibody,
An = analyte (unknown).

If the amount of unknown analyte is relatively low, more label is able to bind to the antibody. If there is a lot of unknown analyte, little label is bound to antibody. Provided free and bound phases can be separated and that the label can be quantified in either of the phases, then the amount of unknown analyte can be determined by reference to the behaviour in the assay of known standards. For reviews see Daussant and Bureau (1984), Morgan (1989), Rittenburg (1989).

3 Immunogen Conjugates

To produce an immunological assay for a mycotoxin, antibodies must be generated that recognize and bind the toxin. This cannot be done directly by exposure of the immune system to mycotoxin alone (even allowing for toxicity) because compounds having atomic masses of less than approximately 1000 are not recognized as "foreign" by the immune system. Consequently no antibodies are produced. "Foreign" proteins will, of course, stimulate the immune system and consequent antibody production and are known as immunogens. If a mycotoxin (or similar hapten, as such low-molecular-weight, non-immunogenic compounds are known) is chemically linked to an immunogen protein and the resulting conjugate administered to a suitable animal, the immune response is stimulated and antibodies produced. Some of the antibodies are capable of binding to free mycotoxin (Daussant and Bureau 1984; Lee and Morgan 1988; Pestka 1988; Morgan 1989; Rittenburg 1990).

According to the literature, most mycotoxin immunoassays have been directed against the aflatoxins, trichothecenes and ochratoxin A. Reviews by Chu (1986), Pestka (1988) and Morgan (1989) list these assays and the chemical strategies employed to synthesize mycotoxin-protein conjugates. These strategies have been adapted from procedures first used in clinical chemistry to develop immunological assays for hormones and drugs, procedures that have been reviewed by Erlanger (1980).

Reactive groups on both the hapten and protein are required for conjugation. For the protein, groups commonly utilized are the amino, carboxylic acid, sulphydryl, phenolic hydroxyl and imidazole functions. Of these, the ε-amino group of lysine residues, an abundant amino acid in most proteins, is most often used for conjugation. Amide bond formation is the favoured covalent linkage between hapten and protein. Three reactions are most commonly used;

1. The mixed anhydride reaction, in which isobutylchlorocarbonate reacts with a carboxylic acid function in the absence of water to form an acid

anhydride intermediate, which readily couples to free amino groups (Erlanger et al. 1959);
2. The water soluble carbodiimide reaction in which amide bond formation proceeds via the formation of an isourea ester between the carbodiimide and carboxylic acid. The isourea ester then reacts with primary amine groups (Egly and Boschetti 1983),
3. The activated ester reaction where a carboxylic acid moiety is activated by the addition of N-hydroxysuccinimide in the presence of dicyclohexylcarbodiimide. Primary amines are then able to react to form an amide bond with liberation of N-hydroxysuccinimide (Egly and Boschetti 1983).

Problems may be encountered with the first two procedures (Gendloff et al. 1986; Hastings et al. 1988) because side reactions may result in undesired compounds being generated and linked to protein subsequently used as immunogen. Antibodies against the undesired hapten can lead to non-specific interference in immunological assay development. However, these problems are not common and many assays have been reported in which these reactions are used. If the interference occurs, it may be overcome by adsorption of antisera with a conjugate formed using the amide bond reaction chemicals and protein alone. This could purify the antisera (extracting undesired antibodies) to leave mycotoxin-specific antibodies for use in the assay. Such an approach has been used in the development of immunoassays to pesticides (Wie and Hammock 1982). Alternatively, if the immunoassay procedure (as in ELISA) requires the synthesis of a mycotoxin-protein conjugate besides the one used as immunogen, then different amide bond formation chemistry can be used for each conjugate.

From the above, it can be seen that it is desirable to have a carboxylic acid group on the hapten. If this is already present, for instance as in ochratoxin A, then the toxin can be conjugated directly to protein. If a carboxylic acid group is not present, then other moieties on the hapten molecule such as hydroxyl or carbonyl groups can be derivatized to allow the introduction of the carboxylic acid functionality into the hapten. Other conjugation methods are, of course, also possible (see Sect. 3.2).

3.1 Antibody Specificity

An initial consideration in the design of an immunological assay for a hapten is the point at which it is linked to protein, because this has a critical effect on the specificity of the antibody response. This was first demonstrated in a classic series of experiments by Landsteiner (1945), whose studies showed that antibodies have a greater specificity for parts of the hapten molecule distal to the point of hapten-protein linkage. This property of the antibody production system can be used to generate assay systems that may be either of broad specificity, capable of recognizing groups or classes of molecules, or they might be highly specific, recognizing one compound amongst others of similar structure (Morgan 1989).

3.2 Conjugation Procedures

More detailed examples will now be given of conjugation procedures with emphasis on actual procedures utilized in our own laboratory, though procedures used by other workers will be discussed to illustrate the variety of chemistries that may be used.

Morgan et al. (1983) developed a double-antibody, non-competitive indirect ELISA for ochratoxin A, a potent nephrotoxin and possible carcinogen that possesses a free carboxylic acid function. Its position on the molecule is such that its use in conjugate synthesis allows recognition of the chlorine atom at C_5 by the generated antibodies, because the chlorine is distal to the point of conjugation. It also allows discrimination of ochratoxin A from non-toxic ochratoxin B which has a hydrogen atom at the C_5 position (see Fig. 1d). Immunogen was synthesized using the mixed anhydride reaction. Ochratoxin A (25 mg) was dissolved in re-distilled dry dioxane (2.0 ml) and the solution cooled to 12 °C. Tri-n-butylamine (0.05 ml) was added and the solution left at 12 °C for 20 min, after which dry, re-distilled isobutylchlorocarbonate (0.02 ml) was added. The solution was left a further 15 min at 12 °C. Bovine serum albumin as immunogen carrier protein (80 mg) was then added in a mixture comprising water (4.3 ml) and dioxane (2.6 ml), at a pH adjusted to 9.5 with 1.0 M sodium hydroxide. The mixture was then allowed to react at 4 °C for 4 h before extensive dialysis to remove low-molecular-weight materials. The second ochratoxin A-protein conjugate used to coat microtitration plate wells (see Sect. 5.3.2) was synthesized using the water soluble carbodiimide method. Ochratoxin A (5 mg) in ethanol (0.12 ml) was added to 0.1 M phosphate buffer (3.0 ml, pH 7.0). Carrier protein (keyhole limpet hemocyanin, 51 mg) in 0.1 M sodium chloride (5.0 ml) was added to the toxin solution followed by 1-ethyl-3-(dimethylpropyl) carbodiimide (25 mg). The mixture was stirred at 20 °C in the dark for 24 h before purification by passage through a Sephadex G-75 column.

A third amide bond-forming reaction has been used by Märtlbauer and Terplan (1988) to synthesize an ochratoxin A-enzyme conjugate. Ochratoxin A (2.0 mg) was mixed with a fivefold molar excess of N-hydroxysuccinimide and tenfold molar excess of dicyclohexylcarbodiimide in dry dimethylformamide (0.13 ml). This mixture was allowed to react at room temperature prior to purification of the desired product.

Aflatoxins and trichothecenes do not have carboxylic acid groups and so have to be derivatized prior to linkage to protein. Figure 2 illustrates some of the conjugation strategies that have been used for aflatoxins. These

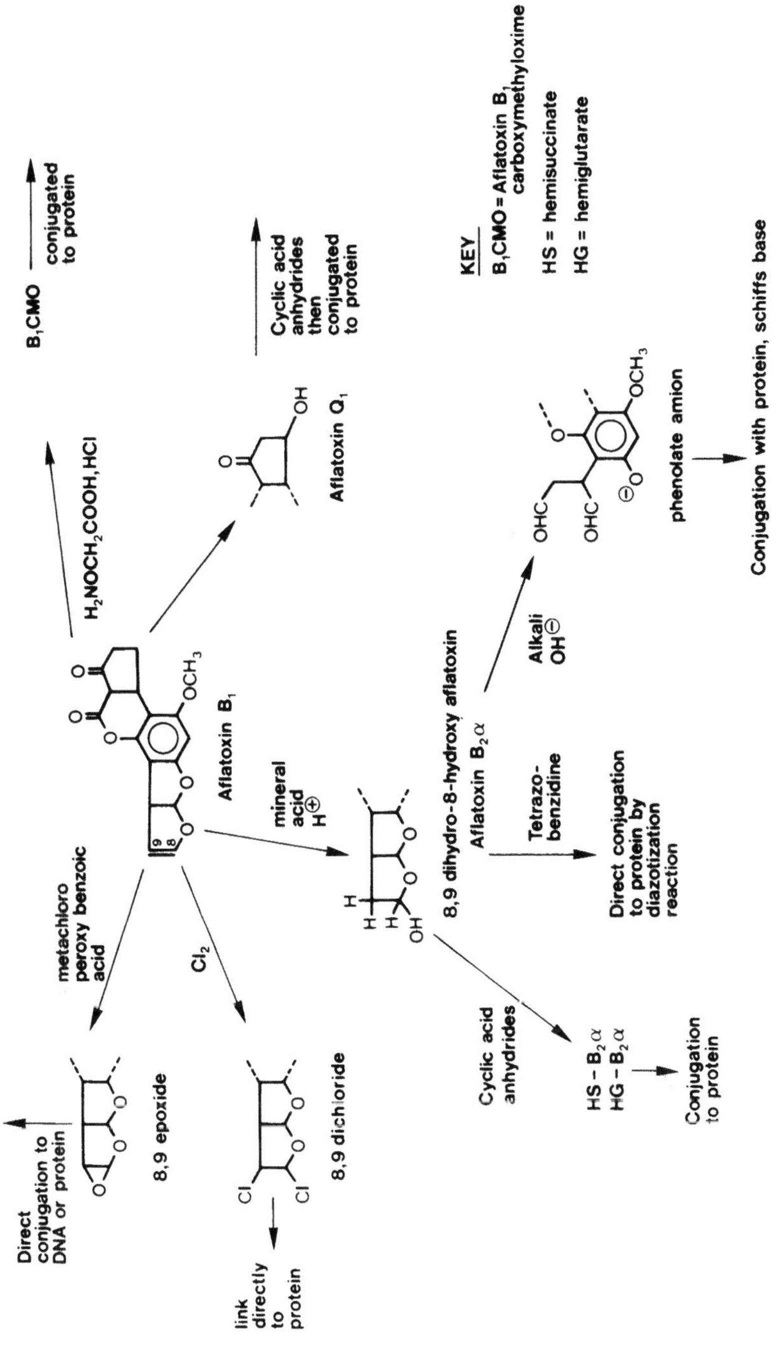

Fig. 2. Strategies used to derivatize aflatoxin B_1 and conjugate to protein

compounds possess reactive groups at both ends of the molecules that can be used for derivatization, but the most successful route in terms of generating antibodies having the desired sensitivity and specificity for immunological assay have involved derivatization of the carbonyl group of the cyclopentanone ring structure to form a carboxymethyl oxime derivative. The introduced carboxylic acid group is then used to link to protein (Morgan 1989).

In a sensitive ELISA developed by Morgan et al. (1986a), aflatoxin B_1 (30 mg) together with O-carboxymethylhydroxylamine (also known as amino oxyacetic acid hemichloride, 45 mg) were dissolved in methanol-water-pyridine (4:1:1, v:v:v, 18 ml) and refluxed for 3 h. After standing at room temperature for a further 12 h, the carboxymethyloxime derivative was purified by silica gel chromatography. The derivative was then conjugated to protein using the mixed anhydride procedure. The antisera generated by this approach can have a wide range of specificities (Ward et al. 1990) some recognizing the most toxic and carcinogenic aflatoxins namely B_1 and G_1, and others recognizing B_1, B_2, G_1 and G_2; B_1 and B_2 and G_1 alone. The antisera of appropriate high affinity sera being used for the rapid and simple determination of aflatoxin B_1 and G_1 in agricultural produce (Morgan et al. 1986a).

Derivatization via the unsaturated 8,9 carbon-carbon bond of the bifuran ring of aflatoxin has also been used to produce immunogens to generate anti-aflatoxin antisera. Three procedures have been used;

1. Reaction of the 8,9 double bond with chlorine gas to form an 8,9 dichloride derivative which can then be linked directly to protein (Sizaret et al. 1980);
2. Generation of the very reactive 8,9 epoxide by reaction of aflatoxin with metachloroperoxybenzoic acid. The epoxide which forms can react immediately with deoxyribonucleic acid (DNA) or protein to form a covalent linkage (Groopman et al. 1984). These aflatoxin-macromolecular adducts can then be used as immunogens and the antisera raised have primarily been used to assess human exposure to aflatoxin non-invasively (Groopman and Donahue 1988);
3. Reaction of aflatoxin with mineral acid causes the formation of the 8,9 dihydro-8-hydroxy derivative, a compound formed in mammalian metabolism of aflatoxin B_1 and which is also known as aflatoxin $B_{2\alpha}$. For immunogen synthesis, aflatoxin $B_{2\alpha}$ is derivatized as follows:
   a) By reaction with succinic and glutaric cyclic acid anhydrides (Lau et al. 1981);
   b) By reaction with alkali to form the phenolate anion. This derivative has free aldehyde moieties that can be used to link to protein amino groups through the Schiffs base reaction. This linkage is stabilized by reduction using sodium borohydride (Gaur et al. 1981). Jackman (1985) also utilized the phenolate anion derivative to synthesize an immunogen

using the Mannisch condensation reaction. The antisera raised was used to develop an ELISA to analyze for the mammalian aflatoxin B_1 metabolite, aflatoxin M_1, also a mutagen and carcinogen.
c) Lawellin et al. (1977) coupled aflatoxin $B_{2\alpha}$ to protein by a diazotization reaction using tetrazobenzidine.

Much effort has been made to develop immunological assays for trichothecenes because they are particularly difficult to analyze by conventional physicochemical techniques as, unlike aflatoxins and ochratoxins, they lack distinctive spectral and fluorescent properties. Gas chromatography-mass spectrometry has been used successfully but requires highly skilled operators and elaborate, expensive equipment. The analytical method is limited further, as are other physicochemical methods, by the necessity to extensively cleanup samples prior to analysis. This limits the number of samples that can be analyzed and increases costs (Mills et al. 1988; Morgan 1989; Morgan and Lee 1990).

There are a number of reports describing immunological assays for T-2 toxin. Nearly all these describe the derivatization of the hydroxyl group of T-2 toxin with either succinic or glutaric acid anhydrides. The derivatives formed possess a carboxylic acid group that is linked to protein using any of the standard amide bond forming reactions described above (Chiba et al. 1988; Morgan 1989).

Cyclic acid anhydrides were used by Mills et al. (1988) in the development of a sensitive ELISA for diacetoxyscirpenol (DAS). This trichothecene (like T-2 toxin) has a single hydroxyl group available for derivatization. DAS (10 mg) was reacted with a 60-fold molar excess of glutaric anhydride in the presence of 1 mol of the catalyst 4, N′,N′-dimethylaminopyridine. Dry, re-distilled tetrahydrofuran (5 ml) was added and the reaction continued for 15 h at 20 °C. The mixture was then lyophilized and the DAS-hemiglutarate derivative purified by silica gel chromatography.

Developing immunological assays to some trichothecenes particularly nivalenol (NIV) and deoxynivalenol (DON, also known as vomitoxin), are challenging to immunochemists because these compounds are multi-hydroxylated, thus straight-forward reaction with cyclic anhydrides, sebacoyl dichloride and similar reagents would lead to poorly defined multiple products. In addition, the fragile nature of the strained ring system only allows mild reactions to be performed without undesirable rearrangement.

Unsuccessful attempts to synthesize toxin-protein conjugates by oxime formation at the α-β unsaturated ketone at C_8 have been reported by Zhang et al. (1986), Ikebuchi et al. (1990) and Teshima et al. (1990a). The lack of success by this approach may be accounted for by the resonance stabilization of the α-β unsaturated ketone which lowers the reactivity of the carbonyl group (Mills et al. 1990).

Greater success has been achieved for both NIV and DON using tetra and tri-acetylated derivatives respectively for immunogen formation

(Zhang et al. 1986; Ikebuchi et al. 1990; Teshima et al. 1990a). Triacetyldeoxynivalenol was reduced using sodium borohydride, the reducing agent causing the formation of hydroxyl groups at C_7 and C_8, which was then reacted with succinic anhydride. The hemi-succinate derivative was conjugated to protein. For NIV, only the C_8 carbonyl was reduced and then reacted with succinic anhydride in the intermediate step of immunogen synthesis. Though radioimmunoassays (DON and NIV) and an ELISA (NIV) have been reported (Zhang et al. 1986; Ikebuchi et al. 1990; Teshima et al. 1990a), the antisera generated recognize only the multi-acetylated derivatives of the respective trichothecenes. The assays developed thus need to include an acetylation step in the assay protocol. This approach only permits an indirect measurement of these trichothecenes and increases analysis time and difficulty.

Reports by Casale et al. (1988) and Mills et al. (1990) describe the synthesis of DON immunogens that allow the direct determination of DON. Casale et al. (1988) protected the C_{15} and C_7 hydroxyl groups with l-butaneboronic acid forming a cyclic boronate ester. This leaves the remaining C_3 hydroxyl available for succinylation and then conjugation to protein. The protecting group – cyclic boronate ester – is easily removed, regenerating the C_{15} and C_7 hydroxyl groups.

Figure 3 illustrates an alternative approach that has been described by Mills et al. (1990). Here the starting material 3-acetyldeoxy-nivalenol (10 mg), was reacted in dry pyridine (3.0 ml) with glutaric anhydride (83 mg). The mixture was warmed to 50 °C and stirred for 36 h. Addition of water (6.0 ml) stopped the reaction. Only the C_{15} hemiglutarate derivative forms, as the C_7 hydroxyl is unreactive because of hydrogen bonding with the C_8 ketone. Following purification, the C_{15} hemiglutarate derivative was dried under a stream of nitrogen and then redissolved in acetone (0.5 ml). The solution was added to 0.1 M sodium phosphate buffer (6.0 ml, pH 6.5) containing 0.02% (w/v) sodium azide. This was then added to an ammonium sulphate suspension of the enzyme acetyl esterase, (36 units) resuspended in phosphate buffer (3.0 ml). The mixture was incubated for one week at 30 °C with gentle agitation. The enzyme specifically hydrolyzes the 3-acetyl group of the 3-acetyl-DON-C_{15} hemiglutarate derivative to form DON C_{15} hemiglutarate. The product was then separated from the non-hydrolyzed 3-acetyl-DON C_{15} hemiglutarate and conjugated to protein using the mixed anhydride procedure.

3.2.1 Miscellaneous Mycotoxin Conjugations

Immunological assays have also been described for zearalenone (Thouvenot and Morfin 1983; Warner et al. 1986; Dixon et al. 1987; Teshima et al. 1990b); sterigmatocystin (Li and Chu 1984; Morgan et al. 1986b; Wei and Chu 1988); rubratoxin B (Davis and Stone 1979); Kojic acid (Abdalla and

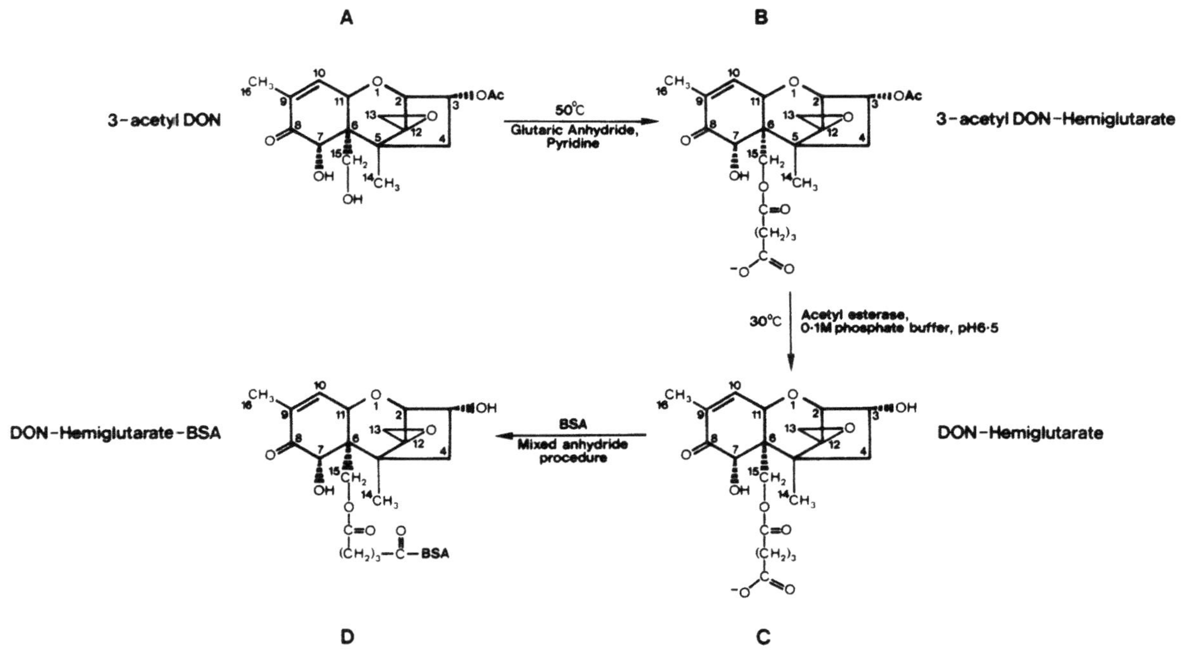

Fig. 3. Synthetic scheme used by Mills et al. (1990) in development of an immunoassay for deoxynivalenol. **A** 3-Acetyl deoxynivalenol; **B** 3-acetyldeoxynivalenol hemiglutarate; **C** deoxynivalenol hemiglutarate; **D** deoxynivalenol-bovine serum albumin conjugate

Grant 1980); HT-2 Toxin (Fan et al. 1987); 3-hydroxy-T-2 toxin (Wei et al. 1986); roridan A (Hack et al. 1988; Märtlbauer et al. 1988b); PR toxin (Wei and Chu 1988); and ergotamine (Shelby and Kelley 1990).

Most assays for zearalenone (for instance those of Thouvenot and Morfin 1983 and Warner et al. 1986) have involved oxime synthesis prior to conjugation. Teshima et al. (1990b) used different conjugation procedures. Immunogen was prepared by dissolving zearalenone (40 mg) in acetic acid (0.5 ml). Concentrated nitric acid (8 μl), water (0.15 ml) and sodium nitrite (0.8 mg) were added and the solution stirred at room temperature for 2 days. The 5-nitrozearalenone which had been synthesized was then reduced to 5-aminozearalenone by dissolving 5-nitrozearalenone (5 mg) in acetone (0.3 ml) then adding water (0.1 ml) and sodium hydro-sulphite. Sodium hydrogen carbonate (saturated solution) was used to bring the mixture to pH 7.0, which was then stirred at room temperature for 2 h. The 5-aminozearalenone was then purified. Conjugation to bovine serum albumin was achieved by glutaraldehyde cross-linking (Erlanger 1980). Bovine serum albumin (5 mg) was dissolved in phosphate buffer (0.42 ml, 0.1 M, pH 6.8). 5-aminozearalenone (2 mg) dissolved in methanol (0.12 ml) was added gradually. A glutaraldehyde solution (2%, w/v: 0.02 ml) was then added and the solution stirred overnight at room temperature. The zearalenone-protein conjugate which was used to coat microtitration plates (see Sect. 5.3.2) was prepared by directly coupling zearalenone to protein using the Mannich reaction (Frankel-Conrat and Olcott 1948; Miwa et al. 1977).

Ovalbumin (100 mg) was dissolved in sodium hydrogen carbonate (0.9 ml, 0.3 M). Zearalenone (2.5 mg) dissolved in methanol (0.1 ml) was added slowly to the protein solution. The mixture was adjusted to pH 6 with sodium acetate (3 M) and formaldehyde (7.5%, v/v; 1.0 ml) was added. The mixture was then stirred under nitrogen at room temperature for 18 h.

Immunoassays for sterigmatocystin have involved raising antisera against the more water soluble hemi-acetal derivative (Li and Chu 1984; Morgan et al. 1986b). The latter reports interesting results relating to heterologous coupling procedures used to generate immunogen and assay solid phase. A novel combination of reagents was employed. Immunogen was synthesized by reaction of sterigmatocystin (10 mg) at 120 °C for 1 h with sebacoyl dichloride (0.01 ml) and pyridine (0.001 ml) in a sealed vial. Excess dichloride was then removed under vacuum. The sterigmatocystin acid chloride product was conjugated directly to protein. The immunogen has an 8 carbon straight-chain link or bridging group separating the hapten from the protein. For the solid phase conjugate, trans 1,4 cyclohexanedioic acid was synthesized to give a cyclic bridging group, which considerably improved the assay limit of detection and gave a reduced background response. This was because the contribution of antibodies specific for the linear bridging group used in immunogen synthesis is negated in the presence of the cyclic bridge.

PR toxin contains several functional groups including an acetoxy, an α-β unsaturated ketone, two stable epoxides and an aldehyde group. The aldehyde group was conjugated to protein using the Schiffs base reaction with the linkage stabilized by reductive alkylation (Wei and Chu 1988). The reaction involves dissolving PR toxin in methanol-pyridine (0.8 ml; 1:1, v:v) and adding protein dissolved in 0.05 M sodium phosphate buffer (pH 7.2) and mixing for 30 min when 0.026 M sodium borohydride (0.5 ml) was added. The mixture was then incubated for a further 30 min at −5 °C and the reaction then terminated by the addition of 0.1 M hydrochloric acid (0.5 ml).

4 Antibody Production

The most vital and fundamental component of an immunoassay is the antibody preparation. It is the antibodies which determine the effectiveness and performance of the assay. Thus some detail on production is required. The production of antibodies in higher animals is initiated by the stimulation of the immune system following invasion by "foreign" material, which may be protein, polysaccharide or lipid in nature. It is understood that recognition of the foreign substance by receptors on the surface of a B lymphocyte in the spleen or lymph nodes stimulates it to proliferate and to secrete an antibody specific to that binding site. Tagging the invading substance in this way identifies it for neutralization by subsequent steps in the defence system.

Compounds of less than approximately 1000 daltons molecular weight are not normally immunogens and must therefore be conjugated to foreign proteins or a synthetic polymer, i.e., poly-L-lysine (see Sect. 3.1). A molecule such as a conjugated protein will possess many antibody stimulating epitopes and each will induce clones of lymphocytes to produce antibodies. Some of these antibodies will be capable of binding to free hapten. These become pooled in the blood to form a polyclonal antiserum (Fig. 4a). Following the discovery of Kohler and Milstein (1975) it is now possible to produce in vitro monoclonal antibodies recognizing a single epitope (Fig. 4b).

4.1 Immunization

The effectiveness of immunization of an animal and thus yield of antibody will depend on a great many circumstances (Paul 1984):

1. Species of animal. The most convenient species are mice, rats, guinea pigs and rabbits, although larger animals such as sheep, goat, donkey or horse may be used, with the benefits of larger serum volumes. There is

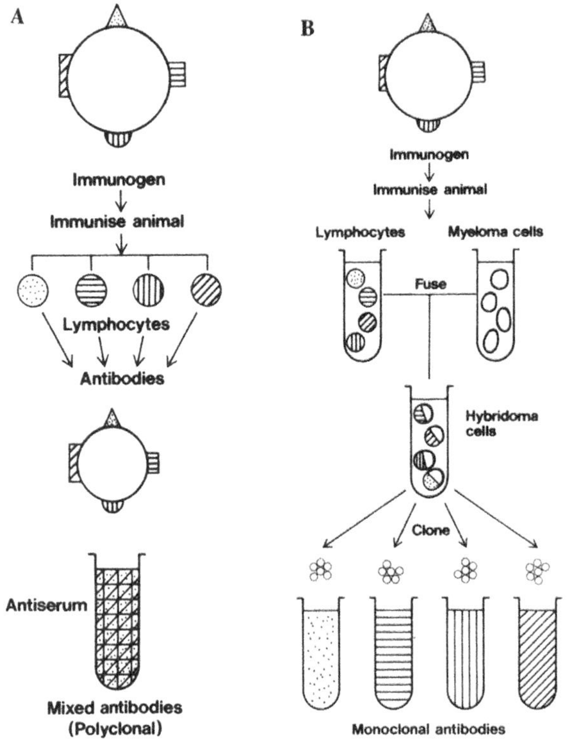

Fig. 4. Schematic representation of antibody production. **A** Polyclonal; **B** monoclonal

some evidence that there can be considerable inter-species variation in response to a single immunogen.

2. Adjuvant. In order to optimize the yield of antibodies, particularly to hapten-conjugates in which an extreme response is sought, an adjuvant should be used with the immunogen for immunization. Adjuvants contain mitogenic compounds derived, in the case of Freunds from heat-killed bacteria in an oil-based dispersion which increase inflammatory reaction and so greatly enhance the yield of specific antibodies. Alternatively a water-miscible saponin like Quil-A is available, but may be less suitable for haptenic immunogens.
3. Site of immunization. There are a number of possible routes to administer the immunogen, but usually some combination of the following is used: intra-muscular injections into the hind leg; intraperitoneal injections; intradermal or subcutaneous injections on the back.
4. Frequency of immunization. The degree of sensitization is increased during a series of booster injections. These have been administered as close as bi-weekly or with breaks of several months or longer. The number of injections depends on the degree of response to previous injections and the practice of the laboratory concerned.

5. Time elapse before bleeding. Following injection there is a lag period before antibodies begin to circulate in the blood. The optimal peak of antibody production may be only a day or two in duration. In the rabbit, the peak is often observed at about 12–14 days and antisera obtained accordingly.
6. Concentration of immunogen. Too large a dose at the first immunization can lead to intolerance. Thus the concentration should be increased in stages if a large dose is necessary. In the case of haptens, the molar ratio of hapten to carrier protein has also been reported to influence immunogenicity (Erlanger 1980).

The optimal criteria for raising antisera to haptenic drugs have recently been investigated in rabbits (Kitagawa et al. 1989) and mice (Hu et al. 1989a,b), though it is generally accepted that inter-animal variation is so great as to make objective study extremely difficult.

4.2 Polyclonal Antisera

There has been a tendency for different research groups to evolve and adopt their own protocols for raising antisera. The methods employed by one laboratory are summarized by Corrie (1983). The following method is used in this laboratory to raise polyclonal antisera in male New Zealand white rabbits aged at least 10 weeks. The immunogen, usually a bovine serum albumin-hapten conjugate, is prepared in Freunds complete adjuvant: sterile saline (7:3; 0.5 mg ml^{-1}). Each rabbit is injected intra-muscularly (0.5 ml) in each hind leg and sub-cutaneously (2×0.25 ml) in each side of the back. After a minimum of 6 weeks, booster injections are administered as before but using incomplete (rather than complete) Freunds adjuvant. The rabbits are bled from a marginal ear vein into heparinized tubes a maximum of three times between 10–16 days after injection. The plasma is removed by centrifugation and stored below −18 °C.

4.3 Monoclonal Antibodies

Rats or mice are the species most usually used for monoclonal antibody production. We prefer Balb/C mice which are given intra-peritoneal injections (0.2 ml) of immunogen (50 µg ml^{-1}) prepared in the same medium as for rabbits. The degree of immunization is tested in antisera obtained from a tail bleed. When a satisfactory response is achieved, the mouse is given a further booster injection and after 3–4 days, the spleen is removed for fusion.

The process of monoclonal antibody production follows well-established procedures (Galfré and Milstein 1981). In essence this involves the fusion of a spleen lymphocyte with a cancerous myeloma cell to produce a hybridoma, which is able to secrete antibody and to divide indefinitely in culture. Non-hybridoma cells are eliminated in selective medium.

Hybridomas secreting monoclonal antibodies of interest are identified by an ELISA screening technique (Sect. 5.6) and a pure cell line obtained by repeated cell dilution and cloning. Monoclonal antibodies can be produced in large amounts in two ways:

1. By injection of the hybridoma cells into the peritoneum of an animal co-specific with the spleen donor, where tumour development leads to the production of ascites fluid containing 5–15 $mg\,ml^{-1}$ of antibody;
2. Alternatively, the hybridomas may be grown in culture, in vessels ranging from 10 ml to 1000 litres in size. Culture supernatant normally contains 5–50 $\mu g\,ml^{-1}$ of monoclonal antibody which can be further purified by a number of techniques as required. Cell lines can be stored in liquid nitrogen for later retrieval.

4.4 Antiserum Titre

The optimal dilution of the antiserum (whether polyclonal or monoclonal in nature) must now be determined for use in the assay format adopted. However, this will be dependent on the concentration of reagents used and the times of incubation periods. In the case of microtitration plates, both protein/toxin coated to the plate and toxin/enzyme in free solution need to be studied by using a two-dimensional "chequerboard". It is set up so that each reagent is assayed against different concentrations of the other. More than one combination of concentrations and incubation times will give a maximal absorbance around the target optical density of 1.5.

With polyclonal antisera, the antibody population present at the highest concentration and/or with the greatest affinity is most likely to determine the specificity and sensitivity. There is the greatest chance, however, of selecting monoclonal antibodies with median affinity. Thus monoclonal antibodies tend to have a lower sensitivity than polyclonals.

5 Immunoassay Formats

Immunoanalysis exploits the ability of antibodies to recognize and bind to an analyte at extremely low concentrations and to differentiate between closely related molecules. The amount of antibody recognition of target analyte can be determined by the use of a label which can be sensitively and readily detected. The labels employed may be quantified accurately, using instrumental techniques, or assessed semi-quantitatively, sometimes using a direct visual approach.

5.1 Radioimmunoassays (RIA)

Sufficient radiolabelled analyte (^{125}I, ^{14}C or ^{3}H) is added to an antibody such that 50% of the label becomes antibody bound. A proportion of this label is then displaced when a known concentration of standard analyte or an unknown amount of sample analyte is added. The antibody-free fraction can be removed by the addition of activated charcoal and from the concentration of isotope present in either the supernatant or the residue, the amount of sample analyte can be calculated by reference to the standard curve.

However, radioisotopes may be labile, are expensive, and together with the hazards inherent in handling isotopes present serious drawbacks to the use of RIA. Although some early methods developed for measuring mycotoxins were by RIA (Langone and Van Vunakis 1976) the most suitable label (and that most frequently adopted) is enzymic in nature.

5.2 Enzyme Immunoassay

Several enzyme systems have been employed in immunoassay, but only two are used widely. Alkaline phosphatase gives yellow reaction products in carbonate buffer, pH 9.6, with p-nitrophenyl phosphate ($1\,mg\,ml^{-1}$) which absorb at 405 nm. Further colour development may be stopped by adding 3 M sodium hydroxide.

The popularity of horse radish peroxidase has increased with the availability of suitable substrates. Early substrates were either carcinogenic or light sensitive. Substrates currently available are 2, 2′ azinobis (3-ethylbenzthiazoline) sulphonic acid (ABTS) and 3, 3′, 5, 5′ tetramethyl benzidiine (TMB).

1. ABTS: in citrate/phosphate buffer pH 4.0 + $10\,\mu l\ 100\,ml^{-1}$ of 30% hydrogen peroxide ABTS ($0.5\,mg\,ml^{-1}$) produces a green colour absorbing at 405 nm. The reaction can be stopped with 1.5% w/v sodium fluoride.
2. TMB: dimethyl sulphoxide (1 ml) is required to dissolve TMB ($10\,mg\,ml^{-1}$). This solution is slowly added to 100 ml of 0.1 M sodium acetate/citric acid buffer (pH 6). After adding $20\,\mu l$ of 30% hydrogen peroxide the buffer should be used immediately. On stopping the reaction with 2 M sulphuric acid, the yellow colour can be read at 450 nm. This substrate is available commercially as a two-component reagent (Cambridge Veterinary Sciences, Ely, UK).

5.3 Phase Separation

In ELISA (enzyme-linked immunosorbent assay) separation of unreacted reagents is brought about by binding either analyte or specific antibodies to a solid surface. Surplus reagents are then removed by washing. It is vital that the antibody or analyte is firmly bound to the solid phase, as leaching during the washing steps will lead to poor assay performance. The surface should not be so reactive as to bind assay reagents non-specifically. In such cases the unoccupied binding sites on the solid phase can be blocked, for instance with a 1% solution of bovine serum albumin. Many different supports for the solid phase have been developed.

5.3.1 Suspended Particles

Small beads of Sepharose, cellulose or polystyrene can be used to bind the reagents to form the solid phase, which are then separated in the assay by centrifugation. Alternatively, magnetic particles can be separated by attraction to a magnetized rack. A major problem of the use of particles is that they tend to settle during incubations and yet may be drawn off with the supernatant. The extra step involved in centrifugation and/or supernatant removal and the difficulty of incorporation into automated procedures are not conducive to the uptake of these techniques. Consequently they have been little used in mycotoxin analysis, although a sensitive screening assay for aflatoxin M_1 has been developed using polystyrol beads (Steiner et al. 1988).

5.3.2 Solid Surfaces

Most frequently the solid phase is the plastic surface of a tube or microtitration plate. Both antibody and analyte (in the form of a protein conjugate) can be readily immobilized passively in 0.05 M carbonate buffer (pH 9.6) at 37 °C for 3 h, or 4 °C for 16 h. Here separation and washing stages are both rapid and simple. Microtitration plates consist of 96 wells which may be sub-divided into strips and are suited to both low- and high-sample throughput operations. All of the formats described can be performed rapidly in minutes, though large sample numbers are more efficiently processed in a batch-wise manner with incubation times of several hours.

Microtitration plate assays have proved highly successful in terms of performance, simplicity and cost, and have been widely used. Most assays are initially validated on microtitration plates, although some may subsequently be modified to other formats. It is for this reason that the microtitration plate format has been covered in greatest detail.

5.4 Direct ELISA: Antibody-Labelled

Analyte-protein conjugate is coated onto the wells of the microtitration plates at concentrations of 0.1–10 $\mu g\,ml^{-1}$. The protein should be unrelated to that used for the immunogen and it is often beneficial to reduce the ratio of hapten:protein (Kemp and Morgan 1986). Sample or standard analyte are then added to the well together with the anti-mycotoxin antibodies conjugated to an enzyme. The antibodies bind either to the bound or free mycotoxin in amounts dependent on the amount of free mycotoxin present. Following incubation and washing, the plate is developed by the addition of a chromogenic substrate for the enzyme. The amount of colour, which can be measured spectrophotometrically, is inversely proportional to the concentration of analyte.

5.5 Direct ELISA: Analyte-Labelled

In this case, the anti-mycotoxin antibody is immobilized on the solid phase. A fixed amount of enzyme-labelled mycotoxin binds to the antibody in amounts dependent on how much free mycotoxin (unknown sample or standard) is also present. The amount of enzyme bound is determined readily by the addition of substrate.

5.6 Indirect ELISA: Antibody-Labelled

Rather than labelling the anti-analyte antibody directly with enzyme, a further indirect step can be introduced. Here an enzyme-labelled anti-species antibody, for instance anti-rabbit or anti-mouse antibodies raised in goat, can be used to label the bound anti-analyte antibody. In this way there is no contact between the enzyme and food matrix or extraction solvent, which may cause interference with the direct assays. Consistent supplies of stable enzyme-labelled anti-species antibody conjugates are available commercially. Hapten-enzyme conjugates have to be produced specifically for each analyte and may have only a short shelf life. It is sometimes possible to incorporate the advantages of both formats by mixing anti-analyte antibody and enzyme labelled second antibody before adding to the microtitration plate. Such a method has been developed for the assay of ochratoxin A (Sidwell et al. 1989).

5.7 Conjugation of Enzyme to Hapten or Antibody

Enzyme-hapten conjugations may use similar methods as for protein-hapten, such as carbodiimide or mixed anhydride (see Sect. 3). However, the optimal molar hapten: enzyme ratio needs to be around 1:1, and care is required to prevent impairment of enzyme activity. In addition, the periodate method of Wilson and Nakane (1978) has been used to conjugate antibody with horseradish peroxidase. Due to the instability of the conjugate, if assays for several mycotoxins are required for only occasional use then the indirect format is preferable.

5.8 Spot Tests

Some ingenious systems have been devised for the rapid detection of mycotoxins using immunoassay. Most use the labelled analyte principle. Antibodies can be immobilized onto cards or discs, as well as microtitration plates. Kits with all reagents supplied ready prepared can be carried out away from the specialized laboratory. These assays are usually only semi-quantitative, but this may be sufficient where rapid screening is required and the number of negative samples is high.

5.9 Immunoaffinity Columns

Antibodies may also be linked to a solid support such as Sepharose to form a powerful chromatographic reagent. Gels pre-activated with cyanogen bromide are available commercially for reaction with the antibody. Made up into a small column, the gel-antibody complex will bind the toxin specifically as the sample extract is passed through. After washing to remove non-specifically bound residues, the antibody binding can be disrupted by changes in pH or use of organic solvent to release the toxin. Liquid samples such as serum, urine or milk can be applied directly or with minimal modification. A column, 1–2 ml in volume, was used to concentrate aflatoxin M_1 from a litre of skimmed milk (Mortimer et al. 1987a). After following suitable washing procedures and regeneration, the columns can be re-used several times. If aflatoxin is the analyte, it can be eluted and trapped in a fluorosil tip (Candlish et al. 1988a) for rapid semi-quantitative measurement by fluorescence in ultraviolet light. Alternatively, elution of free toxin can be followed by quantitative analysis by HPLC (Groopman and Donahue 1988; Sharman et al. 1989).

6 Validation and Application

Before an assay can be applied to analysis of a particular matrix, several validation tests are required. These will need to be repeated if any changes are made to the method, for example, in the matrix or extraction procedure.

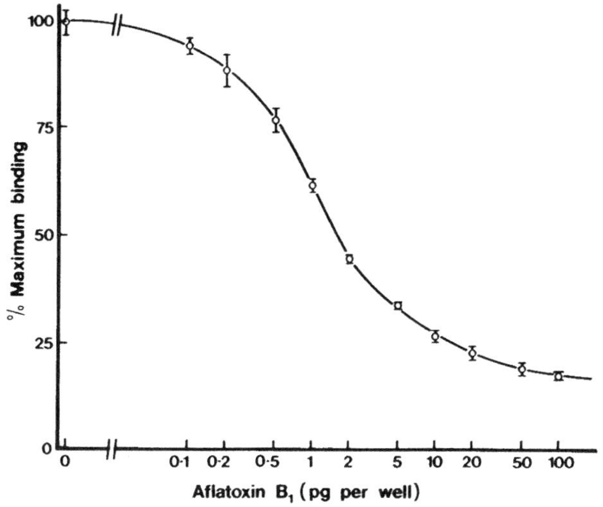

Fig. 5. Standard curve for aflatoxin B_1 showing sensitivity of 0.1 pg/well

6.1 Standards

It is preferable to obtain a source of pure toxin, and many are available commercially. This is not possible for all mycotoxins and it may be necessary to modify a related chemical or to extract and purify the mycotoxin from fungal cultures. Aflatoxins are stable for long periods in methanol solution stored in the deep freeze. Standard solutions of deoxynivalenol are stable at room temperature in ethyl acetate or down to −18 °C in acetonitrile. (Shepherd and Gilbert 1988). Concentrations of standard solutions should be checked by UV absorption and purity by TLC or HPLC.

6.2 Limit of Detection

From the titre curves (see Sect. 4.4), reagent dilutions are selected which produce the desired absorbance after the addition of substrate. A range of standards diluted in assay buffer including a zero value are assayed with an equal volume (usually 50 or 100 µl) of antiserum or hapten/enzyme conjugate depending on assay format. A plot of absorbance against $\log_{10}$ analyte concentration should result in a sigmoidal response curve.

The limit of detection of the assay is often defined as the minimum concentration of analyte which differs by twice the standard deviation of the zero value. Generally speaking, the lowest limits of detection are achieved with the lowest concentration of antibody for given incubation times. A limit of detection of 0.1 pg/well has been achieved (Fig. 5) for aflatoxin B_1 in peanut butter (Morgan et al. 1986a).

6.3 Specificity

To determine the cross-reactivity of an antibody preparation, a collection of compounds structurally related to the hapten mycotoxin are required. Those toxins, metabolites and other structurally related materials likely to occur in the matrix of interest should be tested. Standard curves for each compound are compared in the assay. Cross-reactions are usually calculated as the ratio, expressed as a percentage, of the amount of toxin required to give 50% inhibition of maximum binding to the amount of reference compound giving the same inhibition.

Many compounds will exhibit no antibody recognition. Some, depending on the degree of similarity in the region of the molecule distal to the point of conjugation, may be recognized. It is also possible, particularly with monoclonal antibodies, to have a greater cross-reaction for a non-target molecule. Thus a monoclonal antibody for aflatoxin M_1 had a greater affinity for aflatoxin B_1 (Dixon-Holland et al. 1988) and a monoclonal raised to aflatoxin B_1 preferentially recognized aflatoxin G_1 (Ward et al. 1990).

6.4 Extraction

The extreme sensitivity and specificity of immunoassays frequently ensures that only solubilization and dilution of the analyte is required prior to analysis. Indeed, in the case of foods and biological material which are already liquid, direct measurement may be possible. Milk can be analyzed directly for aflatoxin M_1 (Pestka et al. 1981) though sensitivity was much improved if the sample was subjected to treatment on a reverse phase Sep-Pak C_{18} cartridge prior to ELISA (Frémy and Chu 1984). Alternatively, aflatoxin M_1 can be quantified by HPLC following clean-up on an immunoaffinity column (Mortimer et al. 1987a). Undiluted serum may interfere with an immunoassay. Analysis of human serum for aflatoxin was possible after addition of an equal volume of methanol and centrifugation (Wilkinson et al. 1988b). Few immunoassays have been described for the direct analysis of urine, although an RIA for aflatoxin B_1 in rat urine has been published (Sizaret et al. 1982). More usually, as with human urine, a clean up on an affinity column is required (Wild et al. 1986).

Non-liquid commodities should first be finely ground and extracted with organic solvents prior to analysis of mycotoxins. The trichothecene diacetoxyscirpenol (DAS) in wheat is extracted with methanol and diluted to prevent interference from the solvent in the assay (Mills et al. 1988). The addition of water to the solvent may improve efficiency of aflatoxin extraction, but the ratio of water to solvent and proportion of extractant to sample needs to be optimized (Whitaker and Dickens 1986). Methanolic extraction of aflatoxins from cereal products give consistently lower results than those obtained with chloroform (Friesen 1989). For the extraction of

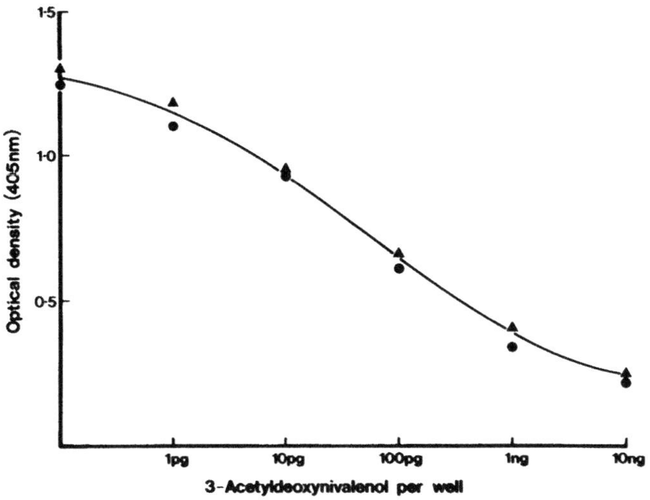

Fig. 6. Superimposability of standard curves for 3-acetyldeoxynivalenol in buffer (▲) and diluted extract (●)

aflatoxins from peanut butter, acetonitrile: water (1:1; v:v) was found most efficient (Morgan et al. 1986a), whereas acidified chloroform was used to extract barley for the determination of sterigmatocystin (Morgan et al. 1986b) and ochratoxin A (Morgan et al. 1986c). Extraction with volatile solvents offers considerable scope for extremely sensitive assays after concentration. Methanol:water, 60:40 was preferred to extract 3-acetyldeoxynivalenol (ADON) and deoxynivalenol (DON) from rice and wheat respectively (Kemp et al. 1986; Mills et al. 1990).

6.5 Superimposability and Linearity

Standards diluted in assay buffer and in a blank matrix extract should produce identical curves which superimpose (Fig. 6). If this is not the case, then it is likely that components of the matrix or the extraction solvent are interfering with antibody binding. This is often expressed by a reduction in extinction of the blank control or displacement of the curve on the concentration of analyte axis. Interference can usually be circumvented by running standards in a blank extract (Wilkinson et al. 1988b). However, it has also been observed that low concentrations of methanol extraction solvent can enhance the performance of some assays. This effect can be utilized by adding a similar concentration of solvent to the standards via the dilution buffer (Mills et al. 1988).

The test for linearity is a more stringent indicator of interference. Different volumes of a positive extract are assayed and results plotted

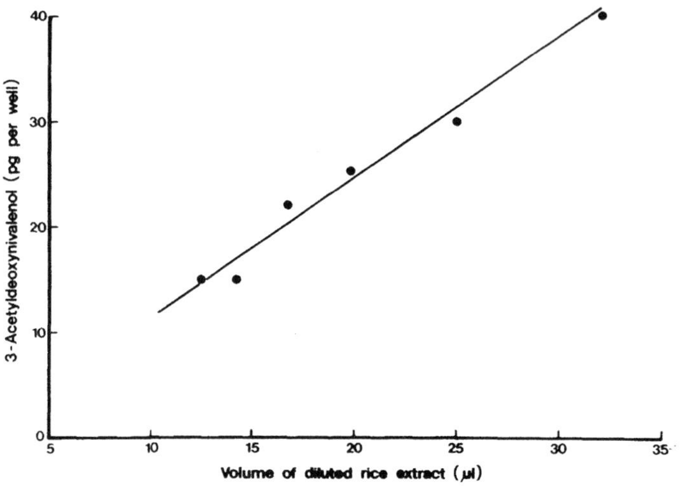

Fig. 7. Linearity between volume of extract and recovery of 3-acetyldeoxynivalenol

against concentration of analyte (Fig. 7). Any interference is indicated by a deviation from linearity over the working range of the assay.

6.6 Recovery and Reproducibility

Recovery of several concentrations of analyte are determined initially by spiking the raw material prior to extraction and clean-up. Dry material is first ground to a fine powder and the mycotoxin added in an organic solvent, thoroughly mixed and allowed to dry. However it is preferable subsequently to use a naturally contaminated sample in which the analyte is integral with the matrix. The Community Bureau of Reference (BCR, Brussels, Belgium) prepares natural products with levels of constituents including mycotoxins which have been determined by exhaustive analysis (Van Egmond and Wagstaffe 1990). Recovery should be close to 100% (±5%) but if due to incomplete extraction or losses in cleanup, lower but consistent recoveries are obtained then correction can be made (Van Egmond and Wagstaffe 1989).

A major advantage of immunoassay is the lack of extensive cleanup required. Fewer steps leads to increased speed of analysis, higher recovery of analyte and improved reproducibility of the assay compared to other techniques.

Having established the recovery of toxin from the matrix, then the reproducibility of the assay performance should be determined. Repeated extractions of the same samples, preferably at more than one concentration of contamination, should be assayed. Both within and between assay variation are required, carried out on several days with different batches of reagents. Between assay plate variation is usually greater than within assay. Coefficients of variation of less than 10% are desirable, and are certainly feasible with immunoassays even when operating at the trace contamination level.

6.7 Precision Profiles

The use of microcomputers in the analytical laboratory is allowing complex statistical analysis of data, thus increasing the assurance of the analyst in the data. The most appropriate statistical assessment of the reliability of the performance of an assay is the precision profile (Ekins 1983). This takes into account the slope of the standard graph and the consequent standard deviation on the dose and response axes. The characteristic profile is U-shaped and any deterioration in assay reagents, assay conditions or operator performance are likely to be reflected in the shape or position of the curve. Comparison of successive precision profiles will reveal any slight but consistent drift in any of these parameters. By selecting an acceptable level of coefficient of variation the working range of the assay can be determined. This identifies the region of the standard curve of highest precision (Ward et al. 1990) and coefficients of variation can be as low as 3% for a well-prepared assay (Ward and Morgan 1991).

7 Correlation Studies

It is important to ensure that results from immunoanalysis techniques correlate with analysis by other established methods. The criteria for the validation of analytical methods including RIA have been described by De Ruig et al. (1989). Comparative studies for non-immunoanalytical techniques for aflatoxins have been reported by Friesen and Garren (1982a), Van Egmond and Wagstaffe (1989 and 1990) and for aflatoxin M_1 by Friesen and Garren (1982b), Shepherd et al. (1986), Van Egmond et al. (1986) and Scott (1989). General methods of mycotoxin analysis are reviewed by Van Egmond and Paulcsh (1986) and Shotwell (1986). Thus methods for RIA and TLC have been compared for recovery of deoxynivalenol from wheat and corn (Xu et al. 1986), ELISA compared with GLC for analysis of T-2 toxin in *Fusarium* culture filtrates (Nagayama et al. 1988); ELISA compared with LC for analysis of aflatoxin M_1 in dairy products (Frémy and Chu 1984); and ELISA compared with HPLC and HPTLC for analysis of aflatoxins from peanut butter (Dell et al. 1990).

In preference, a more complete evaluation should be conducted. For instance, two ELISA screening tests were compared with quantitative measurement of aflatoxins in raw peanuts by liquid chromatography (Dorner and Cole 1989). The EZ-Screen Quick Card (Environmental Diagnostics Inc, Burlington, NC, USA) was shown to have a detection threshold close to 10 rather than the nominal 20 ppb, but both it and the Afla-10 Cup Test (International Diagnostics Systems Corp., IDS, St. Joseph, MI, USA) reliably identified >95% samples containing >10 ppb aflatoxin.

A comparison using methods of extraction approved by the Association of Official Analytical Chemists (AOAC) and TLC analysis of animal feeds with two commercial tests (Trucksess et al. 1990) further confirmed the suitability of ELISA screening methods for rapidly detecting positive samples. Using the Immuno-Dot Screen Cup (IDS) and an affinity column, Aflatest-P (Vicam, Somerville, MA, USA), which was used to clean-up samples for quantitative determination by fluorodensitometry or TLC, was as effective as long-established AOAC methods.

7.1 Screening Trials

In the United States, collaborative trials have concentrated on rapid screening methods. As a result, Agri-Screen (Neogen Corp, Lansing, MI, USA) a semi-quantitative ELISA kit for aflatoxin B_1, has been recommended for adoption as official first action for screening aflatoxins in cottonseed products and mixed feed at $\geqslant 15\,\mu g\,g^{-1}$, and $\geqslant 20\,\mu g\,g^{-1}$ in corn and roasted peanuts (Park et al. 1989a and b). Another method using a filter paper disc in a polyethylene cup (Immuno-Dot Screen Cup, IDS) has been adopted for screening aflatoxins B_1, B_2, G_1 and G_2 at $\geqslant 20\,\mu g\,g^{-1}$ in cottonseed and peanut butter and at $\geqslant 30\,\mu g\,g^{-1}$ in corn and raw peanuts (Trucksess et al. 1989).

Both methods gave extremely low levels of false negatives, but at levels of contamination close to the cut-off point, some samples were falsely determined as positive. Thus the methods were suitable for screening samples to detect those requiring full quantitative analysis by other techniques. As the tests take only a few minutes to perform, they are particularly valuable when rapid decisions are involved.

7.2 Quantitative Trials

Trials have been conducted in the U.K. (Mortimer et al. 1988) of the Quantitox B_1 microtitration plate kit (Rhône-Poulenc Diagnostics Ltd., Glasgow, U.K.) and of three commercial ELISA kits (Patey et al. 1989). A series of trials have investigated the use of immunoaffinity columns (Oxoid, Basingstoke, U.K.) for sample clean-up with quantitation by HPLC or ELISA (Patey et al. 1990a,b, 1991a). Most recently an international trial of the Biokits Aflatoxin Kit (Cortecs) has led to the submission for AOAC approval, of the first quantitative mycotoxin kit (Patey et al. 1991b).

8 Concluding Remarks

The use of antibody-based techniques for quantitative and semi-quantitative analysis of mycotoxins has revolutionised the area within the last decade. In particular, the analysis (in minutes) of samples extracted once without the need for additional clean-up steps has been an exciting development. The simple-to-use aflatoxin ELISA technique that was able to analyze homogeneous samples of peanut butter (mean aflatoxin level 11.7 $ng\,g^{-1}$) on four separate occasions, and two further samples (mean aflatoxin level 5.8 and 24.8 $ng\,g^{-1}$) also on four separate occasions without observing a single false-positive or false-negative result based on a 10 $ng\,g^{-1}$ action level (Ward and Morgan 1991), would not have appeared possible only a few years ago. Equipment needs (except for fully automated, high-sample through-put systems) are minimal, with the imminent prospect of real "in field" testing. Further developments will see an increased range of antibodies with useful specificities. The potential for true broad-specificity antibodies (e.g., to quantify a large number of tricothecenes) becomes increasingly feasible with new techniques of antibody production and manipulation. These probes will become incorporated into even more robust assay systems, with sophisticated control procedures to ensure correct operation and statistical validity. Both quantitative and semi-quantitative immunoassay procedures will operate in tandem, first to screen samples and then to quantify significant positives. It is still possible to envisage the use of biosensors for mycotoxin analysis. Conventional physicochemical procedures will still be needed, of course, to provide important back-up in particularly sensitive cases, to analyze new toxins and those toxins for which no suitable antibody preparations have been developed.

The end purpose of the improved methods of analysis provided by immunoassays will be to gain a clearer insight into the role (if any) of individual mycotoxins in human disease, particularly at the sub-acute dose level, to ensure adequate monitoring of the food supply in association with effective control procedures, and to contribute to the provision of wholesome, nutritious food.

Appendix 1: Commercially Available Kits for the Determination of Mycotoxins

The use of immunoanalysis was initially limited to the specialized laboratory with the expertise and facilities to raise antisera, develop formats and apply the analysis to complex matrices. However, in recent years the technology has become more widely available through the sale of kits. These are usually self-contained including all the reagents required: standards, controls, buffers, substrates etc. Quantitative kits require only the additional use of a spectrophotometer or some other instrument for accurate end-point determination. Semi-quantitative screening assays, in contrast, are very simple to perform even away from the laboratory at the farm or factory.

There has been much interest in kits as a rapid screening procedure. By eliminating negative samples, which are often the majority, detailed analytical work is much reduced. So quick to perform are these assays that a "while-you-wait" service can be operated before accepting or rejecting a batch of raw materials. Now that such kits are receiving official approval, it is clear they will be widely used in quality control in future, displacing existing methods.

In the rush to be first in the market, some kits were released before they had been fully tested under all the likely circumstances of use. Such kits have been withdrawn due to company failure and takeover. Those that remain have the reliability and robustness to give dependable results in the majority of applications. No matter how simple, it is unlikely that an inexperienced operator will obtain accurate, quantitative results on the first use of a kit. Accurate, reproducible pipetting is a necessary skill, but can be quickly acquired. Target standards of performance from an experienced laboratory are illustrated by coefficients of variation as low as 3% (Ward and Morgan 1991).

The Table, whilst undoubtedly incomplete, will serve as a guide to the availability of commercial kits for the analysis of mycotoxins.

Table 1. Suppliers of immunoassay kits for mycotoxin determination

Supplier	Mycotoxin	Principle	Application[a]
Chemopol Ltd	Aflatoxin B_1	RIA	Q
Kodanská 46	Aflatoxin M_1/B_1	RIA	Q
100 10 Prague 10	Ochratoxin A	RIA	Q
Czechoslovakia			
Cortecs Diagnostics	Aflatoxin (B/G)	Microwell	Q
Techbase 1	Aflatoxin	Microwell	S
Newtech Square	Ochratoxin A	Microwell	Q
Deeside Industrial Park			
Deeside, Clwyd			
CH5 2NT, UK			
Environmental Diagnostics Inc	Aflatoxin B/G	Card	S
1238 Anthony Road	Aflatoxin M_1	Card	S
Burlington	Ochratoxin	Card	S
North Carolina 27215, USA	T-2 Toxin	Card	S
	Zearalenone	Card	S
Genego Spa	Aflatoxin B_1	Microwell	Q
Via Ressel	Aflatoxin M_1	Microwell	Q
S. Andrea Zona Ind	Zearalenone	Microwell	Q
34170 Gorizia, Italy			

Table 1. (continued)

Supplier	Mycotoxin	Principle	Application[a]
Idexx 100 Fore Street Portland Maine 04101 USA	Aflatoxin B_1 Aflatoxin M_1	Cup Cup	S S
International Diagnostics Systems Corp St. Joseph Michigan, USA	Aflatoxin B/G	Cup	S
Neogen Corp 620 Lesher Place Lansing Michigan 48912, USA	Aflatoxins Aflatoxin M_1 DON T-2 Toxin Zearalenone	Microwell Microwell Microwell Microwell	S S S S S
Noack Geylinggasse 27/8 A-1130 Wien, Austria	Aflatoxin B/G Aflatoxin M_1 Zearalenone T-2 Toxin Ochratoxin	– – – – –	Q/S S S S S
Oxoid Ltd Wade Road Basingstoke Hants, RG24 OPW, UK	Aflatoxin B/G Aflatoxin M_1	Affinity Affinity	S/C S/C
Penicillin Assays Inc 36 Franklin Street Malden Massachusetts 02148, USA	Aflatoxin B/G Aflatoxin M_1/M_2	RIA	Q Q
Rhône-Poulenc Diagnostics Ltd Montrose House 187 George Street Glasgow, G1 1YT	Aflatoxin B_1 Aflatoxins	Microwell Affinity	Q S/C
Riedel-de Haën Wunstorfer Str 40 D-3016 Seelze 1, Germany	Aflatoxin B_1, M_1 Ochratoxin A Zearalenone	Microwell Microwell Microwell	Q Q Q
Transia Ltd 8, rue Saint-Jean de Dieu 69007 LYON, France	Aflatoxin B_1, M_1 Aflatoxin B/G, M_1 Ochratoxin T-2 Toxin Zearalenone	Microwell Card Card Card Card	Q S S S S

[a] C = Cleanup S = Screen Q = Quantitative

Note: Fukal and Káš (1989) have reported a similar list containing some details not available to the present authors.

Appendix 2: Examples of the Application of Immunological Assays for Mycotoxins in Food and Biological Matrices

Table 2 lists immunological assays applied to mycotoxin analysis in food and biological matrices reported since the review of Pestka (1988).

Table 2. Applications of immunological assays to mycotoxin analysis

Mycotoxin	Application biological matrix	Type of assay	References
Aflatoxin M_1	Milk	Affinity Column	Mortimer et al. (1987a)
Aflatoxin	Peanuts, corn, animal feed	ELISA	Itoh et al (1987)
Aflatoxin	Peanut butter	ELISA	Mortimer et al. (1987b)
Aflatoxin	Peanut products	ELISA	Kawamura et al. (1988)
Aflatoxin B_1 and M_1	Natural contaminated foods	ELISA	Dixon-Holland *et al.* (1988)
Aflatoxin	Peanut products, corn	ELISA	Park et al. (1988)
Aflatoxin B_1	Peanut Butter	ELISA	Mortimer et al. (1988)
Aflatoxin	Dried figs	ELISA	Reichert et al. (1988)
Aflatoxin	Peanut products, corn	ELISA	Park et al. (1989a)
Aflatoxin B_1	Cotton seed, mixed feed	ELISA	Park et al. (1989b)
Aflatoxin B_1, B_2 and G_1	Corn, cottonseed, peanuts, peanut butter, poultry feed	ELISA	Trucksess et al. (1989)
Aflatoxins	Peanuts	ELISA	Dorner and Cole (1989)
Aflatoxin	Peanut butter	ELISA	Patey et al. (1989)
Aflatoxin	Corn	ELISA	Wilcke et al. (1990)
Aflatoxin B_1, T_2-toxin, ochratoxin A	Barley	ELISA	Ramakishna et al. (1990)
Aflatoxin	Peanuts	RIA	Fukal et al. (1987)
Aflatoxin M_1	Human breast milk	ELISA	Wild et al. (1987)
Aflatoxin B_1	Human liver tissue	RIA	Stora and Dvorackova (1987)
Aflatoxin B_1	Rat liver tissue	ELISA	Groopman et al. (1987)

Table 2. (continued)

Mycotoxin	Application biological matrix	Type of assay	References
Aflatoxin B_1, Aflatoxin-protein adducts	Rat tissue, serum	ELISA	Okoye and Neal (1988)
Aflatoxin metabolites, protein and DNA adducts	Human serum, urine, breast milk	ELISA, Affinity columns	Wild et al. (1988)
Aflatoxin – DNA adducts	Human liver	ELISA, Affinity columns	Garner et al. (1988)
Aflatoxin metabolites	Human urine	RIA	Dragsted et al. (1988)
Aflatoxin-DNA	Rat, mice liver, tissues	ELISA	Hsieh et al. (1988)
Aflatoxin-protein adducts	Serum	RIA	Gan et al. (1988)
Aflatoxin B_1, Aflatoxin-DNA, Aflatoxin-protein adducts	Rat and human urine and serum	RIA, ELISA, affinity columns	Groopman and Donahue (1988)
Aflatoxin	Human serum	ELISA	Denning et al. (1988)
Aflatoxin	Human serum	ELISA	Wilkinson et al. (1988a)
Aflatoxin	Human serum	ELISA	Wilkinson et al. (1988b)
Aflatoxin	Foods, human serum	ELISA	Wilkinson et al. (1989)
Aflatoxin metabolites	Human serum, urine	ELISA	Autrup (1989)
Aflatoxin – DNA adducts	Human liver	Immunohist-ochemistry	Lee et al. (1989)
Aflatoxin B_1 and M_1	Foods, human urine	ELISA	Yu et al. (1989)
T-2 toxin	Fusarium cultures	ELISA	Nagayama et al. (1988)
T-2 toxin	Spiked wheat flour	ELISA	Chiba et al. (1988)
T-2 toxin, 3-acetyl-deoxynivalenol, 15-acetyl deoxynivalenol, nivalenol, deoxynivalenol, zearalenone	Corn	Affinity column, RIA	Li et al. (1988)
T-2 toxin, zearalenone,	Cereals, wheat flour	ELISA	Lepschy et al. (1989)
3-acetyl deoxynivalenol	Rice	ELISA	Kemp et al. (1986)

Table 2. (continued)

Mycotoxin	Application biological matrix	Type of assay	References
Diacetoxyscripenol	Wheat	ELISA	Mills et al. (1988)
Nivalenol	Spiked barley	ELISA	Ikebuchi et al. (1990)
Nivalenol	Barley	RIA	Teshima et al. (1990a)
Deoxynivalenol	Corn, wheat	ELISA	Xu et al. (1988)
Deoxynivalenol	Spiked corn	ELISA	Casale et al. (1988)
Deoxynivalenol	Wheat	ELISA	Mills et al. (1990)
Zearalenone, zearalenone metabolites	Swine urine	ELISA	Macdougald et al. (1990)
Ochratoxin A	Barley	ELISA	Candlish et al. (1988b)
Ochratoxin A	Cereals	RIA	Ruprich et al. (1988)
Ochratoxin A	Human milk	ELISA	Gareis et al. (1988)
Ochratoxin A	Chicken meat, wheat flour, porcine plasma, bovine serum	ELISA	Kawamura et al. (1989)
Ochratoxin A	Porcine kidneys	RIA	Rousseau and Van Peteghem (1989)
Sterigmatocystin	Mycotoxin biosynthesis	ELISA	Chung et al. (1989)

References

Abdalla AE, Grant DW (1980) An immunoanalysis for kojic acid. Sabouraudia 18:191–196

Autrup H (1989) Exposure to aflatoxins in Denmark. J Toxicol Toxin Rev 8:53–67

Bhat RV, Beedu SR, Ramakrishna Y, Munshi KL (1989) Outbreak of tricothecene mycotoxicosis associated with consumption of mould-damaged wheat production in Khasmir Valley, India. Lancet 8628:35–37

Bullerman LB (1986) Mycotoxins and food safety. Food Technol: 59–66

Butler WH (1974) Aflatoxin. In: Purchase IFH (ed) Mycotoxin. Elsevier, Amsterdam, p 6

Candlish AAG, Haynes CA, Stimson WH (1988a) Detection and determination of aflatoxins using affinity chromatography. Int J Food Sci Technol 23:479–485

Candlish AAG, Stimson WH, Smith JE (1988b) Determination of ochratoxin A by monoclonal antibody based enzyme immunoassay. J Assoc Off Anal Chem 71: 961–964

Casale WL, Pestka JJ, Hart LP (1988) Enzyme linked immunosorbent assay employing monoclonal antibody specific for deoxynivalenol (vomitoxin) and several analogues. J Agric Food Chem 36:663–668

Chiba J, Kawamura O, Kajii H, Ohtani K, Nagayama S, Ueno Y (1988) A sensitive enzyme linked immunosorbent assay for detection of T-2 toxin with monoclonal antibodies. Food Addit Contam 5:629–639

Chu FS (1986) Immunoassays for mycotoxins. In: Cole RJ (ed) Modern methods in the analysis and structural elucidation of mycotoxins. Acad Press, New York, p 207

Chung DH, Abouzied MM, Pestka J (1989) Immunochemical assay applied to mycotoxin biosynthesis: ELISA comparison of sterigmatocystin production by *Aspergillus versicolor* and *Aspergillus nidulans*. Mycopathologia 107:93–100

Corrie JET (1983) Production of anti-hapten sera in rabbit. In: Hunter WM, Corrie JET (eds) Immunoassays for clinical chemistry. Churchill-Livingstone, London, pp 469–472

Cucullu AF, Lee LS, Mayne RY, Goldblatt LA (1966) Determination of aflatoxins in individual peanuts and peanut sections. J Am Oil Chem Soc 43:89–92

Daussant J, Bureau D (1984) Immunochemical methods in food analysis. In: King RD (ed) Developments in food analysis techniques, vol 3. Elsevier Applied Science, London, p 175

Davis ND, Dickens JW, Freie RL, Hamilton PB, Shotwell OL, Wyllie TD, Fulkerson JF (1980) Protocols for surveys, sampling, post-collection handling and analysis of grain samples involved in mycotoxin problems. J Assoc Off Anal Chem 63:95–102

Davis RM, Stone SS (1979) Production of anti-rubratoxin antibody and its use in radio immunoassay for rubratoxin B. Mycopathologia 67:29–33

Dell MPK, Haswell SJ, Roch OG, Coker RD, Medlock VFP, Tomkins K (1990) Analytical methodology for the determination of aflatoxins in peanut butter: comparison of HPTLC, ELISA and HPLC methods. Analyst 115:1435–1439

Denning DW (1987) Aflatoxin and human disease. Adv Drug React Acute Pois Rev 4:175–209

Denning DW, Onwubalil J, Wilkinson AP, Morgan MRA (1988) Measurement of aflatoxin in Nigerian sera by enzyme linked immunosorbent assay. Trans R Soc Trop Med Hyg 82:169–172

De Ruig WG, Rainer WS, Dijkstra G (1989) Criteria for the detection of analytes in test samples. J Assoc Off Anal Chem 72:487–490

Dixon DE, Warner RL, Ram BP, Hart LP, Pestka JJ (1987) Hybridoma cell line production of specific monoclonal antibody to the mycotoxins zearalenone and α zearalenol. J Agric Food Chem 35:122–126

Dixon-Holland DE, Pestka JJ, Bidigare BA, Casale WL, Warner RL, Ram BP, Hart LP (1988) Production of sensitive monoclonal antibodies to aflatoxin B_1 and aflatoxin M_1 and their application to ELISA of naturally contaminated foods. J Food Prot 51:201–204

Dorner JW, Cole RJ (1989) Comparison of two ELISA screening tests with liquid chromatography for determination of aflatoxins in raw peanuts. J Assoc Off Anal Chem 72:962–964

Dragsted LO, Bull I, Autrup H (1988) Substances with affinity to a monoclonal aflatoxin B_1 antibody in Danish urine samples. Food Chem Toxicol 26:233–242

Egly JM, Boschetti E (1983) Practical guide for use in affinity chromatography and related techniques. Réactifs IBF Societé Chimique, Pointet, France, p 41

Ekins RP (1983) The precision profile: its use in assay design, assessment and quality control. In: Hunter WM, Corrie JET (eds) Immunoassays for clinical chemistry. Churchill-Livingstone, London, pp 76–105

Erlanger BF (1980) The preparation of antigenic hapten-carrier conjugates: a survey. In: Langone JJ, Van Vunakis H (eds) Methods in enzymology, vol 70. Acad Press, London, pp 85–104

Erlanger BF, Borek OF, Beiser SM, Liberman S (1959) Steroid-protein conjugates II. Preparation and characterization of conjugates of bovine serum albumin with progesterone, deoxycorticosterone and estrone. J Biol Chem 234:1090–1094

Fan TL, Xu YC, Chu FS (1987) Simultaneous analysis of T-2 toxin and HT-2 toxin by an indirect enzyme linked immunosorbent assay. J Assoc Off Anal Chem 70:657–661

Frankel-Conrat H, Olcott HS (1948) Reaction of formaldehyde with proteins. VI. Cross-linking of amino groups with phenol, imidazole, or indole groups. J Biol Chem 174:827–843

Frémy JM, Chu FS (1984) Direct enzyme-linked immunosorbent assay for determining aflatoxin M_1 at picogram levels in dairy products. J Assoc Off Anal Chem 67:1098–1101

Friesen, MD (1989) The international mycotoxin check sample programme. J Toxicol Toxin Rev 8:363–373

Friesen MD, Garren L (1982a) International mycotoxin check sample program: Part I. Report on laboratory performance for determination of aflatoxins B_1, B_2, G_1 and G_2 in raw peanut meal, deoiled peanut meal and yellow corn meal. J Assoc Off Anal Chem 65:855–863

Friesen MD, Garren L (1982b) International mycotoxin check sample program: Part II. Report on laboratory performance for determination of aflatoxin M_1 in milk. J Assoc Off Anal Chem 65:864–868

Fukal L, Káš J (1989) The advantages of immunoassay in food analysis. Trends Anal Chem 8:112–116

Fukal L, Prosek J, Sova Z (1987) The occurrence of aflatoxins in peanuts imported into Czechoslovakia for human consumption. Food Addit Contam 4:285–289

Galfré G, Milstein C (1981) Preparation of monoclonal antibodies: Strategies and procedures. In: Langone JJ, Van Vunakis H (eds) Methods in enzymology, vol 73. Academic Press, London, pp 3–46

Gan L, Skipper PL, Peng X, Groopman JD, Chen J, Wogan GN, Tannebaum SR (1988) Serum albumin adducts in the molecular epidemiology of aflatoxin carcinogenesis: correlation with aflatoxin B_1 intake and urinary excretion of aflatoxin M_1. Carcinogenesis 9:1323–1325

Gareis M, Martlbauer E, Bauer J, Gedek B (1988) Determination of ochratoxin A in human milk. Z Lebensm Unters Forsch 186:114–117

Garfield FM (1989) Sampling in the Analytical Scheme. J Assoc Off Anal Chem 72:405–411

Garner RC, Dvorackova I, Tursi F (1988) Immunoassay procedures to detect exposure to aflatoxin B_1 and benzo(a) pyrene in animals and man at the DNA level. Int Arch Occup Environ Health 60:145–150

Gaur PK, Lau HP, Pestka JJ, Chu FS (1981) Production and characterisation of aflatoxin $B_{2\alpha}$ antiserum. Appl Environ Microbiol 41:478–482

Gendloff EH, Casale WL, Ram BP, Tai JH, Pestka JJ, Hart LP (1986) Hapten protein conjugates prepared by the mixed anhydride method cross reactive antibodies in heterologous antisera. J Immunol Methods 92:15–20

Groopman JD, Donahue KF (1988) Aflatoxin, a human carcinogen determination in foods and biological samples by monoclonal antibody affinity chromatography. J Assoc Off Anal Chem 71:861–867

Groopman JD, Trudel LJ, Donahue PR, Marshak-Rothstein A, Wogan GN (1984) High affinity monoclonal antibodies for aflatoxins and their application to solid phase immunoassays. Proc Natl Acad Sci USA 81:7728–7731

Groopman JD, Roebuck BD, Kensler TW (1987) Application of monoclonal antibodies and dietary antioxidant-based animal models to define human exposure to aflatoxin B_1. Prog Exp Tumor Res 31:52–62

Hack R, Märtlbauer E, Terplan G (1988) Production and characterization of monoclonal antibodies to macrocyclic trichothecene roridin A. Appl Environ Microbiol 54:2328–2330

Hastings KL, Talis JJ, Dean JH (1988) Production and characterization of a monoclonal antibody to aflatoxin B_2. J Agric Food Chem 36:404–408

Horwitz W, Albert R (1982) The reliability of aflatoxin assays. Bull Assoc Food Drug Off 46:14–24

Hsieh LL, Hsu SW, Chen DS, Santella RM (1988) Immunological detection aflatoxin B_1-DNA adducts formed in vivo. Cancer Res 48:6328–6331

Hu J-G, Tanimori H, Shibata M, Yokoyama T, Kitagawa T (1989a) Comparative study of complete and incomplete adjuvants for immunisation of a drug immunogen using enzyme immunoassay methods. Chem Pharm Bull 37:1316–1320

Hu J-G, Ide A, Yokoyama T, Kitagawa T (1989b) Studies on the optimal immunisation schedule of the mouse as an experimental animal. The effect of antigen dose and adjuvant type. Chem Pharm Bull 37:3042–3046

Ikebuchi H, Teshima R, Hirai K, Sato M, Ichinoe M, Terao T (1990) Production and characterization of monoclonal antibodies to nivalenol tetra-acetate and their application to enzyme linked immunoassay of nivalenol. Z Biol Chem Hoppe-Seyler 371:31–36

Itoh Y, Hifumi E, Sudoh K, Uda T, Ohtani K, Kawamura O, Nagayama S, Satoh S, Ueno Y (1987) Detection of aflatoxin B_1 by direct ELISA. Proc Jpn Assoc Mycotox 26:31–35

Jackman R (1985) Determination of aflatoxins by enzyme linked immunosorbent assay with special reference to aflatoxin M_1 in raw milk. J Sci Food Agric 36:685–698

Kawamura O, Nagayama S, Sato S, Ohtani K, Ueno I, Ueno Y (1988) A monoclonal antibody base enzyme linked immunosorbent assay of aflatoxin B_1 in peanut products. Mycotoxin Res 4:75–88

Kawamura O, Sato S, Kajii H, Nagayama S, Ohtani K, Chiba J, Ueno Y (1989) A sensitive enzyme linked immunosorbent assay of ochratoxin A based on monoclonal antibodies. Toxicon 27:887–897

Kemp HA, Morgan MRA (1986) Studies on the detrimental effects of bivalent binding in a microtitration plate ELISA and possible remedies. J Immunol Meth 94:65–72

Kemp HA, Mills ENC, Morgan MRA (1986) Enzyme linked immunosorbent assay of 3-acetyldeoxynivalenol applied to rice. J Sci Food Agric 37:888–894

Kitagawa T, Tanimon H, Shibata M, Yoshida K, Hu J-G (1989) Analysis of specific and total antibody responses of rabbits to four kinds of immunogens. Chem Pharm Bull 37:1013–1018

Kohler G, Milstein C (1975) Continuous cultures of fused cells secreting antibody of predefined specificity. Nature 256:495–497

Krogh P (1977) Ochratoxins. In: Rodricks JV, Hesseltine CW, Mehlman MA (eds) Mycotoxins in human and animal health. Pathotox Publishers, Park Forest South, Illinois, USA, pp 489–498

Landsteiner K (1945) The specificity of serological reactions. Harvard Univ Press, Boston

Langone JJ, Van Vunakis H (1976) Aflatoxin B_1: specific antibodies and their use in radioimmunoassay. J Natl Cancer Inst 56:591–595

Lau HP, Gaur PK, Chu FS (1981) Preparations and characterization of aflatoxin $B_{2\alpha}$ hemiglutarate and its use for the production of antibody against aflatoxin B_1. J Food Safety 3:1–13

Lawellin DW, Grant DW, Joyce BK (1977) Enzyme linked immunosorbent analysis of aflatoxin B_1. Appl Environ Microbiol 34:94–96

Lee HA, Morgan MRA (1988) Current developments in immunoassay of food. Anal Proc 25:297–298

Lee HS, Sarosi I, Vyas GN (1989) Aflatoxin B_1 formamido-pyrimidine adducts in human hepatocarcinogenesis: a preliminary report. Gastroenterology 97:1281–1287

Lepschy V, Gleissenthal J, Dietrich R, Märtlbauer E, Schuster M, Suss A, Terplan G (1989) A survey on the occurrence of *Fusarium* mycotoxins in Bavarian cereals for the 1987 harvest. Z Lebensm Unters Forsch 188:521–526

Li Y, Chu FS (1984) Production and characterization of antibody against sterigmatocystin. J Food Safety 6:119–127

Li YS, Wu JL, Lu XQ, Xia QJ (1988) Tricothecenes in staple foods from a high-incidence area of carcinoma of the esophagus and gastric cardia and their carcinogen potential. Chung Hua Chung Liu Tsa Chih 10:4–8

MacDougald OA, Thulin AJ, Pestka JJ (1990) Determination of zearalenone and related metabolites in porcine urine by modified enzyme linked immunosorbent assay. J Assoc Off Anal Chem 73:65–68

Märtlbauer E, Terplan G (1988) Ein enzymimmunologisches Verfahren zum Nachweis von Ochratoxin A in Schweineserum. Arch Lebensmittelhyg 39:143–147

Märtlbauer E, Gareis M, Terplan G (1988) Enzyme immunoassay for macrocyclic trichothecene roridin A; production, properties and use of rabbit antibodies. Appl Environ Microbiol 54:225–230

Mills ENC, Johnston JM, Kemp HA, Morgan MRA (1988) An enzyme linked immunosorbent assay for diacetoxyscirpenol applied to the analysis of wheat. J Sci Food Agric 42:225–233
Mills ENC, Alcock SM, Lee HA, Morgan MRA (1990) An enzyme linked immunosorbent assay for deoxynivalenol in wheat utilising novel hapten derivatisation procedures. Food Agric Immunol 2:109–118
Miwa A, Yoshioka M, Shirahata A, Tamanura Z (1977) Preparation of specific antibodies to catecholamines and L-3, 4-dihydroxyphenylalanine. 1 Preparation of the conjugates. Chem Pharm Bull 25:1904–1910
Morgan MRA (1988) Factors influencing the relative costs of immunoassays and conventional analyses of food constituents and contaminants. In: Morris BA, Clifford MN, Jackman R (eds) Immunoassays for veterinary and food analysis – 1. Elsevier, London, p 221
Morgan MRA (1989) Mycotoxin immunoassays with special reference to ELISAs. Tetrahedron 45:2237–2249
Morgan MRA, Lee HA (1990) Mycotoxins and natural food toxicants. In: Rittenburg JH (ed) Development and application of immunoassay for food analysis. Elsevier Applied Science, London, pp 143–170
Morgan MRA, McNerney R, Chan HWS (1983) The enzyme-linked immunosorbent assay of ochratoxin A in barley. J Assoc Off Anal Chem 66:1481–1484
Morgan MRA, Kang AS, Chan HW-S (1986a) Aflatoxin determination in peanut butter by enzyme-linked immunosorbent assay. J Sci Food Agric 37:908–914
Morgan MRA, Kang AS, Chan HW-S (1986b) Production of antisera against sterigmatocystin hemiacetal and its potential for use in an enzyme linked immunosorbent assay for sterigmatocystin in barley. J Sci Food Agric 37:873–880
Morgan MRA, McNerney R, Chan HW-S, Anderson, PH (1986c) Ochratoxin A in pig kidney determined by enzyme-linked immunosorbent assay (ELISA). J Sci Food Agric 37:475–480
Mortimer DN, Gilbert J, Shepherd MJ (1987a) Rapid and highly sensitive analysis of aflatoxin M_1 in liquid and powdered milks using an affinity column clean-up. J Chromatogr 407:393–398
Mortimer DN, Shepherd MJ, Gilbert J, Morgan MRA (1987b) A survey of the occurrence of aflatoxin B_1 in peanut butters by enzyme linked immunosorbent assay. Food Addit Contam 5:127–132
Mortimer DN, Shepherd MJ, Gilbert J, Clark C (1988) Enzyme linked immunosorbent (ELISA) determination of aflatoxin B_1 in peanut butter: collaborative trial. Food Addit Contam 5:601–608
Nagayama S, Kawamura O, Ohtani K, Ryu JC, Latus D, Sudheim L, Ueno Y (1988) Application of an enzyme linked immunosorbent assay for screening of T-2 toxin producing *Fusarium* species. Appl Environ Microbiol 54:1302–1303
Neal GE (1987) Influences of metabolism: aflatoxin metabolism and its possible relationships with disease. In: Watson DH (ed) Natural toxicants in food progress and prospects. Ellis Horwood Ltd, Chichester, England, pp 125–168
Newberne PM, Rogers AE (1981) Animal toxicity of major environmental mycotoxins. In: Shank R (ed) Mycotoxins and N-nitroso compounds. CRC Press, Boca Raton, pp 51–106
Okoye ZS, Neal GE (1988) Enhanced ethanol-induced changes in disposition and toxic response to dietary aflatoxin B_1 due to *Sacoglottis gabonensis* bark extract, a Nigerian alcoholic beverage additive. Food Chem Toxicol 26:679–689
Park DL, Pohland AE (1989) Sampling and sample preparation for detection and quantification of natural toxicants in food and feed. J Assoc Off Anal Chem 72:399–404
Park DL, Trucksess MW, Lee RC, Chu FS (1988) Evaluation by enzyme linked immunosorbent assay of clean up for thin layer chromatography of aflatoxin B_1 in corn, peanuts and peanut butter. J Assoc Off Anal Chem 71:953–956
Park DL, Miller BM, Nesheim S, Trucksess MW, Vekich A, Bidigare B, McVey JL, Brown LH (1989a) Visual and semiquantitative spectrophotometric ELISA screening

method for aflatoxin B_1 in corn and peanut products: follow up collaborative study. J Assoc Off Anal Chem 72:638–643
Park DL, Miller BM, Hart LP, Yang G, McVey J, Page SW, Pestka J, Brown LH (1989b) Enzyme-linked immunosorbent assay for screening aflatoxin B_1 in cottonseed products and mixed feed: collaborative study. J Assoc Off Anal Chem 72:326–332
Patey AL, Sharman M, Wood R, Gilbert J (1989) Determination of aflatoxin concentrations in peanut butter by enzyme linked immunosorbent assay (ELISA) and study of three commercial ELISA kits. J Assoc Off Anal Chem 72:965–969
Patey AL, Sharman M, Gilbert J (1990a) Determination of aflatoxin B_1 levels in peanut butter using an immunoaffinity clean-up procedure: interlaboratory study. Food Addit Contam 7:515–520
Patey AL, Sharman M, Gilbert J (1990b) Determination of aflatoxin levels in peanut butter using HPLC and ELISA procedures: interlaboratory comparison. Myco Res 6:2–6
Patey AL, Sharman M, Gilbert J (1991a) Liquid chromatographic determination of aflatoxin levels in peanut butter using an immunoaffinity column clean-up procedure: international collaborative trial. J Assoc Off Anal Chem 74:76–81
Patey AL, Sharman M, Gilbert J (1991b) Determination of total aflatoxin levels in peanut butter by enzyme-linked immunosorbent assay (ELISA): international collaborative trial. J Assoc Off Anal Chem (submitted)
Patterson DSP (1982) Mycotoxins. Environ Chem 2:205–233
Paul WE (ed) (1984) Fundamental immunology. Raven, New York
Pestka JJ (1988) Enhanced surveillance of foodborne mycotoxins by immunochemical assay. J Assoc Off Anal Chem 71:1075–1081
Pestka JJ, Li Y, Harder WO, Chu FS (1981) Comparison of RIA and ELISA for determining aflatoxin M_1 in milk. J Assoc Off Anal Chem 64:294–301
Purchase IFH (ed) (1974) Mycotoxins. Elsevier, Amsterdam
Ramakrishna N, Lacey J, Candlish AAG, Smith JE, Goodbrand IA (1990) Monoclonal antibody based enzyme linked immunosorbent assay of aflatoxin B_1, T-2 toxin and ochratoxin A in barley, J Assoc Off Anal Chem 73:71–76
Reichert N, Steinmeyer S, Weber R (1988) Determination of aflatoxin B_1 in dried figs by visual screening, thin layer chromatography and ELISA. Z Lebensm Unters Forsch 186:505–508
Rittenburg JH (1990) Fundamentals of immunoassay. In: Rittenburg JH (ed) Development and application of immunoassay for food analysis. Elsevier, London, pp 29–57
Rousseau DM, Van Peteghem CH (1989) Spontaneous occurrence of ochratoxin residues in porcine kidneys in Belgium. Bull Environ Contam Toxicol 42:181–186
Rousseaux CG (1988) Tricothecene mycotoxins: real or imaginary toxins. Comments Toxicol 2:37–49
Ruprich J, Piskac A, Mala J (1988) Use of the radioimmunologic determination of ochratoxin A mycotoxin in the screening of foods and animal feed of plant origin cereal type. Vet Med (Prague) 33:165–173
Schlatter C (1988) The importance of mycotoxins in foods. Bibl Nutr Dieta 41.55–65
Schmidt FR, Esser K (1985) Aflatoxins medical economic impact and prospects for control. Process Biochem 20:167–174
Schuller PL, Horwitz W, Stoloff L (1976) A review of sampling plans and collaborative studied methods of analysis for aflatoxins. J Assoc Off Anal Chem 59:1315–1343
Scott PM (1989) Methods for determination of aflatoxin M_1 in milk and milk products – a review of performance characteristics. Food Addit Contam 6:283–305
Sharman M, Patey AL, Gilbert J (1989) Application of an immunoaffinity column sample clean-up to the determination of aflatoxin M_1 in cheese. J Chromatogr 474:457–461
Shelby RA, Kelley VC (1990) An immunoassay for ergotamine and related alkaloids. J Agric Food Chem 38:1130–1134
Shepherd MJ, Gilbert J (1988) Long term stability of deoxynivalenol standard reference solutions. J Agric Food Chem 36:305–308

Shepherd MJ, Holmes M, Gilbert J (1986) Comparison and critical evaluation of six published extraction and clean-up procedures for aflatoxin M_1 in liquid milk. J Chromatogr 354:305–315

Shotwell OL (1986) Chemical survey methods for mycotoxins. In: Cole RJ (ed) Modern methods in the analysis and structural elucidation of mycotoxins. Acad Press, London, pp 51–94

Sidwell WJ, Chan HW-S, Morgan MRA (1989) Reduction in assay time for a double antibody, indirect enzyme-linked immunosorbent assay applied to Ochratoxin A. Food Agric Immunol 1:111–118

Sizaret P, Malaveille C, Montesano R, Frayssinet C (1982) Detection of aflatoxins and related metabolites by radioimmunoassay. J Natl Cancer Inst 69:1375–1380

Smith JE, Moss MO (eds) (1985) Mycotoxins: formation analysis and significance. John Wiley & Sons, Chichester

Steiner von J, Hahn G, Heeschen W (1988) On the enzyme-immunological detection of aflatoxin M_1 in milk: development of a rapid aflatoxin M_1 screening test using polystyrol beads as solid phase. Milchwissenschaft 43:772–776

Stora C, Dvorackova I (1987) Aflatoxin, viral hepatitis and primary liver cancer. J Med (Clin Exp Theor) 18:23–41

Teshima R, Hirai K, Sato M, Ikebuchi H, Ichinoe M, Terao T (1990a) Radioimmunoassay of nivalenol in barley. Appl Environ Microbiol 56:764–768

Teshima R, Kawase M, Tanaka T, Hirai H, Sato M, Sawada J, Ikebuchi H, Ichinoe M, Terao T (1990b) Production and characterization of a specific monoclonal antibody against mycotoxin zearalenone. J Agric Food Chem 38:1618–1622

Thouvenot D, Morfin RF (1983) Radioimmunoassay for zearalenone and zearalenol in human serum: production, properties and use of procine antibodies. Appl Environ Microbiol 45:16–23

Trucksess MW, Stack ME, Nesheim S, Park DL, Pohland AE (1989) Enzyme linked immunosorbent assay of aflatoxins B_1, B_2 and G_1 in corn, cottonseed, peanuts, peanut butter and poultry feed: collaborative study. J Assoc Off Anal Chem 72:957–962

Trucksess MW, Young K, Donahue KF, Morris DK, Lewis E (1990) Comparison of two immunochemical methods with thin layer chromatographic methods for determination of aflatoxins. J Assoc Off Anal Chem 73:425–428

Ueno Y (1977) Trichothecenes: Overview address. In: Rodricks JV, Hesseltine CW, Mehlman MA (eds) Mycotoxins in human and animal health. Acad Press, New York, p 189

Van Egmond HP (ed) (1989a) Mycotoxins in dairy products. Elsevier Applied Science, London

Van Egmond HP (1989b) Current situation on regulations for mycotoxins. Overview of tolerances and status of standard methods of sampling and analysis. Food Addit Contam 6:139–188

Van Egmond HP, Paulsch WE (1986) Determination of mycotoxins. Pure Appl Chem 58:315–326

Van Egmond HP, Wagstaffe PJ (1989) Aflatoxin B_1 in peanut meal reference materials: intercomparison of methods. Food Addit Contam 6: 307–319

Van Egmond HP, Wagstaffe PJ (1990) Aflatoxin B_1 in compound-feed reference materials: an intercomparison of methods. Food Addit Contam 7:239–251

Van Egmond HP, Luissink AB, Paulsch WE (1986) The determination of aflatoxin M_1 in milk powder and comparison of four methods of analysis. In: Reference materials and interlaboratory collaborative studies. Bull Int Dairy Fed 207:150–197

Ward CM, Morgan MRA (1991) Reproducibility of a commercially-available kit utilising enzyme linked immunosorbent assay for determination of aflatoxin in peanut butter. Food Addit Contam 8:9–15

Ward CM, Wilkinson AP, Bramham S, Lee HA, Chan HW-S, Butcher GW, Hutchings A, Morgan MRA (1990) Production and characterisation of polyclonal and monoclonal antibodies against aflatoxin B_1 oxime BSA in an enzyme-linked immunosorbent assay. Mycloc Res 6:73–83

Warner R, Ram BP, Hart L, Pestka JJ (1986) Screening for zearalenone in corn by competitive direct enzyme linked immunosorbent assay. J Agric Food Chem 34:714–717
Wei RD, Chu FS (1988) Production and characterization of antibody against PR toxin. J Food Prot 51:463–466
Wei RD, Bischoof W, Chu FS (1986) Production and characterization of antibody against 3-OH-T-2 toxin. J Food Prot 49:267–271
Whitaker TB, Dickens JW (1986) Optimum methanol concentration and solvent/peanut ratio for extraction of aflatoxin from raw peanuts by modified AOAC method II. J Assoc Off Anal Chem 69:508–510
Whitaker TB, Wiser EH (1969) Theoretical investigations into the accuracy of sampling shelled peanuts for aflatoxin. J Am Oil Chem Soc 46:377–379
Wie SI, Hammock D (1982) Development of enzyme-linked immunosorbent assays for residue analysis of diflubenzuron and BAYSIR 8514. J Agric Food Chem 30:949–957
Wilcke WF, Ehrich MR, Ko KW (1990) Using ELISA kits to test corn for aflatoxin. Appl Agric Res 5:32–36
Wild CP, Umbenhauer D, Chapot B, Montesano R (1986) Monitoring of individual human exposure to aflatoxins (AF) and N-nitrosamines (NNO) by immunoassays. J Cell Biochem 30:171–179
Wild CP, Pionneau FA, Montesano R, Mutiro CF, Chetsanga CI (1987) Aflatoxin detected in human breast milk by immunoassay. Int J Cancer 40:328–333
Wild CP, Chapot B, Scherer E, Den Engelse L, Montesano R (1988) Application of antibody methods to the detection of aflatoxin in human body fluids. IARC Sci Publ 88:67–74
Wilkinson AP, Denning DW, Morgan MRA (1988a) Analysis of UK sera for aflatoxin by enzyme linked immunosorbent assay. Hum Toxicol 7:353–356
Wilkinson AP, Denning DW, Morgan MRA (1988b) An ELISA method for the rapid and simple determination of aflatoxin in human serum. Food Addit Contam 5:609–619
Wilkinson AP, Denning DW, Morgan MRA (1989) Immunoassay of aflatoxin in food and human tissue. J Toxicol Toxin Rev 8:69–79
Wilson BM, Nakane PK (1978) Recent developments in the periodate method of conjugating horseradish peroxidase (HRPO) to antibodies. In: Knapp W, Holuber K, Wick G (eds) Immunoflourescence and related staining techniques. Elsevier, Amsterdam, pp 215–224
Xu YC, Zhang GS, Chu FS (1986) Radioimmunoassay of deoxynivalenol in wheat and corn. J Assoc Off Anal Chem 69:967–969
Xu YC, Zhang GS, Chu FS (1988) Enzyme linked immunosorbent assay for deoxynivalenol in corn and wheat. J Assoc Off Anal Chem 71:945–949
Yalow RS, Berson SA (1959) Assay of plasma insulin in human subjects by immunological methods. Nature 184:1648–1649
Yu SZ, Cheng Z, Liu Y, Huang Z, Zhao Y (1989) The aflatoxins and contaminated water in the etiological study of primary liver cancer. In: Natori S, Hashimoto K, Ueno Y (eds) Bioactive molecules, vol 10. Mycotoxins and Phycotoxins '88 7th International IUPAC Symp, Tokyo, 16–19 August 1988, Elsevier, London, pp 37–44
Zhang GS, Li SW, Chu FS (1986) Production and characterization of antibody against deoxynivalenol triacetate. J Food Prot 49:336–339

Multi-Toxin TLC Methods for Aflatoxins, Ochratoxin A, Zearalenone and Sterigmatocystin in Foods

L.M.V. SOARES

1 Introduction

Thin-layer chromatography (TLC) has kept its place as one of the most widely used chromatographic techniques due to its simplicity, speed, ability to handle many samples simultaneously and low cost. Particularly in the case of multitoxin methods, TLC means savings in terms of the analyst's time and an increase in the volume of information obtained.

A fair amount of data on mycotoxin contamination of foods has been collected for the past three decades. Still, only part of the task has been accomplished; an array of different commodities remain to be surveyed in many countries. Food items already identified as frequently contaminated require monitoring. As a consequence, there still is a need for simple, rapid and low-cost methods to be employed in surveys and by food inspection services.

The use of TLC for mycotoxin analysis has been recently reviewed by Nesheim and Trucksess (1986), chromatographic techniques in general by Romer (1984) and multimycotoxin analysis by Steyn (1981).

The aflatoxins B_1, B_2, G_1 and G_2 are hepatotoxic, carcinogenic, mutagenic and teratogenic. They are produced by strains of *Aspergillus flavus* and *Aspergillus parasiticus* and are by far the most studied, surveyed and monitored mycotoxins. Ochratoxin A is nephrotoxic and teratogenic. A number of species of *Aspergillus* and *Penicillium* are capable of producing it. Zearalenone, an estrogenic and anabolic substance, is an occasional metabolite of *Fusarium* species. Sterigmatocystin is originated mainly by *Aspergills* species and has hepatotoxic, nephrotoxic, carcinogenic and teratogenic effects (Fig. 1).

2 Multitoxin Screening and Quantitation on TLC

A low limit of detection is to be preferred when choosing a screening method. High limits mean that a larger number of positive samples will go undetected. However, the final word on the adequacy of the detection limits is given by the task at hand. Information on the foods to which the method has been applied is also helpful. The method's usefulness may be limited

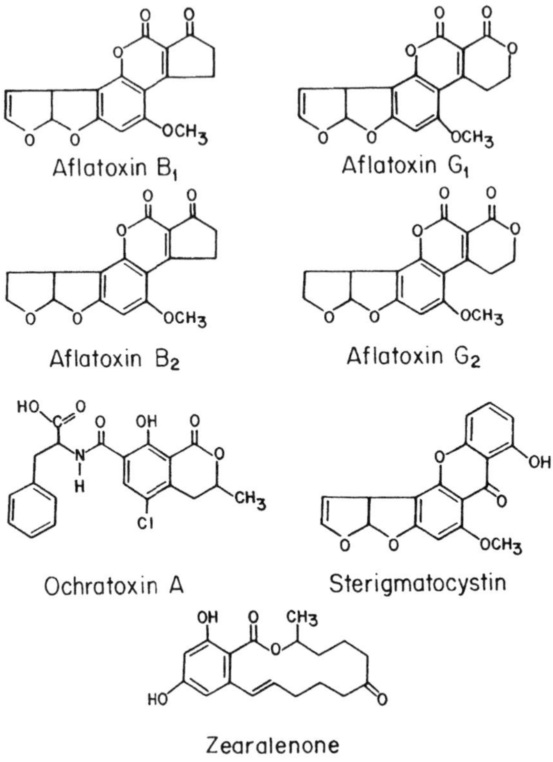

Fig. 1. Structure of aflatoxins B_1, B_2, G_1 and G_2, ochratoxin A, zearalenone and sterigmatocystin

when employed to other commodities than the ones for which it had been originally designed. Finally, recovery and precision should not be overlooked, since they reflect on the method's abilility to provide reliable information.

Many analysts regard multi-toxin TLC methods as purely screening tools. Such belief derives from considering as conflictive the extraction, purification, and quantitation of a chemically diverse group of substances in a single procedure. However, screening methods may be used for quantitative purposes under certain conditions.

In picking a method for quantitation, precision data has a major bearing, along with appropriate recoveries and limits of detection. Ideally, information should be available on relative standard deviations (RSD) of intralaboratory repetitions at different contamination levels (repeatability) and RSD of interlaboratory results of the same test samples (reproducibility). Unfortunately, many multitoxin TLC methods report, at most, their detection limits, and none so far has been submitted to a collaborative study. Table 1 summarizes available data on detection limits and recoveries of

Table 1. Limits of detection and recoveries for aflatoxins, ochratoxin A, zearalenone, and sterigmatocystin in multitoxin TLC methods

References	Aflatoxin B_1 Det. limit (μg/kg)	Aflatoxin B_1 Rec %	Ochratoxin A Det. limit (μg/kg)	Ochratoxin A Rec %	Zearalenone Det. limit (μg/kg)	Zearalenone Rec %	Sterigmatocystin Det. limit (μg/kg)	Sterigmatocystin Rec %
Stoloff et al. (1971)	20	–	45–90	–	200–500	–	60	–
Thomas et al. (1975)	2	–	–	–	100	–	–	–
Seitz and Mohr (1976)	5	–	–	–	200	–	–	–
Wilson et al. (1976)	2	90	20–40	50	200–300	80	–	–
Josefsson and Möller (1977)	5	96–116	10	106–127	35	42–120	10	62–125
Balzer et al. (1978)	2	71	40	87	200	85	–	–
Takeda et al. (1979)	10	88–99	40–60	30–97	300–500	70–91	40–60	30–88
Miguel and de Andres (1982)	2–3	84–93	70	91	150	89	–	–
Johan and Dose (1983)	1	89–99	10	52–78	–	–	20	59–90
Nowotny et al. (1983)	5	80–90	50	50–65	1000	60–80	400	50–70
Le Tutour et al. (1984)	5–7	–	20	–	–	–	–	–
Soares and Rodriguez-Amaya (1989)	2	91–101	5	96–107	55	86–133	15	96–107

these methods. In such a situation, it is left to the analyst to conduct the tests required to provide the necessary assurance regarding the effectiveness of the chosen method. When any extensive work is being planned, it is advisable to conduct such tests. Essays for recovery and detection limit, plus a ruggedness test (Wernimont 1985), can be easily set up and carried out on the types of foods most likely to be analyzed.

2.1 Extraction and Cleanup

Methanol, acetonitrile or chloroform combined with water, salt solution or diluted acid are the solvents described for multitoxin extraction conected with TLC (Table 2). Such a preference points out to the efficiency of a higher polarity mixture for the job. An added benefit is the avoidance to some extent of the extraction of nonpolar lipids and pigments, thus facilitating the cleanup step.

Partition and clarification take considerably less time than in column chromatography. Partition to chloroform can be accomplished in a straightforward way, or by adding acid or base to the aqueous solution (Table 2).

A few methods designed purely for screening use no cleanup. They rely heavily on solvent systems capable of creating appropriate "windows" amidst the interferences.

Methods employing silica-gel column chromatography follow the general outline of the forerunner of such procedures. Eppley (1968) was the first to describe the use of a slurry-packed silica gel column and a sequence of solvents for removing impurities and eluting mycotoxins.

An interesting approach to cleanup is gel filtration (Josefsson and Möller 1977). It has been successfully employed in other areas of residue analysis, such as pesticides, for a long time. So far, surprisingly little use has been found for this technique in mycotoxin methodology.

A word should be said about methods for feeds, which are complex materials. A method developed for feed analysis stands a good chance of performing well for most foods. For that reason, the interested analyst will do well to get acquainted with them. Among the most used of such procedures, an extraction with chloroform or chloroform-water (10 + 1) can be mentioned, followed by elution from a Sep-Pak silica cartridge (Roberts et al. 1981; Howell and Taylor 1981). An extraction with acetonitrile −4% KCl (9 + 1), defatting and dialysis is proposed by Roberts and Patterson (1975). Two other methods start with an acetonitrile −0.1 M phosphoric acid (10 + 1) extraction and a defatting step. One utilizes successive partitions to chloroform in basic and acid conditions; adsorption chromatography on a polyamide stationary phase, coupled with an eluotropic solvent series, is employed by another (Van Egmond et al. 1979).

Table 2. Summary of TLC multi-toxin methods emphasizing aflatoxins, ochratoxin A, zearalenone and sterigmatocystin

Mycotoxins	Extraction	Cleanup	Commodities	Reference
Aflatoxins Ochratoxin A Zearalenone	Chloroform/water (10 + 1)	Silica gel Column chromatography	Wheat, barley, beans, almonds, soybeans, etc.	Eppley (1968)
Aflatoxins Ochratoxin Zearalenone	Chloroform/metanol/hexane (8 + 2 + 1)	None		Vorster (1969)
Aflatoxins Ochratoxin A Zearalenone Sterigmatocystin Patulin	Acetonitrile/4% KCl (9 + 1)	1) Defatting with isoctane 2) Partition to chloroform	Corn, wheat, barley, oats, rye	Stoloff et al. (1971)
Aflatoxins Zearalenone	Methanol/water (6 + 4)	1) Defatting with hexane 2) Partition to chloroform 3) Clarification with cupric carbonate	Corn	Thomas et al. (1975)
Aflatoxins Zearalenone	Methanol/4% NaCl (6 + 4)	1) Defatting with hexane 2) Partition to chloroform	Corn	Seitz and Mohr (1976)
Aflatoxins Ochratoxin A Penicilic acid Citrinin	Cloroform/0.5 NH_3PO_4 (10 + 1)	Silica gel Column chromatography	Corn, beans, peanuts	Wilson et al. (1976)
Aflatoxins Ochratoxin A Zearalenone Sterigmatocystin Patulin	Chloroform/0.1 M H_3PO_4 (10 + 1)	Gel filtration chromatography	Barley, oats, corn	Josefsson and Möller (1977)

Table 2. (continued)

Mycotoxins	Extraction	Cleanup	Commodities	Reference
Aflatoxin B_1 Ochratoxin A Zearalenone	Acetonitrile/water (9 + 1)	1) Defatting with hexane 2) Partitions to chloroform in basic and acid conditions	Corn	Balzer et al. (1978)
Aflatoxins Ochratoxin A Zearalenone Sterigmatocystin and 10 others	20% H_2SO_4/4% KCl/ acetonitrile (2 + 2 + 178)	1) Defatting with isoctane 2) Silica gel column chromatography	Rice, corn, wheat, peanuts	Takeda et al. (1979)
Aflatoxin B_1 Ochratoxin A Zearalenone Sterigmatocystin and 4 others	Acetonitrile/1.25% HCl (10 + 2)	1) Defatting with isoctane 2) Silica gel column chromatography	Corn	Whidden et al. (1980)
Aflatoxins Ochratoxin A Penicilic acid Citrinin	Chloroform/0.5 N H_3PO_4 (10 + 1)	Silica gel column chromatography	Wheat, barley, rye, corn, sorghum, soybean	Miguel and de Andres (1982)
Aflatoxin B_1 Ochratoxin A	Methanol/4% NaCl (6 + 4)	1) Defatting with hexane 2) Clarification with lead acetate 3) Partition to chloroform	Olive oil	Le Tutour et al. (1983)

Aflatoxins Ochratoxin A Zearalenone Sterigmatocystin and 18 others	Methanol	1) Acidification 2) Defatting with hexane 3) Partition to chloroform	Corn, barley, cheese, ryebread	Nowotny et al. (1983)
Aflatoxins Ochratoxin A Sterigmatocystin and 4 others	Acetonitrile/4% KCl/cyclohexane (9 + 1 + 10)	Partitions to dichloromethane under basic and acid conditions	Rice, white bread	Johan and Dose (1983)
Aflatoxin B_1 Ochratoxin A	Methanol/4% NaCl (6 + 4)	1) Defatting with hexane 2) Partition to chloroform	Olives	Le Tutour et al. (1984)
Aflatoxins Ochratoxin A Zearalenone Sterigmatocystin and 7 others	Chloroform/0.1 M H_3PO_4 (10 + 1)	None	Rye, barley, wheat, oats, corn	Grabarkiewicz-Szczesna et al. (1985)
Aflatoxins Ochratoxin A Zearalenone Sterigmatocystin	Methanol/4% KCl (9 + 1)	1) Clarification with ammonium sulfate or cupric sulfate 2) Partition to chloroform	Corn, peanuts, beans, rice, cassava	Soares and Rodriguez-Amaya (1989)

2.2 Conducting the Screening

The simple procedure of spotting each sample extract twice and applying the standards to each second spot (co-chromatography) is most helpful to the analyst during screening. After development, any small difference between the Rf of the suspected spot and a standard can be easily perceived.

Silica gel is the only adsorbent for TLC reported in methods for simultaneous detection of aflatoxins, ochratoxin A, zearalenone and sterigmatocystin. Silica gel type G (calcium sulfate binder) and H (no binder) for laboratory prepared plates and precoated plates (polyamide binder) of any brand can be used. A test to verify the efficiency of the adsorbent is advisable. Differences in performance from batch to batch within a same brand and between brands are commonly observed.

One way to conduct the screening is to apply the sample extracts and standards to one plate and develop it in one solvent system, the purpose being to separate the four aflatoxins, ochratoxin A, zearalenone and sterigmatocystin in the same run. The solvent system most widely used for this purpose is toluene-ethyl acetate-formic acid (6 + 3 + 1) (Roberts and Patterson 1975; Whidden et al. 1980; Johan and Dose 1983; Piskorska-Pliszczynska and Juszkiewicz 1977). The same solvents are also used in the proportions 80 + 20 + 0.5 (Josefsson and Möller 1977), and 60 + 40 + 0.5 (Soares and Rodriguez-Amaya 1989). Some authors recommended a previous development of the plate with benzene-hexane (3 + 1) to wash nonpolar interfering compounds to the top (Le Tutour et al. 1984; Josefsson and Möller 1977). Testing of solvent systems to separate mycotoxin standards has been reported by Steyn (1969) and Gorst-Allman and Steyn (1979). Lists of solvent systems that have been applied to real samples can also help an analyst facing unusual problems (Scott et al. 1970; Gimeno 1980; Miguel and de Andres 1982; Grabarkiewicz-Szczesna et al. 1985; Takeda et al. 1979; Josefsson and Möller 1977).

A more-time consuming approach to screening is to apply the extract to several plates and develop each plate in a solvent system taylored for a specific mycotoxin (Eppley 1968; Stoloff et al. 1971; Wilson et al. 1976; Howell and Taylor 1981; Roberts et al. 1981).

Up to this point, all the discussion assumes one-dimensional development. Two-dimensional development is favored by certain authors, due to the increased possibilities of separating the mycotoxins among themselves and the mycotoxins from interferences. On the other hand, however, a double development causes the spot to grow larger by diffusion. Here one plate will be needed per sample. Likewise, successive developments on the same direction may remove interferences present in extracts lacking adequate cleanup. It may help to separate the toxins. However, a loss in resolution is inherent to the process.

The natural fluorescence of the aflatoxins B_1, B_2, G_1 and G_2 and ochratoxin A is strong enough to allow visualization under longwave UV

(366 nm) light. Zearalenone and sterigmatocystin need a fluorescence enhancer to become visible under radiation of this wavelength. An aluminum chloride solution is the most widely used for this purpose. The plate is sprayed after development and heated for 5–10 min in a 110 °C oven.

For zearalenone, other possibilities exist. Sulfuric acid and p-anisaldehyde sprays increases its fluorescence, as does aluminum chloride. Cesium sulfate, fast violet B salt, 2,4-dinitro phenylhydrazine and chromotropic acid are, in this order, increasingly less effective. Except for aluminum chloride, all others in the list will turn zearalenone detectable under visible radiation. No enhancer is needed under shortwave UV (250 nm) irradiation, where zearalenone's natural fluorescence is stronger.

2.3 Conducting Quantitation

A clean extract is crucial for correct quantitation. Streaking due to impurities may quench the fluorescence of the mycotoxin spot. This leads to underestimating the mycotoxin concentration either by visual or by densitometric means.

Besides quenching, other difficulties may arise when comparing the spots visually. Streaking may, in some cases, fool the eye, by making the spot look larger or more intense, thus leading to an overestimation of the mycotoxin concentration.

As happens with screening, simultaneous quantitation may be performed in a single TLC plate by developing it in a single solvent system. The mycotoxin concentrations may also be evaluated with each in a separate plate, developed in a specific solvent system. The last route may be better for aflatoxins. They exhibit a low Rf in toluene-ethyl acetate-formic acid systems. The area closer to the starting point is more likely to include interferences, and some degree of streaking becoming thus inappropriate for quantitative measurements.

Quantitation by visual evaluation is conducted by finding which standard spot exhibits the same intensity as one of the sample spots. For densitometric evaluation, the scanning for aflatoxins and ochratoxin A may be done simultaneously by setting excitation wavelength (λex.) at 360 nm and emission wavelength (λem.) at 440 nm. Individual best excitation and emission wavelengths may be found in Johan and Dose (1983) or determined experimentally. For zearalenone the use λex. = 254 nm and λem. = 470 nm eliminates the need of fluorescence enhancer. For sterigmatocystin, the plate must first be sprayed with aluminum chloride solution heated, and then scanned with λex. = 360 nm and λem. = 500 nm. A calibration curve for each mycotoxin corelating peak area to mass may be prepared.

2.4 Confirmation

Confirmatory procedures by the way of chemical derivatization reactions or by mass spectrometry are needed to validate the data. However, chemical reactions can be conducted right at the analyst's bench.

For aflatoxin B_1, the addition of water to the double bond connecting carbons 15 and 16 (Fig. 1) is both simple and elegant. It is catalyzed by trifluoracetic acid and performed directly on TLC plate (Przybylski 1975).

Among the derivatization reactions being utilized for sterigmatocystin, the addition of water to the double bond located in the dihydrodifurano moiety (Fig. 1) is akin to the one described for aflatoxin B_1. Shannon and Shotwell (1976) were the first to report its use. It is conducted in the same way as for aflatoxin B_1 except for the need of applying aluminum chloride solution before examining the spots under UV light.

For zearalenone, acetylation performed directly on TLC plate, as proposed by Golinski et al. (1984), is most convenient. A few reactions to be conducted on the TLC plate have been suggested for ochratoxin A without sucess. Methylation of the extract in a vial by a procedure such as the one described by Hunt et al. (1980) can be used.

References

Balzer I, Bogdanic C, Pepeljnjak S (1978) Rapid TLC method for determining aflatoxin B_1, ochratoxin A, and zearalenone in corn. J Assoc Off Anal Chem 61:584–585

Eppley RM (1968) Screening method for zearalenone, aflatoxin, and ochratoxin. J Assoc Off Anal Chem 51:74–78

Gimeno A (1980) Improved method for thin layer chromatographic analysis of mycotoxins. J Assoc Off Anal Chem 63:182–186

Golinski P, Grabarkiewicz-Szczesna J (1984) Chemical confirmatory tests for ochratoxin A, citrinin, penicillic acid, sterigmatocystin and zearalenone performed directly on TLC plates. J Assoc Off Anal Chem 67:1108–1110

Gorst-Allman CP, Steyn PS (1979) Screening methods for the detection of thirteen common mycotoxins. J Chromatogr 175:325–331

Grabarkiewicz-Szczesna J, Golinski P, Chelkolwski J, Szebiotko K (1985) Mycotoxins in cereal grain. Part XI. Simple multidetection procedure for determination of 11 mycotoxins in cereals. Nahrung 29:229–240

Howell MV, Taylor PW (1981) Determination of aflatoxins, ochratoxin A, and zearalenone in mixed feeds, with detection by thin layer chromatography or high performance liquid chromatography. J Assoc Off Anal Chem 64:1356–1363

Hunt DC, McConnie BR, Crosby NT (1980) Confirmation of ochratoxin A by chemical derivatisation and high-performance liquid chromatography. Analyst 105:89–90

Johan H, Dose K (1983) Multianalysenmethode zur routinemässigen Bestimmung der Aflatoxine B_1, B_2, G_1 and G_2 sowie von Citrinin, Ochratoxin A, Patulin, Penicillinsaure und Sterigmatocystin in verschimmelten Nahrungsmitteln. Fresenius Z Anal Chem 314:139–142

Josefsson BG, Möller TE (1977) Screening method for the detection of aflatoxins, ochratoxin, patulin, sterigmatocystin and zearalenone in cereals. J Assoc Off Anal Chem 60:1369–1371

Le Tutour B, Tantaoni-Elaraki A, Aboussalim A (1983) Simultaneous detection of aflatoxin B_1 and ochratoxin A in olive oil. J Assoc Off Anal Chem 60:835–837
Le Tutour B, Tantaoni-Elaraki A, Aboussalim A (1984) Simultaneous thin layer chromatographic determination of aflatoxin B_1 and ochratoxin A in black olives. J Assoc Off Anal Chem 67:611–612
Miguel JA, de Andres V (1982) Simple method for simultaneous determination of aflatoxins, ochratoxin A, and zearalenone in agricultural products. Ann INIA/Ser Ganadera 15:101–110
Nesheim S, Trucksess MW (1986) Thin-layer chromatographic/high-performance thin-layer chromatography as a tool for mycotoxin determination. In: Cole RJ (ed) Modern methods in the analysis and structural elucidation of mycotoxins. Acad Press, New York, pp 239–264
Nowotny P, Baltes W, Kronert W, Weber R (1983) Hochleistungsflüssigkeit chromatographische Bestimmung von Ochratoxin A in Weizenkleie. Chem Mikrobiol Technol. Lebensm 8:29–31
Piskorska-Pliszczynska Y, Juszkiewicz T (1977) Analytical problems and method modifications in mycotoxin determination in cereal and feeds. Arch Inst Pasteur Tunis 3/4:279–288
Przylbylski W (1975) Formation of aflatoxin derivatives on thin layer chromatographic plates. J Assoc Off Anal Chem 58:163–164
Roberts BA, Patterson DSP (1975) Detection of twelve mycotoxins in mixed animal feedstuffs, using a novel membrane cleanup procedure. J Assoc Off Anal Chem 58:1178–1181
Roberts BA, Glancy EM, Patterson DSP (1981) Rapid, economical method for determination of aflatoxin and ochratoxin in animal feedstuffs. J Assoc Off Anal Chem 64:961–963
Romer T (1984) Chromatographic techniques for mycotoxins. In: Lawrence JE (ed) Food constituents and food residues. Marcel Dekker Inc, New York Basel, pp 355–393
Scott PM, Lawrence JW, van Walbeek W (1970) Detection of micotoxins by thin layer chromatography: application to screening of fungal extracts. Appl Microbiol 20:839–842
Seitz LM, Mohr HE (1976) Simple method for simultaneous detection of aflatoxin and zearalenone in corn. J Assoc Off Anal Chem 59:106–109
Shannon GM, Shotwell OL (1976) Thin layer determination of sterigmatocystin in cereal grains and soybeans. J Assoc Off Anal Chem 59:963–965
Soares LMV, Rodriguez-Amaya DB (1989) Survey of aflatoxins, ochratoxin A, zearalenone, and sterigmatocystin in some Brazilian foods. J Assoc Off Anal Chem 72:22–26
Steyn PS (1969) The separation and detection of several mycotoxins by thin layer chromatography. J Chromatogr 45:473–475
Steyn PS (1981) Multimycotoxin analysis. Pure Appl Chem 53:892–902
Stoloff LS, Nesheim S, Yin L, Rodricks JV, Stack M, Campbell AD (1971) A multimycotoxin detection method for aflatoxins, ochratoxins, zearalenone, sterigmatocystin and patulin. J Assoc Off Anal Chem 54.91–97
Takeda Y, Isohata E, Amano R, Uchiyama M (1979) Simultaneous extraction and fractionation and TLC determination of 14 mycotoxins in grains. J Assoc Off Anal Chem 62:573–578
Thomas F, Eppley RM, Trucksess MW (1975) Rapid screening method for aflatoxins and zearalenone in corn. J Assoc Off Anal Chem 58:114–116
Van Egmond HP, Paulsch WE, Sizoo EA, Schuller PL (1979) Rijksintitunt voor der volksgezondheid, Report N° 152/79 LCO. In: Steyn PS (1981) Multimycotoxin analysis. Pure Appl Chem 53:892–902
Vorster LJ (1969) A method for the analysis of cereals and groundnuts for three mycotoxins. Analyst 94:136–142
Wernimont GT (1985) Use of statistics to develop and evaluate analytical methods. Association of Official Analytical Chemists, Arlington, Virginia

Whidden MP, Davies ND, Diener VL (1980) Detection of rubratoxin B and seven other mycotoxins in corn. J Agric Food Chem 28:784–786
Wilson DM, Tabor WH, Trucksess MW (1976) Screening method for the detection of aflatoxin, ochratoxin, zearalenone, penicillic acid, and citrinin. J Assoc Off Anal Chem 59:125–127

Isolation and Characterization of Elicitors

K.-H. KOGEL and B. BEIẞMANN

1 Introduction

1.1 Elicitors as Primary Signals for Induction of Defense Responses in Plants

The discovery of elicitors in the mid-1970s (Anderson-Prouty and Albersheim 1975; Keen and Brügger 1977) has been the subject of considerable attention as a promising and versatile approach to explain the biochemical processes underlying the resistance response of plants upon microbial attack. Since then, it has been realized that elicitors may serve as useful tools both in studies of plant metabolic pathways typically activated during the expression of resistance, and in studies of the primary recognition between a plant and a pathogen. An essential prerequisite for both types of studies is the isolation of well-defined pure elicitor molecules. The modern biochemical, immunological and molecular biological methods are the basis of these attempts and thus have a great effect on the understanding of the determining processes in host-parasite interactions.

The term "elicitor" was originally introduced by Keen et al. (1972) to refer to compounds able to induce phytoalexin production in plants in the absence of a microorganism. Now it is used generally for inducers of various defense responses (Heath 1981; Callow 1984; Ryan 1987). Elicitors have been classified by their biological activities, chemical structures or their biological source. The term "biotic elicitor" is used to distinguish macromolecules of microbial origin from "abiotic elicitors". These are physical and chemical factors like UV light and heavy metal salts which probably do not play a decisive role in host-parasite interaction.

A vast number of fungal and bacterial metabolites, including carbohydrates, glycoproteins, peptides, fatty acids and hydrolytic enzymes have been shown to have elicitor activity (for review see West 1981; Darvill and Albersheim 1984; Boller 1989). Table 1 summarizes highly purified elicitors

---

Abbreviations

CEP: crude elicitor preparation; cf: culture filtrate; EAI: ethyl acetimidate hydrochloride; FAB-MS: fast atomic bombardment-mass spectroscopy; IWF: intercellular washing fluid; MDH: malate dehydrogenase; MPLC: medium pressure liquid chromatography; mw: mycelial wall; PAL: phenylalanine ammonia-lyase; Pgt elicitor: elicitor of *Puccinia graminis* f. sp. *tritici*.

Table 1. Highly purified biotic fungal elicitors of the resistance response of higher plants

Compound	Pathogen : source[a]	Elicitation : plant	Reference
Heptaglucosyl oligosaccharide	*Phytophthora megasperma* var. *sojae*: mw	Glyceollin : soybean	Sharp et al. (1984)
β-1,4 glucosamin polymeres	*Fusarium solani* f.sp. *pisi/phaseoli*: mw	Pisatin : pea	Hadwiger and Beckman (1980)
Glycoprotein	*Puccinia graminis* var *tritici*: mw	Lignin : cereals	Kogel et al. (1988)
Glycoprotein	*Phytophthora megasperma*: mw	Furanocumarins : parsley	Scheel and Parker (1990)
Peptide	*Cladosporium fulvum*: IWF	Rishitin : tomato	Schottens-Toma and De Wit (1988)
Proteins	*Phytophthora cryptogea*: cf	Necrosis : tomato	Ricci et al. (1989)
Epoxystearic acid	*Puccinia graminis* var. *tritici*: uredospores	Lignin : cereals	Kogel (1985)
Arachidonic acid	*Phytophthora infestans*: mw	Rishitin lubimin : potato	Bostock et al. (1981)
Eicosapentaenoic acid			

[a] mw, mycelial wall; IWF, intercellular washing fluid; cf, culture filtrate.

selected from a large number of reports on the isolation of elicitor active materials.

1.2 General Problems of Elicitor Isolation

Although numerous studies have been carried out to find molecules in microorganisms with eliciting activity, only a few resulted in the isolation of single elicitor molecules. With increasing knowledge, the following reasons for the purification problems became clear. (1) Crude elicitor preparations may contain more than one elicitor active component which are different in structure but act synergistically. This in consequence leads to a loss of activity after purification of single compounds (Bostock et al. 1981). (2) Crude elicitor preparations may consist of various elicitoractive components with similar structures. This makes separation difficult and purification of a single component often does not result in an increase of specific elicitor activity (Keen and Legrand 1980). (3) Complex formation of glycoprotein elicitors in extracts may occur. Dissociation only takes place under certain pH and salt conditions (Beißmann and Reisener 1990). (4) The elicitor may be active only in the presence of a cofactor which is lost during purification (Kauss 1984). (5) Yield and quality of elicitors depend on cultivation conditions, growth medium and age of the cultures (De Wit and Kodde 1981). This may result in difficulties in reproducing a purification procedure.

The chapter is focused on the experimentals applied to isolate and characterize the elicitor from *Puccinia graminis*, a crucial pathogen of wheat.

2 Bioassays

The choice of an appropriate bioassay is an important aspect in elicitor identification and characterization. The following requirements should be considered: (1) handling and analysis of bioassay must be fast and easy; (2) it is reasonable to carry out bioassay under controlled green house conditions in order to ensure reproducibility; (3) bioassay should first be done with intact plant material rather than plant parts or organs like cut leaves. Bioassay using cell cultures or protoplasts are meaningful only after having ensured that their reactions can be correlated with those occuring in intact plants. (4) Plant reactions should carefully be analyzed on different levels, e.g., macroscopically, microscopically, by histochemistry and biochemistry (e.g., measurement of enzymes or accumulating products).

The criteria relied upon in bioassays for the identification of elicitors include the determination of phytoalexin accumulation (West 1981), the visual estimation of the reinforcement of plant cell walls through the deposition of callose (Kauss 1984) or cellular necrosis, the measurement of extractable activities of enzymes of the phenylpropanoid pathway or enzymes like chitinases or glucanases (Boller 1987; Hahlbrock and Scheel 1987), as well as determination of membrane lipid peroxidation (Rogers and Anderson 1987) and electrolyte leakage (Dow and Callow 1979).

The elicitor of *Puccinia graminis* f. sp. *tritici* (Pgt elicitor) was identified by its ability to induce the cellular hypersensitive lignification response, the typical inducible defense mechanism in rust-infected wheat (Kogel et al. 1985; Moerschbacher et al. 1986a).

2.1 Measurement of Autofluorescence

Lignified epidermal and mesophyll cells are detected upon elicitor treatment or inoculation in leaves of cereals due to their yellow colour under UV light by epifluorescence microscopy (Zeiss, Oberkochen, FRG) using filters 02 (ultraviolet), 18 (violet), or 09 (blue). Treated leaves are prepared for microscopy by cutting at different time points (15, 25, 35, 45 h posttreatment). They are fixed in 80% ethanol, heated for 3 min, washed with H_2O and tranferred to glycerol. This method is reliable and well suited for screening many samples in a short time.

2.2 Quantitative Enzyme Assays

Correlation of elicitor activity with induction of enzymes of the phenylpropanoid pathway (either on the mRNA or enzyme level) provides a convenient method for quantification of elicitor activity (Hahlbrock et al. 1981; Kuhn et al. 1984; Moerschbacher et al. 1986b). In wheat, phenylalanine ammonia-lyase (PAL), a key enzyme for the synthesis of lignin and phytoalexins, is induced after injection of elicitor into young primary leaves (Fig. 1). PAL activity must be carefully analyzed because the enzyme is not only induced by elicitor treatment, but also by stress factors like wounding, UV light or abiotic chemicals.

3 Isolation of Elicitor-Active Material

3.1 Cell Wall Preparation

Sources for elicitor-active material are the fungal wall, culture filtrates or the intercellular fluids of infected plants. Soluble fragments from isolated mycelial walls have been often obtained by heat treatment or mild acid hydrolysis. For example, elicitor-active material from walls of *Phytophthora megasperma* was either released by autoclaving at 121 °C for 3 h (Anderson-

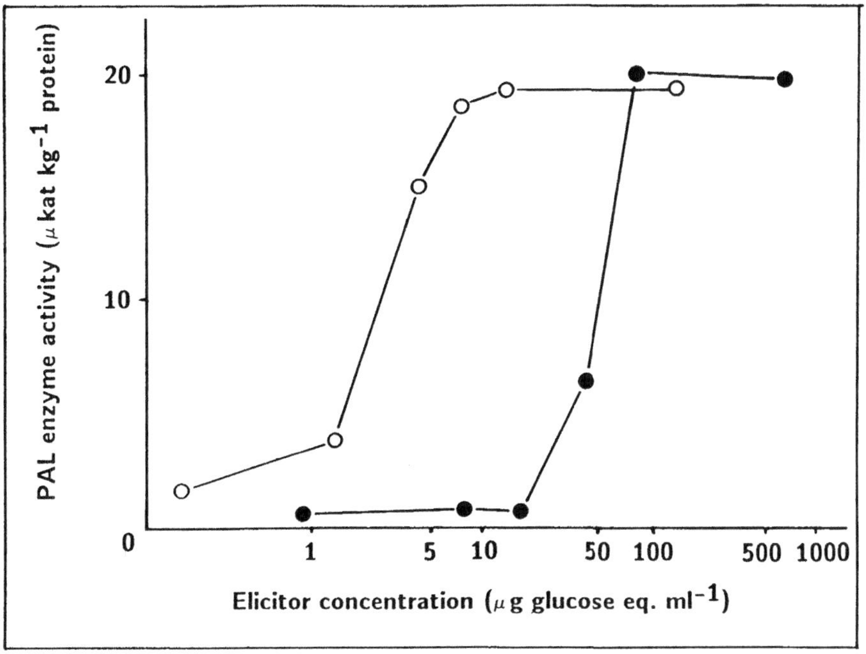

Fig. 1. Measurement of extractable activities of phenylalanine ammonia-lyase as quantitative bioassay for elicitor activity: displayed is a dose-response curve for induction of PAL activity in wheat cultivar Marquis bearing the Sr5 gene conferring rust resistance. Seven-day-old primary leaves of intact plants are injected with varying concentrations of crude elicitor (●) or Pgt elicitor (○). The injection sites are about 5 cm from the tip of the leaf. Treated leaves are placed under growth conditions: illumination at 15000 lx, light/dark circle of 16/8 h, 20 °C, 50% relative humidity. After 24 h, leaf sections (2 cm from the injection site to each side) are cut and immediately frozen in liquid N_2. For enzyme extraction, five leaf segments are ground in a mortar containing 5 ml of cold extraction buffer (0.1 M sodium borate, pH 8.8, 1 mM EDTA, 1 mM DTT), quartz sand and both Dowex 1 × 2 (Cl^-) equilibrated with extraction buffer and polyvinylpyrrolidone. The homogenate is stirred at 4 °C for 20 min, and centrifuged at 20000 × g for 20 min (Knogge and Weissenböck 1986). Measurement of PAL activity: 300 μl of extract are mixed with 0.6 ml extraction buffer and 100 μl of 10 mM phenylalanine in borate buffer. OD_{290} is measured at 40 °C over 15 min. Specific enzyme activity is expressed as μkat kg^{-1} protein

Prouty and Albersheim 1975), by partial acid hydrolysis or by enzymatic hydrolysis with an exo-ß-1,3 glucanase isolated from *Euglena gracilis* (Ayres et al. 1976). The latter yields the smallest fragments retaining elicitor activity. Fragments are heterogeneous in size, ranging from a M_r of 5000 to 200000 and are uncharged and stable even when exposed to extrem conditions (121 °C, pH 2–10).

Table 2, alternatively, depicts a purification scheme for Pgt elicitor isolated from germ tube walls of *P. graminis* uredosporelings. Since heat and acid treatment is not required, this procedure provides a powerful means for the isolation of sensitive and unstable elicitor molecules.

Table 2. Isolation of Pgt elicitor

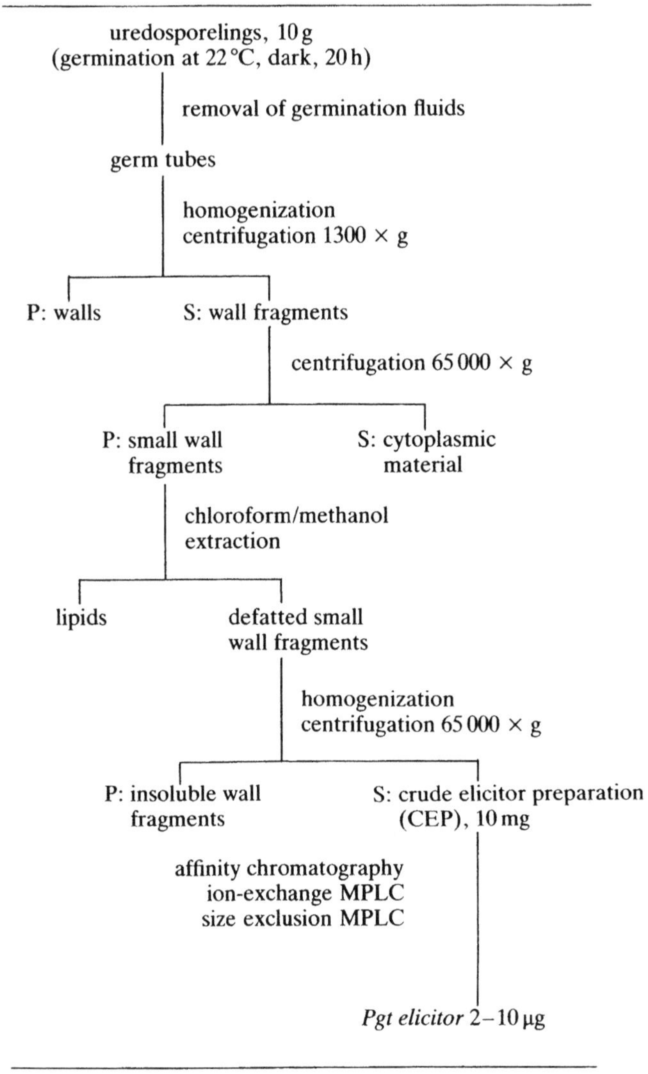

3.2 Intercellular Washing Fluid Preparation

Extracellular components can readily be obtained in large quantities from culture filtrates. However, pathogens may secrete very different components in liquid cultures and in their natural interaction with host plants. For example, filtrates from cultures of various races of *Cladosporium fulvum*, a tomato pathogen, contain potent unspecific high-molecular mass glycoprotein elicitors of rishitin (De Wit and Roseboom 1980; De Wit and Kodde 1981). In contrast, the intercellular fluid produced by *C. fulvum* in its natural interaction with tomato contains higly active race-specific elicitors (De Wit and Spikman 1982). One of these elicitors is a peptide with a M_r of 3200 (Schottens-Toma and De Wit 1988). The work of De Wit's group, based on the method of Klement (1965), opened up the prospect for a distinguished approach in the isolation of interaction relevant molecules of fungus or plant origin. The method was modified for the isolation of interaction specific material from rust-infected cereals (Rohringer et al. 1983) and also provides a source for elicitor-active material (Deverall and Deakin 1985; Kogel et al. 1988).

Elicitor-active intercellular fluids are extracted from rust-infected susceptible wheat primary leaves 1 day before sporulation of the fungus (6 days postinoculation). After cutting 8 cm from the top of the leaves, several leaf segments are bundled and placed in a desiccator filled with cold H_2O. The apoplasts are degazed for 30 sec in vacuo and filled with water by suddenly readmitting air. The infiltration process is repeated twice. Leaf segments are dried between tissue and wrapped in a plastic sheet. The bundle is stabilized by a polyethylene cylinder and placed vertically in a plastic tube with perforated bottom. This is put in a glass tube and centrifuged at 500 × g for 10 min. Six hundred leaf segments yield about 5 ml IWF (intercellular washing fluid).

Electron microscopical examination reveals that the cell walls and plasma membranes of plant and rust fungus are not affected after IWF extraction (Holden and Rohringer 1985). Contamination of apoplastic fluid by cytoplasmic proteins was further determined by measuring the activities of intracellular enzymes such as ribulose-1,5-bisphosphate carboxylase (Holden and Rohringer 1985), glucose-6-phosphate and 6-phosphogluconate dehydrogenase (Deverall and Deakin 1985) or malate- dehydrogenase (MDH). The assay of the latter enzyme is as follows: 10 µl of total leaf extract or IWF are added to 1490 µl phosphate buffer containing 1 mM NADH and 10 mM oxalacetate. The change in absorbance at 366 nm at 30 °C is measured over 15 min. Following the above method, less than 0.8% of total MDH activity of wheat primary leaves are found in IWFs.

3.3 Affinity Chromatography with Immobilized Lectins

The presence of glycoproteins in elicitor preparations can be analyzed in dot-blot binding assays on nitrocellulose filters by using peroxidase-coupled lectins as probe (Kogel et al. 1988). The use of immobilized lectins, especially of Con A Sepharose, for the separation and isolation of glycoproteins and glycopeptides was reviewed (Narasimham et al. 1979).

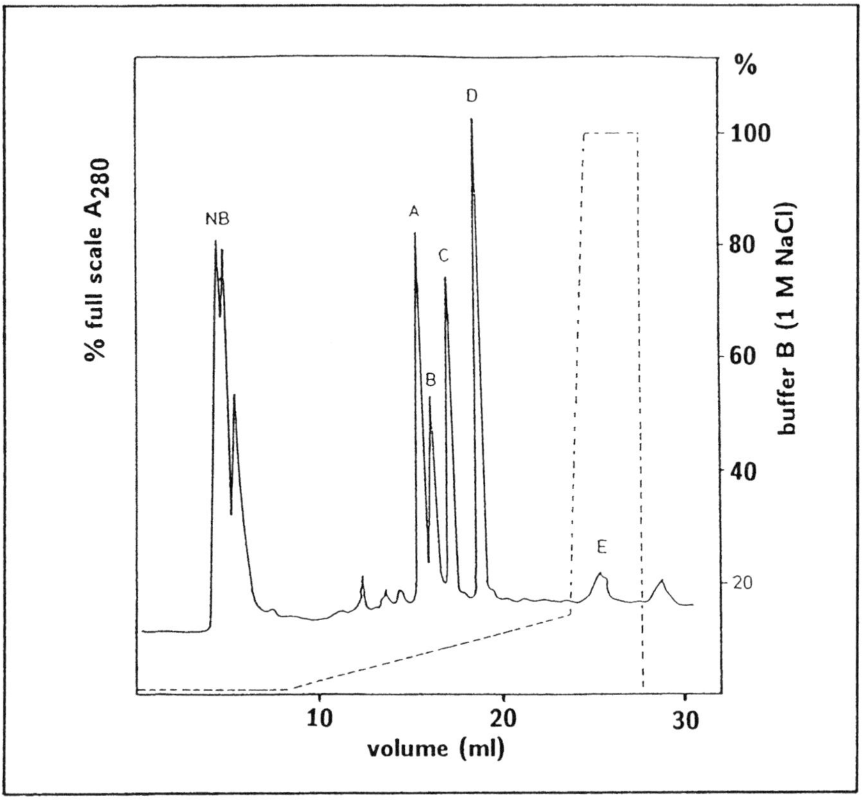

Fig. 2. Elicitor purification from a crude fraction (CEP) isolated from the germ tube walls of *Puccinia graminis* f. sp. *tritici* (see Table 2) by MPLC anion-exchange chromatography on a prepacked MonoQ HR 5/5 column (Pharmacia). 200 µg of CEP (protein basis) in 500 µl of buffer A are applied. The buffer system comprises 25 mM Tris/HCl, pH 8.5, and 4% (w/v) betaine or taurine (buffer A) to minimize ionic interactions of molecules and 1 M NaCl in the same buffer (buffer B). Proteins are eluted at a flow rate of 1 ml min^{-1} with a salt gradient of 0–35% buffer B in 40 ml. The fractions must be desalted on a PD 10 gel filtration column (Pharmacia) before bioassaying. Besides the nonbound fraction, four major peaks are recorded at 280 nm. Peak *C* contains the most active compound, peaks *A*, *B* and *D* exhibit minor activities only and unbound material is inactive

Crude elicitor preparations (CEP) of *P. graminis* are applied to Con A-Sepharose 4B (Pharmacia, Uppsala, Sweden) equilibrated with 20 mM Tris/HCl, pH 7.4, containing 0.5 M NaCl and 1 mM each of $MgCl_2$, $MnCl_2$ and $CaCl_2$ (Con A buffer) at 4 °C. Staining of the unbound material in dot-blot binding assays after Con A/Peroxidase overlay and chloronaphtol/H_2O_2 addition confirms the binding activity of the column-immobilized lectin. Elicitor active glycoproteins are specifically displaced by 0.2 M α-methylmannoside in Con A buffer.

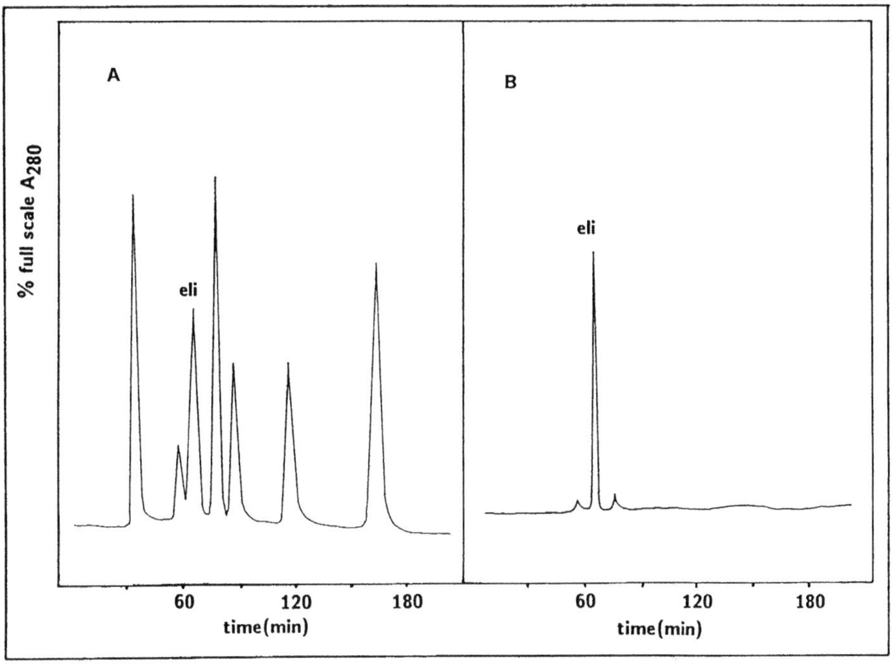

Fig. 3. Medium-pressure size exclusion chromatography of CEP (**A**) and Pgt elicitor purified by MonoQ (**B**) on a Superose 12 HR 10/30 column (Pharmacia). The column is equilibrated with 100 mM acetic acid and calibrated with the following standard proteins: aldolase (M_r 146000), bovine serum albumin (M_r 66000), carbonic anhydrase (M_r 29000). Per run, 200-µl sample containing 60 µg glycoprotein is applied. MPLC is done at a flow rate of 0.3 ml min^{-1}. According to this method Pgt elicitor (eli) has an M_r of 66000

3.4 Medium Pressure Liquid Chromatography (MPLC)

MPLC is a powerful and rapid technique for the isolation and purification of proteins and glycoproteins. This technique was recently adapted to the final purification from elicitors of *P. graminis* and *P. megasperma* (Kogel et al. 1988; Scheel and Parker 1990; Beißmann and Reisener 1990). The effectiveness of the method is enhanced by the combined use of several columns with different separation parameters.

Elicitor molecules of *P. graminis* are highly enriched by MPLC-assisted anion-exchange chromatography (Fig. 2). For final purification, the fraction exhibiting the highest activity is subjected to size exclusion MPLC producing a highly purified glycoprotein (Fig. 3) referred to as Pgt elicitor. Larger amounts of elicitor are obtained by semi-preparative anion exchange MPLC with Q-Sepharose Fast Flow medium packed in a HR 10/10 column (Pharmacia). CEP (50 mg 50 ml^{-1}) is injected onto this column equilibrated with 20 mM $(NH_4)_2CO_3$, pH 9.1. The column is washed for 60 min at a flow rate of 1 ml min^{-1} and eluted with a 220 min gradient from 20 to 200 mM $(NH_4)_2CO_3$. Desalting is achieved by lyophilizing the volatile buffer components without loss of elicitor-active material.

The chromatographic elution profile for CEP on this column resembles the profile obtained with the MonoQ column. The scaled-up procedure provides a high yield of purified elicitor. More than 1.5 mg (MonoQ: 0.3 mg) of Pgt elicitor are obtained from 18 mg CEP.

3.5 Identification of Elicitors in Polyacrylamide Gels and by Western Blot Analysis

Proteinaceous elicitors can be isolated from SDS gels by cutting single bands. Cut pieces are ground with a mortar and pestle followed by extraction with H_2O for 72 h. N-butanol is added to prevent growth of microorganisms. Extracts are filter-sterilized and SDS is removed by precipitation of proteins in 90% (v/v) cold acetone.

Western blot analysis is used to screen the presence of elicitors in crude cell wall preparations or IWFs. Elicitors, thereby, are identified either by polyclonal or monoclonal antibodies or by specific lectins (Fig. 4).

4 Characterization of Elicitors

4.1 Molecular Mass Determination

Ultrafiltration is a convenient method to estimate the molecular mass of elicitors in crude preparations. For a better estimation, gel chromatography or SDS gel electrophoresis is widely used. The molecular size of glycoproteins is often overestimated by size exclusion chromatography and underestimated by SDS PAGE (Freifelder 1982). However, molecular mass determination of Pgt elicitor by size exclusion MPLC using the analytical Superose 12 column (Fig. 3) and by SDS PAGE lead to nearly identical results (Beißmann and Reisener 1990).

4.2 Identification of the Active Site of Elicitors

4.2.1 Heat, Chemical and Enzymatic Treatment

Stability to heat is tested by boiling elicitor-active IWF or Pgt elicitor in a water bath for 1 h.

Periodate oxidation of polysaccharides has been used in structural characterization (Andersson and Gahmberg 1979). Carbohydrate residues

Fig. 4. Identification of Pgt elicitor in intercellular fluids. IWF from compatible interactions as well as Pgt elicitor are separated by SDS PAGE in a 12.5% slab gel using the Laemmli (1970) buffer system. Elicitor detection on Western blots is possible either by immunostaining with polyclonal anti-Pgt elicitor antiserum (*lane 1*, Pgt elicitor; *lane 2*, IWF from rust-infected susceptible wheat leaves; *lane 3*, IWF from unaffected wheat leaves) or by Con A-peroxidase affinostaining (*lane 4*, Pgt elicitor; *lane 5*, IWF from rust-infected susceptible wheat leaves). Electroblotting is carried out by a trans-blot cell at 400 mA for 2 h under extensive cooling at 4 °C. The transfer buffer consists of 25 mM Tris/HCl and 192 mM glycine, pH 8.3, with 20% (v/v) methanol to avoid shrinking of gels. To avoid unspecific binding, filters are blocked by incubation in TBS (50 mM Tris/HCl, 200 mM NaCl, pH 7.9) containing 4% (w/v) BSA for 1 h at room temperature. Con A-binding glycoproteins are detected after lectin/peroxidase overlay and chloronaphtol/H_2O_2 addition (Hawkes 1982). For immunostaining the nitrocellulose filter is incubated with anti-Pgt elicitor antiserum (1:100) in 20 mM Tris/HCl, pH 7.5, containing 0.3 M NaCl (binding buffer) for 2 h at 37 °C. After washing, the filter is incubated with phosphatase conjugated second antibody (1:1000) in binding buffer for 2 h at 37 °C. Antigenic glycoproteins are visualized by adding of 4% chloronaphtol (w/v) and 0.1% H_2O_2 (v/v) in binding buffer. Antiserum preparation: 500 μg Pgt elicitor dissolved in a mixture of 100 μl TBS and 100 μl Freund's complete adjuvant is subcutanously injected into a rabbit. This is followed by two injections each of 150 μg Pgt elicitor together with Freund's incomplete adjuvant 2 and 4 weeks later. The rabbit is bled 2 and 4 weeks after the last injection

containing glycol groups are oxidized to dialdehydes. Thus, periodate oxidation is an appropriate controllable means to destroy native carbohydrate structures. Sensitivity to periodate is checked by incubation of glycoproteins in 0.02 M $NaIO_4$ for 22 h at 20 °C in the dark. Excess of $NaIO_4$ is destroyed by adding 0.2 ml ethylene glycol to 0.8 ml sample, followed by extensive dialysis against H_2O.

For proteolytic digestion, samples are incubated with trypsin [1% (w/w) of the sample protein] in 50 mM $(NH_4)_2CO_3$ at 37 °C for 2 h followed by a second addition of 1% trypsin. Tryptic digestion is terminated 2 h later by the addition of trypsin inhibitor [1% (w/w) of the sample protein]. Samples are extensively dialyzed against H_2O and assayed for elicitor activity.

The structural requirements for elicitor activity may also be deduced from further enzymic or chemical treatment. For example, the elicitor activity of a glycopeptide elicitor from *Cladosporium fulvum* decreases after the release of both mannosyl residues by α-mannosidase and galactofuranosyl residues by mild acid hydrolysis (0.01 N HCl, 100 °C, 1 h; De Wit and Kodde 1981).

4.3 Characterization of Elicitor-Active Glycoconjugates

Detailed structural identification of elicitor-active carbohydrates is only possible by very complex techniques. Gas liquid chromatography (GC) analysis reaveals the monosaccharide composition of the active carbohydrate structures of Pgt elicitor by two ways: (1) Monosaccharides released by hydrolysis in 2 N TFA at 110 °C for 2 h are transferred into alditol acetates by the method of Jones and Albersheim (1972). The sugar derivatives are separated on a 25-m OV-1 capillary column. According to this analysis, Pgt elicitor consists of galactose (50%), mannose (47%) and a small portion of glucose. (2) Methanolysis in 1 M methanol in HCl for 24 h at 85 °C followed by reacetylation and trimethylsilylation (Chaplin 1982) yields trimethylsilylated methylglycosides which were separated on a OV-225 capillary column. This analysis confirms the results obtained by GC of alditol acetates, but additionally revealed a minor amount of N-acetylglucosamine.

Alternatively, monosaccharides liberated by hydrolysis are estimated by anion-exchange HPLC (Hardy et al. 1988). Released sugars are immediately separated on an analytical CarboPAC column (Dionex) in 16 mM NaOH as eluent. Detection is via pulsed amperometric detection which appears even as sensitive as flame ionization detection used in GC. By this technique monosaccharides are resolved even in samples of very complex sugar composition (Beißmann, unpublished). To establish the positions of glycosidic linkages and the ring sizes of carbohydrates, partially methylated alditolacetates are separated by GC and subsequently analyzed by mass spectrometry (for review see Sweeley and Nunez 1985).

Sharp et al. (1984) describe a glycosyl sequence analysis of the primary structures of seven heptaglycans which are available in small amounts only (e.g., 30–50 μg). Permethylated glycans are partially cleaved by formolysis and transferred into partially O-methylated, partially ethylated glucitol fragments. Glucitol derivatives of different size are separated by GC and characterized by mass spectroscopy. The primary structure of the hepta-glucans is deduced from overlapping fragments.

Fast-atom bombardment mass spectroscopy (FAB-MS) has recently become popular to gain further insight into the monosaccharide sequence of oligosaccharides (Sweeley and Nunez 1985). FAB-MS is successfully applied to establish the molecular mass of a 12-residue oligogalacturonide elicitor released by hydrolysis from soybean cell walls and from a pectin fraction (Nothnagel et al. 1983).

Detailed informations on the primary structure, (e.g., position and configuration of glycosidic linkages) can be deduced from NMR spectroscopy. ^{1}H-NMR spectroscopy is about 400 times more sensitive than ^{13}C-NMR (Goux and Unkefer 1987), allowing analysis in the range of 20–100 μg glycoconjugate (Sweeley and Nunez 1985). The efficacy of ^{1}H-NMR spectroscopy applied, for instance, for structural investigation of oligogalacturonic acid elicitors (Nothnagel et al. 1983; Davies et al. 1986) is enhanced by the introduction of 500 MHz ^{1}H-NMR spectroscopy as described by Vliegenthart et al. (1981). The technique allows the determination of the complete primary structures of glycans. On the other hand, ^{13}C-NMR spectra appear in some details less complicated yielding well-resolved signals over a large chemical shift range (Goux and Unkefer 1987). NMR spectroscopy does not require any time consuming derivatization and samples remain unaffected allowing further experiments.

4.4 Identification of Protein-bound Carbohydrate Chains

Table 3 shows a scheme for the analysis of active carbohydrate chains in a glycoprotein elicitor. Asparagine-bound oligosaccharides are released from elicitor by enzymic hydrolysis with peptide N-glycosidase F (PNGase F, EC 3.2.2.18). This enzyme cleaves protein bonds to N-glycans of the high mannose-, complex- and hybride type (Chu 1986; Hirani et al. 1987). Release of N-glycans is indicated by a change of electrophoretic mobility. N-glycans are separated by size exclusion chromatography in 0.1 M acetic acid. The partially deglycosylated glycoprotein is submitted to β-elimination. Under non-reducing conditions (without $NaBH_4$) a constant OD_{240} indicates the end point of the reaction (Neiderhiser et al. 1971). The decrease of serine and threonine is determined by automated amino acid analysis after hydrolysis of the β-eliminated glycopeptide. By this treatment, elicitor-active O-glycans are released from Pgt elicitor greater than M_r 30000.

Table 3. Identification of protein-bound carbohydrate chains

Pgt elicitor (1 mg)

PNGase F,
10–15 units, 0.1 M Na-phosphate buffer, pH 8.6, 20 h, 37 °C

Size exclusion chromatography (Sephadex G 50)

Mr < 30 k

Mr > 30 k (active)

β-elimination,
0.1 M NaOH, 1 M $NaBH_4$,
48 h, 40 °C

Size exclusion chromatography (Sephadex G 50)

Mr < 30 k

Mr > 30 k (active)

5 Functional Characterization of Elicitor by the Identification of Its Cellular Target

Schmidt and Ebel (1987) identified specific binding sites in soybean membrane vesicles for an [^{3}H]ß-1,3 oligoglucan (K_d = 200 nM) with phytoalexin-inducing activity from *Phytophthora megasperma* (Drechs) f. sp. *glycinea*. The [^{3}H]glucan was obtained by reduction of the oligoglucan fraction with sodium boro[^{3}H]hydride to yield [^{3}H]oligoglucosyl glucitols. Reduction did not affect the elicitor activity, as demonstrated for an oligoglucan fraction which was reduced with unlabelled $NaBH_4$. Higher specific binding activity (K_d = 37 nM) was expressed by an oligoglucan which was modified at the reducing end with phenylalkylamine reagent to give [^{125}I] oligoglucosyl tyramine. The [^{125}I]glucan with a specific radioactivity of 15–30 Ci/mmol provided higher sensitivity and lower detection limits for the binding assay.

Specific binding of Pgt elicitor to highly purified plasma membrane vesicles of wheat primary leaf cells has been demonstrated (Kogel et al. 1991). The elicitor is modified with [^{14}C]ethyl acetimidate hydrochloride at the aminogroups of the protein portion. Thus, elicitor activity is retained. Binding is saturable and reversible as shown by Scatchard plot analysis (Fig. 5).

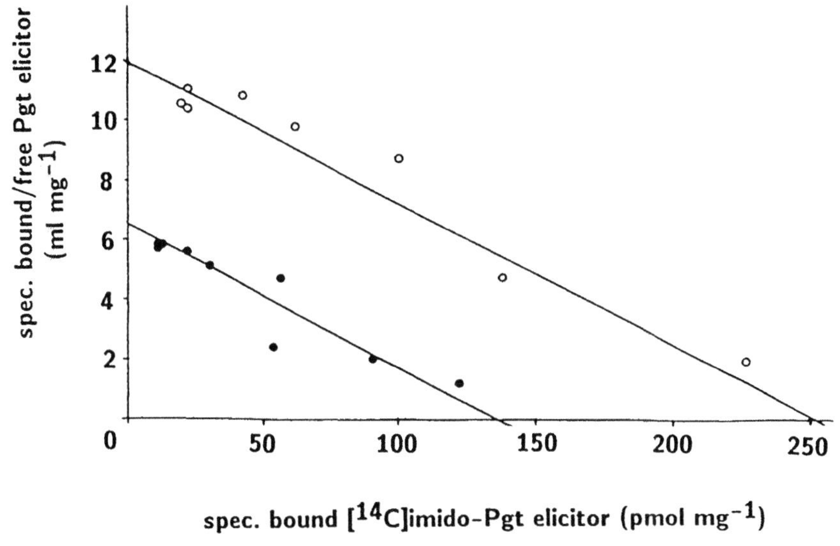

Fig. 5. Scatchard plots analysis of the binding of [^{14}C]imido-Pgt elicitor in the presence of increasing amounts of unlabelled Pgt elicitor. 50 µl (10 µg protein) of pure plasma membrane vesicles (○–○) and plasma membrane vesicles contaminated with intracellular membranes (●–●) prepared according to Kjellbom and Larsson (1984) are immobilized in wells of microtiter plates (Greiner, Solingen, FRG). To avoid unspecific binding free polystyrene matrix is subsequently blocked with 200 µl of 4% BSA in 20 mM Tris/HCl, pH 7.5, containing 0.3 M NaCl (binding buffer). Vesicles are incubated at 4°C with 50 µl of binding buffer containing 1% BSA and 0.2 µM labelled elicitor in the presence of unlabelled elicitor. After 1 h, the supernatant is removed and the wells are washed (5×) with binding buffer. The supernatants are combined and analyzed for radioactivity. Specifically bound [^{14}C]imido-Pgt elicitor/free [^{14}C]imido-Pgt elicitor is plotted against specifically bound [^{14}C]imido-Pgt elicitor. Regression analysis by the least-squares method is used for computation of the dissociation constant (K_d = 2 µM) and number of binding sites (n = 250 pmol mg^{-1}) for pure plasma membrane. The respective values for plasma membrane contaminated with intracellular membranes are K_d = 2.1 µM and n = 135 pmol mg^{-1}. Preparation of labelled elicitor: [^{14}C]imido-Pgt elicitor is prepared by a method of Whiteley and Berg (1974) with modifications. Lyophilized Pgt elicitor (46 µg protein) is dissolved in 946 µl of 0.1 M borate buffer, pH 9, at 25 °C. Then, 54 µl of borate buffer containing ethyl [1-^{14}C]EAI (Amersham-Buchler, FRG) (0.01 mM = 923 MBq) is added. This corresponds to 15 mol [^{14}C]EAI per mol aminogroups of Pgt elicitor. After incubation at 25 °C for 3 h, the reaction is stopped by neutralizing with 0.1 M HCl and freezing in liquid N_2. Unreacted [^{14}C]EAI is removed by a Sephadex G10 column (Pharmacia). Labelled Pgt elicitor is collected from the void volume of this column. [^{14}C]ethylacetate which is an undesired side product of the reaction is removed by evaporation under reduced pressure. [^{14}C]imido-Pgt elicitor has a specific radioactivity of 8.285×10^5 cpm µg^{-1}. Unlabelled imido-Pgt elicitor is prepared by the same method using non-radioactive EAI

Specific binding is verified by ELISA. Subsequent to immobilization and blocking (experimentals see Fig. 5), plasma membrane vesicles are incubated with 5-μg Pgt elicitor. Bound elicitor is detected using rabbit anti-Pgt elicitor IgG and a goat anti-rabbit IgG coupled to phosphorylase. Phosphorylase reaction with nitrophenylphosphate as substrate is recorded by an ELISA reader at 405 nm. In control experiments, anti-Pgt elicitor antibodies are replaced by preimmune serum.

6 Conclusions and Prospect

The combined application of biochemical, immunological and microscopical methods provided a single highly active elicitor of *Puccinia graminis* f. sp. *tritici*. Purity is a prerequisite for an extensive characterization of the elicitor, revealing O-glycan chains of the elicitor glycoprotein as being responsible for activity. Purified elicitor is recognized by the plasma membrane, which is an indication for the mode of elicitor action in rust-infected wheat leaves.

An intention of this chapter is to demonstrate that molecular phytopathology is not simply a repetition or adoption of the tried and true procedures and methods used successfully for plant and microbial science, but requires expertise in handling organisms which due to the coevolution of host and pathogen live in close relationship. The power of the technology has to some degree obscured the elegance of traditional phytopathology. It is, therefore, necessary to ask for the relevance of the purified elicitors in the natural interaction. This is particularly the question for the race – cultivar specificity of elicitors which is beyond the scope of this chapter. It is, however, important to realize that elicitors are one type of molecule which are involved in the early steps of host-parasite interaction. The role of another type of molecules, i.e., suppressors which possibly play a role in this interaction as well, are discussed in another chapter of this volume.

References

Anderson-Prouty AJ, Albersheim P (1975) Host-pathogen interactions: VII. Isolation of a pathogen-synthesised fraction rich in glucan that elicits a defense response in the pathogens host. Plant Physiol 56:286–291

Andersson LC, Gahmberg CG (1979) Surface glycoproteins of resting and activated human T lymphocytes. Mol Cell Biochem 27:117–129

Ayres A, Valent B, Ebel J, Albersheim P (1976) Host-pathogen interactions XI. Composition and structure of wall-released elicitor fractions. Plant Physiol 57:766–774

Beardmore J, Ride JP, Granger JW (1983) Cellular lignification as a factor in the hypersensitive resistance of wheat to stem rust. Physiol Plant Pathol 22:209–220

Beißmann B, Reisener HJ (1990) Isolation and purity determination of glycoprotein elicitors from wheat stem rust using medium pressure liquid chromatography. J Chromatogr 521:187–197

Boller T (1987) Hydrolytic enzymes in plant disease resistance. In: Kosuge T, Nester EW (eds) Plant-microbe interactions: molecular and genetic perspectives, vol 2. Macmillan, New York, pp 385–413

Boller T (1989) Primary signals and second messengers in the reaction of plants to pathogens. In: Boss WF, Morre DJ (eds) Second messenger in plant growth and development. Alan R. Liss, New York, pp 227–255

Bostock RM, Kuc'JA, Laine RA (1981) Eicosapentaenoic and arachidonic acids from *Phytophthora infestans* elicit fungitoxic sesquiterpenes in the potato. Science 212:67–69

Callow JA (1984) Cellular and molecular recognition between higher plants and fungal pathogens. In: Linskens HF, Heslop-Harrison J (eds) Cellular interactions. (Encycl Plant Physiol 17). Springer, Berlin Heidelberg New York, pp 212–237

Chaplin MF (1982) A rapid and sensitive method for the analysis of carbohydrate components in glycoproteins using gas-liquid chromatography. Anal Biochem 123:336–341

Chu FK (1986) Requirements of cleavage of high mannose oligosaccharides in glycoproteins by peptide-N-glycosidase F. J Biol Chem 261:172–177

Darvill AG, Albersheim P (1984) Phytoalexins and their elicitors – a defense against microbial infection in plants. Annu Rev Plant Physiol 35:243–275

Davies KR, Darvill AG, Albersheim P (1986) Host-pathogen interactions: XXX. Characterization of elicitors of phytoalexin accumulation in soybean released from soybean cell walls by endopolygalacturonic acid lyase. Z Naturforsch 41c:39–48

Deverall BJ, Deakin AL (1985). Assessment of Lr20 gene-specificity of symptom elicitation by intercellular fluids from leaf rust-infected wheat leaves. Physiol Plant Pathol 27:99–107

De Wit PJGM, Kodde E (1981) Further characterization and cultivar specificity of glycoprotein elicitors from culture filtrates and cell walls of *Cladosporium fulvum* (syn. *Fulvia fulva*). Physiol Plant Pathol 18:297–314

De Wit PJGM, Roseboom PHM (1980) Isolation, partial characterization and specificity of glycoprotein elicitors from culture filtrates, mycelium and cell walls of *Cladosporium fulvum* (syn. *Fulvia fulva*). Physiol Plant Pathol 16:391–408

De Wit PJMG, Spikman G (1982) Evidence for the occurrence of race and cultivar specific elicitors of necrosis in intercellular fluids of compatible interactions of *Cladosporium fulvum* and tomato. Physiol Plant Pathol 21:1–11

Dow JM, Callow JA (1979) Leakage of electrolytes from isolated leaf mesophyll cells of tomato induced by glycopeptides from culture filtrates of *Fulvia fulva* (Cooke) Ciferri (syn. *Cladosporium fulvum*). Physiol Plant Pathol 15:27–34

Freifelder DM (1982) Physical biochemistry, 2nd edn. W.H. Freeman, San Francisco

Goux WJ, Unkefer CJ (1987) The assignment of carbonyl resonances in ^{13}C-NMR spectra of peracetylated mono- and oligosaccharides containing D-glucose and D-mannose: an alternative method for structural determination of complex carbohydrates. Carbohydr Res 159:191–210

Hadwiger LA, Beckman JM (1980) Chitosan as a component of pea-*F. solani* interactions. Plant Physiol 66:205–221

Hahlbrock K, Scheel D (1987) Biochemical responses of plants to pathogens. In: Chet I (ed) Innovative approaches to plant disease control. John Wiley & Sons, New York, pp 229–254

Hahlbrock K, Lamb CJ, Purwin C, Ebel J, Fautz E, Schäfer E (1981). Rapid response of suspension-cultured parsley cells to the elicitor from *Phytophthora megasperma* var. *sojae*. Plant Physiol 67:768–773

Hardy MR, Townsend RR, Lee YC (1988) Monosaccharide analysis of glycoconjugates by anion exchange chromatography with pulsed amperometric detection. Anal Biochem 170:54–62

Hawkes R (1982) Identification of Concanavalin A-binding protein after sodium dodecyl sulphate-gel electrophoresis and protein blotting. Anal Biochem 123:143–146

Heath MC (1981) A generalized concept of host-parasite specifity. Phytopathology 71:1121–1123

Hirani S, Bernasconi RJ, Rasmussen JR (1987) Use of peptide-N-glycosidase F to release asparagine linked oligosaccharides for structural analysis. Anal Biochem 162:485–492

Holden DW, Rohringer R (1985) Proteins in intercellular washing fluid from noninoculated and rust-affected leaves of wheat and barley. Plant Physiol 78:715–723

Jones TM, Albersheim P (1972) A gas chromatographic method for the determination of aldose and uronic acid constituents of plant cell wall polysaccharides. Plant Physiol 49:929–936

Kauss H (1984) Chitosan elicited synthesis of callose in soybean cells: a rapid and Ca^{++}-dependent process. 6th J Innes Symp Cell surf in plant growth and dev, Norwich, 53 (Abstr)

Keen N, Brügger B (1977) Phytoalexins and chemicals that elicit their production in plants. Am Chem Soc Symp Ser 62:1–26

Keen NT, Legrand M (1980) Surface glycoproteins: evidence that they may function as the race specific phytoalexin elicitors of *Phytophthora megasperma*. Physiol Plant Pathol 17:175–192

Keen NT, Partridge JE, Zaki AI (1972) Pathogen produced elicitor of a chemical defense mechanism in soybeans monogenically resistant to *Phytophthora megasperma* var. *sojae*. Phytopathology 65:768

Kjellbom P, Larsson C (1984) Preparation and polypeptide composition of chlorophyll-free plasma membranes from leaves of light-grown spinach and barley. Physiol Plant 62:501–509

Klement Z (1965) Method of obtaining fluid from the intercellular spaces of foliage and fluid's merit as substrate for phytobacterial substrates. Phytopathology 55:1033–1034

Knogge W, Weissenböck G (1986) Tissue distribution of secondary phenolic biosynthesis in developing primary leaves of *Avena sativa* L. Planta 167:196–205

Kogel G, Beißmann B, Reisener HJ, Kogel KH (1988) A single glycoprotein from *Puccinia graminis* f. sp. *tritici* cell walls elicits the hypersensitive lignification response in wheat. Physiol Mol Plant Pathol 33:173–185

Kogel G, Beißmann B, Reisener HJ, Kogel KH (1991) Specific binding of a hypersensitive lignification elicitor of *Puccinia graminis* f. sp. *tritici* to the plasma membrane from wheat (*Triticum aestivum* L.). Planta 183:164–169

Kogel KH (1985) Untersuchungen zur molekularen Grundlage von Resistenz im Weizen Schwarzrost-System. Thesis, RWTH Aachen

Kogel KH, Heck B, Kogel G, Moerschbacher B, Reisener HJ (1985) A fungal elicitor of the resistance response in wheat. Z Naturforsch 40c:743–744

Kuhn DN, Chappell J, Boudet A, Hahlbrock K (1984) Induction of phenylalanine ammonia-lyase and 4-coumarate: CoA ligase mRNAs in cultured plant cells by UV light or fungal elicitor. Proc Natl Acad Sci USA 81:1102–1106

Laemmli UK (1970) Cleavage of structural proteins during the assembly of the head of bacteriophage T4. Nature 227:680–685

Moerschbacher B, Kogel KH, Noll U, Reisener HJ (1986a) An elicitor of the hypersensitive lignification response in wheat leaves isolated from the rust fungus *Puccinia graminis* f. sp. *tritici*. I. Partial purification and characterization. Z Naturforsch 41c:830–838

Moerschbacher B, Heck B, Kogel KH, Reisener HJ (1986b) An elicitor of the hypersensitive lignification response in wheat leaves isolated from the rust fungus *Puccinia graminis* f. sp. *tritici*. II. Induction of enzymes correlated with the biosynthesis of lignin. Z Naturforsch 41c:839–844

Narasimham S, Wilson JR, Martin E, Schachter H (1979) A structural basis for four distinct elution profiles on concanavallin A sepharose affinity chromatography of glycopeptides. Can J Biochem 57:83–96

Neiderhiser DH, Plantner JJ, Carlson DM (1971) The purification and properties of the glycoproteins of pig gallbladder bile. Arch Biochem Biophys 145:155–163

Nothnagel EA, McNeil M, Albersheim P, Dell A (1983) Host-pathogen interactions. XXII Galacturonic acid oligosaccharide from plant cell walls elicits phytoalexins. Plant Physiol 71:916–926

Ricci P, Bonnet P, Huet JC, Sallantin M, Beauvais-Cante F, Bruneteau M, Billard V, Michel G, Pernollet JC (1989) Structure and activity of proteins from pathogenic fungi *Phytophthora* eliciting necrosis and acquired resistance in tobacco. Eur J Biochem 183:555–563

Rogers K, Anderson AJ (1987) Effects of elicitors on plant membrane functions. In: Alan R (ed) Plant membranes: structure, function, biogenesis. UCLA Symp on Molecular and cellular biology, New Ser vol 63. Alan R. Liss, New York, pp 403–418

Rohringer R, Ebrahim-Nesbat F, Wolf G (1983) Proteins in intercellular washing fluids from leaves of barley (*Hordeum vulgare* L.). J Exp Bot 34:1589–1605

Ryan CA (1987) Oligosaccharide signalling in plants. Annu Rev Cell Biol 3:295–317

Scheel D, Parker J (1990) Elicitor recognition and signal transduction in plant defense gene activation. Z Naturforsch 45c:569–575

Schmidt WE, Ebel J (1987) Specific binding of a fungal glucan phytoalexin elicitor to membrane fractions from soybean *Glycine max*. Proc Natl Acad Sci USA 84:4117–4121

Schottens-Toma IMJ, De Wit PM (1988) Purification and primary structure of a necrosis-inducing peptide from the apoplastic fluids of tomato infected with *Cladosporium fulvum* (syn. *Fulvia fulvum*). Physiol Mol Plant Pathol 33:59–67

Sharp JK, McNeil M, Albersheim P (1984) The primary structures of one elicitor-active and seven elicitor-inactive hexa (β-D-glucopyranosyl)-D-glycitols isolated from the mycelial walls of *Phytophthora megasperma* f. sp. *glycinea*. J Biol Chem 259:11321–11336

Sweeley CC, Nunez HA (1985) Structural analysis of glycoconjugates by mass spectroscopy and nuclear magnetic resonance spectroscopy. Annü Rev Biochem 54:765–801

Vliegenthart JFG, van Halbeek H, Dorland L (1981) The applicability of 500 MHz high-resolution ^{1}H-NMR spectroscopy for the structure determination of carbohydrates derived from glycoproteins. Pure Appl Chem 53:45–77

West CA (1981) Fungal elicitors of the phytoalexin response in higher plants. Naturwissenschaften 68:447–457

Whiteley NM, Berg HC (1974) Amidination of the outer and inner surfaces of the human erythrocyte membrane. J Mol Biol 87:541–561

Identification and Characterization of Suppressors

B. BEIẞMANN and K.-H. KOGEL

1 Introduction

Plants, like other living organisms, have evolved various mechanisms by which they usually can counter the advance of invading microbes. Functionally, these resistance mechanisms may be either constitutive (preformed) or inducible. Induced resistance mechanisms often involve gene activation.

Plant disease depends on the ability of a pathogen to overcome the plant defense response. A pathogen may be successful on this by interfering with the reaction sequence leading to resistance on different levels, e.g., recognition, signal transduction, gene activation or production of antimicrobial compounds. The molecules which inhibit or delay the resistance reaction, thus causing successful infection, are referred to as suppressors (Callow 1984). In contrast to toxins of necrotrophic pathogens which destroy the plant tissue, thus preventing resistance, suppressors allow the pathogen to invade into and to live on intact plant tissue. Hence, suppressors of the plant's resistance response are specific tools of biotrophic and slow-growing necrotrophic pathogens.

The concept that compatibility is due to suppressor action has been substantiated by double inoculation experiments, which show that infection by pathogenic fungi increased the accessibility of the tissues to either non-pathogenic species or avirulent strains of a pathogen. For instance, the preliminary inoculation of barley with a virulent powdery mildew fungus rendered the plant susceptible to an avirulent fungus or even to a non-pathogen (Ouchi et al. 1974). The close proximity of established colonies of the bean rust, cowpea rust, and sunflower rust fungi on their respective hosts increased the number of infection sites at which avirulent rust fungi produce haustoria (Heath 1980). This effect was not restricted to cells

---

Abbreviations

dpi: days post-inoculation; ELISA: enzyme-linked immunosorbent assay; GLC: gas liquid chromatography; HR: hypersensitive response; IWF: intercellular washing fluid; NSE: non-specific elicitor preparation; PAL: phenylalanine ammonia-lyase; Pgt elicitor: elicitor from *Puccinia graminis* f. sp. *tritici*.

Table 1. Biotic fungal suppressors of the resistance response of higher plants

Source	Origin	Chem. nature	Race-specific	Reference
Bean rust/French bean	IWF[a]	n.d.[b]	No	Heath (1981b)
C. fulvum/tomato	IWF	Protein	No	Peever and Higgins (1989)
P. infestans/potato	GF/MY[c]	Glucans	Yes	Garas et al. (1979)
A. rabiei/chickpea	CF[d]	Glycoprotein	n.d.	Kessmann and Barz (1986)
P. megasperma/soybean	CF	Glycoprotein	Yes	Ziegler and Pontzen (1982)
M. pinodes/pea	GF	Glycoprotein	n.d.	Yamamoto et al. (1986)
P. graminis/wheat	IWF/GF	n.d.	No	Beißmann et al. (in press)

[a] IWF: intercellular washing fluid.
[b] n.d.: no data.
[c] GF: germination fluid; MY: mycelium.
[d] CF: culture filtrate.

already containing haustoria, but was accomplished by the diffusion of a compound from the infection sites. The effective compound was isolated from the apolastic space of rust-infected leaves and was of low molecular mass.

These experiments have stimulated the search for suppressors in other pathosystems (Table 1). Since they can interfere with different steps of the reaction sequence, it is not surprising that the known suppressors are not homogeneous in structure and have different properties. The elucidation of single steps of the reaction sequence leading to resistance has recently stimulated more sophisticated attempts to identify suppressors. For instance, it could be demonstrated that the race-unspecific elicitor from *Puccinia graminis* f. sp. *tritici* (Pgt elicitor) specifically binds to a receptor located in the wheat plasma membrane (Kogel and Beißmann 1991, this Vol.; Kogel et al. 1991). The ELISA-based technique used in these binding experiments provides an excellent test system to identify candidates for suppressors of the primary recognition between pathogen and plant. Parallel approaches may be directed towards the identification of biotic inhibitors of enzymes involved in typical resistance pathways (e.g., phytoalexin or lignin biosynthesis).

1.1 Specificity of Suppressors: General Concepts

Several models to account for the specificity of the plant's response to pathogens have been developed based on genetic and biochemical studies. Most systems of cultivar- and race-specific resistance are controlled by one, or at most a few, genes conforming to a "gene-for-gene" relationship (Flor 1956; Ellingboe 1976). It is, however, unlikely that each plant has a specific and different gene to control the recognition of each of a myriad of potential pathogens in a manner analogous to a "gene-for-gene" relationship. It

Table 2. Two models proposing the involvement of suppressors in host-parasite specificity

A	
1. Non-host resistance:	Induction of a hypersensitive reaction (HR) by race-unspecific elicitors
2. Basic compatibility:	Suppressor interferes in the reaction sequence of HR induction by the race-unspecific elicitors
3. Host resistance:	Induction of HR by race-specific elicitors (avirulence gene product) via a cultivar-specific receptor (resistance gene product)
B	
1. Non-host resistance:	Induction of HR by race-unspecific elicitors
2. Basic compatibility:	Supperessor interferes in the reaction sequence of HR induction by the race-unspecific elicitors
3. Host resistance:	Elicitor binding is restored due to changes in suppressor structures which are controlled by the avirulence gene

appears to be more probable that each plant species has evolved the ability to recognize the majority of pathogens by surface components which they have in common. The resulting molecular interaction serves to trigger resistance.

Experimental evidence for this assumption comes from recent investigations on the biochemical mechanism of HR induction by Cosio et al. (1988) as well as by our laboratory (Kogel et al. 1991). Both groups showed that HR is triggered by race-unspecific fungal hyphal wall components. These elicitors bind to membrane-bound receptors of the respective plant. According to the model of Heath (1981a), successful pathogens may establish "basic compatibility" by suppressing general recognition. Cultivar- and race-specificity is then superimposed on the system of basic compatibility. By random mutation under strong selection pressure, the plant may restore the ability to recognize specific components (controlled by avirulence genes) of the pathogen via receptors (controlled by resistance genes). This concept implies that two elicitors may be present on the pathogen, one being race-unspecific, the other being race-specific; only the latter is recognized by the resistance gene product of the host. Furthermore, this concept implies the presence of an unspecific suppressor acting as an antagonist of the unspecific elicitor, thereby establishing basic compatibility (Table 2A). Experimental evidence for this concept comes from investigations on the tomato/*Cladosporium fulvum* interaction. De Wit and Kodde (1981) found an unspecific elicitor in culture filtrates and mycelial walls of *C. fulvum*. De Wit and Spikman (1982) demonstrated that the intercellular fluids of *C. fulvum*-infected tomato leaves contained a race-cultivar specific peptide with necrosis and chlorosis inducing activity. Furthermore, Peever and Higgins (1989) isolated a suppressor of the action of the unspecific elicitor from infected tomato leaves. However, contrary to Heath's concept, this suppressor is also present in non-infected tomato leaves.

Alternatively, Bushnell and Rowell (1981) proposed a model in which a race-cultivar specific suppressor competes with a race-unspecific elicitor for the receptor encoded by the resistance gene. This suppressor is controlled by the avirulence gene of the pathogen. The expression of avirulence is suggested to be due to an alteration of suppressor structure, with the consequence that the suppressor no longer fits the receptor, resulting in a restored recognition of the pathogen by the host (Table 2B). Experimental evidence for this model is furnished by Garas et al. (1979), who isolated a race-specific suppressor from the potato/*Phytophthora infestans* system.

At present it is not possible to make a general decision in favour of one or the other of these models. In contrast, this Chapter intends to demonstrate that with increasing information on suppressor action in different systems, the situation has become even more complex. Clearly, improved models are needed which must include the results obtained with the modern biochemical and molecular biological methods recently introduced into phytopathology.

2 Isolation and Characterization of Suppressors

2.1 Sources for Suppressor Isolation

Suppressors have been found in plants as well as in pathogen-derived material. The main problem in suppressor research is the coincident appearance of suppressors and elicitors in the same sources, e.g., fungal mycelia, spore germination fluids or intercellular washing fluids (IWF). Suppressor activities are indeed often diminished in such crude solutions or can even be completely masked due to high elicitor activity. Hence, the separation of suppressors from elicitors may be the prerequisite to detect suppressor activity. Separation can be achieved by extraction procedures based on differing physicochemical properties of suppressors and elicitors or by separation steps like ultrafiltration or gel filtration upon coextraction. The isolation of some suppressors is described in more detail:

Pycnospores of the pea pathogen *Mycosphaerella pinodes* release both suppressors and elicitors of pisatin synthesis into the germination fluid (Shiraishi et al. 1978). This fluid was fractionated by ultrafiltration (Millipore filter, PTGC 142-05) into a low-molecular mass fraction (<10000 daltons) containing suppressors and a high-molecular mass fraction (>10000 daltons) exhibiting elicitor activity.

Suppressor and elicitor, both present in mycelia of *Phytophthora infestans*, were separated during the extraction procedure (Garas et al. 1979). Frozen mycelia were homogenized in 0.05 M sodium acetate, pH 4.5, for 2 min. The homogenate was sonicated for 15 min and centrifuged at 20000 g for 30 min. Suppressor activity was found in the supernatant, while elicitor-active material was subsequently extracted from the pellet.

Culture filtrates of the fungal pathogen *Ascochyta rabiei*, the causal agent of *Ascochyta* blight on chickpea (*Cicer arietinum* L.), contained both suppressor- and elicitor-active compounds (Kessmann and Barz 1986). For extraction of suppressor-active material, 10- to 12-day-old culture media were filtered, boiled for 10 min and centrifuged (1000 × g, 30 min). Protein was precipitated from the supernatant in three steps by the addition of solid ammonium sulfate to 30, 70 and 100% saturation. The fraction precipitating at 70% ammonium sulfate saturation contained the suppressor-active material, whereas the elicitor-active material could not be precipitated.

In general, IWF from compatible host-parasite interactions are considered to be a potential source for suppressor isolation. In fact, suppressor-active apoplastic fluids were obtained from *Cladosporium fulvum*-infected tomato leaves 10–14 dpi when fungal sporulation was evident (Peever and Higgins 1989). Moreover, IWFs derived from the compatible interactions of rust with bean, sunflower and wheat increased the susceptibility of bean plants to the normally avirulent cowpea rust fungus (Li and Heath 1989).

Recently, a suppressor of the hypersensitive lignification response was identified in IWF of stem rust-infected wheat leaves, and separated by gel filtration on Sephadex G 25 (Fig. 1) from elicitor-active compounds found in the same fluids. Interestingly, in the same manner, suppressor-active compounds were isolated from IWF of healthy wheat leaves as well, although the non-fractionated IWF alone showed neither elicitor nor suppressor activity (Beißmann et al., in press).

2.2 Bioassays for Suppressor Activity

A prerequisite for the detection of suppressor activity is the availability of a sensitive assay to register the elicitation of plant defense responses either by inoculation of resistant host plants with the respective pathogens or, alternatively, by treatment of plants with elicitors. Therefore, most bioassays for suppressor activity are slightly modified assays for elicitor activity and, consequently, must fulfill the same criteria as postulated for elicitor bioassays (Kogel and Beißmann, this Vol.). The principle of most suppressor assays is the comparison of phytoalexin accumulation in plants induced by elicitor alone or by a series of mixtures of elicitor and suppressor. In practice, several important criteria have to be fulfilled to ensure a sufficient sensitivity of bioassays for suppressor detection: First, elicitor must be applied to plant material at optimal concentration. Elicitors showing a sigmoidal dose-response curve of symptom elicitation (Kogel et al. 1988; Kogel and Beißmann, this Vol.) must preferably be used in suppressor assays at concentrations of non-saturating elicitor activity. Otherwise, low suppressor activities may be masked by a high elicitation level. Second, the optimal time of application for each suppressor and elicitor must be determined. Thus, in some cases, pretreatment of plant material with suppressor before elicitor addition may

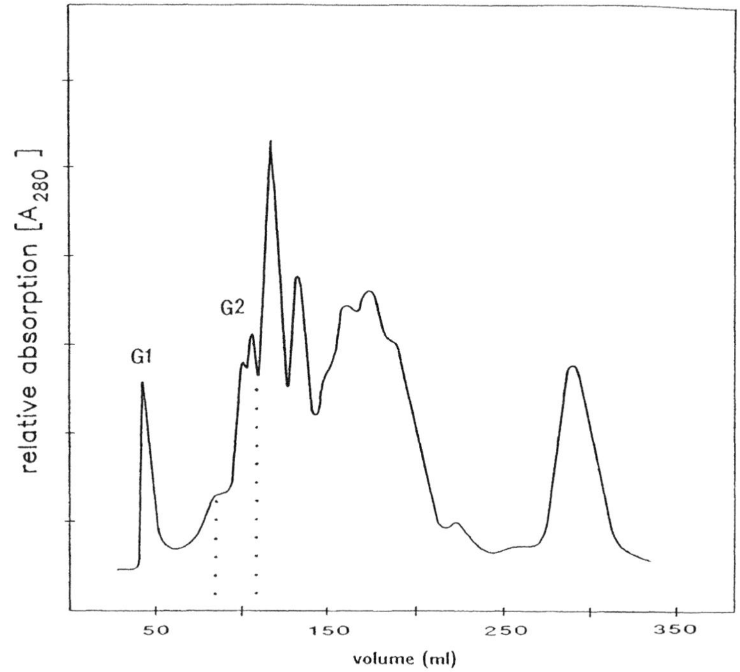

Fig. 1. Identification of elicitor- and suppressor-active components in IWF from stem rust (*Puccinia graminis* f. sp. *tritici*, race 32) infected wheat plants (cv. Prelude). IWF was obtained 6 days post inoculation (dpi) from primary leaves according to Kogel and Beißmann (this Vol.) and applied to a Sephadex G 25 column. Fraction *G1* eluting in the exclusion volume showed elicitor activity, whereas fraction *G2* contained suppressor-active material with an M_r of about 1500 daltons. Elicitor and suppressor activity were determined by the PAL assay described in Section 2.2

result in higher suppressor-activity. Third, an ideal bioassay must be fast and easy to handle to allow the screening of many samples.

The following bioassays are described in more detail with emphasis on the handling of plant material, such that the outlined method should be generally applicable to other plant systems.

Substances of IWF material from *Cladosporium fulvum*-infected tomato leaves as well as from bean-rust infected French bean leaves prevent necrosis-induction and callose formation in tomato leaves upon treatment with a non-specific elicitor (NSE) preparation from *Cladosporium fulvum* (Peever and Higgins 1989). Test solutions consisting of NSE mixed with either IWF or an equivalent volume of water were injected into interveinal panels of tomato leaflets. Samples to be compared were injected into panels on opposite sides of the central vein. The level of necrosis was determined macroscopically 24 h later and leaves were tested qualitatively for callose deposition with the aniline-blue fluorescence method (Lazarovits and Higgins 1979).

IWF from wheat leaves and spore germination fluid from stem rust uredospores contain suppressor-active material which prevented the hypersensitive response in wheat induced by the race-unspecific elicitor from *Puccinia graminis* (Beißmann et al., in press). For suppressor screening, test solutions were injected into the intercellular spaces of 7-day-old wheat primary leaves by a hypodermic syringe. The leaves were analyzed 4 days after injection for hypersensitively reacting cells by fluorescence microscopy. The level of autofluorescing cells in elicitor-treated tissue was significantly higher than in tissue treated with a mixture of elicitor and suppressor.

A quantitative assay for suppressor activity in wheat was established by measuring elicitor-induced activity of PAL, a key enzyme for lignin biosynthesis. In this assay, samples of Pgt elicitor alone and mixed with suppressor were injected into wheat primary leaves and PAL activity was determined 24 h later (Fig. 2) by the method described previously (Kogel and Beißmann, this Vol.).

Suppressors from the spore germination fluid of *Mycosphaerella pinodes* prevented the accumulation of the phytoalexin pisatin upon treatment of pea leaves with an elicitor present in the same solution (Oku et al. 1977; Hiramatsu et al. 1986). The bioassay for suppressor activity was therefore based on the elicitor assay. After removal of the abaxial epidermis, a drop of elicitor solution (500 ppm, 50 µl) was applied onto pea leaf pieces (8 × 8 mm). The leaf segments were subsequently illuminated (3000 lx) at 22 °C for 8–9 h and extracted by heating in 500 µl ethanol to 80 °C for 10 min. The extracts were dried under reduced pressure, redissolved in small amounts of ethanol and separated by TLC (silica gel 60 F_{254}) in toluene/ethylformate/formic acid (60:40:1, v/v/v).

Pisatin amounts in elicitor- or elicitor plus suppressor-treated leaf segments were indirectly quantified by measuring the incorporation of radioactive phenylalanine into pisatin and its biosynthetic precursers. For this, L-[U-^{14}C-phenylalanine] was fed for 1 h at the end of an 8-h incubation period to the leaf pieces with elicitor or elicitor plus suppressor. Pisatin and its precursors were extracted and separated by TLC as described above. The incorporated radioactivity was determined by solubilizing the compounds after visualizing them under UV light by liquid scintillation counting.

A suppressor isolated from culture filtrates of the fungus *Ascochyta rabiei* inhibits the constitutive biosynthesis of isoflavones, as well as the wound- and elicitor-induced accumulation of pterocarpan phytoalexins in slices of chickpea cotyledons (Kessmann and Barz 1986). In the bioassay, isoflavone and pterocarpan amounts were determined. It was very important to conduct the assay under highly aseptic conditions. Therefore, chickpea seeds were surface-sterilized with 70% ethanol and 5% sodium hypochlorite (Kessmann and Barz 1986) and all subsequent steps were carried out with sterilized glassware and fluids. After preincubation on moist filter paper in Erlenmeyer flasks at 25 °C for 24 h in the dark, the chickpea seeds were sliced and again applied on moist paper in Erlenmeyer flasks in the dark. Samples to be tested

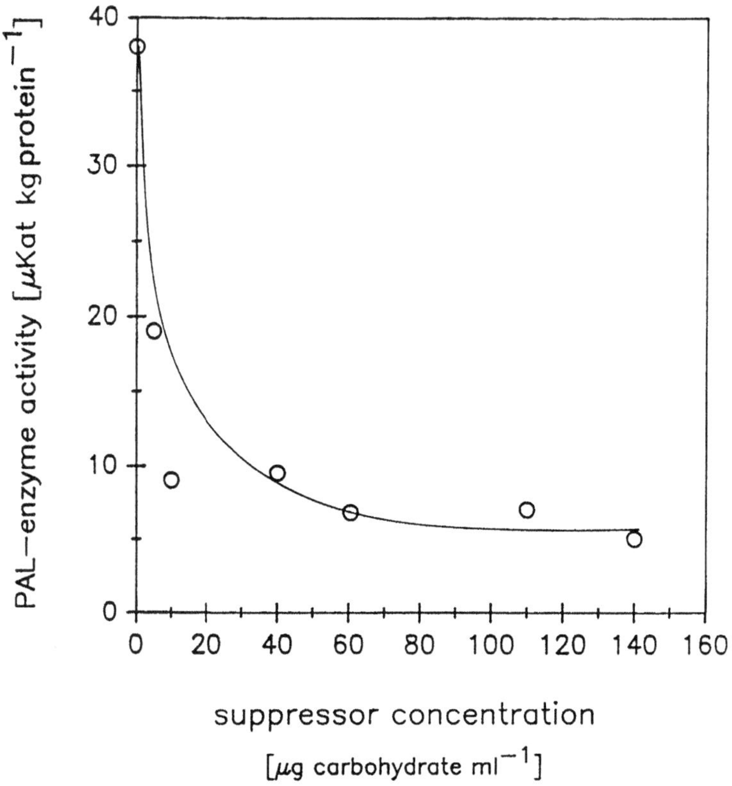

Fig. 2. Influence of different IWF-suppressor concentrations on the activity of phenylalanine ammonia-lyase (PAL). IWF was obtained 6 days post inoculation (dpi) from stem rust (*Puccinia graminis* f. sp. *tritici*) infected wheat primary leaves (cv. Prelude). PAL assays for suppressor activity were performed as described in Section 2.2 after treatment of primary leaves (cv. Prelude) with different concentrations of suppressor from the IWF in the presence of a constant concentration (50 μg) of Pgt elicitor

for elicitor or suppressor activity were diluted in 0.25 M D(-) mannitol, pH 5, autoclaved and applied to the seed slices. After different incubation periods, the induced phenolic compounds were extracted from the seeds with acetone and methanol and quantified by HPLC analysis (Kraft and Barz 1985).

Cytospores of *Phytophthora infestans* release suppressor-active glucans during germination (Doke et al. 1979). This water soluble material prevents browning and accumulation of terpenoid phytoalexins in potato tuber tissue inoculated with an incompatible race of *Phytophthora infestans* or treated with an elicitor preparation from the same pathogen. The inhibition of phytoalexin synthesis was determined by comparing the amounts of synthesized terpenoids in the presence or absence of suppressor. Small discs of potato tubers (18 mm in diameter and 7 mm thick) were placed for 18 h in petri dishes at 18 °C. Ten tuber discs per assay were transferred in 30-ml centrifuge tubes and covered with suppressor solution. The tubes were infiltrated with test solution by applying a 40-psi nitrogen pressure for 10 min, which then was slowly released over a period of 30 min. The infiltrated discs were washed with sterile distilled water and placed on moist filter paper in petri dishes at 18 °C. After 6 h, elicitor solution or zoospore suspension of an incompatible race of *Phytophthora infestans* were applied to the tuber discs and incubated for 72 h at 18 °C. Terpenoid phytoalexins were extracted from the upper 1 mm of the treated discs by soaking the plant material in an Erlenmeyer flask with 20 ml of methanol for 24 h, followed by two additional rinses with 10 ml of methanol for 4 h. The organic phases were combined, filtered, dried under reduced pressure, and dissolved in 1 ml of distilled water. Additional non-polar compounds, among them the phytoalexins, were extracted with 5 ml of hexane, which was subsequently evaporated. For qualitative phytoalexin determination, the extracted compounds dissolved in 0.5 ml of methanol were separated by TLC on silica gel G plates with cyclohexane/ethylacetate (1 : 1, v/v) and stained with the vanilin-sulfuric acid reagent (Garas et al. 1979). The amounts of the phytoalexins rishitin and lubimin were determined by quantitative gaschromatography (Garas et al. 1979).

An alternative bioassay has been developed to determine the suppressor activity of *Phytophthora infestans* glucan fractions, in that glucan suppressors prevent the rupture of potato tuber protoplasts effected by hyphal wall elicitors of the same fungus (Doke and Tomiyama 1980). This bioassay revealed race-cultivar specific differences in suppressor action. The protoplasts of a resistant cultivar were more reactive upon addition of elicitor than those of a susceptible cultivar. Furthermore, the hypersensitive response of such highly reactive protoplasts was inhibited more strongly by glucans isolated from virulent than from avirulent races of the pathogen.

A mannoprotein inhibits the elicitor-induced formation of the phytoalexin glyceollin in a soybean cotyledon assay (Ziegler and Pontzen 1982). Cotyledons from 9- to 10-day-old soybean plants were washed with sterile distilled water. Epidermis and first mesophyll cell layers were removed from the lower side of each cotyledon with a cut section of about 1 mm thick and 6 mm in diameter. Ten cotyledons per assay were placed upside down on moist filter paper in petri dishes. Each wound was covered with a 75-μl drop of test solution and the cotyledons were incubated for 16 h at 26 °C in the dark. The wound droplets were extracted by soaking and shaking the cotyledons in 20 ml of water in a 50-ml Erlenmeyer flask. These extracts were filtered through glass fibre filters (Whatman GF/A). Since the amount of accumulated glyceollin strongly correlated with an increase in UV-absorbance at 285 nm (Ayers et al. 1976), the amount of glyceollin was measured photometrically.

Cell cultures are also suitable systems for suppressor bioassays. A mixture of elicitor and suppressor-active citrate added to cell cultures of cotton or soybean decreased phytoalexin formation measured after solvent extraction, as compared with the amount of phytoalexins synthesized after addition of elicitor alone (Apostol et al. 1987). The suppressor effect of citrate was additionally monitored on cultured cotton and soybean cells by fluorescent dyes. The elicitor-induced fluorescence transitions of pyranine were inhibited by increasing amounts of citrate. In these experiments, suppressor was added to cuvettes containing cell suspension and fluorescent dye about 15 s before elicitor addition. The change in fluorescence was recorded by a spectrofluorometer.

An alternative approach for suppressor identification may consist in the observation of an improved development of the pathogen, e.g., an enhanced susceptibility of normally resistant plants. However, increased pathogen growth may not only be due to suppressor action but also to an alteration of the nutrient conditions. This is especially the case when plants are treated with crude suppressor fractions. For this reason, Heath (1981b) monitored the action of possible suppressors on both the plant and the pathogen side. IWF derived from the compatible interactions of various host plants and rust fungi were injected into bean plants. After infection with the cowpea rust fungus, a pathogen normally avirulent on bean plants, not only an increased formation of haustoria by the fungus, but also the reduced deposition of silica in the walls of adjacent cell walls was detected by microscopy, indicating that characteristic plant defense mechanisms were suppressed (Heath 1981b; Stumpf and Heath 1985).

2.3 The Chemical Nature of Suppressors

The term "suppressor" is a functional description of a species of biological effectors. At present, there appears to be no predictable consensus in the structural determinants of suppressor activity. This is consistent with the finding of different modes of suppressor action (see Sect. 2.4).

Several extracellular glycoproteins have been found to inhibit the elicitation of resistance reactions in plants. Invertase isolated from culture fluids of two compatible races of *Phytophthora megasperma* f. sp. *glycinea* previously characterized to be a mannoprotein (Ziegler and Albersheim 1977) was shown to inhibit the elicitation of glyceollin accumulation in soybean cotyledons (Ziegler and Pontzen 1982). The carbohydrate part of this suppressor, comprising 20% of the whole molecule, consisted mainly of mannose with small amounts of glucose and traces of glucosamine. As the suppressor activity was not affected by heat or pronase treatment but destroyed upon periodate oxidation, the carbohydrate portion was postulated to be the active part of the molecule.

The suppressor from culture filtrates of *Ascochyta rabiei* was characterized as a glycoprotein (Kessmann and Barz 1986). Active glycoproteins bound to a Con A-Sepharose affinity column and eluted with a step gradient of α-D-methylmannopyranoside. Glycoproteins with high inhibiting activity were eluted with 0.1–0.2 M α-D-methylmannopyranoside and subsequently separated by ultrafiltration into a fraction with M_r greater than 30 kDa and a most active fraction with M_r = 10 – 30 kDa. SDS-PAGE of the active fraction revealed one major protein band with a M_r of about 14.5 kDa, besides several minor faintly staining bands as putative candidates for suppressor-active glycoproteins.

The preliminary characterization of the suppressor-active compound F 5 from *Mycosphaerella pinodes* pointed to a mannose containing glycopeptide as the active molecule (Oku, pers. comm; Yoshioka et al., in press). In this case, the biological activity appears to reside in the peptide moiety, which is mainly composed of serine, aspartic acid and glycine since pronase and proteinase K destroyed the suppressor activity (Oku, pers. comm.).

The active portion of suppressors found in IWF from wheat leaves as well as in spore germination fluids from *Puccinia graminis* f. sp. *tritici* are low-molecular mass compounds. These suppressors were separated from elicitor-active compounds in IWF of susceptible stem-rust infected wheat leaves by gel filtration on Sephadex G 25 or isolated from IWF of healthy wheat leaves. (Beißmann et al., in press; Fig. 1). The suppressor-active fraction contains material smaller than 5 kDa and consists of about 4% protein and 96% carbohydrates, including about 12% uronic acids. No Con A-binding proteins could be detected after SDS-PAGE of this fraction. Periodate oxidation, proteolytic digestion, heat treatment and even acid hydrolysis did not affect suppressor activity, indicating that neither carbohydrates nor proteins were the biological active sites. Gel filtration on Biogel P 2 after acid hydrolysis revealed several suppressor-active fractions eluting close to and in the inclusion volume of the column (500 daltons). Hence, it is concluded that the low-molecular mass suppressors released by acid hydrolysis are associated with molecular compounds of higher M_r.

The suppressor from spore germination fluid of *Puccinia graminis* f. sp. *tritici* appears to be very similar to the suppressors from wheat IWF described above. In fact, insensitivity towards heating, periodate, proteolytic digestion and trifluoroacetic acid again points to non proteinaceous and non-carbohydrate determinants of the suppressor activity. Like the suppressors of wheat IWF, gel filtration after acidic hydrolysis of the active compounds from spore germination fluid resulted in the isolation of suppressor-active fractions with low molecular masses.

2.4 Mode of Suppressor Action

Most models assume that the primary target of elicitor action is a receptor located on the plant cell surface (Cosio et al. 1988; Kogel et al. 1991). In contrast, many different ways how suppressors may prevent the elicitor-induced change in cellular metabolism can be suggested. This complicates the attemps to elucidate the mode of suppressor action. Suppressors may (1) impair the primary recognition events, e.g., by interfering with the elicitor-binding to its putative receptor; (2) interrupt the signal transduction, e.g., by preventing the formation of second messengers; (3) suppress gene activation, e.g., by interacting with regulatory proteins; or (4) inhibit the enzyme activities involved in defense responses, e.g., phytoalexin or lignin biosynthesis. So far, little is known about the way suppressors interfere in the resistance pathways and it is at present impossible to propose a general concept for the mode of suppressor action.

IWF from different rust fungus-plant interactions inhibit silica deposition induced after infection of bean plants with the cowpea rust fungus, but not after infection of bean plants with other rust fungi which are also defended by silica deposition (Li and Heath 1989). Silica deposition upon cowpea rust infection is probably not directly affected by suppressor action, but may be the result of impaired recognition.

The suppressor activity of apoplastic fluids from *Cladosporium fulvum*-infected tomato leaves and rust-infected bean leaves was suggested to be due to an enzymatic breakdown of the non-specific elicitor (NSE) from *Cladosporium fulvum* (Peever and Higgins 1989). This was concluded from the observation that suppressor activity was eliminated by heating. The enzymatic elicitor suppression has not yet been further characterized. However, glycohydrolytic and/or proteolytic enzymes in apoplastic fluids are possibly involved in this elicitor inactivation, as the activity of the nonspecific elicitor, previously described as a peptidogalactoglucomannan (De Wit and Kodde 1981), was reduced by in vitro digestion with α-mannosidase and proteolytic enzymes (De Wit and Roseboom 1980).

The activity of the glucan suppressor from *Phytophthora infestans* was suggested to be due to a direct interaction of suppressor and elicitor receptors on the plant side. The elicitor-induced browning and terpenoid formation in potato tuber tissue decreased when elicitor was mixed prior to its application with a microsomal fraction prepared from tubers. It was concluded that the elicitor was bound to and thereby inactivated by microsomal factors which might be receptors for the elicitor. Furthermore, the reduction of the elicitor activity could be reversed by the addition of suppressor-active glucans, leading to the conclusion that the suppressor might affect the binding between microsomal factor(s) and elicitor. In this way, a race-specific effect was monitored in that glucans from a compatible race reestablished the elicitor effect to a higher extent than glucans from an incompatible race (Doke et al. 1979).

Elicitor and suppressor from spore germination fluids of *Mycosphaerella pinodes* affected cellular metabolism in different ways, although both effectors influenced pisatin synthesis in pea plants (Yoshioka et al., in press). The suppressor markedly inhibited the ATPase located in pea plasma membranes. In contrast, the elicitor had no effect on ATPase activity. Furthermore, the suppressor was shown to inhibit the enzymes PAL and cinnamate 4-hydroxylase in vitro (Hiramatsu et al. 1986).

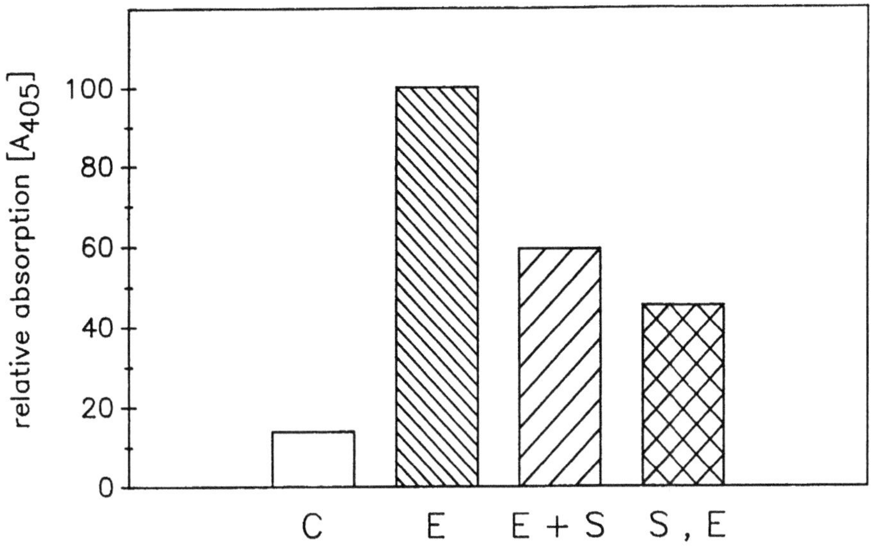

Fig. 3. Influence of suppressor isolated from IWF of stem rust (*Puccinia graminis* f. sp. *tritici*) infected wheat primary leaves (cv. Prelude) on the binding of the Pgt elicitor to wheat plasma membrane-located binding proteins. The binding of Pgt elicitor to immobilized plasma membrane (plm) absorbed onto polystyrene plates was monitored by ELISA using anti-Pgt elicitor monoclonal antibodies and phosphatase labelled anti-mouse antibodies (second antibody). After addition of p-nitrophenylphosphate, OD_{405} was measured by an ELISA reader (Engels pers. comm.). The diagram shows four modifications of the ELISA: *C*, absence of Pgt elicitor and suppressor (control); *E*, incubation of plm with Pgt elicitor, absence of suppressor; *E+S*, incubation of plm with a mixture of Pgt elicitor and suppressor; *S, E*, preincubation of plm with suppressor followed by elicitor addition

Experiments on suppressor-active material isolated from IWF of stem rust-infected and healthy wheat leaves indicated two different ways by which elicitation of HR is prevented (Beißmann et al., in press). (1) The suppressor inhibits the binding of the Pgt elicitor to a putative receptor localized in plasma membranes of wheat. Highly purified plasma membranes of wheat leaves isolated by the method of Kjellbom and Larsson (1984) were analysed for elicitor binding sites by ELISA using anti-Pgt elicitor antiserum to detect bound elicitor (Kogel and Beißmann, this Vol.). The addition of suppressor to the elicitor solution reduced the amount of Pgt elicitor bound to the plasma membranes (Fig. 3). The same result was obtained by pretreatment of the immobilized plasma membranes with suppressor solution. This indicates that the prevention of the HR in wheat plants depends on a receptor-suppressor interaction and not on a direct elicitor-suppressor interference. (2) The suppressor is a competitive inhibitor of PAL, thus preventing the synthesis of cinnamic acid and its derivatives needed for the lignification response of resistant wheat plants. By the method described elsewhere (Kogel and

Beißmann, this Vol.), enzyme extracts of wheat leaves were prepared 24 h after injection of elicitor solution. 550 µl of borate buffer (0.1 M, pH 8.8), 300 µl of enzyme extract, 50 µl of suppressor solution or borate buffer as control and 100 µl of a solution containing various amounts of L-phenylalanine were mixed in a microcuvette at 40 °C. The amount of synthesized cinnamic acid was calculated from the increasing absorption at 290 nm. Km and Vmax of PAL activity were determined in a Lineweaver-Burk plot (Fig. 4).

The double suppression of HR elicitation by inhibition of both elicitor-receptor binding and elicitor-induced enzyme activation is not yet understood. However, this mechanism may contribute to a very effective suppression of the resistance reaction in wheat plants.

3 Conclusions

The identification of suppressors in host parasite interaction is much more complicated than the identification of elicitors. The major difficulty is that suppressors are present in infected plant tissue together with elicitors which may in many cases mask suppressor activity. This situation causes problems in establishing a reliable test system for suppressor detection.

One intention of this review is to demonstrate that in many instances elicitor isolation is a prerequisite for the identification of suppressors.

The models for a specific interaction of host and pathogen presented by Heath (1981a) and Bushnell and Rowell (1981) inspired promising new approaches for suppressor identification. For instance, one aspect of Heath's model concerning the function of suppressors prompted the isolation of suppressors present in germination fluids of the wheat rust fungus as well as in rust-infected wheat plants. These suppressors were identified by a binding assay using the receptor for the race-unspecific Pgt elicitor for suppressor screening (Beißmann et al., in press).

To elucidate the mechanisms by which wheat stem rust is arrested, artificial enzyme inhibitors have been applied to resistant wheat plants and shown to prevent the plant resistance responses. For these experiments, either compounds assumed to interfere with the primary recognition between host and pathogen (Kogel et al. 1985) or inhibitors of enzymes involved in the lignin biosynthesis (Moerschbacher et al. 1990; Tiburzy and Reisener 1990) were used. However, to address specifically the question of how natural resistance is brought about, purified and well-characterized biogenic suppressors should be used. This is especially true for the examination of primary signal perception and transmission. For instance, it appears that purified suppressors are excellent tools to investigate the causal relationship of the induction of the various plant metabolic pathways and the basic resistance response definitely responsible for the restriction of a pathogen's growth.

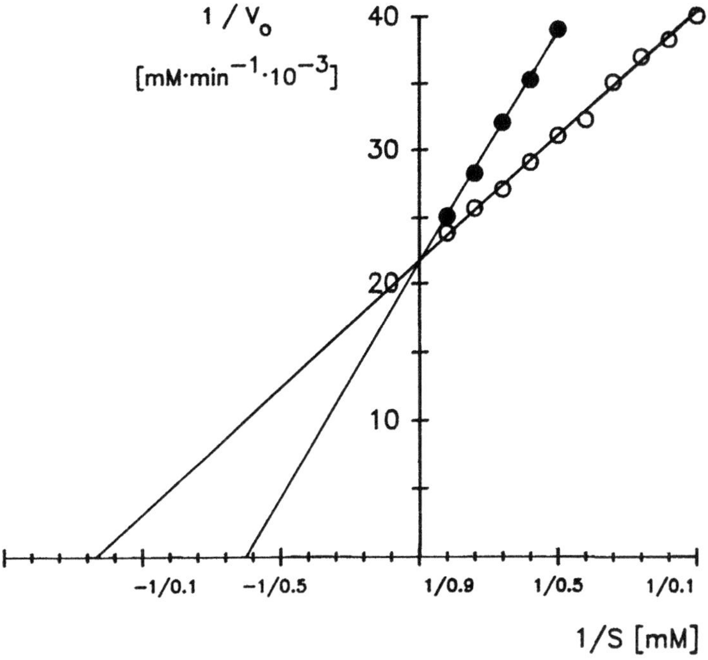

Fig. 4. Influence of suppressor isolated from IWF of stem rust (*Puccinia graminis* f. sp. *tritici*) infected wheat plants (cv. Prelude) on the velocity of the PAL reaction (Lineweaver-Burk plot). PAL assay (see Sect. 2.2) was carried out in the presence of either different concentrations of L-phenylalanine (○–○) or in the presence of different concentrations of L-phenylalanine plus 50 µl of a suppressor solution (●–●). *S* Concentration of phenylalanine; *V* rate of PAL reaction. Data strongly suggest a competitive inhibition of PAL by the suppressor

Whereas evidence has been furnished that at least some elicitors, such as the race-cultivar specific peptide elicitor from *Cladosporium fulvum*, are coded for by fungal avirulence genes (Schottens-Toma and De Wit 1988), nothing is known about the genetic basis of suppressors, e.g., if they are products of parasitic virulence genes.

In the future, application of molecular genetics may solve the problem of how the antagonistic effects of elicitors and suppressors might contribute to race-cultivar specificity in host-parasite interactions. It has already become clear, however, from the results available at present that because of the great variability by which specificity is produced, no general model of the recognition events will be capable of explaining the basis for the outcome of all the many different host-pathogen interactions.

Acknowledgement. We thank Dr. S. Archer, Imperial College, London, Dr. W. Knogge, Max Planck Institut für Züchtungsforschung, Köln and Dr. R. Pontzen, Bayer AG, Monheim for revising the manuscript.

References

Apostol I, Low PS, Heinstein P, Stipanovic RD, Altman DW (1987) Inhibition of elicitor-induced phytoalexin formation in cotton and soybean cells by citrate. Plant Physiol 84:1276–1280

Ayres AR, Ebel J, Finelli F, Berger N, Albersheim P (1976) Host-pathogen interactions. IX. Quantitative assays of elicitor activity and characterization of the elicitor present in the extracellular medium of cultures of *Phytophthora megasperma* var. *sojae*. Plant Physiol 57:751–759

Beißmann B, Engels W, Marticke KH, Reisener HJ (in press) Characterization of elicitor-active glycoproteins in apoplastic fluids of stem rust infected wheat leaves. Physiol Mol Plant Pathol

Bushnell WR, Rowell JB (1981) Suppressors of defense reactions: a model for roles in specificity. Phytopathology 71:1012–1014

Callow JA (1984) Cellular and molecular recognition between higher plants and fungal pathogens. In: Linskens HF, Heslop-Harrison J (eds) Cellular interactions. (Encycl Plant Physiol, New Ser 17). Springer, Berlin Heidelberg New York, pp 212–237

Cosio EG, Pöpperl H, Schmidt WE, Ebel J (1988) High affinity binding of fungal β-glucan fractions to soybean (*Glycine max* L.) microsomal fractions and protoplasts. Eur J Biochem 175:309–315

De Wit PJGM, Kodde E (1981) Further characterization and cultivar specificity of glycoprotein elicitors from culture filtrates and cell walls of *Cladosporium fulvum* (syn. *Fulvia fulva*). Physiol Plant Pathol 18:297–314

De Wit PJGM, Roseboom PHM (1980) Isolation, partial characterization and specificity of glycoprotein elicitors from culture filtrates, mycelium and cell walls of *Cladosporium fulvum* (syn. *Fulvia fulva*). Physiol Plant Pathol 16:391–408

De Wit PJMG, Spikman G (1982) Evidence for the occurrence of race and cultivar specific elicitors of necrosis in intercellular fluids of compatible interactions of *Cladosporium fulvum* and tomato. Physiol Plant Pathol 21:1–11

Doke N, Tomiyama K (1980) Suppression of the hypersensitive response of potato tuber protoplasts to hyphal wall components by water soluble glucans isolated from *Phytophthora infestans*. Physiol Plant Pathol 16:177–186

Doke N, Garas NA, Kuc J (1979) Partial characterization and aspects of the mode of action of a hypersensitivity inhibiting factor (HIF) isolated from *Phytophthora infestans*. Physiol Plant Pathol 15:127–140

Ellingboe AH (1976) Genetics of host parasite interactions. In: Heitefuss R, Williams PH (eds) Physiological plant pathology (Encycl Plant Physiology, New Ser 4). Springer, Berlin Heidelberg New York, pp 761–778

Flor HH (1956) The complementary genic system in flax and flax rust. Adv Genet 8:29–54

Garas NA, Doke N, Kuc J (1979) Suppression of the hypersensitive reaction in potato tubers by mycelial components from *Phytophthora infestans*. Physiol Plant Pathol 15:117–126

Heath MC (1980) Effects of infection by compatible species or injection of tissue extracts on the susceptibility of nonhost plants to rust fungi. Phytopathology 70:356–360

Heath MC (1981a) A generalized concept of host-parasite specificity. Phytopathology 71:1121–1123

Heath MC (1981b) The suppression of the development of silicon-containing deposits in French bean leaves by exudates of the bean rust fungus and extracts from bean rust-infected tissue. Physiol Plant Pathol 18:149–155

Hiramatsu M, Ichinose Y, Shiraishi T, Oku H, Ouchi S (1986) Regulation of pisatin biosynthesis in pea leaves by elicitor and suppressor produced by *Mycosphaerella pinodes*. Ann Phytopathol Soc Jpn 52:53–58

Kessmann H, Barz W (1986) Elicitation and suppression of phytoalexin and isoflavone accumulation in cotyledons of *Cicer arietinum* L. as caused by wounding and polymeric components from the fungus *Ascochyta rabiei*. J Phytopathol 117:321–335

Kjellbom P, Larsson C (1984) Preparation and polypeptide composition of chlorophyll-free plasma membranes from leaves of light-grown spinach and barley. Physiol Plant 62:501–509

Kogel G, Beißmann B, Reisener HJ, Kogel KH (1988) A single glycoprotein from *Puccinia graminis* f. sp. *tritici* cell walls elicits the hypersensitive lignification response in wheat. Physiol Mol Plant Pathol 33:173–185

Kogel G, Beißmann B, Reisener HJ, Kogel KH (1991) Specific binding of a hypersensitive lignification elicitor of *Puccinia graminis* f. sp. *tritici* to the plasma membrane from wheat (*Triticum aestivum* L.). Planta 183:164–169

Kogel KH, Schrenk F, Sharon N, Reisener HJ (1985) Suppression of the hypersensitive response in wheat stem rust interaction by reagents with affinity for wheat plasma membrane galactoconjugates. J Plant Physiol 118:343–352

Kraft B, Barz W (1985) Degradation of the isoflavone biochanine A and its glucoside conjugates by *Ascochyta rabiei*. Appl Environ Microbiol 50:45–48

Lazarovits G, Higgins VJ (1979) Biological activity and specificity of a toxin produced by *Cladosporium fulvum*). Phytopathology 69:1056–1061

Li A, Heath MC (1989) Effect of intercellular washing fluids on the interactions between bean plants and fungi nonpathogenic on beans. Can J Bot 68:934–939

Moerschbacher BM, Noll U, Gorrichon L, Reisener HJ (1990) Specific inhibition of lignification breaks hypersensitive resistance of wheat to stem rust. Plant Physiol 93:465–470

Oku H, Shiraishi T, Ouchi S (1977) Suppression of induction of phytoalexin, pisatin. Naturwissenschaften 64:643–644

Ouchi S, Oku H, Hibino C, Akiyama I (1974) Induction of accessibility to a nonpathogen by preliminary inoculation with a pathogen. Phytopathol Z 79:142–154

Peever TL, Higgins VJ (1989) Suppression of the activity of non-specific elicitor from *Cladosporium fulvum* by intercellular fluids from tomato leaves. Physiol Mol Plant Pathol 34:471–482

Schottens-Toma IMJ, De Wit PM (1988) Purification and primary structure of a necrosis-inducing peptide from the apoplastic fluids of tomato infected with *Cladosporium fulvum* (syn. *Fulvia fulvum*). Physiol Mol Plant Pathol 33:59–67

Shiraishi T, Oku H, Yamashita M, Ouchi S (1978) Elicitor and suppressor of pisatin induction in spore germination fluid of pea pathogen, *Mycosphaerella pinodes*. Ann Phytopathol Soc Jpn 44:659–665

Stumpf MA, Heath MC (1985) Cytological studies of the interactions between the cowpea rust fungus and silicon depleted French bean plants. Physiol Plant Pathol 27:369–385

Tiburzy R, Reisener HJ (1990) Resistance of wheat to *Puccinia graminis* f. sp. *tritici*: association of the hypersensitive reaction with the cellular accumulation of lignin-like material and callose. Physiol Mol Plant Pathol 36:109–120

Yamamoto Y, Oku H, Shiraishi T, Ouchi S, Koshizawa K (1986) Non-specific induction of pisatin and local resistance in pea leaves by elicitors from *Mycosphaerella pinodes*, *M. melonis* and *M. ligulicola* and the effect of suppressor from *M. pinodes*. J Phytopathol 117:136–143

Yoshioka H, Shiraishi T, Yamada T, Ichinose Y, Oku H (in press) Suppression of pisatin production and ATPase activity in pea plasma membranes by orthovanadate, verapamil and suppressor from *Mycosphaerella pinodes*. Plant Cell Physiol

Ziegler E, Albersheim P (1977) Host-parasite interactions. Extracellular invertases secreted by three races of a plant pathogen are glycoproteins which posses different carbohydrate structures. Plant Physiol 59:1104–1110

Ziegler E, Pontzen R (1982) Specific inhibition of glucan elicited glyceollin accumulation in soybeans by an extracellular mannan-glycoprotein of *Phytophthora megasperma* f. sp. *glycinea*. Physiol Plant Pathol 20:321–331

Screening for Plant Antioxidants

W. BORS, M. SARAN, and E.F. ELSTNER

1 Introduction

Chloroplasts of green plants produce oxygen by light-driven reactions and contain photodynamically active molecules which catalyze photo-oxidations. The latter processes and unavoidable leakage of electrons, resulting in transient radical formation, can be considered toxic for plants. It is thus not surprising that it was primarily plants that developed a great variety of protective substances, i.e., antioxidants against chemical insults during metabolism, aside from toxins against predatory attack. The important activity of antioxidants is to regulate, inhibit, avoid or ameliorate damaging effects connected with oxidative processes or during "oxidative stress" conditions (Sies 1991; Rice-Evans and Dormandy 1988; Halliwell and Gutteridge 1989).

Animals (and mankind) during their evolution took advantage of easy availability and established metabolic routes without the need of synthesizing their own antioxidants. This made them dependent on the uptake of such compounds; several low-molecular weight substances of plant cells with antioxidative properties thus are vitamins. Vitamin C (ascorbic acid), E (α-tocopherol) and A (β-carotene) are of special importance (Burton et al. 1985). Certain flavonoids (Cody et al. 1986; Das 1990), which seem to cooperate with ascorbate and tocopherol in cellular detoxification chains, have also been proposed to function as vitamins (vit. "P", Bentsath et al. 1937).

Selenium, as a basic requirement for the hydroperoxide-detoxification enzyme glutathione peroxidase, is also an obligatory component of nutrition with antioxidative capacities that ultimately stems from plant sources. Cofactors of some redox enzymes which activate oxygen, such as dehydrogenases or amine oxidases, e.g., riboflavin (vit. B_2) and pyrroloquinoline-quinone (methoxatin; Gallop et al. 1989) might also be listed in this context. In addition to the examples mentioned above, there is a vast amount of substances which function as antioxidants in plants but do not seem to play an essential role in the metabolism of heterotrophically living mammalian cells. Nevertheless, many of them have been utilized in traditional folk medicine for centuries, especially in China (Huang Liang 1990).

Only in the last few years has more insight been gained into the basic mechanisms underlying the functions of antioxidative compounds in the plant cell or corresponding activities as food additives or drugs. Research activities have been dramatically stimulated by the findings that many serious diseases such as cancer, atherosclerosis, arthritis, organ reperfusion problems, cataract, diabetes and several inborn errors of metabolism are causally related to activated oxygen species. In this context the question asked by Scott "Antioxidants: Can man improve on nature?" (Scott 1988) is highly relevent; screening for plant antioxidants in a wider sense means screening for processes that inhibit or ameliorate deleterious consequences of oxygen radical reactions in all sorts of cells.

The whole field of oxygen radical biochemistry and physiology of activated oxygen species developed explosively after the discovery by McCord and Fridovich (1969) that the copper protein "erythrocuprein" functioned as a superoxide dismutase (SOD). For 20 years, SOD has been utilized by numerous laboratories as a tool for identifying reactions involving superoxide (Halliwell and Gutteridge 1989). In connection with catalase and $^{\cdot}OH$ radical scavengers such as benzoate, formate or mannitol, SOD is part of the standard set for the investigation of oxidative reaction mechanisms operating via the monovalent oxygen reduction, where superoxide dismutation and finally transition metal-catalyzed peroxide reduction forming $^{\cdot}OH$ via the so-called Haber-Weiss cycle takes place (Cohen 1977).

2 Basic Mechanisms of Oxygen Radical Reactions

Processes of "radical leakage", i.e., the accidental transfer of radical intermediates from metabolically active centers to the biochemical environment, or "electron leakage", e.g., the transfer of electrons from redox cycling compounds active in detoxification of xenobiotics to acceptors with suitable redox potentials, are considered as the main sources of radical species, in vivo. Once such a radical state starts to spread through the cell it is open to interception by antioxidative mechanisms. Clearly the fate of the chain reactions depends on the chemical nature of the chain initiator, whether it was an $^{\cdot}OH$ radical, an alkoxyl or peroxyl radical and also depends on the available chain propagators. No uniform reaction pattern can be expected, but altogether the following reactions may play a role and may thus be the basis for an intervention by antioxidants:

1. Monovalent oxygen reduction, yielding superoxide:

   $$O_2 + e^- \rightarrow O_2^{\cdot -};$$

2. Superoxide dismutation:

   $$2O_2^{\cdot -} + 2H^+ \rightarrow H_2O_2 + O_2;$$

3. Peroxide reduction, yielding hydroxyl radicals:

   $$HOOH + D \rightarrow {}^{\cdot}OH + OH^{-} + D^{+};$$

   (where both reduced transition metal ions, hydro- or semiquinone-type organic molecules have been described to act as electron donors, D);

4. Degradation of organic hydroperoxides generating alkoxyl radicals:

   $$ROOH + Me^{2+} \rightarrow RO^{\cdot} + OH^{-} + Me^{3+};$$

5. Production of organic alkyl radicals via hydrogen abstraction by $^{\cdot}OH$ or by transition metal-oxygen complexes (ferryl, perferryl, cupryl compounds):

   $$RH + {}^{\cdot}OH \rightarrow R^{\cdot} + OH^{-};$$

6. Addition of ground-state molecular oxygen to these alkyl radicals and formation of peroxyl radicals:

   $$R^{\cdot} + O_2 \rightarrow ROO^{\cdot};$$

7. Formation of new alkyl radicals and hydroperoxides by chain reactions with organic H-donors:

   $$R_1H + ROO^{\cdot} \rightarrow R_1^{\cdot} + ROOH.$$

8. In addition to the above radical species and peroxides, singlet oxygen is of great importance as a potentially toxic agent. It can be produced by basically three different mechanisms of biological (medical) relevance:
   a) Photodynamic activities mediated by certain dyes, P:

   $$P + hv \rightarrow P^{*},$$
   $$P^{*} + O_2 \rightarrow {}^{1}O_2 + P;$$

   b) Reaction of hydrogen peroxide and hypochlorite in the phagosome of activated leucocytes:

   $$H_2O_2 + OCl^{-} \rightarrow Cl^{-} + {}^{1}O_2 + H_2O; \text{ or}$$

   c) Via intermediary tetroxides formed by bimolecular interaction of peroxyl radicals:

   $$ROO^{\cdot} + {}^{\cdot}OOR \rightarrow ROOOOR \rightarrow {}^{1}O_2 + 2RO^{\cdot}$$

All of the mentioned processes have been shown to occur in aerobic cells and may be detrimental for some of their designed functions; they are, on the other hand, also important for certain metabolic processes such as lipid metabolism (prostaglandin and leucotriene biosynthesis) and thus inevitable for biological functions at certain times and in certain compartments.

3 Basic Mechanisms of Inhibition of Oxidative Processes (Antioxidative Mechanisms)

Changes of reaction velocities of oxidative processes can be brought about by different mechanisms which may be measured by a variety of analytical procedures. Basically four different mechanisms can be shown to operate in vivo, either in intermediary metabolism exerting regulatory functions, or after drug application or poisoning.

3.1 Scavenging of Free Radicals

The reaction of the "scavenger", with an intermediary free radical and formation of more or less stable secondary radicals, usually leads to final products being different from the products in the absence of the scavenger. Radical scavengers mostly donate one electron to the upaired electron of the free radical, and thus reduce it. Polyphenols are very active in this respect and the radical scavenging activities of propylgallate, nordihydroguaiaretic acid, ellagic acid, flavonoids and also ascorbic acid and tocopherol are due to this function.

3.2 Quenching of Singlet Excited States

The physical transfer of energy from the primarily excited molecule to the "quencher", results in energy dissipation by light emission or as heat. It thus avoids single electron transfer reactions or direct reaction of the primarily excited molecule with critical targets. Tocopherol and carotenes are physical quenchers of excited states of pigment molecules as well as of singlet oxygen; they are usually not derivatized during this function and only after every 20th interaction with an excited-state molecule may a chemical change, i.e., an oxidation of the quencher, statistically occur.

3.3 Chelation of Transition Metals

Fenton reactions or analogous processes (reactions 3 or 4 in Sect. 2) depend on the availability of metals in catalyzing amounts. Interference with this process of catalysis by chelation would be expected to have a strong effect on the progress of the radical reaction: transition metal ions may either be activated by the chelator increasing the catalyzing potential or may be rendered inactive. The change brought about depends on the metal and the chelator (Smith et al. 1990). The most frequently discussed reaction in this respect is the iron catalysis during the so-called Haber-Weiss cycle producing the $\cdot OH$ radical (Cohen 1977), and eventually also other strongly oxidizing species such as ferryl or perferryl ions (Koppenol 1985). With most detector systems for the iron-catalyzed production of $\cdot OH$, EDTA stimulates the observable reaction, while desferrioxamine (Desferal) inactivates the iron-mediated catalysis.

3.4 Inactivation/Activation of Enzymes

An example of this process is competition with the native substrate or inactivation of the active center. Flavonoids as the most diverse class of plant antioxidants (Harborne 1988) have been shown to inhibit both prostaglandin synthetase (Moroney et al. 1988) and lipoxygenase (Yamamoto et al. 1990; Lyckander and Malterud 1990). Of particular pharmacological interest are the flavonoids in *Silybium marianum* (Campos et al. 1989; Minonzio et al. 1988).

A number of plant (and synthetic) antioxidants have also been shown to *activate* certain enzymes which divert metabolic processes to non-radical reactions (Bors 1986). A prime example is DT-diaphorase, which reduces quinones in a bivalent reaction directly to hydroquinones, thereby avoiding potentially detrimental semiquinone formation (de Long et al. 1986; Talalay et al. 1988).

Direct induction of antioxidative defense systems in plants by oxidative stress conditions should also be mentioned in this context. As shown for the induction of superoxide dismutase activity in maize after application of the O_2^--producing herbicide paraquat, it involves gene expression and differential isoenzyme synthesis (Matters and Scandalios 1986).

4 Test Systems for Screening Plant Antioxidants

All substances active in the above-mentioned categories of inhibitory processes may be addressed as "antioxidants". When testing for antioxidative properties of unknown substances, essentially two parameters are open to choice for the investigator: (1) the source of the test radicals to be used; and (2) the assay procedure.

The decision in the first case might largely depend on the type of radicals that are suspected to be responsible for certain in vivo processes. The choice of the test procedure might be greatly influenced by the technical equipment of the laboratory, e.g., availability of light measuring devices etc., and furthermore by the decision whether the kinetics of the reaction are to be followed, whether it suffices to analyze for stable end products or whether rapid screening of numerous substances is required.

4.1 Extraction of Plant Material

Antioxidants may be subdivided into water-soluble and lipid-soluble low-molecular weight compounds and proteins. These properties make it virtually impossible to extract all antioxidatively active molecules from plant material with just one procedure. In order to obtain the protein-based activities, the preparation of an acetone dry powder from liquid nitrogen-frozen plant material and subsequent extraction of the centrifugation pellet with phosphate or borate buffer has proven useful. Some plants contain high concentrations of polyphenolic compounds undergoing oxidation during extraction with aqueous media, thus causing inactivation of enzymes. In these cases, extraction in the presence of polyvinyl-pyrrolidone may be helpful. Other enzymes, such as ascorbate peroxidase, require the presence of the substrate (ascorbate) to conserve enzyme activity (Chen and Asada 1989).

For the extraction of water-soluble low-molecular weight antioxidants, pulverization of liquid nitrogen-treated plant material and subsequent exhaustive (repetitive) buffer extraction and testing of the combined extracts (see below) is the standard procedure. In order to test the hydrophobic fraction (tocopherols, carotenes), lipophilic extraction (ether, petrol ether) of the water-extracted and freeze-dried pellets may allow at least qualitative testing and analysis (GC, HPLC) for these substances and activities (Harborne 1973).

4.2 Generation Methods for Activated Oxygen Species

Perusing the following paragraphs, one should be aware of the fact that generally available methods, while being simple, are most likely unspecific. In contrast, highly specific methods such as electrochemistry or pulse radiolysis are complicated and rather uncommon, but yield the most accurate data. In this Chapter, only methods routinely available will be outlined in detail. Special attention will be paid to simple chemical systems incorporating a reductant (ascorbate, cysteine, glutathione, dihydroxyfumarate) and a transition metal chelate (Cu^{2+}, Fe^{2+}/Fe^{3+}; bipyridyl, EDTA, ADP etc.) coupled to a suitable monitoring system.

4.2.1 Physicochemical Methods

1. *Direct Photolysis*. The use of 254-nm low-pressure mercury lamps and H_2O_2 or organic (hydro)peroxides is a readily available source of $^{\cdot}OH$ or alkoxyl ($RO^{\cdot}$) radicals (Bors et al. 1984). When illuminating organic hydroperoxides to obtain $RO^{\cdot}$ radicals, $^{\cdot}OH$ radicals are generated simultaneously by homolytic scission and have to be eliminated by adding

0.1 M *tert*-butanol. In the case of di-*tert*-butylperoxide, addition of *p*-methoxyacetophenone (Griller et al. 1980) has been shown to enhance the yield of t-BuO$^\cdot$ radicals.

2. *Photodynamic Oxidation*. Sensitizer-mediated photolytic oxidation depends on the sensitizer dye used and involves either a type-II reaction (singlet oxygen formation) or belongs to type I (electron or hydrogen transfer reactions generating radicals, e.g., O_2^{-}, Foote 1976). Relatively clean sources of $^1O_2$ are obtained by illumination of 5 μM riboflavin or 5 μM Rose Bengal solutions (with white light from a 30 klx ATRALUX lamp, applied through a 10-cm water filter).

3. *Electrolytic Generation of O_2^{-}*. This provides a clean source of this radical both in aqueous and organic solvents. Counter ions such as tetrabutylammonium stabilize the radical sufficiently to achieve millimolar concentrations (Chevalet et al. 1972; Valentine et al. 1984).

4. *Radiolysis of Aqueous Solutions*. This is the most versatile method to generate any type of radical depending on the type of converting substance (Bors et al. 1984a). Both steady-state (γ-, X-irradiation) and pulse radiolysis are thus important adjuncts to any biochemical assays of antioxidative capacities. The method of pulse radiolysis, combined with kinetic spectroscopy, which is only available at specialized research centers, is probably the most powerful tool to obtain quantitative data for plant antioxidants (Bors et al. 1989, 1990).

4.2.2 Chemical Methods

1. *Fenton Reactions*. Fenton systems contain a reducing agent such as ascorbate or cysteine and a metal ion chelate such as Fe^{3+}-EDTA or Fe^{3+}-ADP; the system may be supplemented with hydrogen peroxide or organic peroxides to alter the velocity of the oxidant-producing reaction. Despite the fact that the Fenton reaction has been known since 1894 (Fenton 1894), even today the mechanism is ambiguous (Cohen 1985; Stadtman and Berlett 1988; Walling 1975) and both free $^\cdot$OH radicals, ferryl species (Koppenol 1985) or "crypto $^\cdot$OH" (Youngman 1984) have been proposed as active species. Since it is the most strongly oxidizing system, it may thus not be the best to discriminate between closely related substances. The following concentrations have been proven useful for comparing antioxidants: ascorbate (0.5 mM); EDTA (0.5 mM); $FeCl_3$ (0.5 mM); H_2O_2 (0.5 mM); in 2 ml reaction volume with 0.2 M phosphate buffer, pH 7.8.

2. *Autoxidation of Dihydroxyfumarate (DHFA)*. Both metal-catalyzed (Autor et al. 1981; Wood and Hess 1987) and horseradish peroxidase-

catalyzed reactions (Berezin et al. 1975; Dordick et al. 1986; Halliwell and de Rycker 1978) are employed. The radicals produced are most likely an unresolved mixture of carbon- and oxygen-centered radicals, capable of abstracting H-atoms from phenols.

Procedure a: DHFA (0.5 mM) is oxidized in the presence of $CuSO_4$ (0.5 mM) in 2 ml reaction volume of 0.2 M phosphate buffer, pH 7.8.

Procedure b (Dordick et al. 1986): the best yield of horseradish peroxidase-catalyzed phenol oxidation in the presence of DHFA was achieved at 4 °C, pH 5 (50 mM sodium acetate buffer) and at DHFA concentration $\leq$1 mM with 0.05 mM peroxidase present. The reaction was carried out in 250-ml Erlenmeyer flasks with a total volume of 25 ml, under continuous stirring and bubbling with O_2. DHFA was determined spectrophotometrically (ε_{314} = 3250 $M^{-1} cm^{-1}$) and catechol or hydrochinone as products of phenol oxidation by HPLC.

3. *Hemin Autoxidation*. This method has been used quite regularly to screen food antioxidants (Berner et al. 1974; Kawashima et al. 1978). "Compound I", which is formed under the described conditions, is an activated iron-oxygen-porphyrin complex, which may initiate lipid peroxidation.

Procedure (Kawashima et al. 1978): hemin solution (0.04 mM, 0.2 ml) is added to an emulsified solution of linoleic acid (0.5 ml) in phosphate buffer, pH 7 (0.1 M, 3 ml), containing 20% Tween 20 (0.5 ml), ethanol (0.1 ml) and water (0.1 ml). With antioxidant samples of 0.1 ml the total volume is 4.5 ml in the respirometer flask. The hemin-catalyzed oxidation of linoleic acid is followed by manometric determination of oxygen consumption – see also 1 and 2 above. Alternatively, 0.05 mM hemin in 10 mM NaOH is oxidized in the presence of 0.5 mM $NaClO_2$ (Elstner 1988).

4. *Azo Initiators*. Thermolysis of a number of azo compounds at 37 °C results in liberation of N_2 and alkyl radicals. The latter attach to O_2 in a diffusion-controlled reaction, forming peroxyl radicals. These in turn initiate fatty acid or lipid peroxidation (Darley-Usmar et al. 1989). Depending on the organic substitutents, both water- and lipid-soluble azo initiators may serve as steady-state sources of peroxyl radicals in the respective environment (Niki et al. 1985; Barclay 1988).

For non-spectroscopic measurements (see No. 5 below), turbid suspensions of linoleic acid (50 mM) in phosphate buffer (50 mM), pH 7.4, can be employed: for clear homogeneous solutions the concentration of linoleic acid should not exceed 1 mM. Azo initiators [azo-*bis*-isobutyronitrile, 2,2′-azo-*bis*-2-amidinopropane hydrochloride, both water-soluble, or the lipid-soluble 2,2′-azo*bis*(2,4-dimethylvaleronitrile)] are added to a final concentration of 10–50 mM into the thermostatted reaction cuvettes.

5. *Phenazine Methosulfate/NADH.* The system has originally been proposed by Nishikimi et al. (1972) as a source of activated oxygen species. Unspecificity of the method (Bors et al. 1979) and finally a detailed study of its mechanism by Richter et al. (1982) have relegated it to an obsolete method of merely historical interest.

4.2.3 Enzymatic Methods

The systems most frequently used comprise the xanthine oxidase and the NAD(P)H oxidase (diaphorase) reaction. Yet both enzymes should be used circumspectly, as they produce an undefined mixture of O_2^-, H_2O_2 and $^{\cdot}OH$ radicals (KMB fragmentation yielding ethylene can be inhibited by both SOD and catalase, whereas H_2O_2 alone cannot oxidize KMB; Elstner 1987).

1. *Xanthine Oxidase (EC 1.2.3.2) Reactions.* As applied to generate activated oxygen species, these surpass by far all other assay procedures. It is also the original test for O_2^- and SOD, introduced by McCord and Fridovich (1969) and has since been optimized (Fridovich 1985). Aside from purines such as hypoxanthine or xanthine, the enzyme also readily oxidizes simple aldehydes. Only under strict conditions (see below) is this reaction a relatively clean source for O_2^-. For screening purposes, where more reactive species than O_2^- are required, such precautions may not be necessary. To optimize O_2^- production and to avoid artifacts, the following procedure is recommended (Fridovich 1985):

The reaction mixture should contain 0.05 mM xanthine (or 10 mM acetal-dehyde) and 0.1 mM EDTA in 50 mM phosphate buffer, pH 7.8 and should be thoroughly saturated with oxygen. The 15% yield of O_2^- at pH 7.8 can be enhanced to 100% by changing the pH to 10.0. The reaction is started with the addition of 0.1 units of xanthine oxidase (350 nM). To minimize $^{\cdot}OH$ production in the adventitious Haber-Weiss reaction, replacement of EDTA by equal concentrations of DTPA or desferrioxamine and addition of catalase (100 U) is advisable.

2. *NADH Oxidases.* From the many other enzymatic reactions producing activated oxygen (Dixon and Webb 1979), autoxidizable and non-autoxidizable NAD(P)H-oxidases (diaphorases, EC 1.6.99.1) have gained some importance (Elstner et al. 1983). This is especially due to the fact that these flavoproteins are quite specific at the electron acceptor site (NADH and/or NADPH) and relatively unspecific at the electron donor site where a vast amount of substances with suitable redox potentials can be reduced. By selecting appropriate acceptors (such as quinones, e.g., anthraquinone sulfonate or juglone; bipyridylium salts, e.g., paraquat; nitroaromatic or

nitroheterocyclic compounds, e.g., nitrofurantoin) and addition of transition metal complexes, practically all above mentioned oxidants can be produced (Youngman and Elstner 1985a).

Procedure (Elstner et al. 1983): NADH dehydrogenase from pig heart or *Clostridium* (2.5 units, Boehringer Mannheim) is added to a reaction mixture, containing in 2 ml: 0.2 M phosphate buffer (pH 7.8) 0.75 mM NADH; 0.5 mM EDTA and 0.05 mM juglone (in 0.1% ethylacetate); or 0.05 mM anthraquinone sulfonate.

3. *Stimulation of Phagocytes*. To study physiological effects of plant antioxidants, this would obviously seem to be the most appropriate method, especially since purification of neutrophils from other blood constituents poses no difficulties (Berkow et al. 1983; Markert et al. 1984). However, stimulation of neutrophils can be achieved by a number of agonists, each resulting in different yields, type and kinetics of formation of the activated oxygen species (McPhail et al. 1984; Dewald et al. 1988). Known agonists are opsonized zymosan or latex beads, calcium ionophore, phorbol esters, chemotactic peptides, complement components, lectins and cytokines. Both O_2^- from the initially stimulated NADPH oxidase and OCl^- from the myeloperoxidase reaction with H_2O_2 and Cl^-, the latter inhibitable by azide, are active species, which are produced under these conditions.

4.3 Monitoring of Radical Reactions and Determination of Scavenger Activities

In principle, three different approaches can be utilized for this purpose:

1. Direct monitoring of one of the reaction intermediates by kinetic methods;
2. Determination of scavenging efficiencies by the competition method;
3. Determination of stable reaction products.

From the great variety of assay procedures available (Bors et al. 1982), kinetic measurements by spectrophotometry or by following chemiluminescence have proven to be rather simple and reliable; for product analysis, the standard techniques of GC and HPLC have been used most often. In the following, only those spectroscopic assay procedures are discussed, which are applicable to large-scale screening tests. The highly informative but not generally available ESR (Borg 1976), combined with spin trapping (Pou et al. 1989; Thornalley 1986) and pulse radiolysis methods (Buxton 1981; Bors et al. 1984a) are merely mentioned in this context and will not be dealt with in detail. What should be kept in mind, however, is the fact that there is generally a trade-off between simplicity and accuracy, when choosing certain assay procedures.

If a postulated mechanism involves the intermediacy of $^{\cdot}OH$ radicals, superoxide anions, organic oxygen radicals or transition metals, as shown by the effects of scavengers or chelators, the validity of these assignments must be verified. Enzyme experiments should be repeated using heat-inactivated or otherwise impaired compounds (e.g., cyanide- or CO-poisoned heme groups), chelators with different properties should be used and a variety of classical $^{\cdot}OH$ scavengers such as formate, benzoate, mannitol, ethanol, etc. should be applied independently.

4.3.1 Spectroscopy

A number of simple spectroscopic or colorimetric methods exist to determine O_2^- generation:

1. Cytochrome c reduction at 550 nm (Kuthan et al. 1982);
2. NBT reduction to the blue formazan (McCord et al. 1977);
3. Adrenaline/epinephrine oxidation at alkaline pH to the red adrenochrome, measurement at 480 nm (McCord and Fridovich 1969) or at 310 nm (Bors et al. 1978; Sun and Zigman 1978);
4. Hydroxylamine oxidation and subsequent determination of nitrite (Elstner and Heupel 1976a,b; Bors et al. 1977; Bielski et al. 1984).

Yet since O_2^- itself is not a very reactive species, it seems of little use to monitor O_2^- reactions when screening for antioxidants. Of the other radicals produced in the previously described methods, alkoxyl, peroxyl radicals or ferryl species are optimal for screening tests, as they are less reactive and thus more discriminating than $^{\cdot}OH$ radicals.

1. *Reactions with Stable Radicals*. The property of antioxidants to donate H atoms or electrons to free radicals is not limited to highly reactive unstable oxygen radicals. Stable radicals such as 2.2-diphenyl-1-picryl-hydrazyl (Ratty et al. 1988) or galvinoxyl (Nishimura et al. 1978; Tsuchiya et al. 1985) react likewise. Depletion or shifts of their absorption bands at 350 nm and at 430 nm, respectively, can easily be monitored even without the need for ESR spectroscopy.

Procedure for galvinoxyl (Bors and Kreileder, unpublished): to aqueous solutions containing 0.2 M *tert*-butanol and 0.1 mM *p*-methoxyacetophenone at a final volume of 3 ml, aliquots of galvinoxyl from a freshly prepared stock solution in 1 M *tert*-butanol are added to a final concentration of 0.01–0.1 mM. As radical sources either hydrogen peroxide (1 mM), *tert*-butyl-hydroperoxide (25 mM), di-*tert*-butylperoxide (1 mM) or 13-hydroperoxylinoleic acid (0.03 mM) can be used. After the addition of the peroxides (1–10 μl samples) and illumination with 254-nm light from a low-pressure mercury lamp, spectral scans at 30-s intervals show a depletion of the 430-nm absorption and an increase at 310 nm.

2. *Bleaching of p-Nitrosodimethylaniline (p-NDA)*. The easily monitored disappearance of the absorption at 440 nm ($\varepsilon = 34.4\,mM^{-1}\,cm^{-1}$) can be achieved both by reducing and oxidizing radicals (Kraljic and Trumbore 1965). It is thus a sensitive but rather unspecific probe and can be combined with a number of radical sources (Bors et al. 1979), even singlet oxygen sources in the presence of imidazole (Kraljic and El Mohsni 1978).

Procedure: Antioxidants, i.e., radical scavengers, are included at various concentrations in 3 ml of a neutral aqueous solution of 0.05 mM *p*-NDA in quartz cuvettes (initial absorbance 1.7 absorption units). The reaction is initiated by the addition of the radical source and the rate of disappearance of the 440-nm absorption is evaluated by competition kinetics (Bors et al. 1985).

3. *Bleaching of Carotenoids*. Inhibition of the bleaching of β-carotin in lipid solutions or in lard was an early test for food antioxidants (Pratt 1965). In these cases, lipid peroxyl radicals were the most likely intermediates and the studies were limited to lipid-soluble antioxidants.

After studying the sensitivity of the water-soluble carotenoid crocin towards different types of biologically (Montalbini et al. 1978), radiolytically, photolytically or chemically produced radicals (Saran et al. 1980; Bors et al. 1982a), we developed a quantitative method for screening antioxidants, using this compound (Bors et al. 1984b). The bleaching of crocin by photolytically produced t-BuO˙ or LO˙ radicals (see Sect. 4.2.2, No. 1) can be competitively inhibited by substances capable of scavenging the alkoxyl radicals. A concentration-dependent plot of the inhibitory effect results in relative reaction rates (Bors et al. 1985) which can be converted into absolute rate constants, since the absolute bleaching rate of crocin by t-BuO˙ is known (Erben-Russ et al. 1987).

Procedure: aqueous solutions of crocin (0.05 mM, $\varepsilon_{440} = 133.5\,mM^{-1}\,cm^{-1}$) in quartz cuvettes below pH 9, in the presence of 0.1 M t-BuOH and 1 mM *tert*-butyl-hydroperoxide (or di-*tert*-butylperoxide) are bleached after photolysis of the peroxides with a low-pressure mercury lamp (5-min exposure). Increasing volumes of aliquots from a concentrated stock solution of the test substance are added and the respective bleaching rates are evaluated by competition kinetics (Bors et al. 1985).

4. *Chemiluminescence Assays*. Due to their high sensitivity, they have been preferred in biological systems, especially for phagocytic cells (Müller-Peddinghaus and Wurl 1987; see also Sect. 4.2.3, No. 3), but also for the xanthine oxidase reaction (Betts 1985) and during lipid peroxidation (Iwaoka et al. 1987; Yamamoto and Ames 1987). Using different sensitizers

such as luminol (rather unspecific), lucigenin (specific for O_2^-) or more recently fluorescent proteins (phycoerythrin for peroxyl radicals, DeLange and Glazer 1989; polynoidin for O_2^-, Colepicolo et al. 1990), radical selectivity and sensitivity can be varied. The sensitivity of the lucigenin assay could be even further enhanced after addition of high concentrations of Triton X-100 (Storch and Ferber 1988).

Because of the differential requirements depending on the radical sources, sensitizer, type of luminometer, etc. no procedural details can be given here. The interested reader is referred to the cited literature.

5. *Fatty Acid/Lipid Peroxidation Assays*. Aside from the rather unreliable, but most frequently used thiobarbituric acid assay for malondialdehyde (and similar products; Pryor et al. 1976) or the even less informative comparison of the peroxide value (POV), which have occasionally been used for screening tests (Pratt 1965; Aoyama et al. 1982), more accurately controlled assays have been developed. Most comprise autoxidizing systems of linoleic acid, either slow uncatalyzed (Tanizawa et al. 1983) or initiated by azo compounds (see 4.2.1). Monitoring either diene conjugation at 234 nm (ε = 28 000 $M^{-1} cm^{-1}$; Niki et al. 1985), O_2 consumption (Darley-Usmar et al. 1989), chemiluminescence (see above), GC determination of ethane (Konze and Elstner 1978) or HPLC analysis of hydroperoxide stereoisomers (see Sect. 4.3.2, No. 2), these assays are sufficiently reliable for standard screening tests.

4.3.2 Chromatography

Such methods may be more time-consuming than spectroscopic methods, but ultimately provide more information (yields and/or identity of products) and are more reliable.

GC analysis of Ethylene Production. The plant hormone ethylene is derived under in vivo conditions from aminocyclopropane carboxylic acid (ACC) via an almost completely resolved pathway (Yang and Hoffman 1984). It can be produced by wounding (Elstner and Konze 1976) or after attack of peroxyl radicals, i.e., in lipid peroxidation (Gardner and Newton 1987).

Under in vitro conditions both ACC itself (Elstner 1988), methionine (Youngman and Elstner 1985b), or the oxidative degradation products ketomercaptobutyrate (KMB) or methional (Cohen 1977) are sources of ethylene (the sensitivity/specificity of radical attack in declining order).

Procedure: as ethylene sources KMB or methionine (2.5 mM) can be used with a 30-min reaction time in Fernbach vessels with gas-tight serum rubber stoppers (Youngman and Elstner 1985b). GC-analysis with flame ionization detector (Varian) is performed with aluminum oxide column 2 feet × 1/8 in.; column temperature, 60 °C; injector temp., 250 °C; detector temp., 250 °C; carrier gas, N_2; retention time for ethylene correlated with authentic standard.

HPLC-Analysis of Fatty-Acid Oxidation Products. This method has been mentioned previously (see 4.3.1, No. 5), yet it is worthwhile to be pointed out again. As used by Porter et al. (1980) or Schieberle et al. (1984), the analysis of stereoisomers of hydroperoxides provides not only a detailed picture of the initial steps in the autoxidation of polyunsaturated fatty acids (PUFA), but also absolute rate constants for H-abstraction reactions.

4.3.3 Other Methods

1. *Polarography*. Oxygen consumption during lipid peroxidation can be conveniently monitored using a Clark electrode (Darley-Usmar et al. 1989). Similarly, H_2O_2 formation as an indication of O_2^- dismutation can be determined polarographically from O_2 uptake and subsequent release after catalase addition. Vice versa the method can be used to determine catalase activity (del Rio et al. 1977).

2. *Recording of Oxygen Pressure*. As an alternative to polarography, the groups of Ingold (Burton and Ingold 1986) and Niki (Niki et al. 1985) have introduced a more sensitive method to determine O_2 consumption during fatty acid or lipid peroxidation. After initiation of the reaction with azo initiators (see Sect. 4.2.1, No. 4) in a closed vessel, pressure-sensitive transducers allow continuous monitoring of oxygen consumption. In the presence of antioxidants, concentration- and/or activity-dependent lag periods can be quantitated.

3. *Measurement of ^{14}C-CO_2*. In turbid solutions, decarboxylation of 1-^{14}C-labelled α-ketoacids after attack of oxidants such as H_2O_2 has proven useful (Elstner and Heupel 1973, 1976a). However, such measurements are by nature quite time-consuming and do not lend themselves to rapid screening tests. They are thus confined to those substances, which because of their strong absorption or insolubility cannot be otherwise investigated.

References

Aoyama T, Nakakita Y, Nakagawa M, Sakai H (1982) Screening for antioxidants of microbial origin. Agric Biol Chem 46:2369–2371

Autor AP, McLennan G, Fox AW (1981) Oxygen free radicals generated by dihydroxyfumarate and ionizing radiation: cytotoxic effect on isolated pulmonary macrophages. In: Bhatnagar RS (ed) Molecular basis of environmental toxicity. Ann Arbor Sci Press, Ann Arbor, pp 51–67

Barclay LRC (1988) The cooperative antioxidant role of glutathione with a lipid-soluble and a water-soluble antioxidant during peroxidation of liposomes initiated in the aqueous phase and in the lipid phase. J Biol Chem 263:16138–42

Bentsath A, Rusznyak S, Szent-Györgyi A (1937) Vitamin P. Nature 139:326–327

Berezin IV, Ugarova NN, Dmitrieva-Getling MP, Kershengol'ts BM (1975) The kinetics and mechanism of action of horseradish peroxidase in the oxidation of dihydroxyfumaric acid by atmospheric oxygen. Biochemistry USSR 40:397–403

Berkow RL, Tzeng DY, Williams LV, Baehner RL (1983) The comparative responses of human polymorphonuclear leukocytes obtained by counterflow centrifugal elutriation and Ficoll-Hypaque density centrifugation. I. Resting volume, stimulus-induced superoxide production, and primary and specific granule release. J Lab Clin Med 102:732–736

Berner DL, Conte JA, Jacobson GA (1974) Rapid method for determining antioxidant activity and fat stability. J Am Oil Chem Soc 51:292–296

Betts WH (1985) Detecting oxy radicals by chemiluminescence. In: Greenwald RA (ed) CRC handbood of methods for oxygen radical research. CRC Press, Boca Raton, pp 197–201

Bielski BHJ, Arudi RL, Cabelli DE, Bors W (1984) Reevaluation of the reactivity of hydroxylamine with O_2^-/HO_2. Anal Biochem 142:207–209

Borg DC (1976) Applications of electron spin resonance in biology. In: Pryor WA (ed) Free radicals in biology, vol I, Acad Press, New York, pp 69–174

Bors W, Lengfelder E, Saran M, Michel C, Fuchs C, Frenzel C (1977) Oxidation of hydroxylamine to nitrite as an assay for the combined presence of superoxide anions and hydroxyl radicals. Biochem Biophys Res Comm 75:973–979

Bors W, Michel C, Saran M, Lengfelder E (1978) The involvement of oxygen radicals during the autoxidation of adrenalin. Biochim Biophys Acta 540:162–172

Bors W, Michel C, Saran M (1979) On the nature of biochemically generated hydroxyl radicals. Studies using the bleaching of p-nitrosodimethylaniline as a direct assay method. Eur J Biochem 95:621–627

Bors W, Saran M, Michel C (1982a) Radical intermediates involved in the bleaching of the carotenoid crocin. Hydroxyl radicals, superoxide anion and hydrated electrons. Int J Radiat Biol 41:493–501

Bors W, Saran M, Michel C (1982b) Assays of oxygen radicals: methods and mechanisms. In: Oberley LW (ed) Superoxide dismutase, vol II. CRC Press, Boca Raton, pp 31–62

Bors W, Saran M, Michel C, Tait D (1984a) Formation and reactivities of oxygen free radicals. In: Breccia A, Greenstock CL, Tamba M (eds) Advances on oxygen radicals and radioprotectors. Lo Scarabeo Edit Scient, Bologna, pp 13–27

Bors W, Michel C, Saran M (1984b) Inhibition of the bleaching of the carotenoid crocin. A rapid test for quantifying antioxidant activity. Biochim Biophys Acta 796:312–319

Bors W, Michel C, Saran M (1985) Determination of kinetic parameters of oxygen radicals by competition studies. In: Greenwald RA (ed) CRC handbook of methods for oxygen radical research. CRC Press, Boca Raton, pp 181–188

Bors W (1986) Bedeutung und Wirkungsweise von Antioxidantien. In: Elstner EF, Bors W, Wilmanns W (eds) Reaktive Sauerstoff Spezies in der Medizin. Springer, Berlin Heidelberg New York, pp 161–183

Bors W, Langebartels C, Michel C, Sandermann H (1989) Polyamines as radical scavengers and protectants against ozone damage. Phytochemistry 28:1589–95

Bors W, Heller W, Michel C, Saran M (1990) Flavonoids as antioxidants. Determination of radical scavenging efficiencies. In: Packer L, Glazer AN (eds) Oxygen radicals in biological systems. B. Oxygen radicals and antioxidants (Methods in Enzymology, vol 186). Academic Press, New York, pp 343–355

Burton GW, Ingold KU (1986) Vitamin E: application of the principles of physical organic chemistry to the exploration of its structure and function. Acc Chem Res 19:194–201

Burton GW, Foster DO, Perly B, Slater TF, Smith ICP, Ingold KU (1985) Biological antioxidants. Philos Trans R Soc Lond B311:565–578

Buxton GV (1981) Pulse radiolysis studies. Inorg React Mechan 7:106–124
Campos R, Garrido A, Guerra R, Valenzuela A (1989) Silybin dihemisuccinate protects against GSH depletion and lipid peroxidation induced by acetaminophen on rat liver. Planta Med 55:417–419
Chen GX, Asada K (1989) Ascorbate peroxidase in tea leaves: occurrence of two isoenzymes and the difference in their enzymatic and molecular properties. Plant Cell Physiol 30:987–998
Chevalet J, Rouelle F, Gierst L, Lambert JP (1972) Electrogeneration and some properties of the superoxide ion in aqueous solutions. J Electroanal Chem 39:201–216
Cody V, Middleton E, Harborne JB (eds) (1986) Plant flavonoids in biology and medicine – biochemical, pharmacological and structure-activity relationships. Alan R Liss, New York
Cohen G (1977) In defense of Haber-Weiss. In: Michelson AM, McCord JM, Fridovich I (eds) Superoxide and superoxide dismutases. Academic Press, London, pp 317–322
Cohen G (1985) The Fenton reaction. In: Greenwald RA (ed) CRC handbook of methods for oxygen radical research. CRC Press, Boca Raton, pp 55–64
Colepicolo P, Camarero VCPC, Nicolas MT, Bassot JM, Karnovsky ML, Hastings JW (1990) A sensitive and specific assay for superoxide anion released by neutrophils or macrophages based on bioluminescence of polynoidin. Anal Biochem 184:369–374
Darley-Usmar VM, Hersey A, Garland LG (1989) A method for the comparative assessment of antioxidants as peroxyl radical scavengers. Biochem Pharmacol 38:1465–69
Das NP (ed) (1990) Flavonoids in biology and medicine. III. Current issues in flavonoid research. Nat Univ Singapore Press, Singapore
DeLange RJ, Glazer AN (1989) Phycoerythrin fluorescence-based assay for peroxy radicals: a screen for biologically relevant protective agents. Anal Biochem 177: 300–306
Dewald B, Thelen M, Baggiolini M (1988) Two transduction sequences are necessary for neutrophil activation by receptor agonists. J Biol Chem 263:16179–84
Dixon M, Webb EC (eds) (1979) Enzymes, 3rd edn. Longman, London
Dordick JS, Klibanov AM, Marletta MA (1986) Horseradish peroxidase catalyzed hydroxylations: mechanistic studies. Biochemistry 25:2946–51
Elstner EF (1987) Metabolism of activated oxygen species. In: Stumpf PK, Conn EE (eds) The biochemistry of plants – a comprehensive treatise. Acad Press, San Diego, pp 253–315
Elstner EF (1988) Heme-activated oxidations by the chlorite-oxygen complex "TCDO" (Oxoferin^R). Z Naturforsch 43c: 893–902
Elstner EF, Heupel A (1973) On the decarboxylation of α-keto acids by isolated chloroplasts. Biochim Biophys Acta 325:182–188
Elstner EF, Heupel A (1976a) Formation of hydrogen peroxide by isolated cell walls from horseradish (*Armoracia lapathifolia* – Gilib.). Planta 130:175–183
Elstner EF, Heupel A (1976b) Inhibition of nitrite formation from hydroxylammoniumchloride: a simple assay for superoxide dismutase. Anal Biochem 70:616–620
Elstner EF, Konze JR (1976) Effect of point freezing on ethylene and ethane production by sugar beet leaf disks. Nature 263:351–352
Elstner EF, Youngman RJ, Oßwald W (1983) Superoxide dismutase. In: Bergmeyer J, Graßl M (eds) Methods of enzymatic analysis, 3rd edn. Verlag Chemie, Weinheim, pp 293–302
Erben-Russ M, Michel C, Bors W, Saran M (1987) Absolute rate constants of alkoxyl radical reactions in aqueous solutions. J Phys Chem 91:2362–65
Fenton HJH (1894) Oxidation of tartaric acid in the presence of iron. J Chem Soc 65: 899–965
Foote CS (1976) Photosensitized oxidation and singlet oxygen: consequences in biological systems. In: Pryor WA (ed) Free radicals in biology, vol II. Academic Press, New York, pp 85–133
Fridovich I (1985) Xanthine oxidase. In: Greenwald RA (ed) CRC handbook of methods for oxygen radical research. CRC Press, Boca Raton, pp 51–53

Gallop PM, Paz MA, Flueckiger R, Kagan HM (1989) PQQ, the elusive coenzyme. Trends Biochem Sci 14:343–346
Gardner HW, Newton JW (1987) Lipid hydroperoxides in the conversion of 1-aminocyclopropane-1-carboxylic acid to ethylene. Phytochemistry 26:621–626
Griller D, Ingold KU, Scaiano JC (1980) The sensitized photodecomposition of di-*tert*-butyl peroxide. A simple method for increasing radical concentrations in certain EPR experiments. J Magn Reson 38:169–171
Halliwell B, de Rycker J (1978) Superoxide and peroxidase-catalysed reactions. Oxidation of dihydroxyfumarate, NADH and dithiothreitol by horseradish peroxidase. Photochem Photobiol 28:757–763
Halliwell B, Gutteridge JMC (1989) Free radicals in biology and medicine, 2nd edn. Clarendon Press, Oxford
Harborne JB (1973) (ed) Phytochemical methods. Chapman & Hall, London
Harborne JB (1988) (ed) The flavonoids. Advances in research since 1980. Chapman & Hall, London
Huang Liang (1990) Biologically active flavonoids in Chinese traditional medicine. In: Das NP (ed) Flavonoids in biology and medicine. III. Current issues in flavonoid research. Nat Univ Singapore Press, Singapore, pp 537–550
Iwaoka T, Tabata F, Takahashi T (1987) Lipid peroxidation and lipid peroxide detected by chemiluminescence. Free Radical Biol Med 3:329–333
Kawashima K, Itoh H, Chibata I (1978) Hemin-catalyzed peroxidation test for rapid evaluation of antioxidant activity. Chem Pharm Bull 26:3558–3562
Konze JR, Elstner EF (1978) Ethane and ethylene formation by mitochondria as indication of aerobic lipid degradation in response to wounding of plant tissue. Biochim Biophys Acta 528:213–221
Koppenol WH (1985) The reaction of ferrous EDTA with hydrogen peroxide: evidence against hydroxyl radical formation. J Free Radical Biol Med 1:281–285
Kraljic I, El Moshni S (1978) A new method for the detection of singlet oxygen in aqueous solutions. Photochem Photobiol 28:577–581
Kraljic I, Trumbore CN (1965) *p*-Nitrosodimethylaniline as a hydroxyl radical scavenger in radiation chemistry. J Am Chem Soc 87:2547–2550
Kuthan H, Ullrich V, Estabrook RW (1982) A quantitative test for superoxide radicals produced in biological systems. Biochem J 203:551–558
de Long MJ, Prochaska HJ, Talalay P (1986) Induction of NAD(P)H: quinone reductase in murine hepatoma cells by phenolic antioxidants, azo dyes and other chemoprotectors: a model system for the study of anticarcinogens. Proc Natl Acad Sci USA 83:787–797
Lyckander JM, Malterud KE (1990) Lipoxygenase inhibiting flavonoids from *Orthosiphon spicatus*. In: Das NP (ed) Flavonoids in biology and medicine. III. Current issues in flavonoid research. Nat Univ Singapore Press, Singapore, pp 469–474
Markert M, Andrews PC, Babior BM (1984) Measurement of O_2^- production by human neutrophils. The preparation and assay of NADPH oxidase-containing particles from human neutrophils. In: Packer L (ed) Oxygen radicals in biological systems (Methods in enzymology, vol 105). Academic Press, New York, pp 358–365
Matters GL, Scandalios JG (1986) Effect of the free radical generating herbicide paraquat on the expression of the superoxide dismutase genes in maize. Biochim Biophys Acta 882:29–38
McCord JM, Fridovich I (1969) Superoxide dismutase. An enzymic function for erythrocuprein (hemocuprein). J Biol Chem 244:6049–55
McCord JM, Crapo JD, Fridovich I (1977) Superoxide dismutase: a review of methodology. In: Michelson AM, McCord JM, Fridovich I (eds) Superoxide and superoxide dismutases. Academic Press, London, pp 11–17
McPhail LC, Clayton CC, Snyderman R (1984) The NADPH oxidase of human polymorphonuclear leukocytes. Evidence for regulation by multiple signals. J Biol Chem 259:5768–5775

Minonzio F, Venegoni E, Ongari AM, Ciani D, Capsoni F (1988) Modulation of human polymorphonuclear leukocyte function by the flavonoid silybin. Int J Tissue React 10:223–231

Montalbini P, Koch F, Burba M, Elstner EF (1978) Increase in lipid-dependent carotene destruction as compared to ethylene formation and chlorophyllase activity following mixed infection of sugar beet (*Beta vulgaris* L.) with beet yellow virus and beet mild yellowing virus. Physiol Plant Pathol 12:211–223

Moroney MA, Alcaraz MJ, Forder RA, Carey F, Hoult JRS (1988) Selectivity of neutrophil 5-lipoxygenase and cyclo-oxygenase inhibition by an anti-inflammatory flavonoid glycoside and related aglycone flavonoids. J Pharm Pharmacol 40:787–792

Müller-Peddinghaus R, Wurl M (1987) The amplified chemiluminescence test to characterize antirheumatic drugs as oxygen radical scavengers. Biochem Pharmacol 36: 1125–32

Niki E, Kawakami A, Yamamoto Y, Kamiya Y (1985) Oxidation of lipids. VIII. Synergistic inhibition of oxidation of phosphatidylcholine liposome in aqueous dispersion by vitamins E and C. Bull Chem Soc Jpn 58:1971–75

Nishikimi M, Rao NA, Yagi K (1972) The occurrence of superoxide anion in the reaction of reduced phenazine methosulfate and molecular oxygen. Biochem Biophys Res Comm 46:849–854

Nishimura N, Okahashi K, Yukutomi T, Fujiwara A, Kubo S (1978) Substituent and solvent effects on the reaction between galvinoxyl and phenols. Aust J Chem 31: 1201–08

Porter NA, Weber BA, Weenen H, Khan JA (1980) Autoxidation of polyunsaturated lipids. Factors controlling the stereochemistry of product hydroperoxides. J Am Chem Soc 102:5597–5601

Pou S, Hassett DJ, Britigan BE, Cohen MS, Rosen GM (1989) Problems associated with spin trapping oxygen-centered free radicals in biological systems. Anal Biochem 177:1–6

Pratt DE (1965) Lipid antioxidants in plant tissue. J Food Sci 30:737–741

Pryor WA, Stanley JP, Blair E (1976) Autoxidation of polyunsaturated fatty acids: II. A suggested mechanism for the formation of TBA-reactive materials from prostaglandin-like endoperoxides. Lipids 11:370–379

Ratty AK, Sunamoto J, Das NP (1988) Interaction of flavonoids with 1,1-diphenyl-2-picrylhydrazyl free radical, liposomal membranes and soybean lipoxygenase-1. Biochem Pharmacol 37:989–995

Rice-Evans C, Dormandy T (eds) (1988) Free radicals: chemistry, pathology and medicine. Richelieu Press, London

Richter HW, Fetrow MA, Lewis RE, Waddell WH (1982) Reactive species produced by the 5-methylphenazinium methyl sulfate/reduced β-nicotinamide adenine dinucleotide/oxygen system in the hydroxylation of benzoic acid. J Am Chem Soc 104:1666–71

del Rio LA, Gomez Ortega M, Leal Lopez A, Lopez Gorge J (1977) A more sensitive modification of the catalase assay with the Clark oxygen electrode. Application to the kinetic study of the pea leaf enzyme. Anal Biochem 80:409–415

Saran M, Michel C, Bors W (1980) The bleaching of crocin by oxygen radicals. In: Bannister JV, Hill HAO (eds) Chemical and biochemical aspects of superoxide and superoxide dismutase. Elsevier/North-Holland, New York, pp 38–44

Schieberle P, Grosch W, Firl J (1984) Photolysis of unsaturated fatty acid hydroperoxides. In: Bors W, Saran M, Tait D. (eds) Oxygen radicals in chemistry and biology. de Gruyter, Berlin, pp 257–265

Scott G (1988) Antioxidants: can man improve on nature? In: Rice-Evans C, Dormandy T (eds) Free radicals: chemistry, pathology and medicine. Richelieu Press, London, pp 103–130

Sies H (1991) (ed) Oxidative stress. Academic Press, London

Smith JB, Cusumano JC, Babbs CF (1990) Quantitative effects of iron chelators on hydroxyl radical production by the superoxide-driven Fenton reaction. Free Radical Res Comm 8:101–106

Stadtman ER, Berlett BS (1988) Fenton chemistry revisited: amino acid oxidation. In: Simic MG, Taylor KA, Ward JF, von Sonntag C (eds) Oxygen radicals in biology and medicine (Basic Life Sciences, vol 49). Plenum, New York, pp 131–136
Storch J, Ferber E (1988) Detergent-amplified chemiluminescence of lucigenin for determination of superoxide anion production by NADPH oxidase and xanthine oxidase. Anal Biochem 169:262–267
Sun M, Zigman S (1978) An improved spectrophotometric assay for superoxide dismutase based on epinephrine autoxidation. Anal Biochem 90:81–89
Talalay P, De Long MJ, Prochaska HJ (1988) Identification of a common chemical signal regulating the induction of enzymes that protect against chemical carcinogenesis. Proc Natl Acad Sci USA 85:8261–65
Tanizawa H, Sazuka Y, Komatsu A, Toda S, Takino Y (1983) A new efficacy test of antioxidants based on air-oxidation of linoleic acid. Chem Pharm Bull 31:4139–43
Thornalley PJ (1986) Theory and biological applications of the ESR technique of spin trapping. Life Chem Rep 4:57–112
Tsuchiya J, Yamada T, Niki E, Kamiya Y (1985) Interaction of galvinoxyl radical with ascorbic acid, cysteine and GSH in homogeneous solution and in aqueous dispersion. Bull Chem Soc Jpn 58:326–330
Valentine JS, Miksztal AR, Sawyer DT (1984) Methods for the study of superoxide chemistry in nonaqueous solutions. In: Packer L (ed) Oxygen radicals in biological systems (Methods in enzymology, vol 105). Academic Press, New York, pp 71–81
Walling C (1975) Fenton's reagent revisited. Acc Chem Res 8:125–131
Wood MA, Hess ML (1987) The effects of dihydroxyfumarate on isolated rabbit papillary muscle function: evidence for an iron dependent non-hydroxyl radical mechanism. Mol Cell Biochem 78:161–167
Yamamoto S, Ueda N, Hada T, Horie T (1990) Flavonoids as 5-lipoxygenase inhibitors. In: Das NP (ed) Flavonoids in biology and medicine. III. Current issues in flavonoid research. Nat Univ Singapore Press, Singapore, pp 435–446
Yamamoto Y, Ames BN (1987) Detection of lipid hydroperoxides and H_2O_2 at picomole levels by an HPLC and isoluminol chemiluminescence assay. Free Radical Biol Med 3:359–361
Yang SF, Hoffman NE (1984) Ethylene biosynthesis and its regulation in higher plants. Annu Rev Plant Physiol 35:155–189
Youngman RJ (1984) Oxygen activation: is the hydroxyl radical always biologically relevant? Trends Biochem Sci 9:280–283
Youngman RJ, Elstner EF (1985a) Generation of active oxygen species by simple enzymatic redox systems. In: Greenwald RA (ed) CRC handbook of methods for oxygen radical research. CRC Press, Boca Raton, pp 105–110
Youngman RJ, Elstner EF (1985b) Ethylene formation from methionine in the presence of pyridoxal phosphate. In: Greenwald RA (ed) CRC handbook of methods for oxygen radical research. CRC Press, Boca Raton, pp 165–168

Methods for the Analysis of Isoprene Emission from Leaves

A.J. Hills, R. Fall, and R.K. Monson

1 Introduction

Isoprenoids have been the subject of numerous studies in the biological sciences due to their role in protecting plants from herbivory, and in the atmospheric sciences due to their participation in many tropospheric chemical reactions. These hydrocarbon molecules are constructed from isoprene units produced in the mevalonic acid (MVA) pathway of plants. The enzymes of this pathway have been found in the cytoplasmic, mitochondrial, and most recently the chloroplastic compartments (Gray 1987; Kleinig 1989). It has been hypothesized that the synthesis of isoprenoid compounds through the MVA pathway is associated with the emission of isoprene from plants (Monson and Fall 1989; Loreto and Sharkey 1990). At this time, it is not known exactly how or why this isoprene is produced, but studies of its emission characteristics may prove to be a useful means of studying dynamics of the MVA pathway in plants as it relates to the synthesis of higher-order isoprenoids.

Studies of isoprene emission from plants also have value in identifying the role of hydrocarbons in catalyzing important reactions of the earth's troposphere. It is known that the tropospheric concentration of isoprene has an important role in regulating the concentration of hydroxyl radical (Zimmerman et al. 1988). Reactions between isoprene and hydroxyl radical can in turn influence tropospheric concentrations of ozone (O_3), carbon monoxide (CO), and methane (CH_4) (Trainer et al. 1987; Chameides et al. 1988; Zimmerman et al. 1988).

Isoprene emissions from plants are known to be linked to the photosynthetic processes (Sanadze 1969; Monson and Fall 1989; Loreto and Sharkey 1990; Monson et al. 1991). Several recent studies have suggested that the link between these processes could provide significant new insight into the physiological and environmental roles of isoprene emission. In this Chapter, we present a detailed discussion of the methods used to detect and quantify isoprene emissions from leaves, as well as ways to link such quantification to the photosynthetic processes of plants.

2 Historical Perspectives on the Analysis of Isoprene Emission

The earliest studies of isoprene emission from leaves were conducted using enclosed leaves or leaf discs in sealed vials (e.g., Sanadze 1969; Rasmussen and Jones 1973). With this technique, a small aliquot of gas (0.2–0.4 cm^3) was withdrawn from the vial headspace using a gas tight syringe. The gas sample was then injected into the sampling loop of a gas chromatograph equipped with a flame ionization detector (FID). Using this method the lower limit of detection was typically 0.5 μmol isoprene per mol of air in a sample of 0.2 cm^3.

In 1974, Rasmussen, Westburg and Holdren published a key paper in improving methods for hydrocarbon analysis. These workers pleaded for better standardization of the techniques used in the chromatographic analysis of atmospheric hydrocarbons. One improvement to arise from this work was the widespread use of a low-temperature (77 K) treatment of the sample gas to concentrate the hydrocarbons before being injected into the gas chromatograph. This concentration step is typically accomplished by drawing the atmospheric sample through a stainless steel loop immersed in liquid nitrogen. The cold loop is then flash-heated by subsequent immersion in boiling water, or by a flow of current through high resistance wire which is wrapped around the loop. Upon heating, the hydrocarbon sample is volatilized onto the gas chromatograph column.

The low-temperature concentration step greatly improved our ability to accurately analyze hydrocarbons at low concentrations in the atmosphere. Much larger volumes of air could be passed through the concentrating loop and analyzed than was possible using direct injections with head-space analysis. One could now accurately analyze hydrocarbon concentrations in the nmol/mol range, rather than the μmol/mol range previously attainable. With such an improvement in accuracy, gas chromatographic analysis of isoprene emissions could now be combined with flow-through gas-exchange systems such as those used in the analysis of photosynthetic traits in individual leaves (see the recent paper in this series by Sharkey 1989 for a discussion of photosynthetic gas-exchange analysis).

Tingey and coworkers (Tingey et al. 1979) were the first to successfully combine gas chromatographic techniques and continuous-flow gas-exchange techniques to analyze isoprene emissions from plants. They used a sample chamber that fit over an entire plant canopy. Canopy chambers have disadvantages in terms of studying the details of how photosynthesis and isoprene emissions respond to environmental variables. These include problems with accurately determining light and temperature gradients within the isolated canopy. Ten years after the studies of Tingey et al., Monson and Fall (1989) successfully used a gas chromatographic system with a chamber that enclosed single aspen leaves, to study details of the relationship between isoprene emission rate and photosynthesis rate. Even more recently, an "on-line" isoprene detector has been developed which allows for continuous monitoring of the isoprene concentration in a continuously flowing air stream (Hills and Zimmerman 1990). This system has also been successfully combined with a gas-exchange system capable of analyzing the photosynthetic traits of single leaves (Monson et al. 1991).

Future innovations in isoprene measurement should come through attempts to make systems portable. This will be especially important as laboratory-based studies of the physiological basis of isoprene emission are integrated with field-based studies aimed at determining the influence of plant isoprene emissions on tropospheric chemistry processes. Innovations in terms of portable isoprene detection systems have already been made. Rasmussen and Khalil (1988) reported that a commercially available, portable gas chromatograph equipped with a capillary column and a photoionization detector has a detection limit for isoprene of approximately 0.02 ppbv; this is much more sensitive detection than can be obtained with a flame ionization detector (FID). J. Greenberg and P. Zimmerman (National Center for Atmospheric Research, Boulder, unpubl.) have used a model RGD-2 mercuric oxide detector developed by Trace Analytical Co. for isoprene detection in remote locations. In addition to being portable, this detector is at least 100 times more sensitive than the FID for isoprene.

3 Techniques for Analysis of Isoprene by Gas Chromatography

Plant isoprene emissions are readily measured by enclosing leaves or branches into a sealed bag or chamber and analyzing the enclosure air by standard gas chromatographic techniques. Various workers have published the details and results of such procedures (see for example, Rasmussen 1970; Zimmerman 1977; Tingey et al. 1979; Ohta 1986; Winer 1989). In addition, a gas chromatographic method for ethylene analysis has recently appeared in this series (Bassi and Spencer 1989), and the sampling procedures for ethylene can be readily adapted for isoprene analysis. The procedures described below are what we have found to work best for our studies of isoprene emission from individual leaves.

3.1 Gas Chromatography with Flame Ionization Detection

In our studies of isoprene emission from leaves, we routinely use a gas chromatograph equipped for capillary column chromatography and flame ionization detection. The gas chromatograph (Hewlett-Packard, model 5790A) has been modified for vacuum inlet collection and cryogenic enrichment of hydrocarbon samples (Fig. 1). These modifications are described

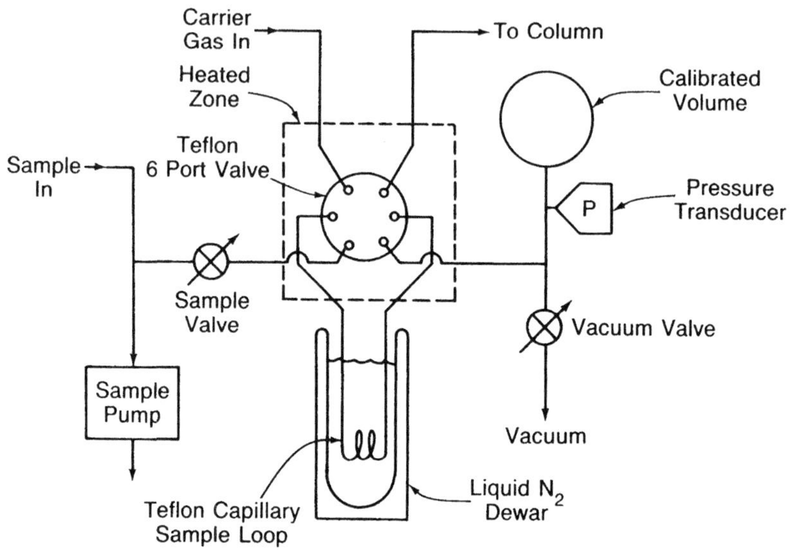

Fig. 1. Schematic diagram of the modifications of the gas chromatograph for vacuum inlet sample collection and cryogenic enrichment. The collection process is initiated by evacuating the entire collection system with the vacuum pump. Then the sample valve is opened. This allows the sample to flow through the sample loop to the calibrated volume. The hydrocarbons are frozen out in the loop. Nitrogen from the sample air goes to the calibrated volume. The Teflon capillary sample loop is housed inside a stainless steel loop. After the concentration step, the 6-port Teflon valve is switched to allow carrier gas to flush the concentrated sample onto the gas-chromatograph column as the loop is flash-heated to 80 °C by a short pulse of electric current through the stainless steel loop (After Goldan et al. 1987)

in detail elsewhere (Goldan et al. 1987; Monson and Fall 1989). Briefly, samples are introduced from a gas stream or gas-tight syringe by opening a stainless steel valve, allowing the sample to flow (1) through a 0.36-mm i.d. Teflon capillary sample loop (housed inside a stainless steel loop) submerged in liquid nitrogen; and (2) into a previously evacuated 7.7-l volume-calibrated stainless steel flask. The sample loop, which allows N_2 and O_2 to pass through while cryogenically enriching the isoprene in the sample, is connected to the head of a capillary column through a Teflon 6-port valve. The 6-port valve is housed in a container heated to 80–90 °C to prevent condensation of analytes. The concentrated hydrocarbon sample is vaporized by flash-heating the loop to 80 °C with a flow of current through the high-resistance stainless steel tubing that surrounds the Teflon capillary loop. The sample is immediately injected onto the column by turning the 6-port valve to the appropriate position.

The capillary column of the gas chromatograph is a 30-m fused silica capillary with a 0.25-mm i.d. and a 1-μm methyl silicone stationary phase

(J & W Scientific, Folsom, CA, USA). Carrier gas is hydrogen at a flow of approximately 1 cm^3/min. A hydrogen rich flame is used to enhance hydrocarbon detection. The column is usually operated isothermally at 30 °C which gives an isoprene retention time of 1.9 min, well resolved from other leaf volatiles (e.g., ethylene, acetaldehyde, and ethanol). When using leaves that emit dimethyl sulfide, it is necessary to use cryogenic cooling of the column and temperature programming to resolve isoprene and dimethyl sulfide.

Some gas chromatographs provide a purge-delay feature with built-in cryogenic cooling. This allows for injection of larger sample volumes without the need for a vacuum inlet system. Thus, isoprene can be "frozen out" and concentrated on the head of a capillary column, and then chromatographed by temperature programming. Another innovation that can be applied to the gas chromatographic analysis of hydrocarbons is the use of a photoionization detector, mentioned above, which is much more sensitive to alkenes than the FID (Driscoll and Duffy 1987; Rasmussen and Khalil 1988). This detector also responds to aromatic, chlorinated aromatic, and chlorinated alkenes found in urban air, and thus should be used with plant samples incubated in purified air when studying isoprene emissions.

3.2 Analysis of Headspace Samples Using Gas Chromatography

Headspace measurements provide a way to quickly and qualitatively test for isoprene emission by a leaf sample. Accumulation of isoprene in the headspace with time can be monitored by gas chromatography to obtain a measure of the isoprene emission rate. However, the latter procedure has a distinct disadvantage, compared to the steady-state techniques described in the next section, in that the analysis is conducted under conditions of decreasing CO_2 concentrations and increasing isoprene concentrations in the atmosphere surrounding the leaf. The latter conditions could, in themselves, influence the isoprene emission rate. Headspace samples are also useful for analysis by gas chromatography-mass spectrometry (GC-MS) to verify the presence of isoprene.

Leaves or leaf discs are placed in glass vials with the cut end of the leaf petiole placed under water, or with a leaf disc floated on water. The vials are sealed with Teflon-lined septa and incubated under an appropriate cool light source. This is best accomplished in a water bath at a fixed temperature with the adaxial surface of the leaves toward the light source. Headspace air samples, typically 0.5–2.0 cm^3, are removed with a gas-tight syringe at various times and can be directly injected onto a capillary column without removal of water. With most isoprene emitters, prolonged incubation (1–2 h) results in accumulation of the hydrocarbon to concentrations in the µmol/mol range, easily providing sufficient isoprene for GC-MS analysis.

3.3 Analysis of Steady-State Isoprene Emission Rates Using Gas Chromatography

If a leaf or branch is enclosed in a gas-exchange system where temperature, light, CO_2 concentration, and water vapor content can be controlled, gas flow rates can be monitored, and the gas exiting the leaf chamber can be analyzed for isoprene content, accurate steady-state isoprene emission rates can be measured. In several studies we have performed such measurements with a leaf cuvette system which has been described in detail elsewhere (Monson et al. 1986; Monson and Fall 1989). In other studies, we have employed a clamp-on leaf cuvette which allows separate gas sampling from adaxial and abaxial leaf surfaces and has a volume small enough (5 cm^3) to allow for rapid gas turnover (6 s; Monson et al. 1991). Rapid gas turnover is essential for studies of isoprene emission transients in response to light-dark transitions (see Monson et al. 1991).

In interfacing the gas-exchange system with the gas chromatograph described above which has been modified for low temperature concentration of the hydrocarbon sample, we have found a need to dry the sample gas stream before it enters the concentrating loop. Due to its narrow diameter, the sample loop will freeze up if more than 5 cm^3 of humidified air is sampled. The drying is accomplished by pulling a part of the gas stream exiting the leaf cuvette at a constant flow of approximately 80 cm^3/min through a glass drying tube (100 cm^3) maintained in a dry ice bath (−80 °C).

The choice of materials for handling gas streams containing hydrocarbons is critical to successful analysis. Many organic compounds are known to adsorb to, or penetrate, surfaces composed of non-polar materials. All tubing that connects the leaf cuvette, dryer, and sample inlet should be PFA Teflon to minimize isoprene loss. Recovery of isoprene in the system, including the leaf cuvette, should be measured frequently and included in calculations of the estimated isoprene emission rate. We use an isoprene calibration standard and mass flow controllers for this purpose. The calibration source contains a weighed amount of isoprene (99+% pure grade, Aldrich Chemical Co., Milwaukee, WI, USA), that is quantitatively transferred into an evacuated, passivated aluminum cylinder (Scott Environmental, Plumsteadville, PA, USA), and pressurized to a predetermined level with purified nitrogen. This gas calibration source is then checked against an NBS traceable standard using FID analysis. Once prepared in this way, the calibration gas provides a stable source of isoprene that can be accurately diluted with carrier nitrogen and flow controllers. The diluted isoprene source can then be used to provide a constant isoprene flow through the gas-exchange system during the calibration procedure to determine the extent of isoprene recovery. For our system an isoprene recovery of 67–75% is typically found.

Care must also be taken to only use the highest grade gases for mixing the air that is used during the gas exchange measurements. In our system, high-purity grade N_2 and O_2 are mixed with 1% CO_2 in nitrogen to the desired proportions using mass flow controllers. This gas is then humidified to the desired level by bubbling the air through a flask of water immersed in a temperature controlled water bath before being sent to the leaf cuvette. With high-purity grade nitrogen and CO_2 we have found no traces of hydrocarbon contaminants during checks with the gas chromatograph. However, in the case of oxygen we routinely find traces of an unknown hydrocarbon contaminant. We now filter the oxygen source through an activated charcoal filter to remove this contaminant.

3.4 Confirmation of Isoprene by GC-MS

In studying isoprene emissions from leaves it is imperative that the substance being detected as isoprene on the gas chromatograph be confirmed using GC-MS analysis. A typical procedure for GC-MS confirmation is described here. Leaves are placed in 14-cm^3 glass vials with the petioles submerged in 1 cm^3 of distilled water. The vials are sealed and incubated as described in the above section on headspace analysis (see Sect. 3.2). An

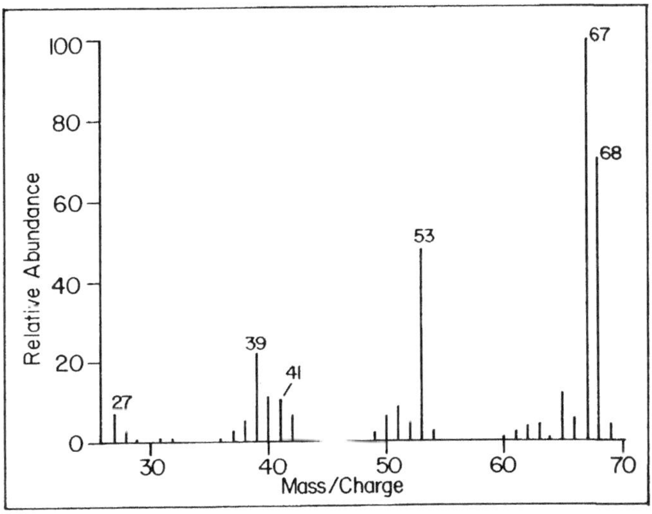

Fig. 2. A typical mass spectrum verifying isoprene emission from an aspen leaf. An aspen leaf (11 cm^2) was incubated in a glass vial as described in the text at 28 °C for 60 min under a cool light source (1000 $\mu mol\,m^{-2}s^{-1}$), and 1 cm^3 of headspace gas was injected for GC-MS analysis as described in the text. Other GC-MS conditions: column temperature was −50 °C for 0.5 min, then increased at a rate of 8 °C per min to a maximum of 100 °C; injector temperature was 250 °C; splitless injection (0.75 min); transfer line temperature was 200 °C; and electron impact voltage was 70 eV. The only significant peak detected has a retention time of 4.1 min and mass spectral fragmentation pattern identical to authentic isoprene

aliquot of headspace gas is withdrawn from the vial after about 90 min of incubation. Headspace samples are analyzed on a Hewlett-Packard 5988A GC-MS system with electron impact ionization, equipped with a 25-m fused silica capillary (0.31-mm i.d., 0.52 μm stationary phase of phenylmethyl silicone) and nitrogen as carrier gas at approximately 1 cm^3/min. Analyses are typically carried out with temperature programming from −50 to 100 °C. A typical mass spectrum for aspen leaf isoprene is shown in Fig. 2. It is identical to the spectrum seen for authentic isoprene. The major fragments are: $(M\text{-}68)^+$ the molecular ion; $(M\text{-}67)^+$ representing loss of a proton; $(M\text{-}53)^+$ loss of a methyl group; $(M\text{-}41)^+$ loss of a vinyl group; $(M\text{-}39)^+$ a $C_3H_3^+$ fragment; and $(M\text{-}27)^+$ a $C_2H_3^+$ fragment.

4 Technique for the Analysis of Isoprene Using Chemiluminescence

The chromatographic method of isoprene analysis outlined above is a sensitive and accurate method for monitoring isoprene levels and emission rates. However, the chromatographic method has limitations, in that it cannot provide a continuous measurement of isoprene emissions nor can it resolve changes in the isoprene emission rate that occur over short time scales (i.e., seconds). Such limitations are especially prohibitive in studies of the photosynthesis/isoprene relationship, since the latter is known to occur on the order of seconds (Monson et al. 1991). Analysis of isoprene by chemiluminescence provides the basis for an instrument that continuously measures isoprene concentrations in a gas stream. When coupled with a gas-exchange system capable of rapid gas turnover in the leaf cuvette, the "on-line" isoprene detector can provide accurate analysis of isoprene dynamics over extremely short time scales.

4.1 Principles of the Detection System

Chemiluminescence, the production of light via a chemical reaction, is a useful analytical technique amenable to the real-time monitoring of certain chemical species. It is selective, owing to the relatively small number of compounds which chemiluminesce upon contact with a given reactant, and very sensitive since the chemiluminescence appears out of a dark background. In principle, a single photon generated from a chemiluminescent reaction can be detected. Useful chemiluminescence systems have been described for the detection of NO and O_3 (Ridley and Howlett 1974; Ridley 1978; Kley and McFarland 1980), F_2 (Nelson et al. 1983; Glinski 1989; Getty and Birks 1979), H_2S, CH_3SH, and OCIO (Spurlin and Yeung 1982), as well

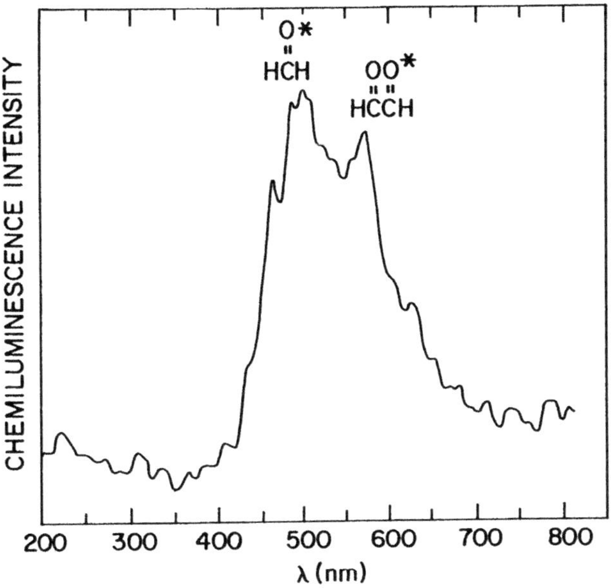

Fig. 3. Chemiluminescence spectrum produced by the reaction of isoprene with ozone. Conditions: pressure was 14.4 kPa; 140 scans (70 signal and 70 background)

as halogenated compounds, and reduced sulfur compounds (Nelson et al. 1983; Glinski 1989; Getty and Birks 1979). The chemiluminescence of alkene/O_3 reactions has been explored in development of ozone monitors and as gas chromatographic detectors (Warren and Babcock 1970; Seila 1976; Rao et al. 1983; Bruening and Concha 1975).

Isoprene can be detected via its chemiluminescent reaction with ozone:

$$\text{isoprene} + O_3 \rightarrow HCHO^* + \text{products}. \quad (1)$$

The chemiluminescence emission spectrum obtained upon reaction of isoprene with ozone is shown in Fig. 3. The largest peak has a maximum at 490 nm and represents emission from excited state formaldehyde (Finlayson et al. 1972; Finlayson et al. 1974). The second peak in the spectrum, which occurs at 550 nm, is probably emission from excited state glyoxal, $(CHO)_2^*$. Excited state glyoxal can be formed in the reaction of ozone with an alkene (such as isoprene), if the alkene possess one or more alkyl groups bonded to the carbon of a carbon-carbon double bond (Toby 1984).

A schematic of the isoprene chemiluminescence detector which we have constructed is shown in Fig. 4. The reaction occurs in a mirror-finish cell constructed of stainless steel. Collinear mixing of ozone and isoprene occurs in the front of the chamber. This "point" reaction region ensures the highest possible O_3/isoprene concentrations during the chemiluminescent reaction, and hence produces maximum photon yield. Gases are evacuated near the

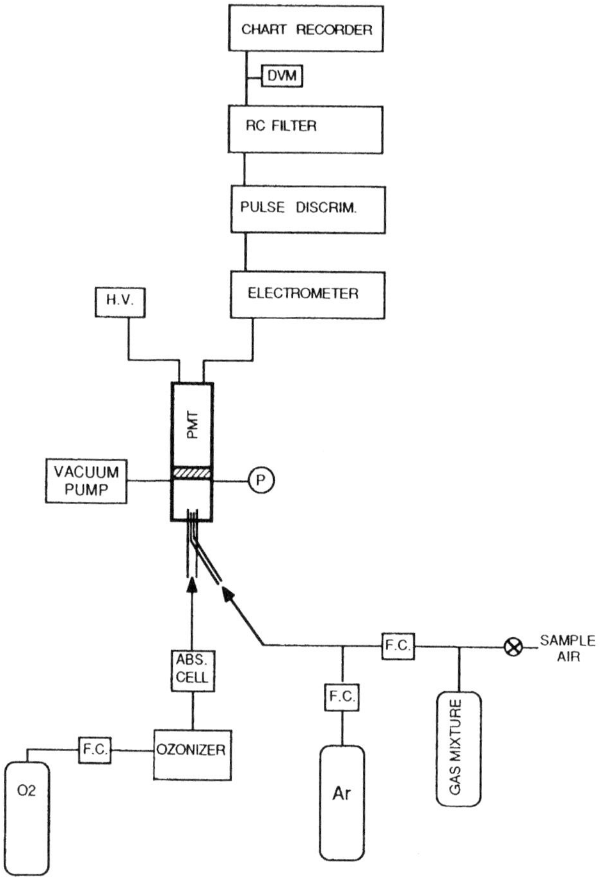

Fig. 4. Schematic of the chemiluminescence detection system. The reaction chamber where the ozone/isoprene reaction takes place is located immediately below the photomultiplier tube (PMT). Ozone which is fed to the reaction chamber is produced with the ozonizer. The electronics that process the signal from the PMT are shown in series in the *upper* portion of the schematic. For more details of the system see text (Hills and Zimmerman 1990)

window at the end of the chamber opposite the reaction point via a concentric slot circling the end of the chamber. Thus, the reaction chamber is continually swept clean and the concentric evacuation slot helps preserve laminar flow conditions, resulting in minimal dead volume. A vacuum pump is used to evacuate the gases. The reaction chamber and photomultiplier tube housing are separated by a glass window through which the chemiluminescent reaction is viewed. The photomultiplier tube that is used has an ultra-low dark current with high quantum efficiency (10–25%) over the emission region (425–550 nm) and low quantum efficiency at red wavelengths (2.5% at 600 nm and 0.1% at 680 nm) where long wavelength chemiluminescence could interfere with isoprene detection.

The electrical signals from the photomultiplier tube are first processed by a fast electrometer and then via a pulse discriminator circuit which eliminates spurious large and wide signal spikes associated with thermally generated dynode currents or other phenomena. The selected pulse train is then fed into a plotter and a personal computer.

4.2 Procedures for Using the Chemiluminescence Isoprene Detector

Calibration of the amount of chemiluminescence in the isoprene detector is accomplished with known dilutions of an isoprene standard. Isoprene standards in the μmol/mol concentration range are prepared by taking vapor of liquid isoprene and performing serial dilutions with helium. Total pressure of the gas standards is 500–650 kPa. These "secondary" standards are calibrated using gas chromatography and are referenced against "primary" standards obtained from the National Bureau of Standards (USA). Concentrations in the range of nmol/mol are prepared in real-time by diluting the μmol/mol standard with argon using mass flow controllers.

The isoprene chemiluminescence instrument can be operated over a wide range of experimental conditions. Since the light detector (photomultiplier) signal is proportional to the amount of light produced in the reaction cell, as long as there is sufficient ozone in the reaction chamber, the signal will increase as the total amount of isoprene in the sampled airstream is increased. System sensitivity is therefore mass dependent as opposed to concentration dependent. In practice, about 150 cm^3/min of sample air flow is directed into the chemiluminescence chamber. Optimal pressure in the reaction chamber (40 kPa) is obtained by balancing increased sensitivity at higher pressures with increased dead time of the reactor. At a typical ozone partial pressure of 3 kPa, one can calculate that 99.99% of the reaction is completed. The detection limit for isoprene is defined as twice the average noise level and is calculated to 400 pmol/mol. Further sensitivity improvements could be realized through more inert and reflective reaction cell materials (i.e., silvered glass). It should be noted that the gold coating of reaction cells that are normally used in NO/O_3 chemiluminescence systems, works poorly for the shorter wavelengths of this system.

At a given pressure and flow rate, the chemiluminescence system provides a linear response with respect to isoprene mass over three orders of magnitude. The chemiluminescence reaction occurs on the order of milliseconds. The system response time is dependent on two factors. First, is the reaction chamber residence time (t), which is defined as:

$$t = V/F, \tag{2}$$

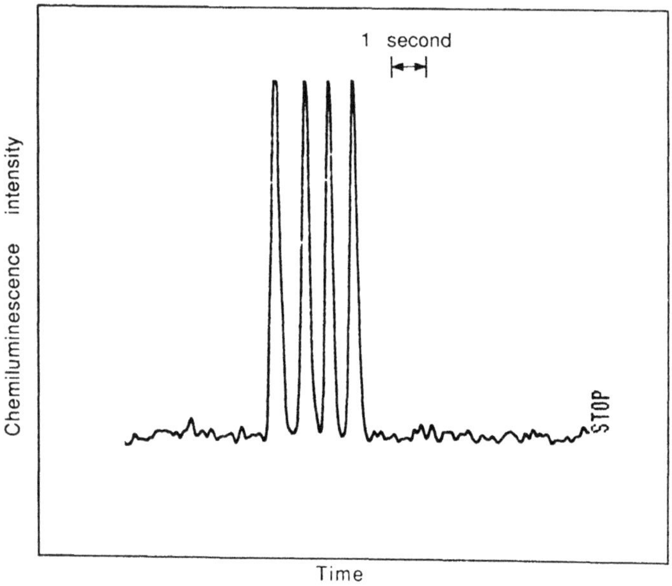

Fig. 5. Time response of the chemiluminescence detection system. The detection peaks are from sequential injections of isoprene separated by 0.7 s. Other conditions: pressure was 19.4 kPa; reaction temperatures was 40 °C; ozonizer oxygen flow was 210 cm^3/min; sample air flow was 1000 cm^3/min

where V is the reaction chamber volume (108 cm^3 in the case of our instrument) and F is the total flow of gas through the chamber (corrected for non-STP pressure and temperatures). The second factor that could affect system response time, is an effect due to adsorption/desorption of isoprene in the reaction zone or tubing leading to this zone (i.e., an instrument "memory" effect). For all of our experiments, memory effects have been minimal and chamber residence time governed the overall system response time. Figure 5 shows an isoprene chemiluminescence trace generated by isoprene injections performed approximately every 0.7 s. The rise time of the signal is 0.1 s, consistent with the calculated residence time for this test.

4.3 Selectivity of the Chemiluminescence System

Selectivity of the chemiluminescence system for isoprene is relatively high (Table 1). In general, one would expect discrimination between isoprene and other alkenes to be poor, since all alkenes chemiluminesce with ozone to some extent producing HCHO*. The factors differentiating chemiluminescence sensitivities for various alkenes are their rate coefficients and emission quantum efficiencies. The instrumental selectivity for isoprene,

Table 1. Response factors for isoprene compared to other compounds as detected by the chemiluminescence system (After Hills and Zimmerman 1990)

Compound	Relative response factor[a]
Alkene compounds	
Isoprene	1.00
Ethene	0.15
Propene	1.00
3-Butene-2-one	0.12
2-Methyl-propenal	0.19
α-Pinene	0.036
β-Pinene	0.039
Nonalkenes	
Acetylene	0.050
Acetaldehyde	0.00027
n-Butanol	0.00027
Cyclohexane	0.000054
Benzene	0.000052
Dimethyl sulfide	0.12[b]
Hydrogen sulfide	0.041[c]
NO	0.025

[a] Relative to isoprene.
[b] Using a Corning GG-475 filter.
[c] Calculated according to Kelly et al. (1983).

relative to many alkenes, is high because isoprene reacts rapidly with O_3 and its relative efficiency for producing light-emitting chemical species in the 500 nm region is high. The high selectivity for isoprene is also facilitated by the fact that there are few alkenes present in most atmospheres that would compete with isoprene for chemiluminescent detection. Selectivities of the instrument to various compounds are presented in Table 1. Chemiluminescence due to ethene, which exhibits a 15% interference on a volume/volume basis, can be further suppressed by adjusting conditions to provide less extent of isoprene reaction, since the rate coefficient of ethene/O_3 is only one-seventh that for isoprene/O_3. The greatest interference is caused by propene. Responses of isoprene and propene are approximately the same and if comparable amounts of each were present the chemiluminescence signals could not be distinguished. Fortunately, biogenic propene fluxes are negligible compared to those of isoprene and no interference problems have been encountered. Monoterpenes are emitted by some species at levels comparable to isoprene fluxes. However, these compounds cause much less chemiluminescence when present with ozone, compared to isoprene and thus cause no problem with isoprene detection (see Table 1). This is somewhat surprising since monoterpenes are known to react rapidly with ozone. Apparently, they either do not produce excited-state species upon reaction with O_3, or the monoterpenes or other reaction products quench these species prior to their fluorescence.

In the lower half of Table 1, we have listed instrument selectivity for nonalkene compounds. Alkanes, alkynes, alcohols and aldehydes give extremely poor chemiluminescence response. The strong red chemiluminescence that characterizes the NO/O_3 reaction is also poorly detected with the isoprene instrument, which is optimized for 500-nm emissions. Reduced sulfur compounds also react with ozone to form excited species, in this case SO_2^*, which emit photons in the visible part of the spectrum and have the potential to interfere with isoprene detection. Dimethyl sulfide (DMS) is among the most sensitively detected of the reduced sulfur compounds and it exhibits a 54% interference relative to isoprene. Fortunately, SO_2^* emission is in the blue wavelength region and consists of an emission band at 310–440 nm (Akimoto et al. 1971). A cutoff filter (Corning GG-475) can be used to block 94% of this emission while passing isoprene chemiluminescence.

5 Use of the Chemiluminescence Detector With a Gas-Exchange System

We have used the chemiluminescence detector along with a traditional open, gas-exchange system in several studies of the relationship between isoprene emission and photosynthesis (Monson et al. 1991). Several precautions and checks must be used in coupling these systems.

First, although water vapor was found to have only a small effect on chemiluminescence sensitivity, it markedly quenches background chemiluminescence signals. Thus, rapid humidity changes can cause baseline shifts. For this reason, a Nafion drier must be placed in the gas line between the leaf cuvette and the isoprene instrument. The drier consists of a tube through which the sample gas flows, and a small countercurrent of dry nitrogen. The dry nitrogen removes H_2O vapor diffusing through the Nafion (see Benner and Lamb 1985). Use of the drier prevents baseline shifts even at high humidity levels. (We tested it up to 90% relative humidity). Second, the measurement of isoprene flux requires an accurate knowledge of instrument sensitivity. Sensitivity is measured daily by recording instrumental response for a standard isoprene mixture. Standards in the nmol/mol range (prepared as described above) are mixed into the analysis gas stream and flowed over the leaf in the cuvette in the dark. In this way any systematic isoprene losses, such as cuvette leakage, are taken into account. Daily sensitivity variation has been found to be at most 5%.

To date, isoprene fluxes have been measured from white oak, aspen, cottonwood, velvet bean, and eucalyptus leaves using the chemiluminescence

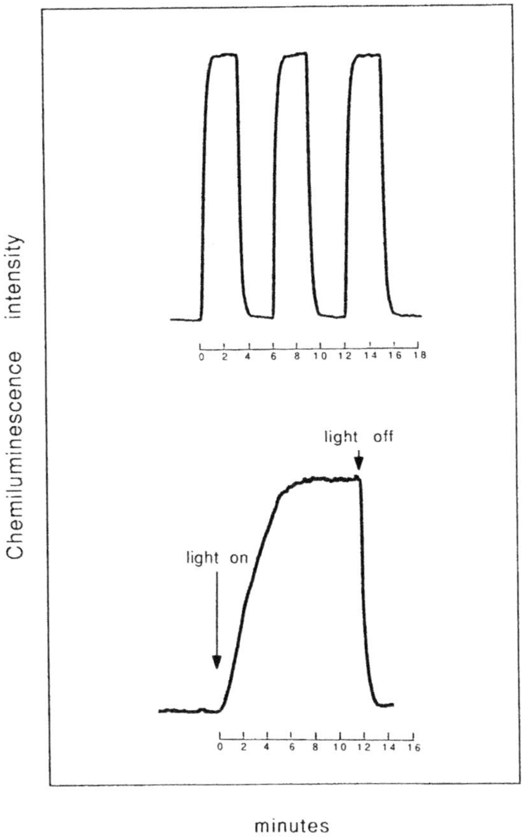

Fig. 6. *Upper plot*: Signal from the chemiluminescence detection system during cycling of an isoprene standard through the gas-exchange cuvette in dark conditions (cycle was 3 min on/3 min off). *Lower plot*: Isoprene flux from a white oak leaf following a dark/light transition and a light/dark transition. Isoprene flux was sampled from the lower leaf surface. The isoprene signal rise time to 98% of the maximum signal was 7 min. The fall time to 2% of the maximum signal was 2 min. Conditions: illuminated leaf area was 7.1 cm^2; photon flux density was 1025 $\mu mol\, m^{-2} s^{-1}$; air flow rate over the leaf was 492 cm^3/min; total gas flow rate through the chemiluminescence reaction chamber was 360 cm^3/min; isoprene instrument cell pressure was 510 kPa

instrument. In Fig. 6 we present representative data from measurements on a white oak leaf (*Quercus alba* L.). In the upper part of the panel a continuous trace is presented of the cycling of an isoprene standard (3 min on and 3 min off). The rise and fall times in this trace are governed by the rate of RC filtering and by gas flow rates through the cuvette system. In the lower part of the panel, a continuous trace is presented of isoprene emissions from a white oak leaf during dark/light and light/dark transitions. The slower rise time in the lower trace compared to the upper traces reflects a real biological response to the dark/light transition. Isoprene was detected in the cuvette air within 6 s after irradiation. This matches the 6 s lag time in flushing the leaf cuvette which was measured using an isoprene standard. The gradual change in isoprene flux rate was validated qualitatively by taking 5 cm^3 aliquots of cuvette air using a gas-tight syringe followed by analysis on a gas chromatograph equipped with a HgO/Hg reductive atomic absorption analyzer (model RGD2, Trace Analytical Co., Menlo Park, CA, USA). The gas chromatographic analysis not only reproduced the isoprene profile shown in the lower panel of Fig. 6 (data not shown), but also demonstrated that isoprene was the only alkene that was being produced by the leaf. The concentration measured in the cuvette was 90 nmol/mol which corresponded to an isoprene flux of 0.86 $\mu mol\, m^{-2} s^{-1}$ on a carbon atom basis. We have found that the best gas flow rates for these measurements are between 300–600 cm^3/min. Environmental conditions in the gas-exchange cuvette are maintained as described previously (Monson et al. 1986).

6 Calculation of the Isoprene Flux from a Leaf

The isoprene flux from a leaf can be calculated using slight modifications of the equations developed for determining the water vapor and CO_2 fluxes from and to leaves. The rate of isoprene efflux from a leaf is calculated as the difference in isoprene flowing into and out of the leaf cuvette:

$$I_s = (u_o i_o - u_e i_e)/s, \quad (3)$$

where I_s is the isoprene emission rate, u_o is the molar flow rate of air out of the cuvette, i_o is the isoprene concentration of the outgoing air, u_e is the molar flow rate of air entering the cuvette, i_e is the isoprene concentration entering the cuvette, and s is the leaf area. Since there is no isoprene in the entering airstream i_e can be ignored. Since the outgoing flow rate is seldom measured in gas-exchanges systems, u_o must be calculated from the flow rate that is measured, u_e. Assuming a well-stirred and well-sealed cuvette, u_o can be calculated from u_e with a correction for the effect of transpirational water loss on the flow rate leaving the cuvette:

$$u_o = u_e(1 - w_e)/(1 - w_o), \quad (4)$$

where w_e is the molar concentration of water vapor in the air entering the cuvette, and w_o is the molar concentration of water vapor in the outgoing air (see von Caemmerer and Farquhar 1981). As a leaf transpires water into the cuvette, the mass of air leaving the chamber per unit volume will be higher and the molar flow rate measured at the ingoing port must be corrected to a greater value if it is to be used as the outgoing flow rate. The correction can be incorporated into Eq. (3) as follows:

$$I_s = i_o/s\ [u_e(1 - w_e)/(1 - w_o)]. \tag{5}$$

In using gas chromatographic analysis of the isoprene concentration in the leaf cuvette, the air stream is often dried as it is passed through a dry-ice condenser before entering the concentrating loop of the modified gas chromatograph (see Sect. 3.3). In this case the isoprene concentration in the measured air stream must be corrected for the influence of removing water. Water vapor removal will act to concentrate the molecular composition of the air stream. The correction can be accomplished by further modifying Eq. (3) to the following:

$$I_s = u_o/s\ [i_o(1 - w_o)/(1 - w_{ice})], \tag{6}$$

where w_{ice} is the molar concentration of water vapor at the temperature of dry-ice (also see von Caemmerer and Farquhar 1981). Note that the relevant flow rate to use in this calculation is that leaving the cuvette (u_o). This value can be determined from Eq. (4). In using the chemiluminescence system, the correction in Eq. (6) must use the molar concentration of water in the air coming out of the Nafion drier. This value is typically so low that it can be ignored and the correction only involves multiplying i_o by $(1 - w_o)$. In using Eq. (6) to calculate the isoprene emission rate, one needs to measure the isoprene concentration of the gas exiting the chamber, the flow rate of gas entering the chamber, the leaf area, and the humidity of the air stream before and after the chamber.

In expressing isoprene emission rates, one can use units of flux based on entire isoprene molecules or the carbon atoms that compose isoprene. In studies relating isoprene emissions to photosynthesis, we prefer to use units based on carbon atoms because they can be more easily related to the photosynthetic assimilation of CO_2. However, in most cases, it is better to use units based on entire isoprene molecules. Fluxes are typically expressed as nmol m^{-2} (leaf area) s^{-1}.

Acknowledgments. The authors gratefully acknowledge useful discussions and interactions with Patrick Zimmerman (National Center for Atmospheric Research, Boulder). The development of techniques described in this chapter were made possible by NSF grant ATM 8719536, EPA grant R-815995, and support from the National Science Foundation to the National Center for Atmospheric Research.

References

Akimoto H, Finlayson BJ, Pitts JN Jr (1971) Chemiluminescent reactions of ozone with hydrogen sulfide, methyl mercaptan, dimethyl sulfide and sulfur monoxide. Chem Phys Lett 12:199–202

Andreae MO, Andreae TW (1988) The cycle of biogenic sulfur compounds over the Amazon basin. 1. Dry season. J Geophys Res 93:1487–1497

Bassi PK, Spencer MS (1989) Methods for the quantification of ethylene produced by plants. In: Linskens HF, Jackson JF (eds) Gases in plant and microbial cells (Modern

methods of plant analysis, New series, vol 9. Springer, Berlin Heidelberg New York, p 309
Becker KH, Schurath U, Wiese A (1976) A new chemiluminescent olefin detector for ambient air. Proc Int Conf on Photochemical oxidant pollution and its control. EPA-600/3-77-001a, 1, PB-264232, 31–40
Benner RL, Lamb B (1985) A fast response continuous analyzer for halogenated atmospheric tracers. J Atmos Ocean Technol 2:582–589
Bruening W, Concha FJM (1975) Selective detector for gas chromatography based on the chemiluminescence of ozone reactions. J Chromatogr 112:253–265
Caemmerer S von, Farquhar GD (1981) Some relationships between the biochemistry of photosynthesis and the gas exchange of leaves. Planta 153:376–387
Chameides WL, Lindsay RW, Richardson J, Kiang CS (1988) The role of bigenic hydrocarbons in urban photochemical smog: Atlanta as a case study. Science 241:1473–1475
Driscoll JN, Duffy M (1987) Photoionization detector: a versatile tool for environmental analysis Chromatogr Mag 1987:21–27
Finlayson BJ, Pitts JN Jr, Akimoto H (1972) Production of vibrationally excited OH in chemiluminescent ozone-olefin reactions. Chem Phys Lett 12:495–498
Finlayson BJ, Pitts JN Jr, Atkinson R (1974) Low-pressure gas-phase olefin-ozone reactions. Chemiluminescence, kinetics, and mechanisms. J Am Chem Soc 96:5356–5367
Getty RH, Birks JW (1979) A chemiluminescence detector for gas chromatography with selectivity for iodine. Anal Lett 12:469–476
Glinski RJ (1989) Chemiluminescence during the reaction of molecular fluorine and carbon diselenide: emission from Se_2, SeF, and an unknown emitter. J Phys Chem 93:1913–1915
Goldan PD, Kuster WC, Albritton DL, Fehsenfeld FC (1987) The measurement of natural sulfur emissions from soils and vegetation: three sites in the eastern United States revisited. J Atmos Chem 5:439–467
Gray JC (1987) Control of isoprenoid biosynthesis in higher plants. Adv Bot Res 14: 25–91
Greenberg JP, Zimmerman PR (1984) Nonmethane hydrocarbons in remote tropical, continental, and marine atmospheres. J Geophys Res 89:4767–4778
Hansen DA, Atkinson R, Pitts JN Jr, Steer RP (1977) Structural effects on the chemiluminescence from the reaction of ozone with selected organic compounds. J Photochem 7:379–404
Hills AJ, Zimmerman PR (1990) Isoprene measurement by ozone-induced chemiluminescence. Anal Chem 62:1055–1060
Jacob DJ, Wofsy SC (1988) Photochemistry of biogenic emissions over the Amazon forest. J Geophys Res 93:1477–1486
Kamens RM, Gery MW, Jeffries HE, Jackson M, Cole EI (1982) Ozone-isoprene reactions: product formation and aerosol potential. Int J Chem Kinet 14:955–975
Kelly TJ, Gaffney JS, Phillips MF, Tanner RL (1983) Chemiluminescent detection of reduced sulfur compounds and gases. Anal Chem 55:135–138
Kleinig H (1989) The role of plastids in isoprenoid biosynthesis. Annu Rev Plant Physiol Mol Biol 40:39–59
Kley D, McFarland M (1980) Chemiluminescence detector for NO and NO_2. Atmos Technol 12:63–69
Loreto F, Sharkey TD (1990) A gas exchange study of photosynthesis and isoprene emission in red oak (*Quercus rubra* L). Planta 182:523–531
Monson RK, Fall R (1989) Isoprene emission from aspen leaves. Influence of environment and relation to photosynthesis and photorespiration. Plant Physiol 90:267–274
Monson RK, Moore RD, Ku MSB, Edwards GE (1986) Co-function of C_3 and C_4-photosynthetic pathways in C_3, C_4 and C_3-C_4 intermediate *Flaveria* species. Planta 168:493–502
Monson RK, Hills AJ, Zimmerman PR, Fall RR (1991) Studies of the relationship between isoprene emission rate and CO_2 or photon-flux density using a real-time isoprene analyzer. Plant Cell Environ 14:517–523

Nelson JK, Getty RH, Birks JW (1983) Fluorine induced chemiluminescence detector for reduced sulfur compounds. Anal Chem 55:1767–1770
Ohta K (1986) Diurnal and seasonal variations in isoprene emission from live oak. Geochem J 19:269–274
Rao AMM, Netravalkar AJ, Arora PK, Vohra KG (1983) Determination of ethylene and other reactive hydrocarbons in the atmospheric air at Trombay, Bombay by gas chromatography using a chemiluminescence detector. Atmos Environ 17:1093–1097
Rasmussen RA (1970) Isoprene identified as a forest-type emission to the atmosphere. Environ Sci Technol 4:667–671
Rasmussen RA, Jones CA (1973) Emission of isoprene from leaf discs of *Hamamelis*. Phytochemistry 12:15–19
Rasmussen RA, Khalil MAK (1988) Isoprene over the Amazon basin. J Geophys Res 93:1417–1421
Rasmussen RA, Westburg HH, Holdren M (1974) Need for standard reference gas chromatograph methods in atmospheric hydrocarbon analysis. Chromatogr Sci Ser 12:80–84
Riba ML, Randrianalimanana E, Mathieu J, Torres L, Namiesnik J (1985) Preconcentration of atmospheric terpenes on solid sorbents. Int J Environ Anal Chem 19:133–143
Ridley BA (1978) Measurements of minor constituents in the stratosphere by chemiluminescence. Atmos Technol 9:27–34
Ridley BA, Howlett LC (1974) An instrument for nitric oxide measurement in the stratosphere. Rev Sci Instrum 45:742–746
Sanadze GA (1969) Light-dependent excretion of molecular isoprene by leaves. In: Metzner H (ed) Progress in photosynthesis research. Proc Int Congr Photosynthesis research, 1968, vol 2. Verlag C. Lichtenstern, Munich, pp 701–706
Schurath U, Wiese A, Becker KH (1976) A chemiluminescence analyzer for unsaturated hydrocarbons in air. Staub-Reinhalt Luft 36:379–85
Seila RL (1976) GC-chemiluminescence method for the analysis of ambient terpenes. US Environmental Protection Agency, Proc Int Conf on Photochemical oxidant pollution and its control. EPA-600/3-77-001a, 1, PB-264232, 41–50
Sharkey TD (1989) Analytical gas exchange measurements of photosynthetic CO_2 assimilation. In: Linskens HF, Jackson JF (eds) Gases in plant and microbial cells (Modern methods in plant analysis, New Series, vol 9). Springer Berlin Heidelberg New York, pp 73–93
Spurlin SR, Yeung ES (1982) On-line chemiluminescence detector for hydrogen sulfide and methyl mercaptan. Anal Chem 54:318–320
Tingey DT, Manning M, Grothaus LC, Burns WF (1979) The influence of light and temperature on isoprene emission rates from live oak. Physiol Plant 47:112–118
Toby S (1984) Chemiluminescence in the reactions of ozone. Chem Rev 84:277
Trainer M, Williams EJ, Parrish DD, Buhr MP, Alwine EJ, Westberg H, Fehsenfeld FC, Liu SC (1987) Models and observations of the impact of natural hydrocarbons on rural ozone. Nature 329:750–757
Warren GJ, Babcock G (1970) Portable ethylene chemiluminescence ozone monitor. Rev Sci Instr 41:280–281
Winer AM (1989) Hydrocarbon emissions from vegetation found in California's central valley. Final report to California Air Resources Board (Contract No A732-155), Riverside, California, Nov 1989
Zimmerman PR (1977) Tampa Bay photochemical oxidant study. Rep EPA 904/9-77-028. Environmental Protection Agency, Research Triangle Park, North Carolina
Zimmerman PR (1979) Testing of hydrocarbon emissions from vegetation, leaf litter and aquatic surfaces and development of a methodology for compiling biogenic emission inventories. EPA 450/4-79-004
Zimmerman PR, Greenberg JP, Westberg CE (1988) Measurements of atmospheric hydrocarbons and biogenic emission fluxes in the Amazon boundary layer. J Geophys Res 93:1407–1416

Analysis of Ecotoxic Agents Using Pollen Tests

P.L. PFAHLER

1 Introduction

The rapid increase in agricultural and industrial technology has been accompanied by a dramatic increase in the number and diversity of potentially toxic agents released into the biosphere. The potential value of pollen systems in assessing the impact of these agents on a wide range of biological organisms has not been adequately evaluated. The production of seed crops depends, to a large extent, on normal fertilization and seed development. Therefore, on a practical level, information as to the effect of these agents on the pollen system and fertilization process is essential to predict the effect of these agents on the production of agricultural crops and the world's food supply.

At this time, no standardized bioassays involving any phase of the pollen system are available, but rapid, effective, and inexpensive methods associated with a higher plant species would be desirable as a screening and monitoring device. The simplest and probably the most reproducible bioassay would involve in vitro germination of mature pollen. However, depending on the form and mode of action of the agent being tested, the use of in vitro germination as a bioassay may be impossible or limited as to its application. Therefore, procedures other than in vitro germination should be considered.

This Chapter presents a detailed description of the methods and procedures which are available to examine the effects of various agents on pollen development, function, and transmission. In the general methods section, recent procedures associated with all areas (microsporogenesis through seed and mature plant development) of the pollen systems in many species, but, at this time, not specifically developed to test the effects of toxic agents, are presented. The information in this area suggests potential areas in which the effects of various agents could be evaluated. In the specific examples section, procedures which have been used with a limited number of species to test various forms of toxic agents are given. A brief overview of past accomplishments, present status, and future challenges in this critical area follows.

2 Biological Considerations

To produce consistent reproducible results, various aspects of the species selected and the manipulation of the reproductive system must be considered and standardized as much as possible. Pollen development, function, and transmission are known to be influenced by species, genotypes within species, and even among anthers on the same plant (Pfahler 1968, 1978; Stanley and Linskens 1974; Shivanna and Johri 1985; Heslop-Harrison 1987; Vasil 1987). In addition, numerous environmental stress factors to which the pollen parent is exposed during growth and microsporogenesis influence pollen characteristics and response patterns (Stanley and Linskens 1974; Pfahler 1978; Heslop-Harrison 1987; Shivanna and Johri 1985; Vasil 1987). Therefore, pollen from a number of plants within the same genotype grown under near optimal conditions should be collected at one time and mixed thoroughly. In most species, pollen viability can be maintained either by desiccation and/or low temperature treatments for extended periods of time (Stanley and Linskens 1974; Barnabas and Rajki 1976, 1981; Shivanna and Johri 1985; Pfahler 1986). Although much less information is available concerning the effect of genetic and environmental factors on the female portion of the reproductive system, the same precautions should be observed when in vivo procedures are used (Pfahler 1967a, 1975; Bergamini-Mulcahy and Mulcahy 1985; Barnabas et al. 1988).

3 General Methods

Potentially toxic agents occur in many forms (gases, liquids, solids), may be responsible for indirect effects, and have varied genetic and physiological activities. Various general techniques, which could be adapted depending on the form of the agent and the type and magnitude of their effect on the pollen system, are available in a number of species. In this section, the pollen system is arbitrarily divided into three phases (development, function, transmission) with recent techniques described for each (Fig. 1).

3.1 Pollen Development

In maize (*Zea mays*), male inflorescences (tassels) were cut from the plant when the first anthers were beginning to dehisce (Pfahler et al. 1986a). Their cut ends were then placed in H_2O in which agents of interest could be introduced. Under these conditions (detached tassels maintained in a low light intensity, high humidity, 25–30 °C environment), all pollen grains from the tassels were normally released over a 2–3-day period (compared to

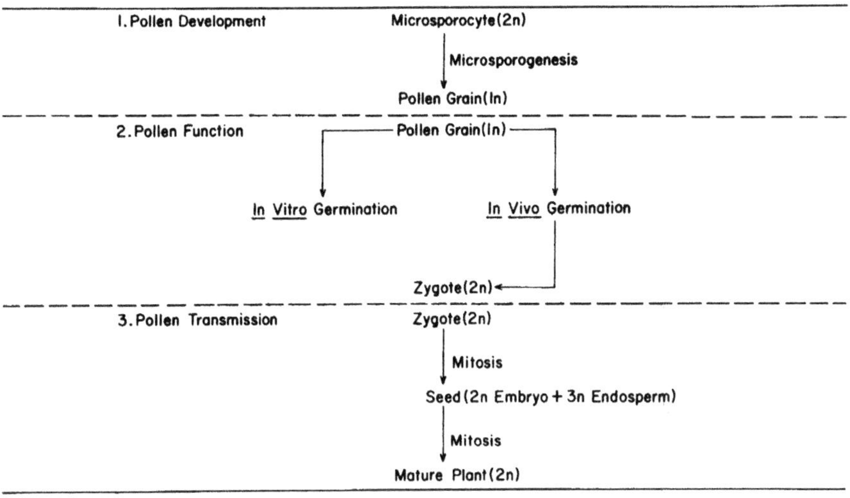

Fig. 1. Diagrammatic representation of the stages in pollen development, function, and transmission which could be altered by various ecotoxic agents

attached tassels which release pollen over an 8- to 10-day period under field conditions), with the most immature pollen grains released later in this time period. Pollen grains obtained with this procedure were capable of in vitro germination and could fertilize in vivo.

A more complex technique in which agents of interest could be introduced early in microsporogenesis has been reported in maize (Polowick and Greyson 1982). Tassel explants (about 1 cm long with about 600 spikelet primordia) were dissected from the plant at about the eighth leaf stage. After removal, the explants were surface sterilized (15% sodium hypochlorite solution) and placed in the following mineral medium (Murashige and Skoog 1962), in which agents of interest could be introduced.

Mineral Medium ($mg\,l^{-1}$) – NH_4NO_3 (1650), KNO_3 (1900), $CaCl_2 \cdot H_2O$ (440), $MgSO_4 \cdot 7H_2O$ (370), KH_2PO_4 (170), Na_2-EDTA (37.3), $FeSO_4 \cdot 7H_2O$ (27.8), H_3BO_3 (6.2), $MnSO_4 \cdot H_2O$ (22.3), $ZnSO_4 \cdot 4H_2O$ (8.6), KI (0.83), $Na_2MoO_4 \cdot 2H_2O$ (0.25), $CuSO_4 \cdot 5H_2O$ (0.025), $CoCl_2 \cdot 6H_2O$ (0.025)

This mineral medium was supplemented with sucrose ($34.2\,g\,l^{-1}$), nicotinic acid ($0.5\,mg\,l^{-1}$), pyridoxine ($0.1\,mg\,l^{-1}$), thiamine ($0.1\,mg\,l^{-1}$), glycine ($3.0\,mg\,l^{-1}$), myo-inositol ($100\,mg\,l^{-1}$), and kinetin ($0.022\,mg\,l^{-1}$). The pH of the final medium was adjusted to 5.8 with NaOH or HCl. The medium was distributed to Erlenmeyer flasks and autoclaved. After introduction of the explants in the medium, the flasks were placed on a rotary shaker ($90\,rev\,min^{-1}$) and exposed to illumination (cool white fluorescent tubes + incandescent bulbs -50 ± 10 microeinsteins $m^2 s^{-1}$) at a 18 h (24 °C)/6 h (20 °C) photoperiod. After 14–20 days, the cultured materials were fixed in a modified paraformaldehyde-glutaraldehyde or formalin-propionic acid-alcohol fixative and embedded in paraffin. For chromosome squashes, anthers were fixed in alcohol-glacial acetic acid (3:1) and stained with propiocarmine. Binucleate and trinucleate pollen grains were present.

A combination of these two techniques in which the stage of tassel development and microsporogenesis could be varied was recently reported in maize (Sari Gorla et al. 1989). In this procedure, the tassels were cut from the plants about 14 days before the beginning of anthesis and the cut ends were placed in an aqueous solution containing the agent of interest for 24 h. Thereafter, the cut ends were put in a supplemented mineral medium described in the last paragraph. The tassels in the medium were placed in a growth chamber until normal dehiscence occurred. Pollen grains obtained in this procedure were released normally and had in vitro germination and fertilization capacity.

A method to determine the mutagenicity of agents applied to the sporophyte prior to pollen development and/or during pollen development is available in maize (Plewa and Wagner 1981). In an aqueous iodine stain ($2.5\,g\,l^{-1}$ KI, $750\,mg\,l^{-1}$ I_2), pollen grains containing the *Wx* allele at the waxy locus stain dark blue while those containing the *wx* allele do not stain and appear brick red. This pollen staining relationship is not influenced by the genotype of the parental sporophyte. In this procedure, homozygous *wx-C/wx-C* sporophytes were exposed during sporophytic growth and/or pollen development to the agent of interest. The pollen grains produced by the treated sporophyte were scored as to the frequency of reverse mutations (*wx-C* to *Wx-C*) after iodine staining. The estimates of mutational events and mutagenic activity of the agent are very precise and accurate because large populations of pollen grains can be easily classified. The homozygous *wx-C/wx-C* sporophyte must be used since the possibility of intralocus recombination exists if the recessive alleles in the homozygote are not identical on the molecular level (Nelson 1962).

3.2 Pollen Function

3.2.1 In Vitro Methods

The most desirable bioassay method would probably involve the introduction of the agent into the in vitro germination medium and observing its effect on the germination percentage and tube growth. From tests (Brewbaker and Kwack 1963) with 86 species representing 39 families, the most suitable aqueous medium for in vitro germination and tube growth contained sucrose ($100\,g\,l^{-1}$), $Ca(NO_3)_2 \cdot 4H_2O$ ($300\,mg\,l^{-1}$), H_3BO_3 ($100\,mg\,l^{-1}$), $MgSO_4 \cdot 7H_2O$ ($200\,mg\,l^{-1}$), and KNO_3 ($100\,mg\,l^{-1}$). The pollen grains from all species did not require all these components in these concentrations for maximum germination and tube growth. The optimal concentration of sucrose is extremely variable among species. Therefore, the components and concentrations should be altered depending on the species. Because of oxygen exchange requirements, pollen grains from certain species such as maize germinate only on a semisolid medium (Pfahler 1967b) with bacto-agar or gelatin used as a solidifying agent. The surface of the semisolid medium must be kept in highly humid conditions to prevent desiccation. For timed studies, germination activity can be stopped by adding an appropriate killing-fixing preservative (ethyl alcohol-glacial acetic acid-formaldehyde) solution with 10–20% glycerin. The presence of glycerin prevents complete drying of the samples during scoring and measuring.

The scoring of the germination percentage and measuring tube length are usually done through a light microscope or using microprojection equipment. This procedure is very time-consuming and, if the pollen tubes are long and tangled, the tube length measurements are very inaccurate. A more rapid and accurate photometric method to estimate pollen tube growth has been developed for pollen grains which can be germinated in a liquid medium (Kappler and Kristen 1987). Pollen of *Nicotiana sylvestris* was collected during early stages of anthesis from plants grown under climate chamber conditions (light phase: 14 h; 25 000 lx; 25 °C; 70% relative humidity; dark phase: 10 h; 20 °C; 70% relative humidity). The pollen was suspended using a stirrer at a concentration of $1\,mg\,ml^{-1}$ in an aqueous medium containing sucrose ($100\,g\,l^{-1}$), $Ca(NO_3)_2 \cdot 4H_2O$ ($700\,mg\,l^{-1}$), and H_3BO_3 ($100\,mg\,l^{-1}$). The medium was buffered with MES (10 mM) and adjusted to pH 5.6 with KOH. The pollen was cultured in 25-ml Erlenmeyer flasks, each containing 1 ml of the aqueous pollen suspension. The conditions were in the dark on a rotary shaker. At various time intervals, the germinating pollen was suspended in 9 ml H_2O and sonicated for 20 s at a setting of 4 using a B12 sonifer cell disruptor (Branson Sonic Power Company, Danbury, Connecticut). After centrifugation (1 min at 1000 g), the pellet with the pollen grains and ruptured tubes was resuspended in 3 ml H_2O. The turbidity of this suspension was photometrically measured at a wavelength of 500 nm with a Beckman Model 34 photometer. A zero-time control was established from cultures killed with 10 mM formaldehyde. The optical density values vs time (up to 23 h after inoculation) were directly proportional (correlation coefficient = 0.994**, regression coefficient = 0.03084** with no significant deviations). With this photometric method, tube growth measurements can be simplified with improved accuracy and precision.

3.2.2 Semivivo and In-Vivo Methods

Semivivo and in-vivo germination can be used to examine the effect of various agents applied during pollen development or to the pistil pre- or postpollination. The easiest and most rapid procedure to observe pollen tubes in the style of many species is the fluorescent method (Shivanna and Johri 1985). Pollen tubes contain callose deposits and plugs which can be stained with aniline blue. In this procedure, pollinated styles were fixed in a glacial acetic acid-alcohol or glacial acetic acid-alcohol-formaldehyde solution at various time periods after pollination and then cleared with 8 *N* NaOH until the tissue became soft. Instead of the NaOH treatment, softening in some species can be accomplished more readily with a 5% sodium sulfite solution at 100 °C for 1 min or by autoclaving in a 5% sodium sulfite solution for 10–15 min. After softening and washing, the stigma and style were cut from the ovary, flattened by gentle pressure under a coverslip, mounted in a 0.005% decolorized aniline blue in 0.05 *M* Na_2HPO_4 solution, and examined with a fluorescent microscope. The callose in the tubes show a bright yellow fluorescence. If the pollen tubes are long, numerous, and tangled, accurate measurement of pollen tube length is difficult.

A quadruple staining method to examine in-vivo pollen tube growth was recently developed and tested successfully in a number of diverse species (Alexander 1987). The staining mixture consisted of the following added in the order and quantity given: lactic acid, 80 ml; 1% aqueous malachite green, 4 ml; 1% aqueous acid fuchsin, 6 ml; 1% aqueous aniline blue, 4 ml; 1% orange G in 50% ethyl alcohol, 2 ml; and chloral hydrate, 5 g. Pollinated pistils were fixed for 6 h in a modified Carnoy's solution (6 parts 100% ethyl alcohol : 4 parts chloroform : 1 part glacial acetic acid), hydrated in descending alcohol concentrations, transferred to the staining mixture, and stained for 24 h at 45 ± 2 °C. The pistils were then removed and placed in a clearing and softening solution (78 ml lactic acid : 10 g phenol : 10 g chloral hydrate : 2 ml 1% orange G in 50% ethyl alcohol) for 24 h at 45 ± 2 °C. Thereafter, the pistils were transferred to a fresh clearing and softening solution and hydrolyzed for 30 min at 58 ± 2 °C. Following hydrolysis, the pistils were washed twice in lactic acid and were either stored in lactic acid for later observation or mounted in a medium (1 part lactic acid : 1 part glycerol) for immediate examination. If, after this treatment, the pistils were not softened and capable of being flattened when light pressure was applied to the coverslip, further softening using the clearing and softening process for an additional 24 h at 55 ± 2 °C was suggested. If chloral hydrate is unavailable, it may be omitted, but the staining and clearing periods should be doubled and the temperature for each process increased from 45 ± 2 to 55 ± 2 °C. Using this method, the pollen tubes were stained dark blue and stylar tissue light green to greenish blue. The pollen tubes were identifiable on the stigmatic region and their growth could be traced along the entire length of the style. The method worked well for species with either hard or soft pistils. The advantages of this method are: (1) applicability to a broad range of species; (2) analysis and photomicrography using a light microscope and/or microprojector; and (3) pollen tubes could be readily traced through the style up to their entry into the micropyle.

A semivivo technique (Bergamini-Mulcahy and Mulcahy 1985) has been developed using petunia (*Petunia hybrida*) and a tobacco species (*Nicotiana alata*). In this procedure, prior knowledge of the rate of pollen tube growth in the style is essential. When the pollen tube front was about one-third down the style, the styles were cut just ahead of the front. The cut ends of the style were then placed in an aqueous medium [$100\,g\,l^{-1}$ sucrose, $300\,mg\,l^{-1}$ $Ca(NO_3)_2 \cdot 4H_2O$, $100\,mg\,l^{-1}$ H_3BO_3]. The styles were then incubated for 8–10 h. The pollen tubes grow out of the cut ends of the style into the medium and average measurements of the tube lengths are possible.

A procedure has been developed in maize in which pollen tube lengths can be accurately measured and embryos capable of producing mature plants can be obtained (Dupuis and Dumas 1989). This method involved the dissection (under sterile conditions) of intact unpollinated pistillate spikelets from the basal half of the cob (female inflorescence) of the plant. These spikelets were then placed in sterile 90 × 15-mm petri dishes containing the mineral medium (Sect. 3.1) supplemented with $0.5\,mg\,l^{-1}$ nicotinic acid, $0.1\,mg\,l^{-1}$ thiamine, $0.1\,mg\,l^{-1}$ pyridoxine, $3.0\,mg\,l^{-1}$ glycine, $100\,mg\,l^{-1}$ myo-inositol, $0.022\,mg\,l^{-1}$ kinetin, $50\,g\,l^{-1}$ sucrose, $500\,mg\,l^{-1}$ casein hydrolate, and $7\,g\,l^{-1}$ bacto-agar. The final pH was adjusted to 5.6 with NaOH or HCl. These petri dishes were not covered directly, but were enclosed in larger sterile petri dishes. The silks were cut to a length of 4.5 cm and the cut ends allowed to extend outside the rim of the smaller petri dishes. Normally dehisced pollen grains were applied to the distal ends of the silks extending outside the smaller petri dishes with a small brush. The larger dishes were then covered and incubated for 24 h at 28 °C (dark) until after fertilization. The silks were then removed by cutting and the smaller dishes were covered, wrapped in parafilm and placed in an incubator (28 °C, dark). Fifteen days after pollination, the developing embryos were removed and placed on the mineral medium (Sect. 3.1) supplemented with $0.5\,mg\,l^{-1}$ nicotinic acid, $0.1\,mg\,l^{-1}$ thiamine, $0.1\,mg\,l^{-1}$ pyridoxine, $40\,g\,l^{-1}$ sucrose and $8\,g\,l^{-1}$ bacto-agar. The plantlets derived from these rescued embryos were grown in the greenhouse until flowering. With this procedure, the growth of the pollen tubes in the silk could be measured using autoradiographic labeling techniques (House and Nelson 1958). On the first day of anthesis, the tassels (male inflorescences) were removed from the plant and the cut ends immediately placed in distilled water containing radioactive ^{32}P in the form of phosphoric acid (activity $-100\,\mu Ci\,ml^{-1}$). Radioactive pollen grains were collected after 24 h and used for pollination as described earlier in this paragraph. At various times (up to 8 h) after pollination, the ovaries with the attached silks were removed, gently rinsed to remove any ungerminated radioactive pollen, and dried under an infra-red lamp at 85 °C for 5 min. The dried pistils were then glued on paper and mounted in autoradiographic frames with X-ray film which was developed after 2 weeks exposure. The white streaks on the photographs could then be measured as an indication of pollen tube growth.

3.3 Pollen Transmission

Pollen transmission encompasses not only the interruption of the fertilization process expressed in reduced seed numbers, but also the differential transmission of various genetic factors. If transmission of genetic factors is to be examined, a species in which extensive genetic information is available must be used.

A method to determine both of these aspects is available in maize (Pfahler et al. 1986b). In this procedure, homozygotes for each endosperm mutant (waxy, sugary, shrunken-2) were pollinated with pollen from the heterozygotes of the same mutant. The agent of interest can be introduced at any stage in the heterozygote up to the mature pollen grain stage or can be applied in aqueous solution to the silk immediately preceding pollination. The seed number and segregation pattern for the endosperm mutant are tabulated for each of the resulting ears to determine the effect of the agent.

4 Specific Examples

The development of any bioassay is greatly complicated by the number and diversity of the agents to be tested, their specific mode of action on the genetic and/or physiological processes, the stage in the life cycle of higher plants and animals where exposure would normally occur, and the stage in the pollen system where their effect is determined. In this section, selected examples of tests involving various forms of known toxic agents introduced at various pollen stages are presented.

4.1 Gases

Various gases (O_3, SO_2, NO_2, CO, CO_2, HN_3, etc.) produced by many agricultural, industrial, and biological activities, are known to influence the vegetative and reproductive growth and development of both plants and animals (Wolters and Martens 1987).

4.1.1 Ozone (O_3)

A test to measure the effect of atmospheric O_3 on the in vitro germination of pollen grains from tobacco (*Nicotiana tabacum*), petunia (*Petunia hybrida*), and tomato (*Lycopersicum esculentum*) is available (Feder 1981). These species were chosen since the sporophytic portion of their life cycle is sensitive to O_3 levels encountered frequently in highly-populated, industrial areas. The pollen grains were lightly dusted on the surface of a semisolid medium [$100\,g\,l^{-1}$ sucrose, $300\,mg\,l^{-1}$ $Ca(NO_3)_2 \cdot 4H_2O$, $100\,mg\,l^{-1}$ H_3BO_3, $200\,mg\,l^{-1}$ $MgSO_4 \cdot 7H_2O$, $100\,mg\,l^{-1}$ KNO_3, $7.5\,g\,l^{-1}$ bacto-agar] which was covered on one side of a #1 coverslip. The coverslip was immediately placed in a well under a microscope. This well was part of a closed recirculating system in which O_3 was generated (General Electric quartz ozone bulb – GE S411) and its concentration controlled (voltage regulator) and monitored (ozone meter). The relative humidity in this closed system was maintained at 100% by bubbling the air stream through H_2O. At appropriate time intervals, photomicrographs of the fields were taken and the germination and tube lengths were determined from the negatives.

4.1.2 Sulfur Dioxide (SO_2)

A test to measure the effect of atmospheric SO_2 on the in vitro and in vivo germination and tube growth in petunia has been reported (Linskens et al. 1985). For the in vitro studies, hydrated pollen grains from clone W166H were placed in a germination medium ($100\,g\,l^{-1}$ sucrose, $100\,mg\,l^{-1}$ H_3BO_3) on slides which were immediately placed in closed petri dishes at 26 °C. The petri dishes contained 20 ppm atmospheric SO_2 generated by mixing appropriate amounts of HCl and $Na_2S_2O_5$ (Adam 1976). At 2 h, germination counts were made after fixing and staining with a 1% aniline blue solution (0.1% aniline blue, 10 ml lactic acid, 10 ml glycerin, 10 g phenol, 10 ml distilled H_2O). At 3 h, pollen tube length measurements were made after 1% aniline blue staining. For the in vivo stylar studies, pollen grains from clone W166H were placed on styles of the compatible clone T2U. The pollinated styles were immediately placed in closed plexiglass containers. Three forms of stylar attachment could be tested: (1) detached from the flowers; (2) attached to the flowers which were detached from the plants; and (3) attached to the flowers and plants. The atmospheric SO_2 level in the chambers was generated by mixing appropriate amounts of HCl and $Na_2S_2O_5$ and the chambers were equilibrated for 17 h prior to the introduction of the pollinated styles. At 17 h after pollination and introduction into the chambers, the styles were removed from the chambers. Using the fluorescent method (Sect. 3.2.2), pollen germination and tube length measurements were made. For the in vivo seed set studies, the same procedure with plexiglass chambers was used. The plants with pollinated styles were removed from the chambers after 17 days and allowed to mature and produce seeds in an environment without enhanced SO_2 levels. Seed number per flower was used as an indication of seed set.

Increased atmospheric SO_2 levels has been associated with acid rain or rainfall with increased acidity. A test to measure the effect of pH on in vitro pollen germination and tube growth in 13 forest species has been reported (Cox 1983). The aqueous germination medium contained $300\,mg\,l^{-1}$

$Ca(NO_3)_2 \cdot 4H_2O$, 100 mg l^{-1} H_3BO_3, 200 mg l^{-1} $MgSO_4 \cdot 7H_2O$, and 100 mg l^{-1} KNO_3. Depending on the osmotic requirements of the species, sucrose concentrations ranged from 6–200 g l^{-1}. The pH of the medium was adjusted to simulate acid rain effects by adding dilute H_2SO_4. The cultures were conducted by placing 80–100 pollen grains in a 50 µl standing drops of this medium. These cultures were supported on 22-mm^2 glass coverslips on four 1-mm-wide black plastic strips above moist filter paper inside 14.5-cm petri dishes. At various times after inoculation, activity was stopped by adding 50 µl of 50% acetic acid. The culture was then mounted on a microscope slide under a coverslip. The germination percentage and tube lengths were determined using a light microscope.

4.2 Electromagnetic Radiation

Most of the agents act directly but, in some instances, unpredictable, indirect effects are possible. For example, chlorofluorocarbons are widely used as solvents, aerosol propellants, and refrigerants, and are released into the biosphere during their manufacture, use, and disposal. These compounds have no direct effect on biological organisms. However, in the stratosphere, these compounds catalytically reduce the ozone layer which, in turn, allows an increase in ultraviolet radiation in the B range (UVB = 280 – 320 nm) to reach the biosphere. UVB is known to produce a wide range of effects in biological organisms (Maugh 1980).

A test to measure the effect of elevated UVB levels on in vitro pollen germination and tube growth in maize has been reported (Pfahler 1981). Pollen grains were inoculated on a semi-solid germination medium [150 g l^{-1} sucrose, 300 mg l^{-1} $Ca(NO_3)_2 \cdot 4H_2O$, 100 mg l^{-1} H_3BO_3, 6 g l^{-1} bacto-agar] in petri dishes. Immediately after inoculation, the dishes were covered with a 5 mil cellulose acetate filter and placed under a FS40 sunlamp. This filter removed radiation below 280 nm and also maintained a 100% relative humidity in the dishes. Activity was stopped at various time periods by flooding the surface with a killing-fixing solution (72 ml H_2O, 5 ml formaldehyde, 3 ml glacial acetic acid, 20 ml glycerin). After flooding, the germination percentage and tube length data was taken immediately or after 2 °C storage using microprojection equipment.

4.3 Phytotoxic Agents

Herbicides are widely used in commercial agriculture to eliminate competing undesirable species. Because of the unpredictable degradation rates of some herbicides, knowledge concerning the possibility of selecting for genetic resistance or tolerance in susceptible succeeding crops is essential.

A method to test the effect of a herbicide on pollen development and function, and the effect of selection for tolerance has been developed in maize (Sari Gorla et al. 1989). Hybrid F_1 plants obtained by crossing tolerant and sensitive inbreds are necessary. In the pollen development portion of the method, tassels were cut from the F_1 plants about 14 days before the beginning of anthesis. The cut ends were immediately submerged in an aqueous solution containing chlorosulfuron and allowed to remain for 24 h. This herbicide is very resistant to degradation under certain environmental conditions and may damage maize when included in the rotation. Also, genetic variation for tolerance has been identified in maize. After removal from the aqueous solution, the cut ends were placed in the supplemented mineral medium described in Sect. 3.1. Normally dehisced pollen grains from these tassels were used to pollinate the sensitive inbred parent, assuming tolerance was at least partially dominant. In the pollen function portion of this method, an aqueous solution of the herbicide was applied to the silks of the sensitive inbred parent immediately prior to pollination with untreated pollen from the F_1 plants. The pregerminated kernels from each portion of this method were planted in sand which contained various levels of chlorosulfuron. After 21 days, height and root length of the seedlings were measured to determine the effect of selecting for genetic tolerance at the pollen level.

4.4 Pesticides and Adjuvants

In commercial agriculture, pesticides with appropriate adjuvants (surfactants, adhesives, etc.) are applied during the life cycle of the crop to control insects, fungi, etc. In many crop species such as maize, the pollen grains and early germination processes on the style are directly exposed to these materials. Any adverse effects on the fertilization process at this stage would be reflected in reduced seed number and yield.

An in vitro method has been reported in maize (Pfahler et al. 1980). This method involved adding the agent of interest to a semisolid medium [$150\,g\,l^{-1}$ sucrose, $300\,mg\,l^{-1}$ $Ca(NO_3)_2 \cdot 4H_2O$, $100\,mg\,l^{-1}$ H_3BO_3, $6\,g\,l^{-1}$ bacto-agar]. The pollen grains were then placed on the medium in closed petri dishes (100% relative humidity). At various time intervals, the germination percentage and tube lengths were measured with microprojection equipment.

An in vivo method associated with pollen function has been developed using endosperm mutants (waxy, sugary, shrunken-2) of maize (Pfahler et al. 1986b). In this procedure, an aqueous solution containing the agent of interest is applied to the silks of the homozygous recessive plant immediately prior to pollination with pollen from heterozygotes of the same mutant. The seed number and segregation pattern for the endosperm mutant was tabulated for each of the resulting ears to determine the effect of the agent on seed number and the pollen transmission at the mutant locus.

4.5 Biological Toxins

Many biological toxins are released which are cytotoxic to plant and animal cells and produce severe and highly complex disruptions of many cell structures and activities.

A semivivo method (Kandasamy and Kristen 1989) to measure the effect of triethyl chloride (TriEL), which is a water-soluble degradation product of the "antiknock" fuel additive, tetraethyllead, has been developed using a tobacco species (*Nicotiana sylvestris*). Plants were grown in a growth chamber (light phase: 14 h, 25 °C, photon fluence rate = 110 μmol $m^2 s^{-1}$, 70% relative humidity. Dark phase: 10 h, 20 °C, 70% relative humidity). The styles were artificially pollinated. Twenty-four hours after pollination, the styles were cut a few millimeters ahead of the pollen tube front and the cut ends were immersed in an aqueous medium [100 g l^{-1} sucrose, 100 mg l^{-1} H_3BO_3, 700 mg l^{-1} $Ca(NO_3)_2 \cdot 4H_2O$, pH = 7.0] containing various TriEL levels. Using fluorescent microscopy (Sect. 3.2.2), the length of the entire tube bundle which emerged from the cut end after various time periods was measured. From a linear dose-response curve, ED_{50} (mean effective dose) values were estimated. For electron microscopy studies, the bundles were fixed in phosphate-buffered 2.5% glutaraldehyde (pH 7.0 at 20 °C for 3 h), washed repeatedly in the buffer, post-fixed in buffered 1% osmium tetraoxide (at 20 °C for 4 h), rinsed in 0.1 *M* sodium acetate, stained for 24 h in 0.5% uranyl acetate, dehydrated in an acetone series, and embedded in Spurr's medium. Ultrathin (60-nm) sections were stained with lead acetate and examined with an electron microscope. For cytochemical localization of the polysaccharides, sections (90 nm) were collected on plastic rings and treated with periodic acid-thiocarbohydrazide-silver proteinate. The sections were then transferred to copper grids and observed with an electron microscope. Studies on cell wall deposits were conducted on semithin (0.5–1.0-μm) sections using fluorescent microscopy (Sect. 3.2.2) after appropriate staining. The stains available include periodic acid-Schiff's reagent, Calcofluor white ST, or aniline blue. This method allows for the determination of the many varied effects associated with toxic agents in the biosphere.

5 Overview

Much research in this specialized area has involved examining the effect of toxic agents on seed production and/or propagation of major agricultural crops (Pfahler and Linskens 1983). As a result, the methods have been conducted on species in which the reproductive system is directly exposed to these agents and seed production and/or propagation is of primary importance. In general, the methods have been very successful in predicting the effects of various toxic agents on the fertilization process and subsequent seed production and/or propagation. In terms of pollen and pistil manipulation, ease of in vitro germination and the availability of extensive genetic information, the species involved (maize, tobacco, tomato, etc.) are well adapted to studies of this nature. Recently, selection for genetic resistance or tolerance to reduce or eliminate the deleterious effects of these agents has been emphasized (Gressel 1985). Since a relationship between the pollen and sporophytic response has been established (Ottaviano and Mulcahy 1989), selection on the pollen level has been advocated as a more effective and efficient method compared to the standard sporophytic selection (Pfahler 1983). The use of these methods in improving these selection programs would be of great practical value and significance.

More basic research is required to use pollen systems tests to predict the effects of toxic agents on higher plants and animals. The great diversity among the agents, the multiple biological effects of the agents on growth, differentiation, and reproduction, and the numerous environmental interactions greatly complicate the development of an appropriate pollen system bioassay. In a limited number of cases, a relationship has been observed suggesting that pollen systems could be used to predict the effect of various agents on many biological processes (Gentile et al. 1973; Kandasamy and Kristen 1987). However, additional information about these relationships and interactions is essential.

References

Adam DF (1976) Sulfur compounds. In: Stern AC (ed) Air pollution: measuring, monitoring, and surveillance of air pollution, vol 3, 3rd edn. Academic Press, New York, pp 214–257

Alexander MP (1987) A method for staining pollen tubes in pistil. Stain Technol 62: 107–112

Barnabas B, Rajki E (1976) Storage of maize (*Zea mays* L.) pollen at −196 °C in liquid nitrogen. Euphytica 25:747–752

Barnabas B, Rajki E (1981) Fertility of deep-frozen maize (*Zea mays* L.) pollen. Ann Bot 48:861–864

Barnabas B, Kovacs G, Abranyi A, Pfahler PL (1988) Effect of pollen storage by drying and deep-freezing on the expression of different agronomic traits in maize (*Zea mays* L.). Euphytica 39:221–225

Bergamini-Mulcahy G, Mulcahy DL (1985) Ovarian influence on pollen tube growth as indicated by the semivivo technique. Am J Bot 72:1078–1080

Brewbaker JL, Kwack BH (1963) The essential role of calcium ion in pollen germination and pollen tube growth. Am J Bot 50:859–865

Cox RM (1983) Sensitivity of forest plant reproduction to long range transported air pollutants: in vitro sensitivity of pollen to simulated acid rain. New Phytol 95:269–276

Dupuis I, Dumas C (1989) In vitro pollination as a model for studying fertilization in maize (*Zea mays* L.). Sex Plant Reprod 2:265–269

Feder WA (1981) Bioassaying for ozone with pollen systems. Environ Health Perspect 37:117–123

Gentile AG, Vaughan AW, Richman SM, Eaton AT (1973) Corn pollen germination and tube elongation inhibited or reduced by commercial and experimental formulations of pesticides and adjuvants. Environ Entomol 2:473–476

Gressel J (1985) Biotechnologically conferring herbicide resistance in crops: the present realities. In: van Vloten-Doting L, Groot GS, Hall TC (eds) Molecular form and function of the plant genome. Plenum, New York, pp 489–504

Heslop-Harrison J (1987) Pollen germination and pollen-tube growth. Int Rev Cytol 107:1–78

House LR, Nelson OE (1958) Tracer study of pollen tube growth in cross-sterile maize. J Hered 49:18–21

Kandasamy MK, Kristen U (1987) Pentachlorophenol affects mitochondria and induces formation of Golgi apparatus-endoplasmic reticulum hybrids in tobacco pollen tubes. Protoplasma 141:112–120

Kandasamy MK, Kristen U (1989) Influence of tetraethyllead on growth and ultrastructure of tobacco pollen tubes. Environ Exp Bot 29:283–292

Kappler R, Kristen U (1987) Photometric quantification of in vitro pollen tube growth: a new method suited to determine the cytotoxicity of various environmental substances. Environ Exp Bot 27:305–309

Linskens HF, van Megen Y, Pfahler PL, Wilcox M (1985) Sulfur dioxide effects on petunia pollen germination and seed set. Bull Environ Contam Toxicol 34:691–695

Maugh TH II (1980) Ozone depletion would have dire effects. Science 207:394–395

Murashige T, Skoog F (1962) A revised medium for rapid growth and bioassays with tobacco tissue cultures. Physiol Plant 15:473–497

Nelson OE (1962) The waxy locus in maize I. Intralocus recombination frequency estimates by pollen and by conventional analyses. Genetics 47:737–742

Ottaviano E, Mulcahy DL (1989) Genetics of angiosperm pollen. Adv Genet 26:1–64

Pfahler PL (1967a) Fertilization ability of maize pollen grains II. Pollen genotype, female sporophyte and pollen storage interactions. Genetics 57:513–521

Pfahler PL (1967b) In vitro germination and pollen tube growth of maize (*Zea mays* L.) pollen I. Calcium and boron effects. Can J Bot 45:839–845

Pfahler PL (1968) In vitro germination and pollen tube growth in maize (*Zea mays*) pollen II. Pollen source, calcium, and boron interactions. Can J Bot 46:235–240

Pfahler PL (1975) Factors affecting male transmission in maize (*Zea mays* L.). In: Mulcahy DL (ed) Gamete competition in plants and animals. North-Holland, Amsterdam, pp 115–124

Pfahler PL (1978) Biology of the maize male gametophyte. In: Walden DB (ed) Maize breeding and genetics. John Wiley & Sons, New York, pp 517–530

Pfahler PL (1981) In vitro germination characteristics of maize pollen to detect biological activity of environmental pollutants. Environ Health Perspect 37:125–132

Pfahler PL (1983) Comparative effectiveness of pollen genotype selection in higher plants. In: Mulcahy DL, Ottaviano E (eds) Pollen: biology and implications for plant breeding. Elsevier Biomedical, New York, pp 361–366

Pfahler PL (1986) Pollen storage effects on early seedling growth in maize. In: Mulcahy DL, Bergamini Mulcahy G, Ottaviano E (eds) Biotechnology and ecology of pollen. Springer, Berlin Heidelberg New York, pp 147–152

Pfahler PL, Linskens HF (1983) Methods for assessing the effects of chemicals on reproduction in higher plants. In: Vouk VB, Sheehan PJ (eds) Methods for assessing the effects of chemicals on reproductive functions (SCOPE PUBL 20). John Wiley & Sons, Chichester, pp 499–514

Pfahler PL, Linskens HF, Wilcox M (1980) In vitro germination and pollen tube growth of maize (*Zea mays*) pollen IX. Pollen source genotype and nonionic surfactant interactions. Can J Bot 58:557–561

Pfahler PL, Mulcahy DL, Barnabas B (1986a) The effect of forced shedding on pollen traits, seedsetting, and transmission at various maize (*Zea mays* L.) endosperm mutant loci. Acta Bot Neerl 35:195–200

Pfahler PL, Mulcahy DL, Barnabas B (1986b) The effect of prepollination stylar treatments on seedset and pollen transmission at various maize (*Zea mays* L.) endosperm mutant loci. Acta Bot Neerl 35:201–207

Plewa MJ, Wagner ED (1981) Germinal cell mutagenesis in specially designed maize genotypes. Environ Health Perspect 37:61–73

Polowick PL, Greyson RI (1982) Anther development, meiosis and pollen formation in *Zea* tassels cultured in defined liquid medium. Plant Sci Lett 26:139–145

Sari Gorla M, Ottaviano E, Frascaroli E, Landi P (1989) Herbicide-tolerant corn by pollen selection. Sex Plant Reprod 2:65–69

Shivanna KR, Johri BM (1985) The angiosperm pollen structure and function. John Wiley & Sons, New York

Stanley RG, Linskens HF (1974) Pollen biology, biochemistry, management. Springer, Berlin Heidelberg New York

Vasil IK (1987) Physiology and culture of pollen. Int Rev Cytol 107:127–174

Wolters JHB, Martens MJM (1987) Effects of air pollutants on pollen. Bot Rev 53:372–414

The Determination of the Allelopathic Potential of Pollen and Nectar

S.D. MURPHY

1 Introduction

Traditionally, studies on allelopathy have been concerned with the effects of chemicals exuded from roots, rhizomes, leaves, glandular hairs and/or litter of one plant on the growth of its neighbours, usually heterospecifics (Rice 1984). The possibility that pollen might also have an allelopathic function on heterospecifics has been largely unexplored. This may be due mainly to the often unwarranted presumption that pollination is a tightly co-evolved, highly efficient process (see discussion in Faegri and van der Pijl 1979).

Pollen allelopathy describes the deleterious effects of chemicals in the pollen of certain species on pollen germination, pollen tube growth, stigma and/or style longevity, sporophytic respiration, leaf chlorophyll content and seed set in other species (Char 1977; Sukhada and Jayachandra 1980a; Thomson et al. 1982; Jimenez et al. 1983; Ortega et al. 1988; Murphy and Aarssen 1989). It is important to recognize that pollen allelopathy is a unique phenomenon (see also Knox et al. 1987). For example, it should not be confused with cross-incompatibility, an immunocytochemical interaction strictly between pollen, stigma and/or style (de Nettancourt 1977; Heslop-Harrison 1982).

"Allelopathy" is a term that has defied succinct definition. In the context of pollen allelopathy, the interaction is normally between species of Anthophyta. Nonetheless, "allelopathy" also may be applied to other types of interactions (plant-animal, plant microorganism and vice versa) (see Cruickshank and Perrin 1964; Towers 1964; Rice 1984; Lovett et al. 1989; Harborne 1990). Thus, such mechanisms as "nectar allelopathy" may also exist, although little research has been done in this area.

Furthermore, the vernacular use of "allelopathy" restricts it to describing the detrimental effects on the neighbours of an allelopathic individual. As Rice (1984) notes, "allelopathic" interactions can be defined so broadly as to include those that are beneficial. This assumes that there is a continuum, where low concentrations of a chemical are stimulatory and high concentrations are toxic (see Rice 1984). In terms of pollen-stigma or pollen-pollen interactions, such a continuum does not always exist. Granted, compounds such as calcium may have concentration-dependent effects (Stanley and Linskens 1974). Pollen allelochemicals, however, have not demonstrated such a trend (Murphy and Aarssen unpub.).

Semantics aside, some studies on the various forms of allelopathy have engendered controversy because of the difficulty in separating the effects of competition from allelopathy (Fuerst and Putnam 1983; Rice 1984; Henn et al. 1988; Weidenhamer et al. 1989). Further, controlling for the many variables encountered when extracting and adding putative allelochemicals to soil media presents enormous problems in designing such experiments (see Henn et al. 1988; Qasem and Hill 1989). In terms of evolutionary theory, the existence of autotoxicity or "auto-allelopathy" is difficult to reconcile. There seems to be little advantage in suppressing neighbours if the allelopathic individual and/or its progeny are similarly affected by the inhibitory compounds (Whittaker and Feeny 1971; Murphy and Aarssen 1989; Thorne et al. 1990).

Research into pollen allelopathy should be expedited by the ability to manipulate and control the stigmatic environment and the pollination process as a whole. Nonetheless, designing methods of analyzing pollen-allelopathic interactions must be considered against all of the issues presented thus far. It is from this perspective that the present review has been written.

2 Selecting Test and Target Species

There are many criteria that may be used to select test species, i.e., those that may be pollen-allelopathic. Approaches may include utilizing genera reported to have some form of allelopathic effect (e.g., litter exudate), empirical evidence of inhibition on sporophytic growth by foreign pollen or random sampling. Species tested thus far include *Agropyron repens*, *Phleum pratense*, *Poa pratensis*, *Zea mays* var. *chalquinoconico* (all Poaceae), *Ambrosia artemisiifolia*, *Erigeron annuus*, *Hieracium aurantiacum*, *H. floribundum*, *H. pratense* (all Asteraceae), *Melilotus alba*, *Vicia cracca* (both Fabaceae) and *Brassica oleracea* (Brassicaceae) (Sukhada and Jayachandra 1980a; Thomson et al. 1982; Hodgkin and Lyon 1983; Jimenez et al. 1983; Ortega et al. 1988; Murphy and Aarssen 1989; unpub.). Extract from the "male flowers" of *Cucurbita pepo* (Cucurbitaceae) is apparently allelopathic and hence, so may its pollen (Itokawa et al. 1982). The target species, those that may be susceptible to allelopathic pollen, may be selected randomly (Murphy and Aarssen 1989). Alternatively, targets may be selected because they are likely to receive allelopathic pollen from other species (Sukhada and Jayachandra 1980a; Ortega et al. 1988; Murphy and Aarssen unpub.).

3 The Extraction of Putative Pollen Allelochemicals

3.1 Overview

Once the test and target species are chosen, the next task is then to extract any allelochemicals from the pollen of the test species. The literature suggests that allelochemicals are mainly phenolics, terpenoids and alkaloids (Rice 1984; Harborne 1990). In addition, the amino acid fraction may be useful as gamma-aminobutyric acid has been found to be allelopathic to herbivores (see Baker and Baker 1975). Therefore, extraction methods may be designed to maximize isolation of phenolics, terpenoids, alkaloids and non-protein amino acids. This is still a broad categorization, however, it does provide some delineation.

3.2 Whole-Pollen Extraction in Polar Solvents

Normally, the initial purpose of the investigation is to maximize the number of pairwise species combinations tested for evidence of pollen-allelopathic interactions. For this, the most convenient method is to extract intact (whole) pollen with a polar solvent (e.g., distilled water; Murphy and Aarssen 1989). This is based on previous studies that have indicated that the allelopathic activity in the Asteraceae and Poaceae is contained mainly in water-soluble fractions (Sukhada and Jayachandra 1980a,b; Thomson et al. 1982; Jimenez et al. 1983; Ortega et al. 1988; Murphy and Aarssen 1989, unpub.).

At anthesis, a known concentration of pollen is harvested, where the sample should be as large as possible (cf. 10–100 million pollen grains). This task is made easier in several ways, the first of which concerns the pollination system itself. With most anemophilous species, the pollen is produced in large quantities and it is quite easy to shake it from the anthers into a collection receptacle. With entomophilous species, the pollen can be removed by vigorous grating of the anthers over a stacked filter-sift system. The top filter should have a mesh between 40 and 90 µ to allow the pollen to fall into the second receptacle. The second filter should have a mesh of 20 µ to permit flow through of fungal spores and other small contaminants.

The harvested pollen is then immersed in distilled water and centrifuged at low speed for 3–6 h. Care must be taken to avoid any contamination (pollen of other species, bacteria, fungi, anthers or stigmatal tissue). After centrifugation, the mixture is assayed for the percentage of disrupted pollen grains in order to ensure that any allelochemicals were exuded rather than ejected (from an undue amount of broken grains). A level of 5% disruption is typical for this procedure (see Murphy and Aarssen 1989). The mixture is then filtered and stored in a sterile, sealed container at 9 °C for at least 72 h (this reduces the risk of loss of chemicals due to extraction-induced volatility). Contamination is avoided easily if proper sterile techniques are used.

This method is useful when the investigation is constrained by time, manpower or budgets. Nonetheless, there is a risk of failing to detect any pollen-allelopathic interactions that may be mediated by less polar compounds (see Sukhada and Jayachandra 1980a,b; Jimenez et al. 1983; Ortega et al. 1988). Further, the strength of the pollen-allelopathic interaction may be reduced by this method if several different classes of compounds are acting synergistically. Finally, this method is not particularly useful if one wishes to isolate and/or characterize the specific pollen allelochemical(s).

3.3 Polar Fractionation of Pollen Extracts

As research into pollen allelopathy has progressed, it has become apparent that a more systematic separation method is needed than simply soaking the pollen in aqueous or organic solvent. Since pollen contains many different classes of compounds, one convenient method is to separate them on the basis of their polarity. The fractions may be used in a bioassay (Sect. 4) or analyzed using more sophisticated methods (Sect. 7).

The procedure outlined by Harborne (1984) has been a useful fractionation method in recent experiments (Murphy and Aarssen unpub.). Several modifications have been made to increase the extraction efficiency. In the initial step, the pollen must be ground with a mortar and pestle for complete extraction. A blender is normally impractical, as the pollen is too small to be ground by the blades. The ground pollen is homogenized in a 5:1 methanol/water solution for 15 min and filtered. The supernatant (filtrate) will contain the more polar extracts, while the residue (bolus) will contain the non-polar compounds. The literature suggests that allelochemicals are somewhat polar, hence the more useful fraction is the supernatant (see Rice 1984). Nonetheless, the residue should be stored at 5 °C, using 0.1% toluene as an fungicide (see also Harborne 1984).

The supernatant is evaporated slowly at temperatures not exceeding 45 °C to reduce loss of the more volatile compounds. Once the supernatant volume is reduced by 90%, 5 N sulfuric acid or 16 N hydrochloric acid is introduced to lower the pH to 2 (see Sukhada and Jayachandra 1980b). Repeated (three to five times) washing of this solution with trichloromethane should separate any moderately polar terpenoids and phenolics. If desired, this trichloromethane extract can be dried and evaporated to leave "pure" terpenoids and phenolics (see Harborne 1984). Air drying is not desirable as there is a greater potential for contamination and some other method (e.g., vacuum pressure) should be used. The acid (aqueous) layer left after the trichloromethane extraction should contain any alkaloids. Jimenez et al. (1983) used a method similar to the general one outlined above in order to fractionate pollen. The pollen was sonicated (instead of macerated) and methylene chloride was used as the initial extraction medium (after several other solvents had been tested; Jimenez et al. 1983). Upon drying and extraction in methylene chloride/methanol, a solid (soluble in organic phase) was obtained, while the aqueous layer was analyzed further (Jimenez et al. 1983). This phase was separated by use of a solvent with low polarity (hexane) and ones of higher polarity (methanol and distilled water). Ortega et al. (1988) abandoned this stepwise extraction process when it was reported that most of the biological activity was found in aqueous layer.

3.4 pH-Based Fractionation: Column Chromatography

One potential difficulty with the polar-fractionation method is that it may introduce bias into the design. For example, the procedure used by Jimenez et al. (1983) is predicated on the implicit assumption that fractions containing phenolics and terpenoids are the most biologically active. Although this may be the case, there have been so few studies on allelopathic pollen that it may be more parsimonious to separate fractions on the basis of a more general character (e.g., pH). The use of column chromatography may be more conducive to examining the sequential loss of biological activity of each fraction. For example, column chromatography has been applied to the study of pollen walls (Baraniuk et al. 1990). The procedures outlined in the present review are useful for generating fractions for both bioassays and characterizations. Reviews of column chromatography are given in Stock and Rice (1974), Deyl et al. (1975) and Poole and Schuette (1984).

The initial extraction procedure may be similar to that used in polar fractionation. The pollen is macerated using a ground-glass mortar and pestle, then extracted with methanol (95%). Use of a polar solvent will promote the extraction of most pollen compounds as they tend to have at least low polarity (see Stanley and Linskens 1974). To ensure that the sample will remain pure and relatively uniform in its properties, the sample may be freeze-dried to −20 °C which will form a pellet. When required, part of the pellet may then be resuspended in distilled water for use in bioassays for pollen-allelopathic interactions (Sect. 4).

The above method may be modified to allow for fractionation of distilled water extracts prior to storage. Sukhada and Jayachandra (1980b) concentrated (10% of original volume) the distilled water extracts under reduced pressure (50 °C). All extracts were then acidified (pH = 2) with concentrated HCl and aliquots (25 ml) equipartitioned against ethyl acetate. These were also concentrated (reduced pressure; 50 °C) and the residue dissolved in a maximum of 15 ml of ethyl acetate. A solution of sodium bicarbonate/sodium carbonate (2 N) was used to produce acidic (carbonate-soluble) and non-acidic (carbonate-insoluble, ethyl acetate-soluble) fractions. Acidic fractions were acidified (pH = 2) with concentrated HCl and re-extracted with ethyl acetate. All fractions were washed (distilled water), concentrated (vacuum pressure) and stored (in a dessicator) for future characterization. The fractions could also have been used in a bioassay, although this was not performed by Sukhada and Jayachandra (1980a,b).

Alumina columns, eluted with petroleum ether:ethyl acetate (70:30), are useful in separating non-acidic allelochemicals such as certain amino acids and alkaloids (see Sukhada and Jayachandra 1980b). Ion-exchange columns eluted with dilute HCl may also be used to separate basic compounds (Murphy and Aarssen unpub.). Neutral compounds are isolated by tandem stacking of AG-50 and AG-1 grade ion-exchange columns eluted with dilute HCl. Silica gel columns (70–120 μ of mesh) may be eluted with benzene:ethyl acetate (90:10 and 90:30) to separate acidic allelochemicals (Sukhada and Jayachandra 1980b). Eluting anion-exchange columns with 1 N HCl will also separate acidic allelochemicals, although there may be a loss of purity in the fractions (Murphy and Aarssen unpub.). Itokawa et al. (1982) used silica gel columns eluted with $CHCl_3$:methanol and *n*-hexane:ethanol (no concentration ratios were given) to separate a mixed fraction of (weakly acidic) phenolics.

4 A Bioassay for Pollen Allelopathy: In Vitro Pollen Culture

4.1 Overview of the Method

In vitro bioassays for pollen-allelopathic interactions are intended as a test of as many species as possible for pollen-allelopathic interactions (Murphy and Aarssen 1989). They may be designed on a large scale to determine which species should be chosen for more rigorous testing or they can be one step in a more linear process of extraction, isolation, characterization, in vitro bioassay and in situ assessment (Sects. 3–7).

Known concentration equivalents of pollen extract (or fraction) are introduced to solid or liquid germination media in a dilution series. Additions begin with an arbitrary concentration of an whole-pollen or fractionated extract (e.g., the equivalent of 50 pollen grains/μl) and are followed by a stepwise dilution to a control concentration (0 pollen grains/μl) (Murphy and Aarssen unpub.).

Pollen from the target species is germinated on a control medium versus one with the pollen extract added. Any reduction in percentage germination of the target pollens by varying the concentrations of extracts from whole pollen or fractions may then be recorded. The percentage pollen germination of the target pollen is then plotted against the concentration of test pollen extract (or fraction) added to the germination medium. This represents the standard toxicity curve for a pollen-allelopathic species versus a targeted species.

4.2 Designing Pollen Germination Media for Bioassays

Studies on pollen allelopathy in *Phleum pratense* have used both liquid media and solid agar media (Murphy and Aarssen 1989, unpub.). Through several pilot experiments, it has been found that a liquid medium containing 300 ppm $CaCl_2$, 100 ppm KH_2PO_4, 100 ppm H_3BO_4, 100 ppm lactose and 2000 ppm sucrose will germinate the pollen of most species (Murphy and Aarssen 1989). A buffer (acetic acid/sodium acetate) is added to maintain a constant pH of 5.8, a value conducive to germination of most pollens (see Stanley and Linskens 1974; Murphy and Aarssen 1989). Of 159 species tested to date, the only ones that have failed to germinate in this medium are two apomicts, *Erigeron annuus* and *Taraxacum officinale* (both Asteraceae) (Murphy and Aarssen 1989, unpub.). Alternatives to this germination medium are reviewed in Vasil (1987).

The use of a standard solution eliminates the logistical problems of providing a pollen germination medium for each species and it also standardizes the response. Since it is relative pollen germination values that are of interest, the fact that a medium may be more conducive to one species over another is irrelevant. Nonetheless, if some species have unusual requirements (e.g., for pH) or if only a few species interactions are to be tested, then an experiment where each targeted species is provided with its optima would be desirable (see Stanley and Linskens 1974).

Well slides are most suited to a rapid, qualitative assay of many pollen-pollen interactions (Murphy and Aarssen 1989). Problems with this method include a greater chance of contamination, the delicate nature of the slides, the chance that the pollen may sink to an anaerobic pocket, difficulty in sealing the slides without excluding all oxygen and the possibility that the pollen will clump, forming anaerobic pockets or creating local saturation of innate growth promoters (Stanley and Linskens 1974; Vasil 1987; Murphy and Aarssen 1989). The pollen grains should be separated with a fine probe to prevent artificially high concentrations of any germination promotors (Brewbaker and Majumder 1961; Stanley and Linskens 1974; Vasil 1987; Murphy and Aarssen 1989). If the cultures are carefully prepared and monitored, the standard errors (percent pollen germination) between replicates may be reduced to 2% or less (versus values as high as 8%; see Stanley and Linskens 1974; Murphy and Aarssen 1989).

To examine the pollen-pollen interactions of fewer species combinations, it may be easier to use solid (agar) media (Murphy and Aarssen unpub.). The composition of the medium is similar to that of the liquid, save for the omission of the sucrose under most circumstances (Murphy and Aarssen unpub.). Known concentrations of a putative pollen-allelopathic extract are introduced into agar more easily than into liquid media. Normally, the extract equivalents are added to give a final volume of extract/μl of medium. This (μl) unit of volume is reasonable considering the small size of most stigmatic surfaces. If mixed carefully, the agar will contain a homogeneous concentration of allelopathic pollen extract (Murphy and Aarssen unpub.). Such a homogeneous profile may not exist in situ but is necessary in order to quantify the pollen-allelopathic effect in vitro. A similar method has been used in assessing the allelopathic potential of sporophytic extracts on the root length and seed germination and growth (Pederson 1986; Dornbos and Spencer 1990). In these studies, the aqueous extracts were layered on the surface of the agar rather than mixed with the agar itself.

Agar is also advantageous in that it is easier to manipulate for contamination control, moisture content, pH, osmotic pressure, aeration, local pollen concentrations, photography and permanent mounting (Bishop 1949; Kubo 1955, 1960; Eigsti 1966; Sukhada and Jayachandra 1980a; Murphy and Aarssen unpub.). Normally, the agar is spread as thinly as possible to prevent pollen from sinking into anaerobic pockets. Kubo (1960) reports sectioning to the scale of microns, however, the method is too cumbersome to be used in large scale experiments. In most cases, one must rely on estimations of thickness but this is attained easily with practice.

Other methods of in vitro culture (Van Tieghem hanging drop tests, spot tests, cellophane membranes) are valid germination tests but these are less conducive to the introduction of a solution containing potentially pollen-allelopathic compounds (see Stanley and Linskens 1974). Staining of pollen grains for viability is not an attractive option as the stains may be inaccurate for certain species and may react with any pollen allelochemicals (see also Hauser and Morrison 1964; Alexander 1969; Heslop-Harrrison and Heslop-Harrison 1970; Parfitt and Ganeshan 1989).

4.3 Pollen Rupture: Causes and Prevention

The primary reason for such meticulous designs of pollen germination media is to avoid artifacts by providing a carefully controlled environment. One of the most obvious indications that the controls have failed is the rupture of the pollen grain or tube of the target species. In no case has this been found to indicate a pollen-allelopathic interaction (Murphy and Aarssen unpub.).

Rupturing of pollen may be caused by an inappropriate pH. This may be a function of the medium itself or chemicals released from the pollen extract (Murphy and Aarssen 1989). Although the alteration of the germination environment by the extract may be an important phenomenon unto itself, any pH effects must be avoided in order to detect pollen-allelopathic interations (see Kubo 1955; Sukhada and Jayachandra 1980a; Ganeshaiah and Shaanker 1988; Murphy and Aarssen 1989). If a buffer was added to the germination media (Sect. 4.3), it should prevent fluctuations of pH and provide optimal or tolerable pH ranges (Sukhada and Jayachandra 1980a; Murphy and Aarssen 1989). Addition of buffers (e.g., sodium-based) may cause bursting due to toxic concentrations of cations (see Stanley and Linskens 1974). This might be avoided by employing several different types of buffers and testing for a significant germination X buffer treatment interaction effect. In experiments on the allelopathic potential of pollen of *Phleum pratense* and *Hieracium* species, there has been no such interaction effect despite the use of an acetic acid/sodium acetate buffer (Murphy and Aarssen 1989, unpub.).

Pollen bursting may also be caused by aniso-osmotic conditions (Murphy and Aarssen unpub.). Adding sucrose to the germination medium normally promotes iso-osmotic conditions and this can be monitored with an osmometer (Jimenez et al. 1983; Murphy and Aarssen unpub.). Unfortunately, adding sucrose to a germination medium increases the risk of bacterial or fungal growth (Bishop 1949). It is imperative that such growth be avoided during the experiment as this affects pollen germination and often causes pollen to burst (Laughnan and Gabay 1973; Murphy and Aarssen unpub.). In some cases, the contaminant may be borne by the pollen in situ. For example, the nectar of *Asclepias syriaca* may harbour a yeast (*Metschinkowia reukaufii*) that may be vectored by the pollen (Eisikowitch et al. 1990).

Lactose may be added as a retardant or substituted for sucrose to prevent such contamination (Bishop 1949; Murphy and Aarssen 1989). It is *not* recommended that fungicides be added, as the sulfur in many of these compounds may interfere with the experiment (Stanley and Linskens 1974; Murphy and Aarssen unpub.). One way to avoid contaminating flora and/or fauna borne by pollen is to surface-sterilize the pollen grains with sodium hypochlorite followed by a brief (1 min) washing with distilled water immediately after harvesting (Tulecke 1954, 1957). Recent experiments have shown that such surface-sterilizing of pollen from the test and/or target species does not significantly alter the results of in vitro bioassays (Murphy and Aarssen unpub.).

4.4 Germinating Pollen from the Target Species

The pollen should be collected when mature (i.e., at anther dehiscence) and germinated as quickly as possible. Speed may be important as tri-nucleate pollen (e.g., Asteraceae and Poaceae) metabolizes so rapidly that germination values can be too low to give useful results (see Mulcahy and Mulcahy 1988). Brewbaker (1967) gives an extensive list of species with bi- and trinucleate pollen. Pollen collected from more than one plant introduces some degree of randomness into the design but this is often necessary as one individual rarely produces enough pollen to provide for sufficient replication (see also Stanley and Linskens 1974). Unfortunately, this will eliminate the detection of any intraspecific variation in pollen-allelopathic strength or vulnerability.

Counting the percentage germination of the target pollen in vitro has been an expedient and repeatable bioassay for allelopathic pollen. There is, however, an alternative means of assessing the effect in vitro. This would be to measure the growth (length) of the pollen tube in the germination medium. Although this is feasible (e.g., Newcomer 1938), the process of measuring tube length is considerably slower than counting the percentage germination. Further, a pilot experiment done in 1987 revealed that the lengths of tube growth in vitro bore little relation to actual length in situ (Murphy and Aarssen unpub.). Conversely, the percentage germination counts in vitro and in situ were not significantly different (by a two-way analysis of variance) (Murphy and Aarssen unpub.). Thus, measuring tube length in vitro might allow the detection of certain putative pollen-allelopathic interactions otherwise overlooked, however, there is a far greater risk of artifacts.

4.5 Replication and Statistical Analyses

In terms of statistical design and logistics, five to ten replicate plates or slides (with controls) are desirable (Murphy and Aarssen unpub.). Within these, it is recommended that five to ten fields of view be chosen randomly for examination under a constant microscopic magnification (e.g., 200×) (Murphy and Aarssen 1989). Using this design, statistical analyses may be kept simple yet informative, as a two-way analysis of variance (ANOVA) is quite suitable to detect significant pollen-allelopathic effects (Murphy and Aarssen 1989).

4.6 The Standard Toxicity Curve

The inhibition of target pollen germination by a varying concentrations of whole pollen or fractionated extracts will produce a quantifiable toxicity curve (Fig. 1). This will show which concentrations of extracts from whole pollen and/or fractions cause the greatest incremental decreases in target pollen germination. The curve obtained when a sequence of diluted extract concentrations are introduced should be non-linear (resembling an inverse cube function; Fig. 1). This reflects a dose-dependent response and demonstrates that the effect is most likely to be pollen-allelopathic in nature. A linear curve may indicate that there is a dilution of some mutually-stimulating pollen germination factor (see Brewbaker and Majumder 1961; Jennings and Topham 1971; Thomson et al. 1982). A curve having a sharp truncation is indicative of some form of contamination or media effect (e.g., toxic pH levels; Murphy and Aarssen unpub.). As a second control, the extraction medium (in the absence of any putative allelochemicals) should

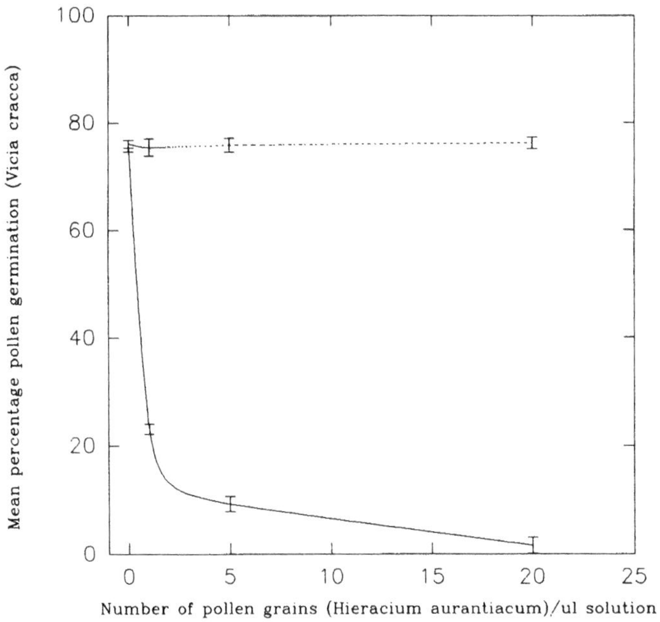

Fig. 1. A sample "standard toxicity curve" for the pollen of the test species, *Hieracium aurantiacum* (Asteraceae). The *curve* shows the dose-dependent response (decrease) of the percentage germination of pollen of *Vicia cracca* (Fabaceae) to increasing concentration equivalents of extract from pollen of *H. aurantiacum*. The *dashed line* represents a second control where solvent, in absence of any pollen allelochemicals, was added to the germination medium

be added to the pollen germination medium. A linear, horizontal curve will re-confirm that there is no "medium effect" (Fig. 1).

The standard toxicity curve for pollen extracts not only indicates whether the effect is truly pollen-allelopathic but also if relevant concentrations of allelochemicals may be toxic enough to cause effects in situ. Ecologically realistic values of concentration equivalents from the putative pollen-allelopathic species are of the most interest. The stigmae of most wind-pollinated species may have up to 25 foreign grains of a given species (versus a similar number of its own pollen) per unit area or volume (Murphy and Aarssen unpub.). The stigmae of animal-pollinated species may be less susceptible to foreign pollen but similar values are still quite valid (see Thomson et al. 1982; Murphy and Aarssen unpub.).

Given these parameters for ecological relevance, most of the in vitro activity of the pollen-allelopathic extract should occur in the concentration range of 1 to 25 pollen grain equivalents/μl. It is cautioned that the range of values for ecological relevance may be more variable than indicated here. An individual of *Parthenium hysterophorus* actually releases less pollen than one of *P. pratense* but the pollen of the former is released in aggregates of 600–800 grains (see Raynor et al. 1971, 1972; Sukhada and Jayachandra 1980a; Sawada 1989; Murphy and Aarssen unpub.). Thus, an ecologically relevant value may be in the tens of thousands of grains for *P. hysterophorus*.

5 An Alternative Bioassay for Pollen Allelopathy Using TLC Plates

Hodgkin and Lyon (1983) germinated pollen on TLC plates in order to assess the inhibitory capacity of extracts from tissues, including pollen, of *Brassica oleracea*. The extraction procedure began with the maceration of the pollen in a homogenizer (25 ml of 70% redistilled ethanol). The slurry was then centrifuged (10 min at 11 000 g) and the supernatant dried in a (rotary) evaporator. After drying, the supernatant was dissolved in distilled water (10 ml) and partitioned (three times) with redistilled ethyl acetate (10 ml). All three partitioned fractions were combined and dissolved in ethyl acetate (0.5 ml), thereby yielding the organic extract.

The extracts were chromatographed on TLC plates in a chloroform: methanol (94:6%) solution and the plates air-dried. A mixture of germination media and the pollen of the target species (*Petunia hybrida* and *Lilium lankongense*) was sprayed onto the TLC plates with an airbrush and the plates incubated in sealed, moist chambers (22 °C).

After 6 h, the TLC plates were air-dried, sprayed with aniline blue (0.1%) in K_3PO_4 (0.1 M) and observed under UV light (350 nm) or under fluorescent microscopes. The germination counts showed distinct bands of inhibition corresponding to various Rf values. In this manner, it may be easier to determine what type of chemical is responsible for the inhibitory effect. It must be noted, however, that the Rf values must be replicated and overlapping zones of inhibition might render the interpretation nebulous. Hodgkin and Lyon (1983) were able to determine Rf values for the inhibitory fractions of pollen of *B. oleracea* but did not identify the compounds. Nonetheless, this method appears useful and should be investigated further with the purpose of refining it.

6 Assessing Effects of Pollen Allelochemicals on Vegetative Tissue

Pollen allelochemicals may not only affect the reproductive capacity of the targeted species; they may also affect the vegetative tissues (Sukhada and

Jayachandra 1980a; Jimenez et al. 1983; Ortega et al. 1988). Several bioassays have been developed to detect such effects.

Sukhada and Jayachandra (1980a) deposited an ecologically relevant number (12 000–20 000) of pollen grains from *Parthenium hysterophorus* on the surface of leaves of *Phaseolus vulgaris*. The treatments (two plants per pot exposed to foreign pollen) were replicated (×10) and grown in a greenhouse with the appropriate control pots (Sukhada and Jayachandra 1980a). After 20 days the chlorophyll content was measured and the plants subsequently dried (Sukhada and Jayachandra 1980a). Both chlorophyll content and dry weight of the shoots were reduced significantly by exposure to pollen from *P. hysterophorus* (Sukhada and Jayachandra 1980a).

The effect of pollen extracts of *Zea mays* var. *chalquinoconico* on the radicle growth of *Bidens pilosa*, *Cassia jalapensis* and *Rumex crispus* was examined by Jimenez et al. (1983). The extracts were added in a 1:1 ratio with agar (2%) and the seeds of target species germinated for at least 72 h (Jimenez et al. 1983). Actual pollen grains were sprinkled on the seeds of the targets in ascending concentrations (0–200 mg; Jimenez et al. 1983). Most treatments appear to have caused inhibition of seed germination and/or seedling growth, although the statistical evidence was rather nebulous (see Jimenez et al. 1983). The standard toxicity curve for various concentrations of pollen was apparently non-linear, but the lack of information on the standard deviations across treatments precludes any conclusions (see Jimenez et al. 1983).

Ortega et al. (1988) used a bioassay based on germination of the seeds of the target species (*Citrullus lanatus*) as well as in vitro studies on the cell division and respiration of the hypocotyl of the target. A methanolic extract from pollen of *Z. mays* var. *chalquinoconico* was introduced to sterile petri plates on which the seeds of *C. lanatus* were germinated (using filter paper as the substrate) (Ortega et al. 1988). The radicle lengths of the plants exposed to the extracts were reduced significantly but the number of replicates (six) seems somewhat inadequate (see Ortega et al. 1988). The standard toxicity curve appeared to be non-linear, indicative of a pollen-allelopathic effect, although standard deviations were not included in this study (Ortega et al. 1988).

The pollen extract significantly inhibited both cell division and mitochondrial respiration in the target plants, where the site of allelochemical interference was postulated to be near cytochrome c (Ortega et al. 1988). The results seem convincing, although some reservations about the isolation techniques of the mitochondria have been expressed to this author (D.B. Layzell pers. comm.). If the site of inhibition is in the electron transport chain, this might have an analogous effect on the photophosphorylation in chloroplasts (see Ortega et al. 1988). It is not clear if this could be related to reduction in chlorophyll content observed in *Phaseolus vulgaris* after exposure to pollen of *P. hysterophorus* (see Sukhada and Jayachandra 1980a).

7 Methods for the Characterization of Pollen Allelochemicals

7.1 Overview

Once the relevant fractions are isolated and tested in in vitro bioassays, it is important that the allelochemicals within the fractions be characterized. In some cases, this may require further isolation techniques to obtain a fraction pure enough to be useful. Details of the various methods useful to the analysis of pollen allelochemicals have been outlined elsewhere (Harborne 1984; Linskens and Jackson 1986a,b, 1987). The use of column chromatography in characterization is similar to that outlined previously (Sect. 3.4). The present review will restrict itself to examples of various techniques applied specifically to the characterization of pollen allelochemicals.

7.2 TLC

Thin layer chromatography (TLC) is one of the more traditional methods of isolation and characterization (Harborne 1984). As demonstrated by Hodgkin and Lyon (1983), it remains a useful analytical tool. Sukhada and Jayachandra (1980b) made extensive use of TLC in the analysis of the allelochemicals found in pollen. Here, 250 μ silica gel (grade H) plates were developed in benzene:ethyl acetate (7:3; 1.5 h) and analyzed under UV light (Sukhada and Jayachandra 1980b).

Normally, the use of TLC includes chemical reagents that react with specific chemical classes or compounds. For example, Seikel (1964) and Harborne (1984) provide lists of general reagents used to detect subclasses of phenolic compounds. Sukhada and Jayachandra (1980b) sprayed TLC plates with alcoholic ferric chloride and diazotized *p*-nitroaniline. The former reagent detected caffeic acid and chlorogenic acid (both indicated by a green colour), while the latter reagent was important in the detection of vanillic acid (see Seikel 1964; Sukhada and Jayachandra 1980b).

7.3 Gas (Liquid) Chromatography and Mass Spectrometry

The GC-MS apparatus has been employed in the analysis of whole-flower extracts and soil residues thought to be allelopathic (Itokawa et al. 1982; Thorne et al. 1990). It has yet to be used successfully for characterization of pollen allelochemicals. One disadvantage is that the high temperature used in GLC may give artifactual results if the allelochemicals are volatile (see Harborne 1984). In some cases, the available GC-MS apparatus may not sophisticated enough to handle the small samples of extract that may contain several dozen compounds (H.G. Weger pers. comm.).

7.4 HPLC

As Harborne (1984) noted, HPLC has an advantage over GLC in that the former may be operated at room or ambient temperature, thereby reducing the potential for thermal-catalyzed reactions and artifacts. Both HPLC and reverse-phase HPLC have been employed to isolate such potentially allelopathic compounds as alkaloids, flavones, flavonoids, phenols, phenolic acids, phenylpropanoids, quinones and sesquiterpenes (see Harborne 1984). In all of these cases, however, the extracts were obtained from sporophytic tissues, rather than pollen (or nectar). It may be difficult to employ HPLC in isolation of these potentially allelopathic compounds simply because the apparatus requires a large and rather pure sample (Harborne 1984).

With respect to pollen and nectar analyses, the typical use of HPLC is to assess and characterize the larger molecular weight compounds such as proteins. For example, HPLC has proven useful for the characterization of allergens in pollen (notably *Phleum pratense*; Gjesing et al. 1985). The study by Grunfeld et al. (1989) is one of the few examples of the current use of HPLC in pollen and nectar analysis (of *Fragaria* × *ananassa*). Both the pollen and nectar samples of the early experimental runs were often contaminated and rendered inadmissible. These problems were resolved by using freeze-drying to concentrate the samples and adding a preservative (25 mg each of cyclohexamide and chloramphenicol in 20 ml distilled water) (Grunfeld et al. 1989). The samples were further purified by running them through a 3 ml disposable chromatographic column and filtration through a 45 μ filter pack (Grunfeld et al. 1989).

Grunfeld et al. (1989) analyzed only the larger molecular weight compounds (common amino acids and carbohydrates). Hence, it is still unclear if this method will be useful in identification of allelochemicals other than non-protein amino acids. The mixed acetate extracts from the "male flowers" (not necessarily including pollen or nectar) of *Cucurbita pepo* were analyzed by HPLC (*n*-hexane:ethyl acetate 5:4) but the active compounds were identified only as "benzyl glycosides" (Itokawa et al. 1982).

7.5 Spectroscopy

UV spectroscopy has been most useful in characterizing pollen allelochemicals. It is compatible with the common extraction media used (water, 95% ethanol and methanol) and requires only a very small sample (0.3 ml) (Sukhada and Jayachandra 1980b; Harborne 1984). In one case, IR spectroscopy was also used but this was for a whole-flower extract rather than one from pollen (Itokawa et al. 1982).

UV spectroscopy is not useful if chloroform is used and some other extraction medium (e.g., petroleum ether) should be used as an organic solvent (see Harborne 1984). For the most part, UV spectroscopy is used in conjunction with chromatography (normally thin-layer) and is most useful for alkaloids and phenolics, although sesquiterpene lactones are detectable (Sukhada and Jayachandra 1980b; Harborne 1984). It is often possible to obtain commercial samples of various compounds with which to compare fractions isolated by column or thin-layer chromatography, thereby expediting the analyses (see Sukhada and Jayachandra 1980b).

7.6 NMR

The first attempt at determining the NMR of allelochemicals was for the non-acidic fractions obtained from tissues of *Parthenium hysterophorus* (Sukhada and Jayachandra 1980b). One such non-acidic compound, parthenin, was identified using NMR but the study did not specify whether this compound was obtainable from pollen in sufficient quantities (Sukhada and Jayachandra 1980b). Similarly, NMR was used in characterization of phenolic allelochemicals from male flowers of *Cucurbita pepo*, but the study does not specify if the pollen was included as part of the assay (Itokawa et al. 1982).

8 Assessing Pollen-Allelopathic Interactions in Situ

The in vitro bioassays are an excellent means of elucidating the relevant pollen-allelopathic interactions and the characterization protocols can identify the active compounds. Nonetheless, the study is somewhat proximate until the ecological importance of the interaction is determined. This requires an in situ approach similar to that of Sukhada and Jayachandra (1980a).

While it is possible to harvest stigmae that have been pollinated naturally, it is difficult to identify the pollen to species (Murphy and Aarssen unpub.). This is especially true of the Poaceae and quantifying the pollen-allelopathic effect in situ can be quite exacting (see also Dreissen et al. 1989). In some cases, however, there may only a few species pollinating at a given time and there may be measurable differences in the pollen diameters to allow for rapid assessments (Murphy and Aarssen unpub.).

In situ methods often involve the hand pollination of stigmae with foreign pollen from a putative pollen-allelopathic donor. A camel hair brush or fine-point probe is quite efficacious in such artificial pollination. The absolute numbers of pollen from both the allelopathic species and from the species compatible with the stigmata should be ecologically realistic (see also Sect. 4.5). Pollen from individuals conspecific to the target species should be added in a parallel series. That is, the ratio of foreign: conspecific pollen should range from 100:0% to 0:100%. Ideally, at least 100 plants of each treatment should be used but this can vary with the amount of time and effort required for hand-pollination.

Once this is accomplished, there are several approaches to quantify any resultant treatment effects. One may stain and examine the stigmae under fluorescence for evidence of inhibition of pollen germination and/or pollen tube growth (see Newcomer 1938; Adams 1953). It is here that one may wish to measure the length of the pollen tubes after a standard time interval (see also Cruzan 1989). Another measure of the pollen-allelopathic effect would be to compare the seed set among and between treatments (see Sukhada and Jayachandra 1980a). Alternatively, one may examine the treatment effect on ovule development or stigmatal (pistil, floral) longevity (Sukhada and Jayachandra 1980a; Thomson et al. 1982).

Regardless of the attributes used to measure treatment effect, there must be a means of accounting for any possible site effects. These can be eliminated if the plants are grown in controlled situation (e.g., a greenhouse) but it imposes an artificial environment on the plants. This is not necessarily unacceptable, it is merely suggested that a field study be attempted to compliment the greenhouse study. In the field, controlling for microhabitat differences is difficult at best. Hence, one often must settle for measuring site characters (nutrient content, water content, slope, height, pH) encountered by the target plants. Reciprocal transplants of target species may be used to test for any site effects on the reproductive capacity of the target species. In many cases, the microsite differences between local populations may be small enough such that the control individuals will be provide adequate buffers against design errors (Murphy and Aarssen unpub.).

One may also attempt to predict how much pollen is released from a pollen-allelopathic individual and the probability of various quantities alighting on foreign stigmae. Rigorous analyses of the aerodynamic dispersion and deposition of wind-borne pollen can be quite complex, but approximate values for pollen concentrations may be obtained using greased slides or a spore trap (see Ogden et al. 1969; Raynor et al. 1971, 1972; Harrington 1979; Mason 1979; Raynor 1979; Sukhada and Jayachandra 1980a; Niklas 1985 for further discussion). For animal-pollinated species, studies of pollinator behaviour and morphological comparisons of the test and target species are required (see Levin and Kerster 1967a,b; Macior 1968; Waser 1986). In both biotic and abiotic pollination systems, the seasonal and diurnal phenologies of the test versus target species are crucial to the interpretation (see Feinsinger 1987; Murphy and Aarssen 1989).

9 Allelochemicals Toxic to Animals: Pollen and Nectar Analyses

Thus far, the discussion has focused on plant-plant interactions and pollen allelopathy. Any allelochemicals found in nectar would be unlikely to affect the reproductive capacity of other species. It is possible, however, that substances ("allelochemicals") within the nectar may be vectored via pollination if pollen and nectar are intimately associated (see Beattie et al. 1985; Sandu and Waraich 1985; Eisikowitch et al. 1990). More commonly, there appear to be examples of allelochemicals in pollen and nectar that affect only animals and prevent ineffective pollinators from utilizing the plant's resources. It may be difficult to separate the effective agent (i.e., pollen or nectar), although the importance of contamination of one substance by the other has been debated (Gottsberger et al. 1984; Erhardt and Baker 1990). Further, instances of injury or death to pollinators attributed to toxins within pollen or nectar may be caused by predators or pesticide use (Todd and Reed 1969; Stanley and Linskens 1974).

Examples of pollen toxic to animals may include *Aesculus*, *Andromeda*, *Asclepias*, *Astragalus*, *Cornyocarpus*, *Hyoscyamus*, *Polygonum*, *Rhododendron*, *Scolypoda*, *Tilia*, *Veratrum* and *Zigadenus* (Pryce-Jones 1944; Stanley and Linskens 1974; Tepedino et al. 1989). In *Hyoscyamus* and *Zigadenus*, the allelochemicals are likely alkaloids (Stanley and Linskens 1974; Tepedino et al. 1989). An extensive list of alkaloidal plants is given by Willaman and Li (1970) and the methods for characterizing alkaloids are reviewed in Harborne (1984). The toxin in both *Aesculus* and *Tilia* was identified as a saponin (see Stanley and Linskens 1974).

The non-protein amino acid gamma-aminobutyric acid, along with others, is released from germinating pollen of *Petunia* (60.6 nmol/mg pollen after 8 h) (Linskens and Schrauwen 1969). It is not clear if gamma-aminobutyric acid affects the pollen germination of other plant species, but it appears to be toxic to animals (Baker and Baker 1975). Gamma-aminobutyric acid has also been isolated from the nectar of *Liriodendron tulipifera* and may occur in other species (Baker and Baker 1975). It is cautioned that pollen may contaminate nectar samples and the toxicity may be in fact be pollen-mediated.

Despite its initial report as being a pollen toxin, the polyphenolic compound arbutin was isolated from nectar (of *Arbutus unedo*) (Pryce-Jones 1944). This substance is found in the sporophytes of *Docynia* and *Pyrus* (both Rosaceae) and in *Triticum* (Poaceae) (Yamaha and Cardini 1960; Harborne and Simmonds 1964). Such phenolic compounds may be present in the nectar and/or pollen of these species and may be isolated and identified following the methods of Seikel (1964).

The nectar of *Zigadenus* may contain allelopathic alkaloids, although evidence for this is equivocal (Hitchcock 1959; Tepedino 1981, 1982; Tepedino et al. 1989). As noted previously, it is known that the pollen is highly toxic due to the presence of alkaloids. The extraction of alkaloids from the nectar has been demonstrated, however, there was a possibility of contamination of the sample (i.e., the pollen alkaloids may have seeped into the nectar sample) (see Hitchcock 1959). The specific identities of the alkaloids in *Zigadenus* (pollen and/or nectar) have yet to be determined.

Sophora microphylla has been reported to contain alkaloidal nectar (Clinch et al. 1972). This report should be viewed with caution as the purportedly toxic pollen in the related *Sophora japonica* proved to be false (the pollinators were killed by the "bee wolf", *Philanthus triangulum*) (see Stanley and Linskens 1974).

In the most systematic study to date, Baker and Baker (1975) tested the nectar of 126 species for the presence of allelopathic alkaloids. The initial screen employed the iodoplatinate test (blue indicated tertiary amines; white indicated primary and/or secondary amines; Baker and Baker 1975). The 19 positive tests were then subjected to a solution of bismuth nitrate, acetic acid and potassium iodide aqueous; i.e. "Dragendorff's reagent") where a change in color from (pale) yellow to orange indicates presence of alkaloids (Baker and Baker 1975). The nectar of *Campanula rapunculoides*, *Cucurbita pepo*, *Cuscuta salina*, *Iris pseudoacorus*, *Lotus corniculatus*, *Mimulus moschatus*, *Nymphoides peltatum* and *Rhododendron ponticum* all tested positive for alkaloids (Baker and Baker 1975).

10 Conclusions

The identification of potential pollen-allelopathic species and the subsequent quantitative and qualitative analysis of the pollen fractions is a protracted process. Indeed, the initial study by Sukhada and Jayachandra (1980a; 1980b) required some 8 years to complete. The preceding discussion should serve to hasten the pace of future research into pollen allelopathy.

It is cautioned that pollen allelopathy may be but one of several mechanisms that operate during heterospecific pollen transfer (HPT) (see Waser 1978, 1983). Previous studies have demonstrated that pollen allelopathy is not the only mechanism (if any) that operates when foreign pollen alights on a stigma (Griffiths 1950; Lewis and Crowe 1958; Martin 1970; Hogenboom 1975; Sedgley and Blessing 1982; Heslop-Harrison et al. 1984; Kinsman 1984; Campbell 1985; Campbell and Motten 1985; Kohn and Waser 1985; Waser and Fugate 1986; Kinsman et al. 1987; Galen and Gregory 1989; Kinsman et al. 1989; Randall and Hilu 1990; McGuire and Armbruster 1991). The determination of the relative importance of pollen allelopathy awaits more detailed studies.

Such studies should investigate the regulation of the biochemical pathways manufacturing pollen allelochemicals. At present, this is unknown, although there is relevant literature on the sporophytic pathways of phenolics, terpenoids, alkaloids and non-protein amino acids (Neish 1964; Thomson 1964; Towers 1964; Charlwood and Banthorpe 1978; Loomis and Croteau 1980; Fowden 1981; Robinson 1981). Research into the site of synthesis of pollen allelochemicals, whether they are produced by pollen or loaded via the tapetum, is in progress (see also Itokawa et al. 1982; Khan and Jahan 1988; Murphy and Aarssen 1989). Whether allelochemicals in pollen may target the same enzymes as those in sporophytic tissues is also of interest (see Rice 1984).

Other aspects currently under investigation include the taxonomic distribution of pollen allelopathy, the localization of pollen allelochemicals, the transport of allelochemicals from pollen to external environment and the importance of the stigmatic environment in mediating the pollen-allelopathic effect. From an economic perspective, pollen allelochemicals may interfere with mentor pollen treatments used in agriculture, horticulture and silviculture (see Lane 1983).

Regardless of which aspect of pollen allelopathy is of interest, all future research will require careful experimental design for the isolation and characterization of the allelochemicals. It is hoped that the preceding discussion will aid in expediting such research.

Acknowledgements. The author would like to thank Lonnie W. Aarssen, David B. Layzell and Daniel D. Lefebvre for their critical analyses of many of the techniques discussed herein. Susan Aiken, Pierre Gagnon, Harold G. Weger, Sharon Kinsman and V.J. Tepedino have also provided comments and/or information relevant to this discussion. Andrea L. Corbett has provided technical assistance during 1989 and 1990.

References

Adams JD (1953) Observing pollen tubes within the styles of *Zea mays* L. Stain Technol 28:295–298

Alexander MP (1969) Differential staining of aborted and non-aborted pollen. Stain Technol 44:117–122

Baker HG, Baker I (1973) Studies of nectar-constitution and pollinator-plant coevolution. In: Gilbert LE, Raven PH (eds) Coevolution of plants and animals. University of Texas Press, Austin, pp 100–140

Baraniuk JN, Bolick M, Buckley CE (1990) Pollen grain chromatography: a novel method for separation of pollen wall solutes. Ann Bot 66:321–329

Beattie AJ, Turnbull C, Hough T, Jobson S, Knox RB (1985) The vulnerability of pollen and fungal spores to ant secretions: evidence and some evolutionary implications. Am J Bot 72:606–614

Bishop CJ (1949) Pollen tube culture on a lactose medium. Stain Technol 24:47–52

Brewbaker JL (1967) The distribution and phylogenetic significance of binucleate and trinucleate pollen in the angiosperms. Am J Bot 54:1069–1083

Brewbaker JL, Majumder SK (1961) Cultural studies of the pollen population effect and self-incompatibility inhibition. Am J Bot 48:457–464

Campbell DR (1985) Pollinator sharing and seed set of *Stellaria pubera*: competition for pollination. Ecology 66:544–553
Campbell DR, Motten AF (1985) The mechanism of competition for pollination between two forest herbs. Ecology 66:554–563
Char MBS (1977) Pollen allelopathy. Naturwissenschaften 64:489–490
Charlwood BV, Banthorpe DV (1978) The biosynthesis of monoterpenes. Prog Phytochem 5:65–126
Clinch PG, Palmer-Jones T, Forster IW (1972) Effect on honeybees of nectar from the yellow kowhai (*Sophora microphylla* Ait.). NZ J Agric Res 15:194–201
Cruickshank IAM, Perrin DR (1964) Pathological function of phenolic compounds in plants. In: Harborne JB (ed) Biochemistry of phenolic compounds. Academic Press, New York, pp 511–544
Cruzan MB (1989) Pollen tube attrition in *Erythronium grandiflorum*. Am J Bot 76: 562–570
Deyl Z, Macek K, Janak J (eds) (1975) Liquid column chromatography: a survey of modern techniques and applications. Elsevier, New York
Dornbos DL Jr, Spencer GF (1990) Natural products phytotoxicity. A bioassay suitable for small quantities of slightly water-soluble compounds. J Chem Ecol 16:339–352
Driessen MNB, Willemse MTM, van Luijn JAG (1989) Grass pollen grain determination by light- and UV-microscopy. Grana 28:115–122
Eigsti OJ (1966) Pollen tube cultures for teaching and research. Turtox News 44:162–166
Eisikowitch D, Kevan PG, Lachance M-A (1990) The nectar-inhibiting yeasts and their effect on pollen germination in common milkweed, *Asclepias syriaca*. Israel J Bot 39:217–225
Erhardt A, Baker I (1990) Pollen amino acids – an additional diet for a nectar feeding butterfly? Plant Syst Evol 169:111–121
Faegri K, van der Pijl L (1979) The principles of pollination ecology, 3rd revised edn. Pergamon Press, Toronto
Feinsinger P (1987) Effects of plant species on each other's pollination: is community structure influenced? Tree 2:123–126
Fowden L (1981) Non-protein amino acids. In: Conn EE (ed) Biochemistry of plants, vol 7. Secondary plant products. Academic Press, New York, pp 215–248
Fuerst EP, Putnam AR (1983) Separating the competitive and allelopathic components of interference: theoretical principles. J Chem Ecol 9:937–943
Galen C, Gregory T (1989) Interspecific pollen transfer as a mechanism or competition: consequences of foreign pollen contamination for seed set in the alpine wildflower, *Polemonium viscosum*. Oecologia 81:120–123
Ganeshaiah KN, Shaanker RU (1988) Regulation of seed number and female incitation of mate competition by a pH-dependent proteinaceous inhibitor of pollen grain germination in *Leucaena leuconocephala*. Oecologia 75:110–113
Gjesing B, Jager L, Marsh DG, Lowenstein H (1985) The international collaborative study establishing the first international standard for timothy (*Phleum pratense*) grass pollen allergenic extract. J Allergy Clin Immunol 75:258–267
Gottsberger G, Schrauwen J, Linskens, H-F (1984) Amino acids and sugars in nectar, their putative evolutionary significance. Plant Syst Evol 145:55–77
Griffiths DJ (1950) The liability of seed crops of perennial ryegrass (*Lolium perenne*) to contamination by wind-borne pollen. J Agric Sci 40:19–38
Grunfeld E, Vincent C, Bagnara D (1989) High-performance liquid chromatography analysis of nectar and pollen of strawberry flowers. J Agric Food Chem 37:290–294
Harborne JB (1984) Phytochemical methods: a guide to modern techniques of plant analysis, 2nd edn. Chapman & Hall, New York
Harborne JB (1990) Constraints on the evolution of biochemical pathways. Biol J Linn Soc 39:135–151
Harborne JB, Simmonds NW (1964) The natural distribution of the phenolic aglycones. In: Harborne JB (ed) Biochemistry of phenolic compounds. Academic Press, New York, pp 77–127

Harrington JB (1979) Deposition: principles of deposition of microbiological particles. In: Edmonds RL (ed) Aerobiology: the ecological systems approach. Dowden, Hutchinson and Ross, Stroudsburg, pp 111–137

Hauser EJP, Morrison JH (1964) The cytochemical reduction of nitro blue tetrazolium as an index of pollen viability. Am J Bot 51:748–752

Henn H, Petit D, Vernet P (1988) Interference between *Hieracium pilosella* and *Arrhenatherum elatius* in colliery spoils of north of France: allelopathy or competition? Oecologia 76:268–272

Heslop-Harrison J (1982) Pollen-stigma interaction and cross incompatibility in the grasses. Science 215:1358–1364

Heslop-Harrison J, Heslop-Harrison Y (1970) Evaluation of pollen viability by enzyme induced fluorescence: intracellular hydrolysis of fluorescin diacetate. Stain Technol 45:115–120

Heslop-Harrison Y, Reger BJ, Heslop-Harrison J (1984) The pollen-stigma interaction in the grasses. 6. The stigma ("silk") of *Zea mays* L. as host to the pollens of *Sorghum bicolor* (L.) Moench. and *Pennisetum americanum* (L.) Leeke. Acta Bot Neerl 33:205–227

Hitchcock JD (1959) Poisoning of honey bees by death camas blossoms. Am Bee J 99:418–419

Hodgkin T, Lyon GD (1983) Germination of *Lilium* pollens on TLC plates and their inhibition by extracts from *Brassica oleracea* tissues. In: Mulcahy DL, Ottaviano E (eds) Pollen: biology and implications for plant breeding. Elsevier, New York, pp 343–349

Hogenboom NG (1975) Incompatibility and incongruity: two different mechanisms for the non-functioning of intimate partner relationships. Proc R Soc Lond Ser B 188: 361–375

Itokawa H, Oshida Y, Ikuta A, Inatomi H, Adachi T (1982) Phenolic plant growth inhibitors from the flowers of *Cucurbita pepo*. Phytochemistry 21:1935–1937

Jennings DL, Topham PB (1971) Some consequences of raspberry pollen dilution for its germination and for fruit development. New Phytol 70:371–380

Jimenez JJ, Schultz K, Anaya AL, Hernandez J, Espejo O (1983) Allelopathic potential of corn pollen. J Chem Ecol 9:1011–1025

Khan MI, Jahan B (1988) Allelopathic potential of senesced anthers of *Bombax ceiba* L., inhibiting germination and growth of lettuce seeds. Pak J Bot 20:205–212

Kinsman S (1984) The effect of heterospecific pollen on female reproductive function in hummingbird-pollinated plants. Bull Ecol Soc Am 65:61

Kinsman S, Mulder CP, Haber WA (1987) Interspecific pollen transfer in ten neotropical moth-flowered plant species. Bull Ecol Soc Am 68:340

Kinsman S, Taylor DJ, Bonnevie NL (1989) Pollination intensity and the effect of heterospecific pollen in two introduced herbs. Bull Ecol Soc Am 70:169

Knox RB, Gaget M, Dumas C (1987) Mentor pollen techniques. Int Rev Cytol 107: 315–332

Kohn JR, Waser NM (1985) The effect of *Delphinium nelsonii* on seed set in *Ipomopsis aggregata*, a competitor for hummingbird pollination. Am J Bot 72:1144–1148

Kubo A (1955) On the germination of the pollen grains of the Ericaceae. Jpn J Bot 15: 15–27

Kubo A (1960) On the germination of pollen grains of *Brassica napus* L. Bot Mag 73: 453–457

Lane WD (1983) Fruit-set after pretreatment with foreign compared with killed compatible pollen. Can J Bot 62:1678–1681

Laughnan JR, Gabay SJ (1973) Reaction of germinating maize pollen to *Helminthosporium maydis* pathotoxins. Crop Sci 13:681–684

Levin DA, Kerster HW (1967a) An analysis of interspecific pollen exchange in *Phlox*. Am Nat 101:387–399

Levin DA, Kerster HW (1967b) Natural selection for reproductive isolation in *Phlox*. Evolution 21:679–687

Lewis D, Crowe LK (1958) Unilateral incompatibility in flowering plants. Heredity 12: 233–256

Linskens H-F, Jackson JF (eds) (1986a) Modern methods of plant analysis (New Series, vol 2). Nuclear magnetic resonance. Springer, Berlin Heidelberg New York

Linskens H-F, Jackson JF (eds) (1986b) Modern methods of plant analysis (New Series, vol 3). Gas chromatography/mass spectrometry. Springer, Berlin Heidelberg New York

Linskens H-F, Jackson JF (eds) (1987) Modern methods of plant analysis (New Series, vol 5). High performance liquid chromatography. Springer, Berlin Heidelberg New York

Linskens H-F, Schrauwen J (1969) The release of free amino acids from germinating pollen. Acta Bot Neerl 18:605–614

Loomis WD, Croteau R (1980) Biochemistry of terpenoids. In: Stumpf PK (ed) Biochemistry of plants, vol 4. Lipids: structure and function. Academic Press, New York, pp 364–419

Lovett JV, Ryuntyu MY, Liu DL (1989) Allelopathy, chemical communication, and plant defence. J Chem Ecol 15:1193–1202

Macior LW (1968) Pollination adaptation in *Pedicularis canadensis*. Am J Bot 55:1031–1035

Martin FW (1959) Staining and observing pollen tubes in the style by means of fluorescence. Stain Technol 34:125–128

Martin FW (1970) Pollen germination on foreign stigmas. Bull Torrey Bot Club 97: 1–6

Mason CJ (1979) Atmospheric transport: principles of atmospheric transport. In: Edmonds RL (ed) Aerobiology: the ecological systems approach. Dowden, Hutchinson and Ross, Stroudsburg, pp 85–95

McGuire AD, Armbruster WS (1991) An experimental test for reproductive interactions between two sequentially blooming *Saxifraga* species (Saxifragaceae). Am J Bot 78:214–219

Mulcahy GB, Mulcahy DL (1988) The effect of supplemented media on the growth in vitro of bi- and trinucleate pollen. Plant Sci 55:213–216

Murphy SD, Arassen LW (1989) Pollen allelopathy among sympatric grassland species: *in vitro* evidence in *Phleum pratense* L. New Phytol 112:295–305

Neish AC (1964) Major pathways of biosynthesis of phenols. In: Harborne JB (ed) Biochemistry of phenolic compounds. Academic Press, New York, pp 295–359

de Nettancourt D (1977) Incompatibility in angiosperms. Springer, Berlin Heidelberg New York

Newcomer EH (1938) A procedure for growing, staining, and making permanent slides of pollen tubes. Stain Technol 13:89–91

Niklas KJ (1985) The aerodynamics of wind pollination. Bot Rev 51:328–387

Ogden WC, Hayes JV, Raynor GS (1969) Diurnal patterns of pollen emission in *Ambrosia*, *Phleum*, *Zea* and *Ricinus*. Am J Bot 56:16–21

Ortega RC, Anaya AL, Ramos L (1988) Effects of allelopathic compounds of corn pollen on respiration and cell division of watermelon. J Chem Ecol 14:71–86

Parfitt D, Ganeshan S (1988). Comparison of procedures for estimating viability of *Prunus* pollen. Hort Sci 24:354–356

Pederson GA (1986) White clover seed germination in agar containing tall fescue leaf extracts. Crop Sci 26:1248–1249

Poole CF, Schuette SA (1984) Contemporary practice of chromatography. Elsevier, New York

Pryce-Jones J (1944) Some problems associated with nectar, pollen and honey. Proc Linn Soc Lond 1944:129–174

Qusem JR, Hill TA (1989) On difficulties with allelopathy methodology. Weed Res 29:345–347

Randall JL, Hilu KW (1990) Interference through improper pollen transfer in mixed stands of *Impatiens capensis* and *I. pallida* (Balsaminaceae). Am J Bot 77:939–944

Raynor GS (1979) Sampling techniques in aerobiology. In: Edmonds RL (ed) Aerobiology: the ecological systems approach. Dowden, Hutchinson and Ross, Stroudsburg, pp 151–172

Raynor GS, Hayes JV, Ogden EC (1971) Experimental data on dispersion and deposition of timothy and corn pollen from known sources. Brookhaven Natl Lab Monogr Ser 50266 (T-595):1–32

Raynor GS, Ogden EC, Hayes JV (1972) Dispersion and deposition of timothy pollen from experimental sources. Agric Meteorol 9:347–366

Rice EL (1984) Allelopathy, 2nd edn. Academic Press, New York

Robinson T (1981) The biochemistry of alkaloids, 2nd edn. Springer, Berlin Heidelberg New York

Sandu DW, Waraich MK (1985) Yeast associated with pollinating bees and flower nectar. Microb Ecol 11:51–58

Sawada S (1989) Pollen yield and its components in eight temperate grasses. J Jpn Grass Sci 35:61–62

Sedgley M, Blessing MA (1982) Foreign pollination of watermelon (*Citrullus lanatus* [Thunb.] Matsum and Nakai). Bot Gaz 143:210–215

Seikel MK (1964) Isolation and identification of phenolic compounds in biological materials. In: Harborne JB (ed) Biochemistry of phenolic compounds. Academic Press, New York, pp 33–76

Stanley RG, Linskens H-F (1974) Pollen biology, biochemistry and management. Springer, Berlin Heidelberg New York

Stock R, Rice CBF (1974) Chromatographic methods, 3rd edn. Chapman & Hall, New York

Sukhada KD, Jayachandra (1980a) Pollen allelopathy – a new phenomenon. New Phytol 84:739–746

Sukhada KD, Jayachandra (1980b) Allelopathic effects of *Parthenium hysterophorus* L. Part IV. Identification of inhibitors. Plant Soil 55:67–75

Tepedino VJ (1981) Notes on the reproductive biology of *Zigadenus paniculatus*, a toxic range plant. Great Basin Nat 41:427–430

Tepedino VJ (1982) Effects of defoliation on reproduction of a toxic range plant, *Zigadenus paniculatus*. Great Basin Nat 42:524–528

Tepedino VJ, Knapp AK, Eickwort GC, Ferguson DC (1989) Death camas (*Zigadenus nuttallii*) in Kansas: pollen collectors and a florivore. J Kans Entomol Soc 62: 411–412

Thomson JD, Andrews BJ, Plowright RC (1982) The effect of a foreign pollen on ovule development in *Diervilla lonicera* (Caprifoliaceae). New Phytol 90:777–783

Thomson RH (1964) Structure and reactivity of phenolic compounds. In: Harborne JB (ed) Biochemistry of phenolic compounds. Academic Press, New York, pp 1–32

Thorne RLZ, Waller GR, McPherson JK, Krenzer EG, Young C-C (1990) Autotoxic effects of old and new wheat straw in conventional tillage and no-tillage wheat soil. Bot Bull Acad Sin (Taipei) 31:35–49

Todd FE, Reed CB (1969) Pollen gathering of honey bees reduced by pesticide sprays. J Econ Entomol 62:865–867

Towers GHN (1964) Metabolism of phenolic glycosides in higher plants and microorganisms. In: Harborne JB (ed) Biochemistry of phenolic compounds. Academic Press, New York, pp 249–294

Tulecke WR (1954) Preservation and germination of the pollen of *Ginkgo* under sterile conditions. Bull Torrey Bot Club 81:509–512

Tulecke WR (1957) The pollen of *Ginkgo biloba*: in vitro culture and tissue formation. Am J Bot 44:602–608

Vasil IK (1987) Physiology and culture of pollen. Int Rev Cytol 107:127–174

Waser NM (1978) Interspecific pollen transfer and competition between co-occurring plant species. Oecologia 36:223–236

Waser NM (1983) Competition for pollination and floral character differences among sympatric plant species: a review of the evidence. In: Jones CE, Little RJ (eds)

Handbook of experimental pollination biology. Van Nostrand Reinhold, New York, pp 277–293
Waser NM (1986) Flower constancy: definition, cause and measurement. Am Nat 127: 593–603
Waser NM, Fugate ML (1986) Pollen precedence and stigma closure: a mechanism of competition for pollination between *Delphinium nelsonii* and *Ipomopsis aggregata*. Oecologia 70:573–577
Weidenhamer JD, Hartnett DC, Romeo FT (1989) Density-dependent phytotoxicity: distinguishing between resource competition and allelopathic interference in plants. J Appl Ecol 26:613–624
Whittaker RH, Feeny PP (1971) Allelochemics: chemical interactions between species. Science 171:757–770
Willaman JJ, Li HL (1970) Alkaloid-bearing plants and their contained alkaloids. Lloydia (Suppl) 33:1–286
Yamaha T, Cardini C (1960) The biosynthesis of plant glycosides. I. Monoglucosides. Arch Biochem Biophys 86:127–132

Determination of Cyanide and Cyanogenic Glycosides from Plants

A.M. BRINKER and D.S. SEIGLER

1 Introduction

Many plants have the capacity to release hydrogen cyanide (HCN) when the tissues are damaged. The entire plant or any plant part may exhibit this property. Most commonly the compounds responsible are cyanogenic glycosides [*O*-β-glycosides of α-hydroxynitriles (cyanohydrins)], relatively polar, water-soluble compounds, usually accompanied by β-glycosidases that cleave sugars from the glycosides (Fig. 1). Associated hydroxynitrile lyases catalyze decomposition of the resultant cyanohydrins to carbonyl compounds and free hydrogen cyanide; this last step also may occur spontaneously. About 60 cyanogenic glycosides are known from higher plants (Nahrstedt 1987; Seigler 1991); probably the best known of these is amygdalin, which is widespread in seeds of plants such as apples, peaches, cherries, and apricots. Examples of the different structural classes of cyanogenic compounds are given in Fig. 2. The ability to produce and accumulate cyanogenic glycosides is found in all major vascular plant groups (Hegnauer 1977; Saupe 1981; Seigler 1991). Several major reviews are available (Seigler 1977, 1981a,b; Conn 1979, 1980a,b, 1981a,b; Nahrstedt 1981, 1985, 1987; Vennesland et al. 1981; Evered and Harnett 1988). The

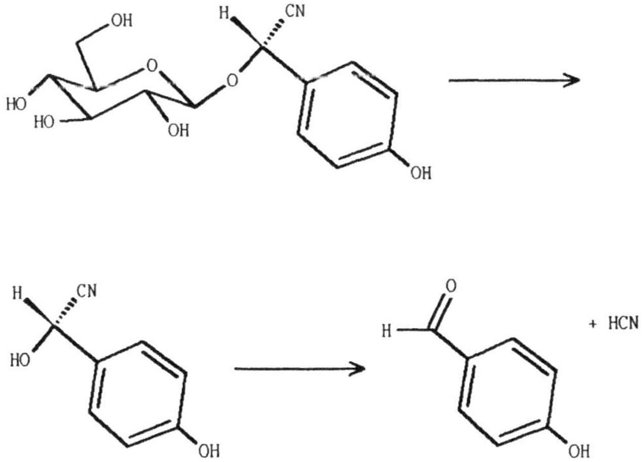

Fig. 1. Enzymatic hydrolysis of cyanogenic glycosides (After Seigler 1991)

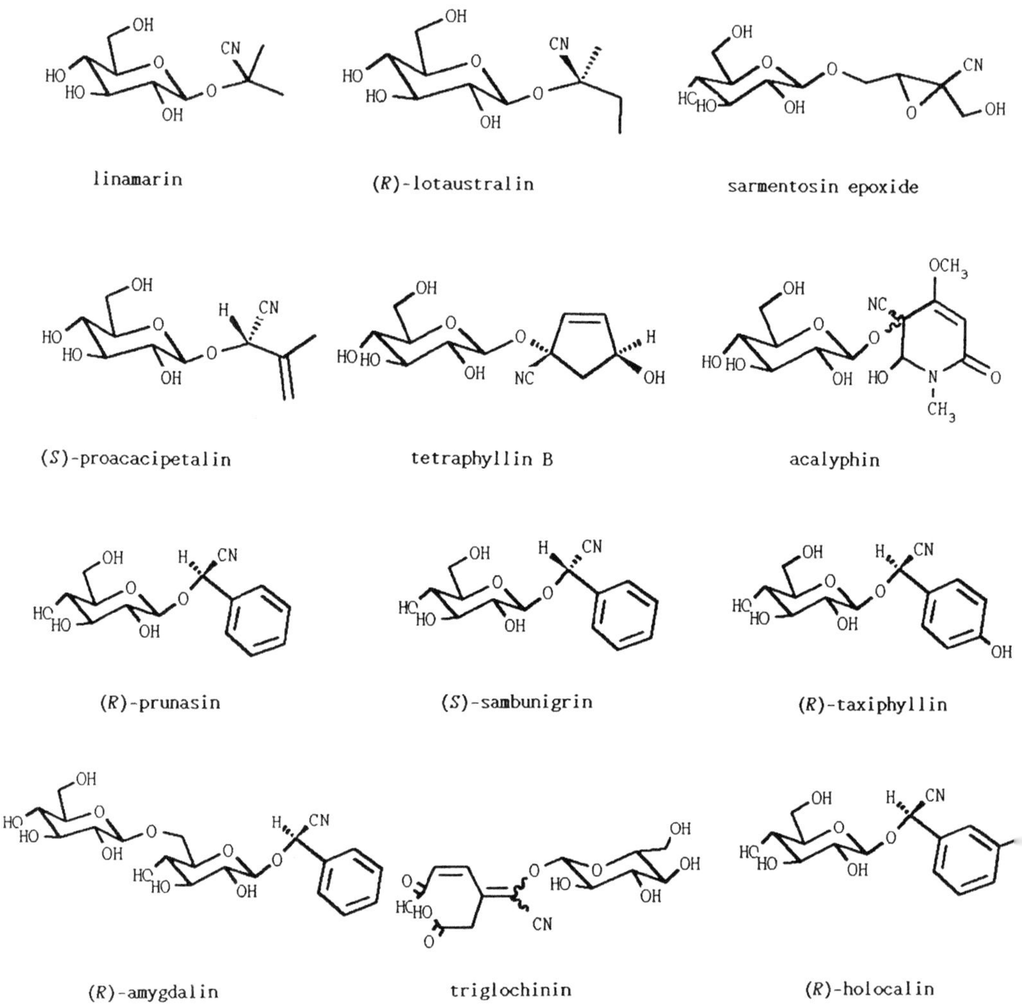

Fig. 2. Examples of the different types of cyanogenic glycosides

carbon bearing the sugar, cyanide and (usually) a proton generally is chiral; both (*R*)- and (*S*)-epimers are known for many series of compounds and in some instances epimers co-occur (Nahrstedt 1987; Seigler 1991). Enantiomeric mixtures of cyanogenic glycosides, especially those of the cyclopentenoid series, also co-occur (Spencer et al. 1983, 1985; Jaroszewski and Olafsdóttir 1987; Jaroszewski et al. 1987a,c, 1988) as do both monoglycosides and diglycosides in flax (Smith et al. 1980), rubber (Selmar et al. 1987, 1988) and some *Passiflora* species (Spencer et al. 1986; Olafsdóttir et al. 1989a). Linamarin and lotaustralin usually co-occur and probably are found in more plant species than any other cyanogenic glycosides (Nahrstedt

1987). Complex mixtures of di-, tri- and even tetraglycosides are found in members of the family Asteraceae (Compositae; Hübel et al. 1982; Nahrstedt et al. 1983; Schwind et al. 1990).

Cyanogenic compounds are readily epimerized by dilute base (Nahrstedt 1975, 1981; Ettlinger et al. 1977), whereas under mildly acidic conditions they are hydrolyzed to HCN, a sugar and a carbonyl compound.

In this Chapter, we will discuss how HCN is detected qualitatively and quantitatively, and how intact cyanogenic compounds are isolated, purified and quantified.

2 Plant Material

In general, plant material should be either fresh, lyophilized, or air dried (Nahrstedt 1981). Air drying at less than 50 °C usually is the most practical method of preservation. At higher temperatures, the hydrolytic enzymes often are denatured and the compounds themselves may be modified in some cases (Nahrstedt 1981). Both cyanogenic glycosides and β-glycosidases are usually preserved in herbarium specimens, even those more than 100 years old.

Lyophilization is an effective method of preservation, but frozen plant samples must be maintained below −18 °C as freezing causes rupture of cells and liberation of both β-glucosidases and cyanogenic glycosides. Once the plant material is dehydrated, little decomposition of cyanogens occurs.

3 Qualitative Determination of Cyanide

Either Feigl-Anger or sodium picrate paper provides an easy and reliable method for qualitative or semiquantitative determination of HCN (Hegnauer 1986). These qualitative methods are very useful for screening large numbers of samples because they are simple, fairly specific, and require no equipment other than vials and corks. Additional details are given in Brinker and Seigler (1989) from which much of the following is taken:

Sodium picrate paper is prepared by dipping filter paper into an aqueous solution of 0.5% picric acid and 5% Na_2CO_3, then allowing the paper to dry. The paper may be stored dry but should be moistened just before use (Guignard 1906; Kingsbury 1964; Horwitz 1980). The detection limit of picrate paper with plant samples is 0.001–0.002% HCN by weight which corresponds to about 0.01–0.02% prunasin (Hegnauer 1986).

Sodium picrate paper is especially useful in situations where water is present, but false positive tests sometimes are observed (Nahrstedt 1980; Hegnauer 1986; Brinker and Seigler 1989), for example with breakdown products of glucosinolates found in members of the Brassicaceae and other families. Many false-positive tests can be recognized because the color produced is different from that given by cyanide, but may also be distinguished by adding a drop of 10% acetic acid to the paper. Disappearance of the red color indicates a false positive test; the color produced by hydrogen cyanide is stable under these conditions (Hegnauer 1986). Another disadvantage of picrate paper is that it entails the use of picric acid, which can explode if mishandled (Brinker and Seigler 1989).

Feigl-Anger paper is made by dipping pieces of filter paper into a solution of equal volumes of freshly prepared solutions of tetrabase (4,4′-tetramethyldiaminodiphenylmethane) and copper ethylacetoacetate. The solutions are made by dissolving 1 g solute in 100 ml $CHCl_3$. The test strips are then dried in a fume hood (Feigl and Anger 1966; Tantisewie et al. 1969). If stored in the light, the paper will fade or discolor but still give a positive test. Feigl-Anger paper is somewhat more sensitive than picrate paper (Nahrstedt 1980) although it also gives some false-positive tests (Brinker and Seigler 1989; Hegnauer 1986). We routinely use the Feigl-Anger test.

The test for cyanogenesis is carried out in the same way with either test paper. The material to be tested is crushed in a test tube or small vial. If the plant material has been dried, a few drops of water or pH-6.8 phosphate buffer should be added, but a minimum of liquid should be used as HCN is extremely water-soluble. A strip of test paper is suspended over the material in the vial and held in place by the cork used to stopper the vial. A color change signals that cyanide is being released: picrate paper turns from yellow to orange or red; Feigl-Anger paper turns from pale blue-green to bright blue or purple. If picrate paper is used, the reaction can be accelerated by warming the vial. The amount of cyanide present can be estimated approximately by the extent of the color change. Moderately to strongly cyanogenic species will give positive tests within a few minutes to several hours.

Residual cyanide from corks or rubber stoppers may give a positive test with either of the above reagents (Brinker and Seigler 1989; Hegnauer 1986).

Negative tests can occur for several different reasons. The plant may lack either cyanogenic compounds or β-glucosidases. Some β-glucosidases are not stable to drying. Plant specimens that have been dried above 60 °C usually require addition of exogenous enzyme and those treated with preservatives such as formalin frequently do not give positive cyanide tests. Finally, tannins can inhibit the hydrolysis of cyanogenic glycosides and be responsible for negative tests (Goldstein and Spencer 1985).

If no positive response is observed after 24 h, a fresh test should be set up and a few drops of an appropriate β-glucosidase added; very slow (24–48 h) positive reactions should be confirmed with a new test to which a drop or two of dilute (1–2 mM) antibiotic solution is added to prevent growth of cyanogenic bacteria (Saupe et al. 1982).

4 Enzymatic Preparations and Hydrolysis of Cyanogens

If it appears that addition of exogenous enzyme is necessary, there are two commercially available enzymes that will work in many cases. Emulsin, the β-glucosidase from almonds, is frequently used and will hydrolyze several cyanogenic compounds (some very slowly) including: cardiospermin 4-hydroxybenzoate (Nahrstedt 1976b); cardiospermin sulfate (glucose, HCN, and sulfate) (Hübel and Nahrstedt 1979); dhurrin (Erb et al. 1981; Halkier and Møller 1989); heterodendrin (Hübel and Nahrstedt 1975; Nahrstedt and Scheid 1981); holocalin (*m*-hydroxybenzaldehyde and glucose) (van Valen 1978a); 3-hydroxyheterodendrin (Brimer et al. 1981); linustatin (glucose and linamarin) and neolinustatin (glucose and lotaustralin) (Smith et al. 1980); passicapsin (Fischer et al. 1982); proacaciberin (Nartey et al. 1981); proteacin (Swenson et al. 1989); prunasin (benzaldehyde, HCN and glucose) (Aritomi et al. 1985); sarmentosin (glucose) (Fang et al. 1982); sutherlandin (Swenson et al. 1987); and triglochinin (slowly) (Jaroszewski and Ettlinger 1981). For earlier reports see Seigler (1977).

Snail or snail gut acetone powder, a complex mixture of enzymes (β-glucosidase, β-glucuronidase, sulfatase, and β-D-mannosidase), is probably the most general method for hydrolysis of cyanogenic glycosides (Brimer et al. 1983). This enzyme preparation appears to be especially useful for the hydrolysis of cyclopentenoid cyanogenic glycosides such as epivolkenin and taraktophyllin (Jaroszewski et al. 1987a); tetraphyllin B and volkenin (Jaroszewski and Olafsdóttir 1986; Jaroszewski et al. 1987c).

In contrast to snail gut enzymes and emulsin, most other β-glucosidases have relatively specific substrate requirements (Nishizawa and Hashimoto 1970; Seigler 1977; Hösel 1981; Hösel et al. 1987; Spencer 1987, 1988; Poulton 1988; Seigler and Brinker 1991). Because the endogenous enzyme or mixture of enzymes is often the most effective system for hydrolysis of the co-occurring cyanogenic compounds, it may be desirable to isolate a β-glucosidase preparation from an aliquot of the plant being studied. A method for effective enzyme preparations has been described (Brinker and Seigler 1989).

5 Quantitative Determination of Cyanide

Methods for quantitation of cyanide in plant tissues involve hydrolysis of the cyanogenic glycosides and release of HCN, which is trapped in a basic solution and determined colorimetrically (Nahrstedt et al. 1981; Brinker and Seigler 1989). The amount of plant material required depends on the amount of cyanogenic compounds present; this often can be estimated by a preliminary Feigl-Anger test. For many plants a 0.5–1.0 g sample is adequate (Brinker and Seigler 1989).

The sample must be ground to a fine powder; yields of cyanide are highest if the material is ground in liquid nitrogen, lower if ground with dry ice, and lowest if ground in buffer at room temperature in a Waring blendor or mortar and pestle (Torres et al. 1988). Some researchers have ground samples in acidic (H_3PO_4; Cooke 1978) or basic (Na_2HPO_4; Lieberei et al. 1985, 1986; Pereira et al. 1981) media to inhibit β-glucosidases, then adjusted the pH to a value at which the enzyme is active just before or after sealing the reaction flasks. Alternatively, samples can be added to the β-glucosidase solution and frozen and thawed several times to release cyanogenic compounds (Halkier and Møller 1989). Methods for grinding and extraction of plant material have been reviewed (Nahrstedt 1981; Brimer 1988; Brinker and Seigler 1989).

In the usual hydrolysis procedure (frequently called distillation), plant tissue is incubated in Warburg flasks or small Erlenmeyer flasks with a central well (Pereira et al. 1981; Brinker and Seigler 1989). The ground plant material is added to the flask outside the center well along with 0.1 M phosphate buffer (pH 6.8) and, if necessary, exogenous β-glucosidase. The volume of liquid should be kept to a minimum as cyanide is very soluble in water. A known amount (0.5–1 ml, depending on the size of the well) of fresh 1 M NaOH solution is added to the center well, and the flask stoppered tightly. Samples should be prepared as quickly as possible to minimize loss of HCN. The flasks are heated at 37 °C for 6–8 h (occasionally longer incubation times are required).

During incubation, HCN released from the plant material is converted to nonvolatile CN^- in the NaOH solution and is trapped in the center well. In our experience, the best method for determination of cyanide is that of Lambert et al. (1975). This method is relatively specific for cyanide ion and can detect as little as 5 μg/l. Merck Spectroquant kits and the ABS procedure for cyanide determination (based on the method of Aldridge 1944) have been used by many workers (Nahrstedt 1977; Lieberei et al. 1985). Other methods for the quantification of cyanide have been reviewed by Nahrstedt et al. (1981) and Seigler et al. (1989).

The procedure for the Lambert method is as follows: aliquots of the NaOH solution from the well of each hydrolysis flask are transferred to test tubes. Enough distilled water is added to adjust the NaOH concentration to 0.1 M. Typically, 0.1 ml aliquots of 1 M NaOH solution are removed and diluted to 1 ml; if the total sample volume is less, 0.1 M NaOH is added to bring the volume up to 1 ml. The following reagents are then added, in order, to each tube: 0.5 ml 1.0 M acetic acid, 5 ml succinimide/*N*-chlorosuccinimide reagent (2.5 g succinimide and 0.25 g *N*-chlorosuccinimide in 1 l water), and 1 ml barbituric acid/pyridine reagent (12 g barbituric acid is mixed with a small amount of water to make a paste, then 120 ml pyridine is added with enough water to give a final volume of 400 ml). The tubes are shaken or vortexed vigorously, allowed to sit for 10 min, and the absorbance measured at 580 nm. The color is stable for at least 30 min (Lambert et al.

1975; Brinker and Seigler 1989). For preparation of a standard curve, aliquots of a standardized NaCN solution (specific instructions are given in Brinker and Seigler 1989) between 0 and 0.2 ml are made up to 1 ml with 0.1 M NaOH. Concentrations are determined by comparison with this standard curve. Phenylpyridine, which is odorless, may be used in place of pyridine (Markley et al. 1987).

6 Isolation and Preliminary Purification of Cyanogenic Glycosides

In order to isolate intact cyanogenic glycosides, plant material must be carefully ground and extracted. Ground air dried or freeze dried material should be extracted with a non-polar solvent (petroleum ether, chloroform, etc.) to remove lipids (Nahrstedt 1981) and reextracted with a more polar solvent (often 50–80% methanol) to remove the cyanogenic glycosides. All fractions and the marc should be tested for the presence of cyanide as some cyanogens are soluble in solvents of lower polarity [e.g., cardiospermin 4-hydroxybenzoate is extractable in EtOAc (Nahrstedt 1976b)].

Workup of undried material is complicated by the presence of β-glucosidases. Decomposition of cyanogens can be minimized by grinding with liquid nitrogen. Fresh material also may be ground in a tissue homogenizer or Waring blendor, but cyanide recovery is lower (Torres et al. 1988). Ground plant tissue should be boiled in 80–100% methanol for at least 10 min to denature the β-glucosidases and extract the glycosides (Seigler 1977). Samples should then be filtered through cheesecloth and subsequently through coarse filter paper, or centrifuged to remove fine materials. Methanol is then removed under vacuum and the resultant aqueous sample extracted with chloroform. Each fraction should be checked for cyanogens. The remaining aqueous sample, which will usually contain the cyanogenic glycosides, is concentrated to dryness under vacuum or frozen and lyophilized.

Careful selection of extraction conditions can minimize contaminants, and in some instances may result in separation of cyanogenic compounds (Nahrstedt 1981).

After extraction, the cyanogenic glycosides must be separated from many types of contaminants (Nahrstedt 1981; Seigler 1977). This may be done by passing the extract over a column of one of the following solid phases or by a batch method, i.e., adding one of the solid phases to the extract, mixing well, and then removing the solid phase by filtration. Suggested materials include: diatomaceous earth (Celite) (Towers et al. 1964; Fikenscher and Hegnauer 1977); activated charcoal (Towers et al. 1964; Nahrstedt 1976a,b; Fang et al. 1982; Tober and Conn 1985); Amberlite

XAD-2 (Aritomi et al. 1985; Nagumo et al. 1985); Sephadex G-10 (Spencer and Seigler 1980, 1987; Seigler et al. 1982); Sephadex LH20, cellulose and polyamide (Nahrstedt 1981; Nahrstedt et al. 1982a). Phenols and acidic substances can be removed by treatment with 10% PbOAc followed by H_2S (Swenson et al. 1987, 1989). Tannins may be removed by precipitation with casein, caffeine, or gelatin (Goldstein and Spencer 1985). One should determine the amount of cyanide in the sample before and after treatment to insure that excessive loss of cyanogenic compounds does not occur.

7 Purification by Chromatography

Cyanogenic glycosides still are contaminated with flavonoid glycosides, tannins, and saponins, which have similar physical and chemical properties. After removal of lipids and pretreatment as above, the cyanogenic fraction may be further purified by chromatography on a variety of substrates (Brimer 1988). Feigl-Anger and picrate tests are useful for monitoring purification steps. In general, the presence of chromatography paper and stationary phases does not affect these tests (Brinker and Seigler 1989). Other methods for detecting cyanogenic substances on paper chromatograms or cellulose TLC plates are given in Seigler (1977) and Nahrstedt (1981).

7.1 Chromatography on Cellulose

Paper chromatography often has been used to purify cyanogenic glycosides (Nahrstedt 1981; Seigler 1977). This method does not require expensive equipment, but resolution is relatively low. Compounds purified in this manner include: deidaclin (Spencer and Seigler 1980); dhurrin and taxiphyllin (Fikenscher and Ruijgrok 1977; Tjon Sie Fat 1978; Erb et al. 1981; van Valen 1978b,c,d, 1979a); epiheterodendrin (Erb et al. 1979); 4-glucosyloxymandelonitrile (Tjon Sie Fat 1979); gynocardin and epivolkenin (Spencer et al. 1982); heterodendrin (Hübel and Nahrstedt 1975); holocalin (van Valen 1978a); linamarin and lotaustralin (van Valen 1979b; Fikenscher and Hegnauer 1981); linustatin and neolinustatin (Spencer et al. 1986); passibiflorin, epipassibiflorin, and passitrifasciatin (Spencer and Seigler 1985); passisuberosin (Spencer and Seigler 1987); proacaciberin and proacacipetalin (Nartey et al. 1981); prunasin (Fikenscher and Hegnauer 1977, 1981; Poulton and Shin 1983; Spencer and Seigler 1984); tetraphyllin A and volkenin (Spencer et al. 1983); tetraphyllin B (Spencer and Seigler 1982, 1984); tetraphyllin B sulfate (Seigler et al. 1982); triglochinin (Fikenscher and Ruijgrok 1977; Tjon Sie Fat 1978, 1979; Tjon Sie Fat and van

Valen 1978; van Valen 1978c,d,e; Jaroszewski and Ettlinger 1981); and zierin (Fikenscher and Hegnauer 1981).

Chromatography on cellulose columns is useful for removing impurities, but not highly selective, i.e., mixtures of cyanogens are often not resolved (Seigler 1977; Nahrstedt 1981). Compounds purified by this method are: acalyphin (Nahrstedt et al. 1982a); cardiospermin sulfate (Hübel and Nahrstedt 1979); deidaclin, tetraphyllin A and B, and volkenin (Spencer et al. 1983); gynocardin, prunasin, epivolkenin, and tetraphyllin B (Spencer and Seigler 1982, 1984; Spencer et al. 1982); linamarin, lotaustralin and epilotaustralin (Michely et al. 1983; Pitsch et al. 1984); linustatin and neolinustatin (Spencer et al. 1986); passicapsin (Fischer et al. 1982); passitrifasciatin (Spencer and Seigler 1985); proacaciberin (Nartey et al. 1981); sarmentosin epoxide (Nahrstedt et al. 1982b); sutherlandin (Swenson et al. 1987); tetraphyllin B sulfate (Seigler et al. 1982); triglochinin (Jaroszewski 1979); and vicianin (Lizotte and Poulton 1985).

7.2 Chromatography on Silica Gel

Separations can be carried out on silica gel columns with high resolution and at relatively low cost, making this one of the most effective methods for purification of cyanogenic compounds. For example, a series of difficultly separable cyclopentenoid cyanogenic glycosides was purified on silica gel columns with EtOAc-Me_2CO-CH_2Cl_2-MeOH-H_2O (40:30:12:10:8) (Jaroszewski et al. 1987a,c, 1988; Olafsdóttir et al. 1989a,b). Other compounds purified on silica gel are: amygdalin, prunasin, and vicianin (Nagumo et al. 1985); *Anthemis* glycosides A and B (Nahrstedt et al. 1983); cardiospermin 4-hydroxybenzoate and cardiospermin 4-hydroxy-(*E*)-cinnamate (Nahrstedt 1976b; Nahrstedt and Scheid 1981); deidaclin and tetraphyllin A (Jaroszewski and Jensen 1985); dhurrin (van Valen 1979a; Erb et al. 1981; Nahrstedt and Limmer 1982; Nicollier et al. 1983); epiheterodendrin (Erb et al. 1979); epiproacacipetalin and proacacipetalin (Jaroszewski and Ettlinger 1987); gynocardin (Jensen and Nielsen 1986); 3-hydroxyhet erodendrin (Brimer et al. 1981); linamarin, lotaustralin, and epilotaustralin (Pitsch et al. 1984); linustatin and neolinustatin (Smith et al. 1980); passitrifasciatin (Olafsdóttir et al. 1991); proacaciberin (Nartey et al. 1981); prunasin (Aritomi et al. 1985); prunasin and prunasin 6′-malonate (Nahrstedt et al. 1989); 6-*O*-(α-L-rhamnopyranosyl)-epivolkenin, 6-*O*-(α-L-rhamnopyranosyl)-taraktophyllin, epivolkenin, and taraktophyllin (Jaroszewski et al. 1988); sambunigrin and holocalin (Brimer and Dalgaard 1984); sarmentosin epoxide (Nahrstedt et al. 1982b); taxiphyllin (Nicollier et al. 1983); tetraphyllin B and volkenin (Jaroszewski et al. 1987c; Jaroszewski and Olafsdottir 1986); triglochinin (Jaroszewski 1979); vicianin (Kofod and Eyjólfsson 1969); and xeranthin (Schwind et al. 1990).

7.3 Chromatography on Other Supports

A number of other supports have been used for particular separations. Polyamide columns are used to purify cyanogens that hydrogen bond strongly, usually those with phenolic or carboxylic groups. Frequently the only major impurities after treatment with polyamide are sugars (Nahrstedt 1981). Some compounds purified by this method are: amygdalin, prunasin, and vicianin (Nagumo et al. 1985); cardiospermin derivatives (Nahrstedt 1976b); dhurrin (Erb et al. 1981; Nahrstedt et al. 1987); dhurrin and proteacin (Swenson et al. 1989); epiheterodendrin (Erb et al. 1979); heterodendrin (Hübel and Nahrstedt 1975); linamarin (Michely et al. 1983); sarmentosin epoxide (Nahrstedt et al. 1982b); sutherlandin (Swenson et al. 1987); taxiphyllin (Nahrstedt 1979); and triglochinin (Jaroszewski 1979).

Compounds purified on other types of columns include: amygdalin 6″-(4-hydroxybenzoate) and amygdalin 6″-(4-hydroxycinnamate) (Sephadex LH-20) (Nahrstedt et al. 1990); gynocardin (alumina) (Jensen and Nielsen 1986); 3-hydroxyheterodendrin and proacaciberin [silanized silica gel 60 (RP-2)] (Brimer et al. 1981; Nartey et al. 1981); prunasin and prunasin 2′-glucoside (Diaion HP20AG) (Aritomi et al. 1985); tetraphyllin B sulfate (Dowex 50W) (Jaroszewski and Fog 1989); and triglochinin [prepurified on a polycaprolactam column and further purified on an Amberlite IR-45 (weakly basic) column] (Jaroszewski and Ettlinger 1981).

7.4 Thin-Layer Chromatography

Thin-layer chromatography (TLC) is one of the most effective techniques for work with cyanogenic compounds. Retention values (R_f) are helpful for preliminary identification of cyanogenic glycosides (Brimer et al. 1983) and prediction of solvents and conditions for column, PC, or HPLC separations. Compounds can be detected on TLC plates with UV light, spray reagents (Seigler 1977; Nahrstedt 1981), naphthoresorcinol (sugar-specific) (Jaroszewski and Jensen 1985), and iodine; also by the sandwich method (Butler and Butler 1960; Brimer et al. 1983) or by desorbing material from the plates and subsequent cyanide tests with Feigl-Anger or picrate paper (this method requires less enzyme than the sandwich method). Cyanogenic glycosides usually can be desorbed from silica gel or cellulose with methanol-water mixtures (50–80%), although some compounds, such as triglochinin, only desorb in water. TLC also is useful for rapid monitoring of separation and fractionation methods, and for isolation and purification of the compounds of interest.

One system is capable of separating at least 16 cyanogens of several biosynthetic classes: amygdalin, cardiospermin sulfate, cardiospermin, dhurrin, gynocardin, holocalin, 3-hydroxyheterodendrin, linamarin, lotaustralin, lucumin, proacaciberin, proacacipetalin, prunasin, sambunigrin, and triglochinin [EtOAc-Me_2CO-$CHCl_3$-MeOH-H_2O (40:30:12:10:8)] (Brimer et al. 1983).

Other compounds purified by TLC include: acalyphin [Me_2CO-MeOH-H_2O (5:1:0.5)] (R_f 0.23), [MeCOEt-EtOAc-MeOH-H_2O (5:3:1:1)] (R_f 0.60), [EtOAc-MeOH-H_2O-HOAc (70:20:5:5)] (R_f 0.51) (silica) (Nahrstedt et al. 1982a); amygdalin (R_f 0.55), (*S*)-neoamygdalin (R_f 0.55), mandelonitrile (R_f 0.91), and mandelonitrile-β-glucuronide (R_f 0.50) [*n*-ButOH-C_6H_6-H_2O-MeOH (2:1:1:1.25)] (silica gel 60) (Fenselau et al. 1977); amygdalin, amygdalin 6″-(4-hydroxybenzoate) and amygdalin 6″-(4-hydroxycinnamate) [EtOAc-Me_2CO-CH_2Cl_2-MeOH-H_2O (20:15:6:5:4)] (silica) (Nahrstedt et al. 1990); *Anthemis* glycoside A (R_f 0.17) and B (R_f 0.42) and epilucumin (R_f 0.5) [EtOAc-MeOH-H_2O-HOAc (11:3:1:1)]; also [EtOAc-MeOH-H_2O (79:11:10)] (silica) (Nahrstedt et al. 1983); cardiospermin and cardiospermin 4-hydroxybenzoate [EtOAc-MeOH-H_2O-HOAc (80:10:2:0.1 and 80:8:2:1)] (Nahrstedt 1976b); cardiospermin 4-hydroxybenzoate, cardiospermin 4-hydroxy-(*E*)-cinnamate and heterodendrin [EtOAc-MeOH-HOAc (10:2:0.2 and 99:1:0.1)] (Nahrstedt and Scheid 1981); cardiospermin sulfate [EtOAc-MeOH-H_2O (6:3:1)] (Hübel and Nahrstedt 1979); deidaclin and tetraphyllin A [EtOAc-Me_2CO-MeOH (4:4:1)] (silica) (Jaroszewski and Jensen 1985); dhurrin [*i*-PrOH-H_2O (7:3)], [$CHCl_3$-MeOH (75:25)], [MeCOEt-Me_2CO-H_2O (15:5:3)], [EtOAc-MeOH-H_2O (93:5:2)], [EtOAc-MeOH-HOAc-H_2O (70:20:2:8)], [ButOH-EtOH-H_2O (40:11:19)] (silica gel 60) (Erb et al. 1981), [EtOAc-MeCOEt-H_2O-HCOOH (7:5:1.5:0.2)] (silica) (Nahrstedt et al. 1987); dhurrin (R_f 0.30) and prunasin (R_f 0.43) [EtOAc-MeOH-H_2O (8:1:1)] (silica) (Nahrstedt and Limmer 1982); dhurrin and taxiphyllin [C_6H_6-EtOAc (1:1)] (silica) (Nicollier et al. 1983); dhurrin (only) [MeCOEt-EtOAc-HCO_2H-H_2O (5:3:2:1)] (Rf 0.7) (silica) (Towers et al. 1964); dhurrin, taxiphyllin, and triglochinin (R_f values for dhurrin) [EtOAc-MeOH-H_2O (79:11:10)] (R_f 0.39), [$CHCl_3$-EtOH (4:3)] (R_f 0.60), [H_2O satd. $CHCl_3$-MeOH (2:1)] (R_f 0.54), [Me_2CO-H_2O (9:1)] (R_f 0.74), [Me_2CO-H_2O (7:3)] (R_f 0.93), [EtOAc-C_6H_6-EtOH (4:1:1)] (R_f 0.29) (silica gel 60) (van Valen 1978b,d, 1979a); epiheterodendrin and heterodendrin [*n*-PrOH-$CHCl_3$-H_2O (85:10:5)], [EtOAc-MeOH-H_2O (93:5:2)] (silica) (Erb et al. 1979; Zinsmeister et al. 1981); epilotaustralin, linamarin, and lotaustralin [*i*-PrOH MeOH (7:3)], (*i*-PrOH-$CHCl_3$-H_2O (85:10:5)], ($CHCl_3$-MeOH (75:25)], [*i*-PrOH-MeOH (9:1)] (silica gel 60), [$CHCl_3$-MeOH (75:25)], [*n*-ButOH-H_2O (50:9)], [EtOAc-MeOH-HOAc-H_2O (60:15:15:10)] (microcrystalline cellulose) (Pitsch et al. 1984); epilucumin (R_f 0.45), sambunigrin (R_f 0.6), xeranthin (R_f 0.3), zierin (R_f 0.6), and zierin xyloside (R_f 0.45) [EtOAc-Me_2CO-CH_2Cl_2-H_2O-MeOH (20:15:6:4:5)] (silica) (Schwind et al. 1990); epivolkenin, deidaclin, taraktophyllin, tetraphyllin A and B, and volkenin [EtOAc-Me_2CO-CH_2Cl_2-MeOH-H_2O (20:15:6:5:4)] (Jaroszewski et al.

1987c; Olafsdóttir et al. 1989a); 4-glucosyloxymandelonitrile (1), dhurrin (2), nandinin (3), and triglochinin (4) [H_2O satd. $CHCl_3$-MeOH (2:1)] [R_f (1) 0.86, (2) 0.67, (3) 0.60, (4) 0.04], [EtOAc-HOAc-H_2O(8:1:1)] [R_f (2) 0.46, (3) 0.36], [EtOAc-$CHCl_3$-MeOH (10:2:3)] [R_f (1) 0.83, (2) 0.46, (3) 0.36] (silica gel PF) (Tjon Sie Fat 1979); gynocardin [$CHCl_3$-MeOH (3:1)] (silica) (Jensen and Nielsen 1986); heterodendrin [MeCOEt-Me_2CO-H_2O (4:5:1)], [i-PrOH-$CHCl_3$-H_2O (85:10:5)] (silica), [EtOAc-Me_2CO-H_2O (4:5:1)], [EtOH-*n*-ButOH-H_2O (7:2:2)], [MeOH-H_2O (1:1)] (cellulose) (Hübel and Nahrstedt 1975); holocalin [EtOAc-MeOH (9:1)] (R_f 0.34) (silica), [H_2O satd. $CHCl_3$-MeOH (3:1)] (R_f 0.43), [Me_2CO-H_2O (9:1)] (R_f 0.90), [EtOAc-C_6H_6-EtOH (4:1:1)] (R_f 0.63) (cellulose) (van Valen 1978a); 3-hydroxyheterodendrin [Me_2CO-$CHCl_3$-H_2O (17:2:1)], [*n*-ButOH-C_6H_5N-$C_6H_5CH_3$-H_2O (5:3:3:1 upper phase)] (silica or cellulose), [EtOAc-HOAc-H_2O (3:3:1)] (cellulose) (Brimer et al. 1981); linamarin (15% HOAc) (silica gel 60) (Michely et al. 1983); linamarin and linustatin [MeOH-$CHCl_3$-15% NH_4OH (2:2:1)] (silica gel 60) (Selmar et al. 1988); linamarin and lotaustralin [EtOAc-MeOH-H_2O (79:11:10)], [$CHCl_3$-MeOH (5:1)], [$CHCl_3$ (sat d. with H_2O)-MeOH (2:1)] (silica) (van Valen 1979b); linustatin and neolinustatin [$CHCl_3$-MeOH-17% NH_4OH (2:2:1)] (silica) (Smith et al. 1980); passicapsin and linamarin [EtOAc-MeCOEt-H_2O-HCO_2H (50:10:10:5)] (silica) (Fischer et al. 1982); passitrifasciatin [EtOAc-Me_2CO-CH_2Cl_2-MeOH-H_2O (20:15:6:5:4)] (Olafsdóttir et al. 1991); proacacipetalin [*n*-ButOH-C_6H_5N-$C_6H_5CH_3$-H_2O (5:3:3:1 upper phase)], [Me_2CO-$CHCl_3$-H_2O (17:2:1)], [MeCOEt-Me_2CO-H_2O (15:5:3)] (silica), [EtOAc-HOAc-H_2O (3:3:1)] (cellulose) (Nartey et al. 1981); prunasin [EtOAc-HOAc-MeOH-H_2O (12:3:3:2)] (silica) (Nahrstedt 1976a), [MeCN-H_2O (85:15)] (silica) (Lizotte and Poulton 1986), [EtOAc-Me_2CO-MeOH (4:3:1)] (silica) (Nahrstedt et al. 1989); prunasin and prunasin 2′-glucoside (15% HOAc) (microcrystalline cellulose) (Aritomi et al. 1985); prunasin 6′-malonate [EtOAc-Me_2CO-MeOH-H_2O (20:15:5:4)] (silica) (Nahrstedt et al. 1989); sarmentosin [$CHCl_3$-MeOH (8:3)], [EtOAc-Me_2CO (3:2)] (silica) (Fang et al. 1982); sarmentosin epoxide [EtOAc-MeOH-H_2O-HOAc (8:2:0.1:0.1)] (silica) (Nahrstedt et al. 1982b); taxiphyllin [Me_2CO-$CHCl_3$-H_2O (85:10:5)] (silica) (Brimer and Dalgaard 1984); taxiphyllin (R_f 0.54) and triglochinin (R_f 0.24) [MeCOEt-EtOAc-H_2O-HCO_2H (50:50:10:5 and 5:7:1:0.5)] (silica) (Nahrstedt et al. 1979, 1984); tetraphyllin A and B, deidaclin, and volkenin [Me_2CO-H_2O (5:1)] (cellulose) (Tober and Conn 1985); tetraphyllin B sulfate [Me_2CO-H_2O (4:1)] (R_f 0.2), [MeCOEt-Me_2CO-H_2O (15:5:3)] (microcrystalline cellulose) (Seigler et al. 1982); triglochinin [Me_2CO-H_2O (7:3)] (silica) (Tjon Sie Fat and van Valen 1978), [MeCOEt-EtOAc-HOAc-H_2O (5:3:2:1)] (silica), [EtOAc-HCO_2H-H_2O (16:3:2)] (cellulose) (Jaroszewski 1979), [H_2O sat d. $CHCl_3$-MeOH (2:1)] (silica) (van Valen 1978e); and zierin xyloside [EtOAc-hexane (5:2) and EtOAc-hexane-MeOH (19:7:4) by centrifugally accelerated radial TLC] (Schwind et al. 1990).

7.5 Other Purification Methods

Droplet counter current chromatography (DCCC) is a highly promising method for the isolation and intermediate purification of cyanogenic compounds. It is especially useful for sensitive compounds, as no solid phase is involved. Compounds that have been purified by DCCC include: dhurrin and taxiphyllin (Nicollier et al. 1983; Nahrstedt et al. 1987); prunasin and prunasin 2′-glucoside (Aritomi et al. 1985); and xeranthin (Schwind et al. 1990).

8 Quantitation of Cyanogenic Compounds

Methods for the quantitation of individual cyanogenic glycosides have been reviewed (Nahrstedt et al. 1981; Brimer and Dalgaard 1984; Brimer 1988). Particularly important techniques include gas liquid (GLC), thin layer (TLC) (see above) and high-performance liquid chromatography (HPLC).

8.1 Gas Chromatography

Cyanogenic compounds (usually as their trimethylsilyl derivatives) can be separated and quantitatively measured by gas liquid chromatography (GLC) (Nahrstedt 1981; Seigler 1977). This method is fast and reliable in instances where the identity of the compounds is known (van Valen 1979b). Compounds separated (as TMS ethers) and liquid phases are: acalyphin (OV-101) (Nahrstedt et al. 1982a); amygdalin and neoamygdalin (OV-1) (Nahrstedt 1975); amygdalin, epilucumin, lucumin, prunasin, and sambunigrin (OV-225) (Nahrstedt et al. 1983); cardiospermin (OV-1, OV-17, OV-225) (Nahrstedt 1976b; Nahrstedt and Scheid 1981); cardiospermin 4-hydroxybenzoate (OV-17) (Nahrstedt 1976b); dhurrin (OV-225) (Nahrstedt et al. 1979; Erb et al. 1981; Nahrstedt and Limmer 1982); epiheterodendrin (OV-1) (Erb et al. 1979); (*S*)-epilotaustralin, linamarin and (*R*)-lotaustralin (OV-101) (Pitsch et al. 1984; Lieberei et al. 1986; Spencer et al. 1986; Nahrstedt and Davis 1986); epilucumin, holocalin, prunasin, sambunigrin, zierin, and zierin xyloside (DB-5, DB-225) (Schwind et al. 1990); heterodendrin (OV-1, OV-225) (Hübel and Nahrstedt 1975; Nahrstedt and Scheid 1981); linamarin (OV-17) (Michely et al. 1983); linustatin (DB-5) (Selmar et al. 1987, 1988); prunasin and sambunigrin (DB-5) (Nahrstedt et al. 1989), (Dexsil 300) (Poulton and Shin 1983); sarmentosin epoxide (OV-225) (Nahrstedt et al. 1982b); taxiphyllin (OV-225) (Nahrstedt et al. 1979; Nahrstedt and Limmer 1982); and tetraphyllin A and B, gynocardin, volkenin, and deidaclin (SP-2250) (Seigler et al. 1982; Tober and Conn 1985).

The configuration of samples of (*R*)-lotaustralin and (*S*)-epilotaustralin can be established by GLC of the tetraacetates on Silar-10C (Pitsch et al. 1984; Lieberei et al. 1986; Spencer et al. 1986) or ECNSS-M (Pitsch et al. 1984; Spencer et al. 1986).

8.2 High-Performance Liquid Chromatography

High-performance liquid chromatography (HPLC) is a rapid means for analysis of cyanogenic glycosides and their derivatives (Nahrstedt 1981). This technique is usually used for either qualitative or quantitative analysis, but also is a valuable preparative tool. Representative compounds analyzed by HPLC include (RP is reversed phase; more precise conditions of separation are given in the articles cited): acalyphin [MeCN-H_2O (4:96)] (RP-18) (Nahrstedt et al. 1982a); amygdalin 6″-(4-hydroxybenzoate) and amygdalin 6″-(4-hydroxycinnamate) [MeCN-H_2O (3:7)], [MeOH-H_2O (9:11)] (RP-18) (Nahrstedt et al. 1990); *Anthemis* glycoside A and B [H_2O-MeCN-HOAc (88:12:0.1)] (RP-18) (Nahrstedt et al. 1983); cardiospermin 4-hydroxybenzoate and cardiospermin 4-hydroxy-(*E*)-cinnamate [MeCN-HOAc-H_2O (25:0.1:75)] (RP-18) (Nahrstedt and Scheid 1981); dhurrin and taxiphyllin (dioxane) (C8) (Nicollier et al. 1983); dhurrin [H_2O-MeCN-HOAc (95:4.5:0.5)] (RP-18) (Nahrstedt 1979; Erb et al. 1981), [H_2O-MeOH-MeCN-HOAc (85:5:9.5:0.5)] (Nahrstedt et al. 1987); dhurrin, proteacin and sutherlandin [MeCN-H_2O (1:19)] (RP-18), [MeOH-H_2O (5:95)] (Astec Cyclobond I) (Swenson et al. 1987, 1989); epiheterodendrin and heterodendrin [EtOAc-MeOH (95:5)] (silica) (Erb et al. 1979); epilucumin [MeCN-0.1% TFA in H_2O (1:19 and 3:7)] (RP-8), [MeOH-0.1% TFA in H_2O (1:4)] (RP-18) (Schwind et al. 1990); epiproacacipetalin and proacacipetalin [EtOAc-MeOH-H_2O (97:2:1)] (silica) (Jaroszewski and Ettlinger 1987); epivolkenin [H_2O-MeOH (10:1, 87:13, 9:1 and 4:1)] (RP-18), [EtOAc-MeOH-H_2O (85:13:2 and 90:8:2)] (silica) (Jaroszewski and Olafsdóttir 1987; Jaroszewski et al. 1987a, 1988; Olafsdóttir et al. 1989a,b); gynocardin [H_2O-MeOH (10:1)] (RP-18), [EtOAc-MeOH-H_2O (85:13:2)] (silica) (Jaroszewski and Olafsdóttir 1987); heterodendrin [MeCN-MeOH (85:15)] (RP-18) (Nahrstedt and Hübel 1978), [MeOH-H_2O (3:7)] (RP-18) (Zinsmeister et al. 1981); holocalin and sambunigrin [MeOH-H_2O (15:85)] (RP-8) (Brimer and Dalgaard 1984); holocalin, zierin and zierin xyloside [MeCN-H_2O (1:9)] (RP-18) (Hübel et al. 1982); 3-hydroxyheterodendrin [H_2O-MeCN (97:3)], [H_2O-MeOH (17:3)] (RP-18) (Brimer et al. 1981); isotriglochinin, taxiphyllin, and triglochinin [H_2O-MeCN-HOAc (95:4.5:0.5)] (RP-18) (Nahrstedt 1979); linamarin and lotaustralin [H_2O-MeCN (19:1)] (RP-18) (Lieberei et al. 1986; Nahrstedt and Davis 1986), [H_2O-MeCN-HOAc (95:4.5:0.5)] (RP-18) (Michely et al. 1983; Pitsch et al. 1984), [MeOH-H_2O (3:7 and 1:4)] (RP-18) (Jaroszewski et al. 1987b; Olafsdóttir et al. 1989a); linustatin [H_2O-MeCN (97:3)] (RP-18) (Selmar et al. 1987);

nandinin [H_2O-MeCN-HOAc (17:3:0.1)] (Whatman Partisil M9 10/50) (Olechno et al. 1984); passibiflorin, passicapsin, epivolkenin and taraktophyllin [H_2O-MeOH (7:3)] (RP-18) (Olafsdóttir et al. 1989b); passisuberosin [MeCN-H_2O (85:15)] (Alltech NH_2) (Spencer and Seigler 1987); passitrifasciatin [MeOH-H_2O (1:4)] (RP-18) (Olafsdóttir et al. 1991); prunasin [MeCN-H_2O (4:1)] (Alltech NH_2) (Spencer and Seigler 1984); 6-O-(α-L-rhamnopyranosyl)-epivolkenin and 6-O-(α-L-rhamnopyranosyl)-taraktophyllin [H_2O-MeOH (10:1)] (RP-18) (Jaroszewski et al. 1988); sambunigrin and zierin [MeOH-0.1% TFA in H_2O (1:3)] (RP-18) (Schwind et al. 1990); sarmentosin epoxide [H_2O-MeOH-HOAc (97.5:2.5:0.2)] (RP-18) (Nahrstedt et al. 1982b); taraktophyllin [H_2O-MeOH (10:1, 4:1 and 9:1)] (RP-18), [EtOAc-MeOH-H_2O (85:13:2 and 90:8:2)] (silica) (Jaroszewski and Olafsdóttir 1987; Jaroszewski et al. 1987a, 1988; Olafsdóttir et al. 1989a); tetraphyllin A and deidaclin [H_2O-MeOH (7:3 and 4:1)] (RP-18), [EtOAc-MeOH-H_2O (93:5:2)] (silica) (Jaroszewski and Jensen 1985; Olafsdóttir et al. 1989a), [MeCN-H_2O (1:9)] (RP-18) (Tober and Conn 1985); tetraphyllin A and B [H_2O-MeOH (85:15)] (RP-18) (Jaroszewski et al. 1987c); tetraphyllin B and volkenin [EtOAc-MeOH-H_2O (78:20:2)] (silica) (Jaroszewski and Olafsdóttir 1986; Jaroszewski et al. 1987c), [H_2O-MeOH (9:1 and 4:1)] (RP-18) (Olafsdóttir et al. 1989a; Jaroszewski et al. 1987c), [MeCN-H_2O (1:9)] (RP-18) (Tober and Conn 1985); tetraphyllin B [MeCN-H_2O (4:1)] (Alltech NH_2) (Spencer and Seigler 1984); tetraphyllin B sulfate (water followed by 0.5 M NH_4OAc) (Nucleosil NH_2) (Jaroszewski and Fog 1989); volkenin [H_2O-MeOH (9:1)] (RP-18) (Olafsdóttir et al. 1989a); epilucumin, xeranthin, zierin, and zierin xyloside [MeCN-0.1% TFA in H_2O (7:93)], [H_2O-MeOH (7:3, 17:3 to 3:7)], [MeCN-H_2O (17:83)] (RP-18), [MeCN-0.1% TFA in H_2O (1:9)] (RP-8) (Schwind et al. 1990).

Cardiospermin sulfate was not retained and not separable from other compounds at the solvent front on reversed phase columns. Use of ion-pair chromatography with tetramethylammonium as the cation permitted separation of cardiospermin and its sulfate (Nahrstedt 1981).

A highly specific and sensitive system which involves separation on a C-8 reversed phase column, enzymatic post-column cleavage of cyanogenic glycosides, and electrochemical detection of the cyanide formed has been described (Brimer and Dalgaard 1984).

9 Hydrolysis with Acid or Base

Determination of the products of hydrolysis with either acid or base has been widely employed for quantitation of purified cyanogenic glycosides (Nahrstedt 1981). Procedures for determination of the resulting aglycones or their derivatives, sugars, aromatic acids, sulfate and malonate are given in reviews (Seigler 1977; Nahrstedt 1981).

Most cyanogenic glycosides yield a sugar and cyanohydrin when heated with dilute aqueous acid. The cyanohydrin dissociates nonenzymatically to produce HCN and an aldehyde or ketone (Seigler 1977; Conn 1979). With more concentrated acid, a different series of hydrolysis products may result (Seigler 1977).

Trifluoroacetic acid has been used often, largely because it is a strong acid, yet easily removed from hydrolysis mixtures by concentration under vacuum. Compounds hydrolyzed in this manner include: amygdalin 6″-(4-hydroxybenzoate) (amygdalin, prunasin, gentiobiose, glucose, 4-hydroxybenzoic acid) and amygdalin 6″-(4-hydroxy-(*E*)-cinnamate) (amygdalin, prunasin, gentiobiose, glucose, 4-coumaric acid) (Nahrstedt et al. 1990); *Anthemis* glycoside A and B (both gave benzaldehyde, glucose, xylose, 4-coumaric acid, epilucumin) (Nahrstedt et al. 1983); dhurrin (glucose, 4-hydroxybenzaldehyde) (Nahrstedt and Limmer 1982); proacaciberin (arabinose and proacacipetalin) (Nartey et al. 1981); vicianin (glucose, arabinose, prunasin) (Lizotte and Poulton 1985); xeranthin (*m*-hydroxybenzaldehyde, xylose, glucose, apiose, and caffeic acid; milder conditions gave zierin and zierin xyloside) (Schwind et al. 1990); and zierin xyloside (*m*-hydroxybenzaldehyde, glucose, xylose) (Hübel et al. 1982).

Hydrochloric acid is also useful, but more difficult to remove completely from hydrolysis mixtures. Compounds hydrolyzed with HCl include: acalyphin (Nahrstedt et al. 1982a); cardiospermin 4-hydroxybenzoate (4-hydroxybenzoic acid, HCN, glucose) and cardiospermin 4-hydroxy-(*E*)-cinnamate (4-hydroxycinnamic acid, HCN, glucose) (Nahrstedt 1976b; Nahrstedt and Scheid 1981); cardiospermin sulfate (glucose, cardiospermin, sulfate) (Hübel and Nahrstedt 1979); dhurrin and taxiphyllin (glucose, *p*-hydroxybenzaldehyde) (Towers et al. 1964); grayanin [methyl caffeate, methyl (*R*)-mandelate, glucose] (MeOH, conc. HCl) (Shimomura et al. 1987); heterodendrin (glucose and hydrogen cyanide) (Hübel and Nahrstedt 1975), (2-hydroxyisovaleric acid) (Nahrstedt and Hübel 1978); linustatin (glucose) (Selmar et al. 1987); neolinustatin (glucose) (5% HCl), (lotaustralin, glucose, β-gentiobiose) (0.5 N HCl) (Smith et al. 1980); passibiflorin (epivolkenin and antiarose) (Olafsdóttir et al. 1989b); passicapsin (epivolkenin, boivinose) (Olafsdóttir et al. 1989b); passisuberosin (glucose, HCN) (Spencer and Seigler 1987); passitrifasciatin (epivolkenin and 6-deoxy-D-allose) (Olafsdóttir et al. 1991); prunasin [(*R*)-mandelic acid and glucose] and prunasin 2′-glucoside (glucose, prunasin) (Aritomi et al. 1985); and vicianin (prunasin) (Nagumo et al. 1985).

Other acid hydrolyses include: epilucumin (benzaldehyde, xylose, glucose, HCN) (mild acidic hydrolysis with UV light) (Nahrstedt et al. 1983); nandinin (caffeic acid, glucose, 4-hydroxybenzaldehyde, 4-hydroxymandelic acid) (strong acid hydrolysis) (Olechno et al. 1984); sarmentosin (2% H_2SO_4) (Fang et al. 1982); sarmentosin epoxide (glucose, HCN)

(1 N H_3PO_4) (Nahrstedt et al. 1982b); tetraphyllin B (4-hydroxy-2-cyclopentenone) (0.5 N H_2SO_4) (Morah 1988); and vicianin (glucose, arabinose) (5% H_2SO_4, reflux) (Nagumo et al. 1985).

Many cyanogenic compounds also are hydrolyzed under basic conditions; the products obtained vary with pH. Under strongly basic conditions, the nitrile group of most cyanogens is converted to the amide and subsequently to the carboxylic acid without cleavage of other bonds. Compounds hydrolyzed by base include: amygdalin 6″-(4-hydroxybenzoate) (isomerized amygdalin, 4-hydroxybenzoic acid) and amygalin 6″-(4-hydroxycinnamate) (isomerized amygdalin, 4-coumaric acid) (0.5 M Na_2CO_3) (Nahrstedt et al. 1990); *Anthemis* glycoside A (epilucumin, a compound which contained 4-coumaric acid, glucose and xylose) (0.07 M NH_4OH) (Nahrstedt et al. 1983); dhurrin and taxiphyllin (HCN, 4-hydroxybenzaldehyde, glucose) (Mao and Anderson 1965; Schwarzmaier 1976); nandinin (cyanide, the corresponding carbonyl compound) (Tris buffer, pH 8.5) (Olechno et al. 1984); prunasin 6′-malonate (prunasin, malonate) (0.5 M Na_2CO_3) (Nahrstedt et al. 1989); and sarmentosin (isosarmentosin) (Fang et al. 1982). Hydrolysis of xeranthin in 0.5 M NaOH gave two main components, one of which yielded caffeic acid and glucose on further hydrolysis, and the second gave apiose, xylose and glucose and was later shown to be zierin xyloside attached to an apiose unit (Schwind et al. 1990).

Acknowledgments. The senior author acknowledges support from a National Science Foundation Graduate Fellowship, a U.S. Public Health Service Cell and Molecular Biology Training Grant, a Francis M. and Harlie M. Clark Research Support Grant, a Graduate College Dissertation Grant, and a research assistantship provided by the School of Life Sciences, University of Illinois. We also acknowledge research support by the National Science Foundation (NSF-82-15274).

References

Aldridge WN (1944) A new method for the estimation of microquantities of cyanide and thiocyanide. Analyst (London) 69:262–265

Aritomi M, Kumori T, Kawasaki T (1985) Cyanogenic glycosides in leaves of *Perilla frutescens* var. *acuta*. Phytochemistry 24:2438–2439

Brimer L (1988) Determination of cyanide and cyanogenic compounds in biological systems. In: Evered D, Harnett S (eds) Cyanide compounds in biology, Ciba Symposium vol 140. Wiley, Chichester, pp 177–196

Brimer L, Dalgaard L (1984) Cyanogenic glycosides and cyanohydrins in plant tissues. Qualitative and quantitative determination by enzymatic post-column cleavage and electrochemical detection, after separation by high-performance liquid chromatography. J Chromatogr 303:77–88

Brimer L, Christensen SB, Jaroszewski JW, Nartey F (1981) Structural elucidation and partial synthesis of 3-hydroxyheterodendrin, a cyanogenic glucoside from *Acacia sieberiana* var. *woodii*. Phytochemistry 20:2221–2223

Brimer L, Christensen SB, Mølgaard P, Nartey F (1983) A densitometric method for quantification of cyanogenic glycosides, employing enzyme preparations (β-

glucuronidase) from *Helix pomatia* and picrate-impregnated ion-exchange sheets. J Agric Food Chem 31:789–793
Brinker AM, Seigler DS (1989) Methods for the detection and quantitative determination of cyanide in plant materials. Phytochem Bull 21:24–31
Butler GW, Butler BG (1960) Biosynthesis of linamarin and lotaustralin in white clover. Nature 187:780–781
Conn EE (1979) Cyanide and cyanogenic glycosides. In: Rosenthal GA, Janzen DH (eds) Herbivores. Academic Press, New York, pp 271–307
Conn EE (1980a) Cyanogenic glycosides. In: Bell EA, Charlwood BV (eds) Secondary plant products. Springer, Berlin Heidelberg New York, pp 461–492
Conn EE (1980b) Cyanogenic compounds. Annu Rev Plant Physiol 31:433–451
Conn EE (1981a) Cyanogenic glycosides. In: Conn EE (ed) Secondary plant products, vol 7. In: Stumpf PK, Conn EE (eds) The biochemistry of plants. Acad Press, New York, pp 479–500
Conn EE (1981b) Biosynthesis of cyanogenic glycosides. In: Vennesland B, Conn EE, Knowles CJ, Westley J, Wissing F (eds) Cyanide in biology. Acad Press, London, pp 183–196
Cooke RD (1978) An enzymatic assay for the total cyanide content of cassava (*Manihot esculenta* Crantz). J Sci Food Agric 29:345–352
Erb N, Zinsmeister HD, Lehmann G, Nahrstedt A (1979) A new cyanogenic glycoside from *Hordeum vulgare*. Phytochemistry 18:1515–1517
Erb N, Zinsmeister HD, Nahrstedt A (1981) Die cyanogenen Glykoside von *Triticum, Secale* und *Sorghum*. Planta Med 41:84–89
Ettlinger MG, Jaroszewski JW, Jensen SR, Nielsen BJ, Nartey F (1977) Proacacipetalin and acacipetalin. J Chem Soc Chem Commun 952–953
Evered D, Harnett S (eds) (1988) Cyanide compounds in biology, Ciba Symposium 140. Wiley, Chichester, 261 pp
Fang S, Yan X, Li J, Fan Z, Xu X, Xu R (1982) Studies on the chemical constituents of *Sedum sarmentosum* Bunge. IV. The structures of sarmentosin and isosarmentosin. Acta Chim Sin (Chin Ed) 40:273–280
Feigl G, Anger V (1966) Replacement of benzidine by copper ethylacetoacetate and tetra base as spot-test reagent for hydrogen cyanide and cyanogen. Analyst 91:282–284
Fenselau C, Pallante S, Batzinger RP, Benson WR, Barron RP, Sheinin EB, Maienthal M (1977) Mandelonitrile β-glucuronide: synthesis and characterization. Science 198: 625–627
Fikenscher LH, Hegnauer R (1977) *Chaptalia nutans*, eine stark cyanogene Pflanze Brasiliens. Planta Med 31:266–269
Fikenscher LH, Hegnauer R (1981) Die Verbreitung der Blausäure bei den Cormophyten. Weitere Beobachtungen bei den Leguminosae-Papilionoideae (Galegeae, Genisteae, Loteae, Phaseoleae). Planta Med 43:323–335
Fikenscher LH, Ruijgrok HWL (1977) Die cyanogenen Glucoside von *Platanus*-Sippen. Planta Med 31:290–293
Fischer FC, Fung SY, Lankhorst PP (1982) Cyanogenesis in Passifloraceae. II. Cyanogenic compounds from *Passiflora capsularis, P. warmingii* and *P. perfoliata*. Planta Med 45:42–45
Goldstein WS, Spencer KC (1985) Inhibition of cyanogenesis by tannins. J Chem Ecol 11:847–858
Guignard L (1906) Le Haricot à acide cyanhydrique (*Phaseolus lunatus*). Étude historique, botanique et chimique. Nouveau procédé pour déceler l'acide cyanhydrique. Bull Sci Pharmacol 13:401–419
Halkier BA, Møller BL (1989) Biosynthesis of the cyanogenic glucoside dhurrin in seedlings of *Sorghum bicolor* (L.) Moench and partial purification of the enzyme system involved. Plant Physiol 90:1552–1559
Hegnauer R (1977) Cyanogenic compounds as systematic markers in Tracheophyta. In: Kubitzki K (ed) Plant systematics and evolution, Supplementum 1. Springer, Berlin Heidelberg New York, pp 191–209

Hegnauer R (1986) In: Chemotaxonomie der Pflanzen, vol. 7. Birkhäuser Verlag, Basel

Horwitz W (ed) (1980) Official methods of analysis of the association of official analytical chemists, 13th edn. Assoc Off Anal Chem, Washington, DC, p 433

Hösel W (1981) The enzymatic hydrolysis of cyanogenic glycosides. In: Vennesland B, Conn EE, Knowles CJ, Westley J, Wissing F (eds) Cyanide in biology. Acad Press, London, pp 217–232

Hösel W, Tober I, Eklund SH, Conn EE (1987) Characterization of β-glucosidases with high specificity for the cyanogenic glucoside dhurrin in *Sorghum bicolor* (L.) Moench seedlings. Arch Biochem Biophys 252:152–162

Hübel W, Nahrstedt A (1975) Ein neues Cyanglykosid aus *Heterodendron oleaefolium*. Phytochemistry 14:2723–2725

Hübel W, Nahrstedt A (1979) Cardiospermin sulfate – a sulphur containing cyanogenic glucoside from *Cardiospermum grandiflorum*. Tetrahedron Lett 4395–4396

Hübel W, Nahrstedt A, Fikenscher LH, Hegnauer R (1982) Zierinxylosid, ein neues cyanogenes Glykoside aus *Xeranthemum cylindraceum*. Planta Med 44:178–180

Jaroszewski JW (1979) Concerning acacipetalin and triglochinin. PhD Thesis, University of Copenhagen

Jaroszewski JW, Ettlinger MG (1981) Ring cleavage of phenols in higher plants: biosynthesis of triglochinin. Phytochemistry 20:819–821

Jaroszewski JW, Ettlinger MG (1987) ^{1}H and ^{13}C NMR spectra of derivatives of 2-hydroxy-3-methylbut-3-enenitrile and 2-hydroxy-3-methylbut-2-enenitrile. Magn Reson Chem 25:555–557

Jaroszewski JW, Fog E (1989) Sulphate esters of cyclopentenoid cyanohydrin glycosides. Phytochemistry 28:1527–1528

Jaroszewski JW, Jensen B (1985) Deidaclin and tetraphyllin A, epimeric glucosides of 2-cyclopentenone cyanohydrin, in *Adenia globosa* ssp. *globosa* Engl. (Passifloraceae). Crystal structure of deidaclin tetraacetate. Acta Chem Scand B39:867–875

Jaroszewski JW, Olafsdóttir ES (1986) Natural glycosides of cyclopentenone cyanohydrins: revised structure of so-called epitetraphyllin B. Tetrahedron Lett 27:5297–5300

Jaroszewski JW, Olafsdóttir ES (1987) Monohydroxylated cyclopentenone cyanohydrin glucosides of Flacourtiaceae. Phytochemistry 26:3348–3349

Jaroszewski JW, Andersen JV, Billeskov I (1987a) Plants as a source of chiral cyclopentenes: taraktophyllin and epivolkenin, new cyclopentenoid cyanohydrin glucosides from Flacourtiaceae. Tetrahedron 43:2349–2354

Jaroszewski JW, Jensen PS, Cornett C, Byberg JR (1987b) Occurrence of lotaustralin in *Berberidopsis beckleri* and its relation to the chemical evolution of Flacourtiaceae. Biochem Syst Ecol 16:23–28

Jaroszewski JW, Olafsdóttir ES, Cornett C, Schaumburg K (1987c) Cyanogenesis of *Adenia volkensii* Harms and *Tetrapathaea tetrandra* Cheeseman (Passifloraceae) revisited: tetraphyllin B and volkenin. Optical rotatory power of cyclopentenoid cyanohydrin glucosides. Acta Chem Scand B41:410–421

Jaroszewski JW, Bruun D, Clausen V, Cornett C (1988) Novel cyclopentenoid cyanohydrin rhamnoglucosides from Flacourtiaceae. Planta Med 54:333–337

Jensen SR, Nielsen BJ (1986) Gynocardin in *Ceratiosicyos laevis* (Achariaceae). Phytochemistry 25:2349–2350

Kingsbury JM (1964) Poisonous plants of the United States and Canada. Prentice-Hall, Englewood Cliffs, NJ, p 25

Kofod H, Eyjólfsson R (1969) Cyanogenesis in species of the fern genera *Cystopteris* and *Davallia*. Phytochemistry 8:1509–1511

Lambert JL, Ramasamy J, Paukstelis JV (1975) Stable reagents for the colorimetric determination of cyanide by modified Konig reactions. Anal Chem 47:916–918

Lieberei R, Selmar D, Biehl B (1985) Metabolization of cyanogenic glucosides in *Hevea brasiliensis*. Plant Syst Evol 150:49–63

Lieberei R, Nahrstedt A, Selmar D, Gasparotto L (1986) Occurrence of lotaustralin in the genus *Hevea* and changes of HCN-potential in developing organs of *Hevea brasiliensis*. Phytochemistry 25:1573–1578

Lizotte PA, Poulton JE (1986) Identification of (*R*)-vicianin in *Davallia trichomanoides* Blume. Z Naturforsch 41C:5–8
Mao CH, Anderson L (1965) Cyanogenesis in *Sorghum vulgare*. II. Mechanism of the alkaline hydrolysis of dhurrin (*p*-hydroxymandelonitrile glucoside). J Org Chem 30:603–607
Markley B, Meloan CE, Lambert JL, Chiang YC (1987) A sensitive, selective, and rapid colorimetric method to determine cyanide contamination in containers, capsules, and liquids. Anal Lett 20:1225–1236
Michely D, Zinsmeister HD, Roth E, Nahrstedt A (1983) Das Cyanglykosid von *Avena sativa*. Z Lebensm Unters Forsch 177:350–352
Morah FNI (1988) Tetraphyllin B from *Adenia cissampeloides*. Phytochemistry 27:2985–2986
Nagumo S, Imamura K, Inoue T, Nagai M (1985) Cyanogenic glycosides and 4-hydroxycoumarin glycosides from *Gerbera jamesonii hybrida*. Chem Pharm Bull 33:4803–4806
Nahrstedt A (1975) Die Isomerisierung von Amygdalin und Homologen. Arch Pharm (Weinheim) 308:903–910
Nahrstedt A (1976a) Prunasin in *Holocalyx balansae*. Phytochemistry 15:1983–1984
Nahrstedt A (1976b) Ein neues cyanogenes Glykosid aus der Rosacee *Sorbaria arborea*. Z Naturforsch 31C:397–400
Nahrstedt A (1977) Ersatz karzinogener und allergener Amine in der Cyanid-Bestimmung nach Aldridge. Dtsch Apoth Ztg 127:1357–1359
Nahrstedt A (1979) Qualitative und quantitative Bestimmung der Cyanglucoside Triglochinin, Taxiphyllin und Dhurrin mittels Hochdruckflüssigkeitschromatographie. J Chromatogr 177:157–161
Nahrstedt A (1980) Absence of cyanogenesis from Droseraceae. Phytochemistry 19: 2757–2758
Nahrstedt A (1981) Isolation and structure elucidation of cyanogenic glycosides. In: Vennesland B, Conn EE, Knowles CJ, Westley J, Wissing F (eds) Cyanide in biology. Academic Press, New York, pp 145–181
Nahrstedt A (1985) Cyanogenic compounds as protecting agents for organisms. Plant Syst Evol 150:35–47
Nahrstedt A (1987) Recent developments in chemistry, distribution and biology of the cyanogenic glycosides. In: Hostettmann K, Lea PJ (eds) Biologically active natural products. Clarendon Press, Oxford, pp 213–234
Nahrstedt A, Davis RH (1986) Uptake of linamarin and lotaustralin from their foodplant by larvae of *Zygaena trifolii*. Phytochemistry 25:2299–2302
Nahrstedt A, Hübel W (1978) Die cyanogenen Verbindungen von *Heterodendron oleaefolium*. Phytochemistry 17:314–315
Nahrstedt A, Limmer S (1982) Dhurrin, the cyanogenic glucoside of *Cercocarpus ledifolius*. Phytochemistry 21:2738–2740
Nahrstedt A, Scheid H (1981) Leucine-derived cyanogenic glucosides in the Rosaceae-Spiraeoideae. Phytochemistry 20:1309–1310
Nahrstedt A, Hösel W, Walther A (1979) Characterization of cyanogenic glucosides and β-glucosidases in *Triglochin maritima* seedlings. Phytochemistry 18:1137–1141
Nahrstedt A, Erb M, Zinsmeister H (1981) Methods of liberating and estimating hydrocyanic acid from cyanogenic plant material. In: Vennesland B, Conn EE, Knowles CJ, Westley J, Wissing F (eds) Cyanide in biology. Academic Press, New York, pp 461–471
Nahrstedt A, Kant JD, Wray V (1982a) Acalyphin, a cyanogenic glucoside from *Acalypha indica*. Phytochemistry 21:101–105
Nahrstedt A, Walther A, Wray V (1982b) Sarmentosin epoxide, a new cyanogenic compound from *Sedum cepaea*. Phytochemistry 21:107–110
Nahrstedt A, Wray V, Grotjahn L, Fikenscher LH, Hegnauer R (1983) New acylated cyanogenic diglycosides from fruits of *Anthemis cairica*. Planta Med 49:143–148
Nahrstedt A, Kant JD, Hösel W (1984) Aspects on the biosynthesis of the cyanogenic glucoside triglochinin in *Triglochin maritima*. Planta Med 50(5):394–398

Nahrstedt A, Proksch P, Conn EE (1987) Dhurrin, (-)-catechin, flavonol glycosides and flavones from *Chamaebatia foliolosa*. Phytochemistry 26:1546–1547
Nahrstedt A, Jensen PS, Wray V (1989) Prunasin-6'-malonate, a cyanogenic glucoside from *Merremia dissecta*. Phytochemistry 28:623–624
Nahrstedt A, Sattar EA, El-Zalabani SMH (1990) Amygdalin acyl derivatives, cyanogenic glycosides from the seeds of *Merremia dissecta*. Phytochemistry 29:1179–1181
Nartey F, Brimer L, Christensen SB (1981) Proacaciberin, a cyanogenic glycoside from *Acacia sieberiana* var. *woodii*. Phytochemistry 20:1311–1314
Nicollier GF, Pope DF, Thompson AC (1983) Biological activity of dhurrin and other compounds from Johnson grass (*Sorghum halepense*). J Agric Food Chem 31:744–748
Nishizawa K, Hashimoto Y (1970) Glycoside hydrolases and glycosyl transferases. In: Pigman W, Horton D (eds) The carbohydrates: chemistry and biochemistry, 2nd edn, vol 2A. Academic Press, New York, pp 241–300
Olafsdóttir ES, Andersen JV, Jaroszewski JW (1989a) Cyanohydrin glycosides of Passifloraceae. Phytochemistry 28:127–132
Olafsdóttir ES, Cornett C, Jaroszewski JW (1989b) Cyclopentenoid cyanohydrin glycosides with unusual sugar residues. Acta Chem Scand 43:51–55
Olafsdóttir ES, Jaroszewski JW, Seigler DS (1991) Cyanohydrin glycosides with unusual sugar residues: revised structure of passitrifasciatin. Phytochemistry 30:867–869
Olechno JD, Poulton JE, Conn EE (1984) Nandinin: an acylated free cyanohydrin from *Nandina domestica*. Phytochemistry 23:1784–1785
Pereira JF, Seigler DS, Splittstoesser WE (1981) Cyanogenesis in sweet and bitter cultivars of cassava. HortScience 16:776–777
Pitsch C, Keller M, Zinsmeister HD, Nahrstedt A (1984) Cyanogene Glykoside aus *Triticum monococcum*. Planta Med 50:388–390
Poulton JE (1988) Localization and catabolism of cyanogenic glycosides. In: Evered D, Harnett S (eds) Cyanide compounds in biology. Ciba Symposium vol 140. Wiley, Chichester, pp 67–80
Poulton JE, Shin SI (1983) Prunasin biosynthesis by cell-free extracts from black cherry (*Prunus serotina* Ehrh.) fruits and leaves. Z Naturforsch 38C:369–374
Saupe SG (1981) Cyanogenic compounds and angiosperm phylogeny. In: Young DA, Seigler DS (eds) Phytochemistry and angiosperm phylogeny. Praeger, New York, pp 80–116
Saupe SG, Seigler DS, Escalante-Semerena JC (1982) Bacterial contamination as a cause of spurious cyanide tests. Phytochemistry 21:2111–2112
Schwarzmaier U (1976) Über die Cyanogenese von *Bambusa vulgaris* und *B. guadua*. Chem Ber 109:3379–3389
Schwind P, Wray V, Nahrstedt A (1990) Structure elucidation of an acylated cyanogenic triglycoside, and further cyanogenic constituents from *Xeranthemum cylindraceum* Phytochemistry 29:1903–1911
Seigler D (1977) The naturally occurring cyanogenic glycosides. In: Reinhold L, Harborne JB, Swain T (eds) Prog Phytochem, vol 4. Pergamon, Oxford, pp 83–120
Seigler DS (1981a) Recent developments in the chemistry and biology of cyanogenic glycosides and lipids. Rev Latinoam Quim 12:39–48
Seigler DS (1981b) Cyanogenic glycosides and lipids: structural types and distribution. In: Vennesland B, Conn EE, Knowles CJ, Westley J, Wissing F (eds) Cyanide in biology. Academic Press, London, pp 133–143
Seigler DS (1991) Cyanide and cyanogenic glycosides. In: Rosenthal GA, Berenbaum MR (eds) Herbivores. Acad Press, San Diego (in press)
Seigler DS, Brinker AM (1991) Characterization of cyanogenic glycosides, cyanolipids, nitroglycosides, organic nitro compounds and nitrileglycosides from plants. In: Waterman PW (ed) Alkaloids and sulphur compounds, vol 8. In: Dey PM, Harborne JB (eds) Methods in plant biochemistry. Acad Press, London (in press)
Seigler DS, Spencer KC, Statler WS, Conn EE, Dunn JE (1982) Tetraphyllin B and epitetraphyllin B sulphates: novel cyanogenic glucosides from *Passiflora caerulea* and *P. alato-caerulea*. Phytochemistry 21:2277–2285

Seigler DS, Maslin BR, Conn EE (1989) Cyanogenesis in the Leguminosae. In: Stirton CH, Zarucchi JL (eds) Advances in legume biology. Monogr Syst Bot Missouri Bot Gard 29:645–672

Selmar D, Lieberei R, Biehl B, Nahrstedt A, Schmidtmann V, Wray V (1987) Occurrence of the cyanogen linustatin in *Hevea brasiliensis*. Phytochemistry 26:2400–2401

Selmar D, Lieberei R, Biehl B (1988) Mobilization and utilization of cyanogenic glycosides. Plant Physiol 86:711–716

Shimomura H, Sashida Y, Adachi T (1987) Cyanogenic and phenylpropanoid glucosides from *Prunus grayana*. Phytochemistry 26:2363–2366

Smith CR Jr., Weisleder D, Miller RW, Palmer IS, Olson OE (1980) Linustatin and neolinustatin: cyanogenic glycosides of linseed meal that protect animals against selenium toxicity. J Org Chem 45:507–510

Spencer KC (1987) Specificity of action of allelochemicals: diversification of glycosides. In: Waller GR (ed) Allelochemicals: role in agriculture and forestry. ACS Symp Ser 330, Am Chem Soc, Washington, DC, pp 275–288

Spencer K (1988) Chemical mediation of coevolution in the *Passiflora-Heliconius* interaction. In: Spencer K (ed) Chemical mediation of coevolution. Acad Press, Orlando, pp 167–240

Spencer KC, Seigler DS (1980) Deidaclin from *Turnera ulmifolia*. Phytochemistry 19: 1863–1864

Spencer KC, Seigler DS (1982) Tetraphyllin B from *Adenia digitata*. Phytochemistry 21:653–655

Spencer KC, Seigler DS (1984) Cyanogenic glycosides of *Carica papaya* and its phylogenetic position with respect to the Violales and Capparales. Am J Bot 71:1444–1447

Spencer KC, Seigler DS (1985) Passibiflorin, epipassibiflorin, and passitrifasciatin: cyclopentenoid cyanogenic glycosides from *Passiflora*. Phytochemistry 24:981–986

Spencer KC, Seigler DS (1987) Passisuberosin and epipassisuberosin: Two cyclopentenoid cyanogenic glycosides from *Passiflora suberosa*. Phytochemistry 26:1665–1667

Spencer KC, Seigler DS, Fikenscher LH, Hegnauer R (1982) Gynocardin and tetraphyllin B from *Carpotroche brasiliensis*. Planta Med 44:28–29

Spencer KC, Seigler DS, Domingo JL (1983) Tetraphyllins A and B, deidaclin and epitetraphyllin B from *Tetrapathaea tetrandra* (Passifloraceae). Phytochemistry 22: 1815–1816

Spencer KC, Seigler DS, Fraley SW (1985) Cyanogenic glycosides of the Turneraceae. Biochem Syst Ecol 13:433–435

Spencer KC, Seigler DS, Nahrstedt A (1986) Linamarin, lotaustralin, linustatin and neolinustatin from *Passiflora* species. Phytochemistry 25:645–647

Swenson W, Dunn JE, Conn EE (1987) Cyanogenesis in *Acacia sutherlandii*. Phytochemistry 26:1835–1836

Swenson WK, Dunn JE, Conn EE (1989) Cyanogenesis in the Proteaceae. Phytochemistry 28:821–823

Tantisewie B, Ruijgrok HWL, Hegnauer R. (1969) Über cyanogene Verbindungen bei den Parietales und bei einigen weiteren Sippen. Pharm Weekbl 104:1341–1355

Tjon Sie Fat L (1978) Cyanogenesis in some grasses III. Proc K Ned Akad Wet 81C:347–354

Tjon Sie Fat L (1979) Cyanogenesis in some Ranunculaceae. Proc K Ned Akad Wet 82C:197–209

Tjon Sie Fat L, van Valen F (1978) Cyanogenesis in some grasses. II. Proc K Ned Akad Wet 81C:204–210

Tober I, Conn EE (1985) Cyclopentenylglycine, a precursor of deidaclin in *Turnera ulmifolia*. Phytochemistry 24:1215–1218

Torres E, Pereira JF, Brinker AM, Seigler DS (1988) Determination of total nitrogen and cyanide nitrogen in cassava. J Sci Food Agric 42:149–156

Towers GHN, McInnes AG, Neish AC (1964) The absolute configurations of the phenolic cyanogenetic glucosides taxiphyllin and dhurrin. Tetrahedron 20:71–77

van Valen F (1978a) Cyanogenesis in Liliaceae. Proc K Ned Akad Wet 81C:132–140
van Valen F (1978b) Cyanogenesis in *Trochodendron aralioides* Sieb. & Zucc. Proc K Ned Akad Wet 81C:198–203
van Valen F (1978c) Cyanogenesis in some Magnoliidae. Proc K Ned Akad Wet 81C: 355–362
van Valen F (1978d) Cyanogenesis in Papaverales. Proc K Ned Akad Wet. 81C:492–499
van Valen F (1978e) Cyanogenesis in Euphorbiaceae. Planta Med 34:408–413
van Valen F (1979a) Cyanogenesis in Boraginaceae. Proc K Ned Akad Wet 82C:171–176
van Valen F (1979b) Cyanogenesis in some Papilionaceae. Planta Med 35:141–149
Vennesland B, Conn EE, Knowles CJ, Westley J, Wissing F (eds) (1981) Cyanide in Biology. Acad Press, New York, 548 pp
Zinsmeister HD, Erb N, Nahrstedt A, Hegnauer R, Tjon Sie Fat L (1981) Das cyanogene Glykosid von *Holcus mollis*. Z Naturforsch 36C:336–337

Subject Index

GPSR Compliance

The European Union's (EU) General Product Safety Regulation (GPSR) is a set of rules that requires consumer products to be safe and our obligations to ensure this.

If you have any concerns about our products, you can contact us on ProductSafety@springernature.com

In case Publisher is established outside the EU, the EU authorized representative is:

Springer Nature Customer Service Center GmbH
Europaplatz 3
69115 Heidelberg, Germany

Batch number: 10755053

Printed by Printforce, the Netherlands